Integrated Pest Management

Potential, Constraints and Challenges

Integrated Pest Management

Potential, Constraints and Challenges

Edited by

Opender Koul

Insect Biopesticide Research Centre
Jalandhar, India

G.S. Dhaliwal

Department of Entomology
Punjab Agricultural University
Ludhiana, India

and

G.W. Cuperus

Department of Entomology and Plant Pathology
Oklahoma State University, Stillwater, Oklahoma, USA

CABI Publishing

CABI is a trading name of CAB International

CABI Head Office
Nosworthy Way
Wallingford
Oxfordshire OX10 8DE
UK

Tel: +44 (0)1491 832111
Fax: +44 (0)1491 833508
Email: cabi@cabi.org
Web site: www.cabi.org

CABI
745 Atlantic Avenue
8th Floor
Boston, MA 02111
USA

Tel: +1 617 682-9015
Email: cabi-nao@cabi.org

A catalogue record for this book is available from the British Library, London, UK

Library of Congress Cataloging-in-Publication Data
Integrated pest management: potential, constraints and challenges/ edited by Opender Koul, G.S. Dhaliwal, G.W. Cuperus.
p. cm.
Includes bibliographical references and index.
ISBN: 0-85199-686-8 (alk. paper)
1. Pests—Integrated control. I. Koul, Opender. II. Dhaliwal, G.S.
III. Cuperus, Gerrit W. IV. Title.
SB950.I4577 2004
632'.9—dc22

2003015419

ISBN-13: 978-0-85199-686-8

First published 2004
Transferred to print on demand 2017

Printed and bound in the UK by Marston Book Services Ltd, Oxfordshire.

Contents

About the Editors

Opender Koul, Fellow of the National Academy of Agricultural Sciences and the Indian Academy of Entomology, is an insect toxicologist/physiologist/chemical ecologist and currently the Director of the Insect Biopesticide Research Centre, Jalandhar, India. After obtaining his PhD in 1975 he joined the Regional Research Laboratory (CSIR), Jammu and then became Senior Group Leader of Entomology at Malti-Chem Research Centre, Vadodara, India (1980–1988). He has been a visiting scientist at the University of Kanazawa, Japan (1985–1986), the University of British Columbia, Canada (1988–1992), and the Institute of Plant Protection, Poznan, Poland (2001). His extensive research experience concerns insect–plant interactions, spanning toxicological, physiological and agricultural aspects. Honoured with an Indian National Science Academy (INSA) medal and the Kothari Scientific Research Institute award, he has authored over 140 research papers and articles and is the author/editor of the books *Insecticides of Natural Origin*, *Phytochemical Biopesticides*, *Microbial Biopesticides* and *Predators and Parasitoids*. He has also been an informal consultant to the Board of Science and Technology for International Development (BOSTID), the National Research Council (NRC) of the USA and at the International Centre for Insect Physiology and Ecology (ICIPE), Nairobi.

G.S. Dhaliwal, a Fellow of the National Environmental Science Academy (NESA), Society of Plant Protection Sciences and Society of Pesticide Sciences, India, is Professor of Ecology in the Department of Entomology at the Punjab Agricultural University, Ludhiana, India. Having completed his PhD in Entomology at the Indian Agricultural Research Institute (IARI), New Delhi, in 1972, he was awarded the Gurprasad Pradhan Gold Medal and became a postdoctoral fellow at the International Rice Research Institute, Manila, for 2 years. He has authored/edited more than 30 books on different aspects of pest management and the environment. Honoured with the Best Scientist Award of NESA, he is the founding President of the Indian Society for the Advancement of Insect Science and the Society of Biopesticide Sciences, India, and President of the Indian Ecological Society as well as Vice-President of the Indian Society of Allelopathy and the Society of Pesticide Science, India. He is a member of the World Food Prize Nominating Academy, The World Food Prize Foundation, Des Moines, Iowa.

Gerrit W. Cuperus, was a Regent's Professor and Integrated Pest Management Coordinator at Oklahoma State University for over 20 years. Dr Cuperus obtained his PhD in 1982, joined the Department of Entomology at Oklahoma State University and has since been involved in national IPM programmes of the USA aiming at an interdisciplinary focus to solve management issues. Dr Cuperus has chaired and served in different capacities in various national committees on food safety and pest management. He has made specific contributions in extension/research and has won distinguished service awards from the US Department of Agriculture (USDA). His research efforts, focused on stored-product pest management, have helped to build the Stored Product Research and Education Center (SPREC) at Oklahoma State University. He has authored over 60 research papers and articles and is an editor of *Successful Implementation of IPM for Agriculture Crops* (1992) and *Stored Product Management* (1995).

Contributors

Ramesh Arora, *Department of Entomology, Punjab Agricultural University, Ludhiana 141 004, India*

Waheed I. Bajwa, *Department of Entomology and Integrated Plant Protection Center, Oregon State University, Corvallis, OR 97331, USA*

R.C. Berberet, *Department of Entomology and Plant Pathology, Oklahoma State University, Stillwater, OK 74078, USA*

David Bergvinson, *International Maize and Wheat Improvement Center (CIMMYT), El Batán, Mexico CP 56130, Mexico*

Julio S. Bernal, *Department of Entomology, Biological Control Laboratory, Texas A&M University, College Station, TX 77843–2475, USA*

William M. Coli, *Department of Entomology, University of Massachusetts, Amherst, MA 01003, USA*

Richard S. Cowles, *Connecticut Agricultural Experiment Station, Valley Laboratory, PO Box 248, Windsor, CT 06095, USA*

G.W. Cuperus, *Department of Entomology and Plant Pathology, Oklahoma State University, Stillwater, OK 74078, USA*

G.S. Dhaliwal, *Department of Entomology, Punjab Agricultural University, Ludhiana 141 004, India*

Paul Guillebeau, *Department of Entomology, University of Georgia Cooperative Extension Service, Athens, GA 30602, USA*

Larry J. Gut, *Department of Entomology, Michigan State University, East Lansing, MI 48824, USA*

K.M. Heinz, *Department of Entomology, Biological Control Laboratory, Texas A&M University, College Station, TX 77843–2475, USA*

Craig S. Hollingsworth, *Department of Entomology, University of Massachusetts, Amherst, MA 01003, USA*

Marcos Kogan, *Department of Entomology and Integrated Plant Protection Center, Oregon State University, Corvallis, OR 97331, USA*

Opender Koul, *Insect Biopesticide Research Centre, 30 Parkash Nagar, Jalandhar 144 003, India*

David E. Legg, *Department of Renewable Resources, University of Wyoming, Laramie, WY 82071, USA*

James R. Miller, *Department of Entomology, Michigan State University, East Lansing, MI 48824, USA*

R.T. Noyes, *Department of Biosystems and Agricultural Engineering, Oklahoma State University, Stillwater, OK 74078, USA*

Imre S. Otvos, *Natural Resources Canada, Canadian Forest Service, Pacific Forestry Centre, 506 West Burnside Road, Victoria, BC, V8Z 1M5, Canada*

Jarrad Prasifka, *Department of Entomology, Biological Control Laboratory, Texas A&M University, College Station, TX 77843–2475, USA*

M. Sétamou, *Department of Entomology, Biological Control Laboratory, Texas A&M University, College Station, TX 77843–2475, USA*

C. Michael Smith, *Department of Entomology, Kansas State University, Manhattan, KS 66506–4004, USA*

Lukasz L. Stelinski, *Department of Entomology, Michigan State University, East Lansing, MI 48824, USA*

Donald R. Thomson, *DJS Consulting Services, LLC, 3015 SW 109 Street, Seattle, WA 98146, USA*

Robert H.J. Verkerk, *Department of Biological Sciences, Imperial College London, Silwood Park, Ascot, Berkshire SL5 7PY, UK*

Preface

The concept of integrated pest management (IPM) excelled during the mid-1970s when environmental, health and production problems associated with dependence on large-scale use of synthetic organic pesticides came into the limelight. Following some large-scale successes with IPM based on biological control systems, improved profitability and pesticide reduction IPM has moved from a peripheral to the central stage of pest suppression. Today it is considered to be the springboard to sustainable crop management. Over the years more than 70 definitions of IPM (including related terms) have been proposed. Although some new terms such as 'biointensive IPM' and 'ecologically based pest management' have been suggested, the essence of all the definitions is the promotion of compatibility of management tactics to ensure economic and ecological sustainability. There have been many success stories, particularly in the developed world, and many bottlenecks, more so in developing countries. The availability of modern tools and transgenic crop-protection technology has opened new opportunities and challenges. All these issues form the focus of the book, where they have been discussed by world authorities in their respective areas of specialization.

With the growing interest in IPM, opportunities and challenges have come to the fore and it is necessary to understand the potential in such programmes. To begin with, the book outlines the historical perspective of IPM in the first chapter and sets the stage for the discussion on potential, constraints and challenges involved in IPM. It covers the era of traditional approaches, from ancient times to 1938, the era of pesticides, from 1939 to 1975 and the current era of IPM, from 1976 onward. The significance of the 'farmer first' concept in IPM development and implementation is stressed. The potential of different management tactics in future IPM programmes has been discussed in various chapters, such as cultural practices, the relevance of modelling in the successful implementation of IPM, the manipulation of tritrophic interactions for the systems and the role of behaviour-modifying chemicals.

There are several challenges and opportunities for transgenic insecticidal cultivars, which have been discussed in Chapter 6, and the role of *Bacillus thuringiensis* (*Bt*) insect-resistant transgenes, molecular markers, cloning and sequencing plant resistance genes in host-plant resistance to pests has been highlighted in Chapter 7. The risk–benefit analysis of different groups of pesticides (i.e. insecticides, fungicides and weedicides) with respect to biological controls is an important component of IPM implementation. An effort has been made to make pesticides more compatible with IPM by improving pesticide selectivity via manipulating various spray parameters such as placement, timing and formulations, or through official policies and regulations and these are comprehensively discussed.

Chapter 9 focuses on the manipulation of host finding and acceptance behaviours to shift highly mobile and discriminating insect populations to plants or traps outside the valued crops. Potential tools for implementing the 'push–pull strategy' have been explained and several models have been proposed to demonstrate the application of behavioural manipulation in trap crops. The consumer response to IPM has been discussed in Chapter 11 and the various constraints and bottlenecks have been highlighted.

The forest ecosystem is much more complex, resilient and longer term than that of agriculture, and the threshold level of damage caused by insects or pathogens is much higher in forestry than what most consumers are willing to accept on or in their fruits or vegetables. Therefore, IPM has an important role to play in silviculture pest management and this has been discussed comprehensively in Chapter 10.

The role of IPM in sustaining productivity in future has been discussed in Chapter 12. The contribution of IPM in meeting economic, environmental and social mandates has been elaborated. The role of diagnostic tools, weather forecasting, transgenic plants, biological control and chemical pesticides in future IPM programmes has been highlighted and the strategies to meet the challenges of pest adaptation have been outlined. The need for improved information transfer among all groups involved in the development, implementation and application of IPM has been stressed. Finally, it is essential to know the status of IPM in developed versus developing countries. Therefore, the potentials and the constraints between the two worlds have been compared extensively in the last chapter of this book.

We received a tremendous response and support from all the authors for preparing their chapters in tune with the theme of the book, for which we express our gratitude to them. We are also thankful to Tim Hardwick at CABI Publishing for his cooperation and help at various stages in the preparation of this volume. Through this IPM book we also want to pay homage to Prof. Bill Brown, who could not complete his chapter for this volume due to his untimely demise in January 2003. Prof. Brown, worked in Nigeria, Thailand, South Korea and Bolivia after becoming a Professor of Plant Pathology and Cooperative Extension IPM Coordinator at Colorado State University. He had a passion for teaching plant pathology and the philosophy of IPM.

We hope the book will prove useful to all those interested in promoting the cause of IPM in formal and informal applications in both developed and developing countries, so that sustainability in the agricultural system and environmental protection for future generations is achieved.

Opender Koul
G.S. Dhaliwal
Gerrit W. Cuperus

1 Integrated Pest Management: Retrospect and Prospect

G.S. Dhaliwal,[1] Opender Koul[2] and Ramesh Arora[1]
[1]*Department of Entomology, Punjab Agricultural University, Ludhiana 141 004, India;*
[2]*Insect Biopesticide Research Centre, Jalandhar 144 003, India*
E-mail: gsd251@redifmail.com; koul@jla.vsnl.net.in

Introduction

In the history of the world, no century can match the population growth of the one that has just come to a close. We entered the 20th century with fewer than 2 billion people and left it with more than 6 billion. At the current pace, when the population increases by about 78 million per year, the global population is projected to rise to over 8 billion by 2025 (Hinrichsen and Robey, 2000). Thus, the population may grow even more in the 21st century, but in a very different way. The new century's growth will occur almost exclusively in the developing countries – among people with limited financial resources. The developed countries, which almost doubled their population in the 20th century, will grow slowly or not at all. Over half of the world population growth will occur in Asia and one-third will be in Africa. The world farmers will thus need to produce enough food for the expanding population (Dhaliwal and Arora, 2001).

The future of world food supply is closely linked with the pattern of population growth. In many countries over the past two decades, growth in food supply has lagged behind population growth. Worldwide, the grain harvest increased by 1% annually between 1990 and 1997, a rate of growth substantially slower than the average population growth rate of 1.6% in the developing world (Hinrichsen and Robey, 2000). Since humans started practising agriculture about 10,000 years ago, the increased need for food that resulted from increasing population or rising living standards was met by bringing a greater area of land under cultivation. However, during the last century, larger needs for food were mostly met by increasing productivity per unit of land. This trend is likely to continue as the populations increase and the land is required for purposes other than agriculture (FAO, 2001).

Food production per unit of land is limited by many factors, including fertilizer, water, genetic potential of the crop and the organisms that feed on or compete with food plants. These organisms that interfere significantly with the productivity or quality of plants considered useful for humans are called pests. Broadly speaking, a pest is any organism that is in competition with humans for some resource. In agriculture, pests occur in many groups of organisms, including plants, arthropods (insects, mites), fungi, bacteria, nematodes, viruses, viroids, mycoplasma-like organisms, rodents, birds and other vertebrates. Food plants of the world are damaged by 10,000 species of insects, 30,000 species of weeds, 100,000 dis-

 Integrated Pest Management: Potential, Constraints and Challenges (eds O. Koul, G.S. Dhaliwal and G.W. Cuperus)

eases (caused by fungi, viruses, bacteria and other microorganisms) and 1000 species of nematodes (Hall, 1995).

The global losses due to various categories of pests vary with the crop, the geographical location and the weather. Total yield losses from different pests of all crops have been estimated to be US$500 billion worldwide. Despite the plant-protection measures adopted to protect the principal crops, 42.1% of attainable production is lost as result of attack by pests. However, if no control measures were used to protect crops, the figure would be 69.8%. Animal pests account for 15.6% loss of production, pathogens 13.3% and weeds 13.2% (Oerke *et al.*, 1994).

Evolution of Management Tactics

During ancient times, humans had to live with and tolerate the ravages of crop pests, but they gradually learned to improve their condition through trial-and-error experiences. Over the centuries, farmers developed a number of mechanical, cultural, physical and biological control measures to minimize the damage caused by phytophagous insects. Synthetic organic insecticides developed during the mid-20th century provided spectacular control of these pests and resulted in the abandonment of traditional pest-control practices. Thus, the evolution of the concept of pest management spans a period of more than a century (Table 1.1). Many components of pest management were developed in the late 19th and early 20th centuries. Rapidly developing technologies and changing societal values had their impact on pest-control tactics also. The history of agricultural pest control thus has three distinct phases, namely, the era of traditional approaches, the era of pesticides and the era of integrated pest management (IPM) (Metcalf, 1980; Dhaliwal *et al.*, 1998).

Era of traditional approaches (ancient–1938)

Cultural and mechanical practices, such as crop rotation, field sanitation, deep ploughing, flooding, collection and destruction of damaging insects/insect-infested plants, etc., developed by farmers through experience were among the oldest methods developed by humans to minimize the damage caused by insect pests (Smith *et al.*, 1976). These were followed by the use of plant products from neem, chrysanthemum, rotenone, tobacco and several other lesser-known plants in different parts of the world. The Chinese were probably the pioneers in the use of botanical pesticides as well as biological control methods for the management of insect pests of stored grains and field crops (Dhaliwal and Arora, 1994a).

However, systematized work on many important tactics of pest control, including the use of resistant varieties, biological control agents and botanical and inorganic insecticides, was done in the USA from the end of the 18th to the end of the 19th century. Remarkable success was achieved in the management of grape phylloxera, caused by *Viteus vitifoliae* (Fitch), by the grafting of European grapevine scions to resistant North American rootstocks during the 1880s. At around the same time, cottony cushion scale, *Icerya purchasi* Maskell, which was causing havoc in the citrus industry in California, USA, was successfully controlled by release of the vedalia beetle, *Rodolia cardinalis* (Mulsant), imported from Australia (DeBach, 1964).

A number of synthetic inorganic insecticides containing arsenic, mercury, tin and copper were also developed towards the end of the 19th and the beginning of the 20th century. With the development of these insecticides, the focus of research in pest control slowly shifted from ecological and cultural control to chemical control, even before the development of synthetic organic insecticides (Perkins, 1980).

Era of pesticides (1939–1975)

The synthetic inorganic insecticides were broad-spectrum biocides and were highly toxic to all living organisms. These were followed in due course by the synthetic organic insecticides, such as alkyl thiocyanates, lethane, etc. The era of pesticides, however, began with the discovery of the insecticidal

Table 1.1. Landmarks in the history of agricultural pest management (modified after Dhaliwal *et al.*, 1998).

Period	Landmark(s)
Ancient	The Chinese used wood-ash for the control of insect pests in enclosed spaces and botanical insecticides for seed treatment. They also used ants for biological control of stored grain as well as foliage-feeding insects. In India, neem leaves were placed in grain bins to keep away troublesome pests. In the Middle and Near East, powder of chrysanthemum flowers was used as an insecticide
1762	The myna (a bird) from India was imported for the control of locusts in Mauritius
1782	'Underhill' variety of wheat reported resistant to Hessian fly in the USA
1831	'Winter Majetin' variety of apple reported resistant to woolly apple in the USA
1855	A. Fitch reported the role of ladybird beetles, green lacewings and other predacious insects in the control of insect pests of crops
1858	Pyrethrum first used for insect control in the USA
1889	Biological control of cottony cushion scale on citrus in the USA by use of the vedalia beetle imported from Australia
1890	Control of grape phylloxera in Europe by grafting of European grapevine scions to resistant North American rootstocks
1923	Multiple-component suppression techniques involving the use of resistant varieties, sanitation practices and need-based application of insecticides developed for the control of boll-weevil in the USA
1939	• Insecticidal properties of DDT reported by Paul Muller in Switzerland • *Bacillus thuringiensis* Berliner first used as a microbial insecticide
1941	Insecticidal activity of hexachlorocyclohexane (HCH) discovered in France
1946	Parathion, the first organophosphatic insecticide developed
1948	'Doom' based on *Bacillus popilliae* Dutky and *Baciilus lentimorbus* registered in the USA for the control of Japanese beetle larvae on turf
1951	• R.H. Painter published his classic book *Insect Resistance in Crop Plants* • Introduction of first carbamate insecticide, isolan
1959	• Concept of integrated control involving integration of chemical and biological control introduced • Concept of economic injury level and economic threshold developed by V.M. Stern and co-workers
1962	Publication of the book *Silent Spring*, by Rachel Carson, which dramatized the impact of the misuse and overuse of pesticides on the environment
1964	Publication of the book *Biological Control of Insect Pests and Weeds*, by Paul DeBach, which established biological control as a separate discipline in entomology
1975	• Elcar (*Helicoverpa nucleopolyhedrovirus* (NPV)) registered for the control of boll-worm and tobacco budworm on cotton • First insect growth regulator (Methoprene) registered for commercial use in USA • Publication of the book *Introduction to Insect Pest Management* by R.L. Metcalf and W.H. Luckmann, which was the first comprehensive treatise on IPM and established the concept on a firm footing
1980	The interest in botanical pesticides revived and the First International Conference on Neem was held at Rottach-Egern, Germany
1987	Development of first transgenic plant, reported by M. Vaeck and co-workers of Belgian biotechnology company, Plant Genetic System, by transferring *B. thuringiensis* δ-endotoxin gene to tobacco for the control of *Manduca sexta* (Johannsen)
1989	An IPM Task Force was established to garner international support for the development and implementation of IPM programmes. A team of consultants appointed by the Task Force reviewed the status of IPM and made recommendations. The Task Force was later reconstituted as the Integrated Pest Management Working Group (IPWG) in 1990
1992	• Concept of environmental economic injury levels proposed by L.P. Pedigo and L.G. Higley • Dr Edward F. Knipling and Dr Raymond C. Bushland were awarded the World Food Prize for developing sterile-insect technique

Continued

Table 1.1. *Continued.*

Period	Landmark(s)
	• United Nations Conference on Environment and Development (Rio de Janeiro, Argentina) assigned a pivotal role to IPM in the agricultural programmes and policies envisaged as part of its Agenda 21
1994	A Task Force consisting of FAO, the World Bank, UNDP and UNEP co-sponsored the establishment of the Global IPM Facility with the Secretariat located at FAO, Rome
1995	Dr Hans R. Herren was awarded the World Food Prize for developing and implementing the world's largest biological control project for cassava mealy bug, which had almost destroyed the entire cassava crop of Africa
1996	Insect-resistant transgenic (*Bt*) cotton, maize and potato were commercialized in the USA
1997	Dr Ray F. Smith and Dr Perry L. Adkisson were awarded the World Food Prize for their pioneering work in the development and implementation of the IPM concept
2002	*Bt* cotton approved for commercialization in India

DDT, dichlorodiphenyltrichloroethane; FAO, Food and Agriculture Organization; UNDP, United Nations Development Programme; UNEP, United Nations Environment Programme; *Bt*, *Bacillus thuringiensis*.

properties of 2,2-*bis*(p-chlorophenyl)-1,1, 1-trichloroethane (DDT) by Paul Muller in 1939. The impact of DDT on pest control is perhaps unmatched by any other synthetic substance and Muller was awarded a Nobel Prize for this work in 1948.

DDT was soon followed by a number of other insecticides, such as hexachlorocyclohexane (HCH), chlordane, aldrin, dieldrin, heptachlor (organochlorine group), parathion, toxaphene, schradan, O-ethyl O-4(nitrophenyl) phenyl phosphonothionate (EPN) (organophosphorus group) and allethrin (synthetic pyrethroid), during the 1950s and a large number of other popularly used organophosphates and carbamates in the ensuing decade.

Due to their efficacy, convenience, flexibility and economy, these pesticides played a major role in increasing crop production. The success of high-yielding varieties of wheat and rice, which ushered in the 'green revolution', was partially due to the protective umbrella of pesticides (Pradhan, 1983). The spectacular success of these pesticides masked their limitations. The intensive and extensive use, misuse and abuse of pesticides during the ensuing decades caused widespread damage to the environment. The most serious effects of pesticides are those on human health and life. It has been reported that at least 3 million, and perhaps as many as 25 million, agricultural workers are poisoned each year by pesticides, and some 20,000 deaths can be directly attributed to agrochemical use (Meerman *et al.*, 1997). Nearly three-quarters of the documented deaths take place in the Third World, even though it consumes only 15% of the global pesticide output. In addition, long-term teratogenic, carcinogenic and mutagenic effects of pesticides are well documented. In a landmark report, the National Resource Defence Council (NDRC) of the USA reported that one out of every 3400 children between 1 and 5 years of age could one day get cancer because of the pesticides they consumed as young children (NDRC, 1989).

Only a small amount (< 1%) of the pesticide applied to a crop reaches target pests and the remaining (> 99%) enters different components of the environment to contaminate soil, water, air, food, feed, forage and other commodities. Nearly 100% of the human population has been found to contain some residues of pesticides, such as DDT and HCH (Dhaliwal and Arora, 2001). In developing countries, such as India, a large proportion of market samples of nearly all types of food commodities have been found to contain pesticide residues above the legal maximum residue limits (MRLs). In addition to human beings, the whole range of living organisms, including natural enemies, pollinators, domestic and wild animals, birds, fish and other aquatic organisms and even soil fauna are affected by the use of insecticides in agriculture (Dhaliwal and Singh, 2000). In addition, pest problems in some crops increased following the continuous application of pesticides. This, in turn, further increased the consumption of pesticides, resulting in the phenomenon of the pesticide treadmill (Altieri, 1995). The combined impact of all these problems, together with the rising cost of pesticides, provided the necessary feedback for limiting the use of the chemical control strategy and led to the development of the IPM concept.

Era of IPM (1976 onwards)

Although many IPM programmes were initiated in the late 1960s and early 1970s in several parts of the world, it was only in the late 1970s that IPM gained momentum. The first major IPM project in the USA, commonly called the Huffaker project, spanned 1972–1978 and covered six crops, i.e. lucerne, citrus, cotton, pines, pome and stone fruits, and soybean. This was followed by another large-scale IPM project called the Consortium for Integrated Pest Management (CIPM) (1979–1985), which focused on lucerne, apple, cotton and soybean. The average adoption of IPM for four crops was claimed to be about 66% over 5.76 million ha (Frisbie and Adkisson, 1985a,b). In 1993, the US government set up the National IPM Initiative and submitted that implementing IPM practices on 75% of the nation's crop area by 2000 was a national goal (Sorenson, 1994). In a recent accounting of the progress by the US Department of Agriculture (USDA) in achieving this goal, it has been estimated that some level of IPM has been implemented on about 70% of the US crop acreage (Baron, 2002).

National IPM programmes were launched in the late 1980s and early 1990s in several developing countries. The most outstanding success has been the Food and Agriculture Organization (FAO) IPM programme for rice in South-east Asia. By the end of 1995, 35,000 trainers and 1.2 million farmers had been exposed to IPM through this programme (FAO, 1995). A recent development at FAO in support of IPM is the establishment of the Global IPM Facility, co-sponsored by the United Nations Development Programme (UNDP), the United Nations Environment Programme (UNEP) and the World Bank. The concept is in response to the United Nations Conference on Environment and Development, held in Rio de Janeiro, Brazil, in 1992, which assigned a central role for IPM in agriculture as part of Agenda 21. The Facility will serve as a coordination, consulting, advising and promoting agency for the advancement of IPM worldwide (Kogan, 1998).

Major efforts in implementing IPM in irrigated rice have been carried out in Asia by the FAO through the Inter-Country Programme for the Development and Application of Integrated Pest Control in Rice in South-east Asia. This programme remains one of the best examples of IPM implementation in the tropical region. It involves purposeful, direct efforts to change farmers' practices, in contrast to some more indirect routes of IPM technology diffusion in many industrialized, temperate environments. The programme itself has evolved into its present transnational form from a relatively small project supported by Australia in the late 1970s, following the large-scale pest outbreaks in several South-east Asian countries (NRI, 1992).

The first phase of the FAO programme (1980–1986) focused on developing and testing the technical aspects of the IPM concept in its seven participating countries, namely, Bangladesh, India, Malaysia, the Philippines, Sri Lanka and Thailand. More recently, the project has been directed towards enhancing farmers' adoption of IPM. Australia, The Netherlands and the Arab Gulf fund have supported the programme. One significant accomplishment of the programme has been to cause policy changes within several governments, in the form of official support of IPM as the means for national plant protection in the Philippines, Indonesia, India, Sri Lanka and Malaysia (NRI, 1992).

A case study of the National IPM Programme in Indonesia as a part of the regional programme during 1989–1991 provides an interesting scenario. Following research findings showing the relation between brown planthopper outbreaks and high pesticide use, the Indonesian government banned the use of 57 broad-spectrum pesticides and declared IPM as the national pest-control strategy. These measures, together with the gradual abolition of the high subsidies on pesticides, created a favourable climate for the large-scale implementation of IPM (Fig. 1.1).

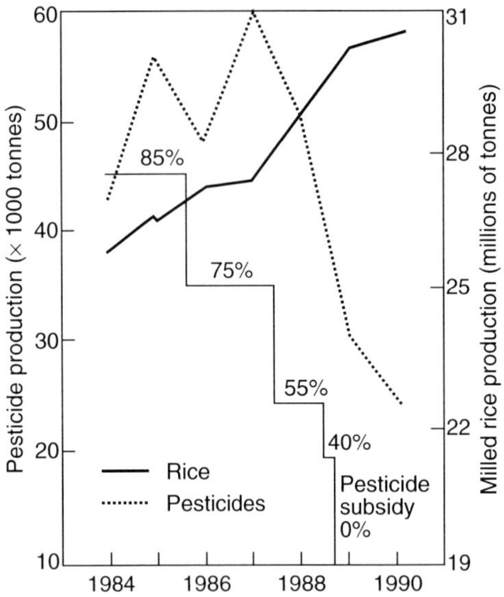

Fig. 1.1. Impact of pesticide subsidy on rice production in Indonesia (after FAO, 1990).

To assess the impact of IPM on pest control and yield, the latter were compared with those of non-IPM farmers (van de Fliert, 1993). IPM farmers really decreased their frequency of pesticide sprays during and after the training to a level consistently lower than that of non-IPM farmers (Table 1.2). Consequently, IPM farmers' expenditure on insect chemical control decreased. The number of farmers not applying pesticides was also significantly higher among IPM farmers. In spite of this, IPM farmers obtained higher yields than non-IPM farmers, which, in addition to lower pest-control expenditure, resulted in higher returns from the crop. Additionally, there were fewer variations in fields among IPM farmers than among non-IPM farmers, indicating less risky farm management under the IPM regime. Timeliness and adequacy of various cultivation and pest-control practices, resulting from better monitoring and decision-making skills, seem to be most important in this respect (van de Fliert, 1998).

During the second phase of the National IPM Programme (1994/95), 310,550 farmers participated in rice IPM field schools. IPM farmers applied pesticides, on average, 1.25 times while non-IPM farmers made an average of 3.62 applications during the season. Similarly, IPM farmers used fewer formulations (cocktails) per spray event, with an average of 2.6 for IPM farmers versus 7.2 for non-IPM farmers.

China, which has been experimenting with IPM for the control of rice pests since the early 1980s, was also invited to join the FAO project in 1989. During 1989/90 alone, nearly 160,000 farmers from over 2000 villages received IPM training. Compared with untrained farmers, IPM-trained farmers saved roughly a third of the pesticides in rice cultivation and still obtained a 7% higher yield. It was estimated that the investment in IPM training generated a return of more than

Table 1.2. Insect pest-control practices, expenditures and yields of IPM and non-IPM farmers in two villages in Central Java, before, during and after IPM training (from van de Fliert, 1993).

	Before training		During training		After one season		After two seasons	
Parameter	IPMF	NIPMF	IPMF	NIPMF	IPMF	NIPMF	IPMF	NIPMF
Frequency of spray applications (no./season)	1.4	1.5	0.7	1.5	0.8	1.3	0.3	0.8
Frequency of granular applications (no./season)	0.4	0.3	0.7	0.9	0.7	0.8	0.6	0.7
Farmers not using pesticides (%)	26	31	41	19	46	24	50	43
Average insect control cost (Rs 1000/ha)	32	22	18	31	18	21	9	19
Average yield (t/ha)	5.38	5.11	5.77	4.64	3.70	3.11	6.38	5.68

IPMF, IPM farmers; NIPMF, non-IPM farmers.

400%. Encouraged by these results, the Ministry of Agriculture has set up a National Steering Committee for the comprehensive prevention and control of diseases and insect pests to protect the nation's rice crop and increase profits. The Committee conducts IPM tests, gives demonstrations and makes appraisals (Quinghua, 1995).

In the Philippines, a National IPM Programme was launched in 1993 and a total of 40,024 farmers were trained during 1993–1995, among whom 36,024 are rice farmers. Of 1632 farmers' field schools (FFSs), 1470 were devoted to rice farmers. The increase in rice yield obtained by IPM farmers varied from 4.7 to 62%. The expenditure on pesticide use (15% of total cost) was almost eliminated in the case of IPM farmers (Kenmore, 1997).

In Vietnam, during 1992–1995, 5941 FFSs were organized in 3095 villages (out of 9274 villages in the country), with a total of 173,650 farmers trained. The number of insecticide applications by IPM farmers was reduced by 80–90% and, in some agroecological regions, there was almost no use of insecticides. IPM fields produced 150–460 kg/ha more rice than non-IPM fields. One study of over 1300 villages in Vietnam showed a 4% yield increase in rice and over 20% increase in profits (Kenmore, 1997).

In Bangladesh, the Cooperative for American Relief to Everywhere (CARE)-Bangladesh began introducing IPM activities into an ongoing rice irrigation project in the late 1980s. Most of the farmers that CARE worked with were very small, owning less than half an acre of land. The farmers were trained in rice–fish culture and IPM. A survey among the farmers trained in rice–fish culture and IPM revealed that, during 1992/93, virtually all of them stopped using pesticides and satisfactory control of insect pests was being provided by fish and naturally occurring parasitoids and predators (Table 1.3).

Farmers cultivating fish in their rice fields totally eliminated the use of pesticides even without training in IPM. However, only 20% of the farmers could cultivate fish, mainly due to lack of adequate water, the poor water-holding capacity of the soil and the risk of flooding. Those farmers who could not cultivate fish still reduced their pesticide use by 76% with IPM training. By harvest time, some farmers were able to harvest quantities of fish worth as much as their rice. These results underline the need for undertaking an IPM programme as part of overall rural development programmes (Kamp *et al.*, 1993).

During 1995, farmers at IPM schools applied an average of 0.19 spray/season as compared with 1.65 sprays/season during the pretraining period. The average yield increased by 17% from 3.76 t/ha during the pretraining period to 4.4 t/ha after IPM training. The average number of pests per ten rice hills decreased from 20 to 11 and the number of natural enemies increased from 18 to 31.

In India, FFSs were organized at locations covering all the major rice growing areas of the country during 1994/95. Pesticide use in IPM fields decreased by 50–100% as compared with non-IPM areas. There was an increase in yield between 6.2 to 42.1% in IPM fields as compared with non-IPM areas (Rajak *et al.*, 1997).

The success of the rice programme persuaded FAO to launch similar IPM Intercountry Programmes for vegetables (1996) and cotton (1999). Moreover, in the

Table 1.3. Effect of training in rice–fish culture and IPM on pesticide use by rice growers during 1992/93, Rangpur, Bangladesh (from Kamp *et al.*, 1993).

Parameter	Group I	Group II	Group III	Group IV
Number of farmers	121	58	972	60
Type of training	Rice–fish culture only	Rice–fish culture and IPM	IPM only	Nil
Reduction in pesticide use (%)	100	100	76	–
Increase in rice yield (%)	6	13	10	–

individual countries, FFS activities were started in a range of other crops, sometimes with external donor support, sometimes without. The Asian model of FFS IPM training has been applied in other continents as well, albeit with adjustments for different cropping and socio-economic conditions (Eveleens, 2002).

Origin of IPM Concept

The basic tactics of IPM were proposed and used to protect crop plants against the ravages of pests long before the term was coined. Most discussions of IPM include the concept of economic threshold, implying that whatever cost one applies to the pest control techniques should be returned from the production of the crop (McNeal, 1988). Another concept is that of team effort, where IPM strives to bring together as many disciplines and areas of interest as possible. The concern over pesticides as a main source of pest control is that it makes one totally reliant on such technology and, if that technology starts to fail, the only way out is a new replacement that promises an improvement over the old one, even if the transfer may also cause difficulties. Thus, in the absence of modern synthetic pesticides, crop protection specialists during the late 19th and early 20th centuries relied on pest biology and cultural practices to propose multitactical approaches, which could be considered as precursors of modern IPM systems. To be precise, the concept of technology packages assembled in IPM can help us to avoid some of the pitfalls inherent in reliance on a single technology (McNeal, 1988).

Integrated control

According to McNeal (1988), one of the activities most responsible for the genesis of IPM was the work done in cotton entomology in Arkansas in the 1920s. This research was overshadowed in the late 1940s as pesticides came along, but the work continued into the 1950s with the cotton scouting programme in Arkansas. However, apparently the idea of integrated control was first conceived by Hoskins *et al.* (1939) when they said:

> Biological and chemical control are considered as supplementing to one another or as the two edges of the same sword … nature's own balance provides the major part of the protection that is required for the successful pursuit of agriculture … insecticides should be used so as to interfere with natural control of pests as little as possible.

The credit for using the term 'integrated control' for the first time goes to Michelbacher and Bacon (1952), who, while working on the control of codling moth, *Cydia pomonella* (Linnaeus), stressed 'the importance of considering the entire, entomological picture in developing a treatment for any particular pest … All effort was directed towards developing an effective integrated control program of the important pests of walnut.' Subsequently, Smith and Allen (1954) stated that 'integrated control … will utilize all the resources of ecology and give us the most permanent, satisfactory and economical insect control that is possible'. Following this was a series of papers that established integrated control as a new trend in entomology (Kogan, 1998).

Stern *et al.* (1959) were the first to define integrated control as 'applied pest control, which combines and integrates biological and chemical control'. This definition remained in place through the late 1950s and early 1960s, but began to change soon in the early 1960s as the concept of pest management gained acceptance among crop protection specialists.

Pest management

The idea of managing insect-pest populations was proposed by Geier and Clark (1961), who called this concept 'protective population management', which was later shortened to 'pest management' (Geier, 1966). By the mid-1970s, integrated control and pest management coexisted essentially as synonyms. However, a synthesis of the two expressions had already become available when Smith and van den Bosch (1967) wrote: 'The determination of insect numbers is broadly under the influence of the total agroecosystem and a background role of the principal elements is essential to integrated pest population management'.

Integrated pest management

It was, however, in 1972 that the term 'integrated pest management' was accepted by the scientific community, after the publication of a report under the above title by the Council on Environmental Quality (CEQ, 1972). In creating this synthesis between integrated control and pest management, no obvious attempts seemed to have been made to advance a new paradigm. Much of the debate had already taken place during the 1960s and by then there was substantial agreement on the following issues (Kogan, 1998):

- 'Integration' means the harmonious use of multiple methods to control single pests as well as the impacts of multiple pests.
- 'Pests' are any organisms detrimental to humans, including invertebrate and vertebrate animals, pathogens and weeds.
- 'Management' refers to a set of decision rules based on ecological principles and economic and social considerations. The backbone for the management of pests in an agricultural system is the concept of economic injury level (EIL).
- 'IPM' is a multidisciplinary endeavour.

Alternative paradigms

Although the success of IPM has been accepted worldwide, some new terms have been proposed to lay emphasis on particular strategies. Frisbie and Smith (1991) proposed 'biologically intensive IPM' or 'biointensive IPM', which would rely on host-plant resistance, biological control and cultural control. In fact, the utilization of biological control and other non-chemical methods has been amply stressed in all IPM programmes. Recently, a special committee of the National Research Council's Board of Agriculture (NRC, 1996) proposed 'ecologically based pest management' (EBPM), emphasizing that it was:

- safe for farmers and consumers;
- cost-effective and easy to adopt and integrate with other crop protection practices;
- durable and without adverse environmental and safety consequences; and
- used with ecosystems as the ecological focus.

All these goals have been well taken care of by the concept of IPM.

Thus IPM is here to stay and to provide suitable solutions to future pest problems. Recently, Benbrook (2002) proposed a new term: the IPM continuum. According to the author, IPM systems exist in almost limitless variety along an IPM continuum. It includes four major zones/levels: no, low, medium and high or biointensive IPM. Farmers in the 'non-IPM' zone manage pests with routine pesticide applications. Low-end IPM depends on basic field sanitation, scouting and pesticide applications linked to thresholds. Medium-level IPM shifts a portion of the control burden to largely preventive measures and requires farmers to bypass most applications of pesticides because of the greater degree of reliance on beneficial organisms. High-level IPM systems manage pests largely through multitactic prevention-based interventions. Biointensive IPM (or Bio IPM) lessens pest pressure through management of ecological and biological processes and interactions.

Defining IPM

Since the first definition of integrated control (Stern *et al.*, 1959), more than 65 definitions of integrated control, pest management or IPM have been proposed. A broader definition was adopted by FAO Panel of Experts (FAO, 1967):

> Integrated pest control is a pest management system that, in the context of associated environment and population dynamics of the pest species, utilizes all suitable techniques and methods in as compatible a manner as possible and maintains pest populations at levels below those causing economic injury.

It is not simply the juxtaposition or superimposition of two control techniques but the integration of all suitable management techniques with the natural regulating and limiting elements of the environment. According to the National Academy of Sciences, IPM refers to an ecological approach in pest management in which all available necessary techniques are consolidated in a unified programme, so that pest populations can be managed in such a manner that economic damage is avoided and adverse side effects are minimized (NAS, 1969).

Most other contemporary definitions perpetuate the perception of an entomological bias in IPM because of the emphasis on pest populations and EIs, of which the former is not always applicable to plant pathogens and the latter is usually attached to the notion of an action threshold that is often incompatible with pathogen epidemiology or many weed-management systems. Smith (1978) defined IPM as a multidisciplinary ecological approach to the management of pest populations, which utilizes a variety of control tactics compatibly in a single coordinated pest-management system. In its operation, integrated pest control is a multitactical approach that encourages the fullest use of natural mortality factors, complemented, when necessary, by artificial means of pest management. In other words, IPM seeks to integrate multidisciplinary methodologies to develop pest-management strategies that are practical, effective, economical and protective of both public health and the environment (Smith *et al.*, 1976). IPM has also been defined as a pest population management system that utilizes all suitable techniques in a compatible manner to reduce pest populations and maintain them at levels below those causing economic injury (Frisbie and Adkisson, 1985a,b). Dr Ray F. Smith and Dr Perry Adkisson were awarded the 1997 World Food Prize for their pioneering work in the development and implementation of the IPM concept. However, in 1998, USDA came up with a definition that IPM is a sustainable approach that combines the use of prevention, avoidance, monitoring and suppression strategies in a way that minimizes economic, health and environmental risks (www.reeusda.gov/nipmn).

IPM is systematic approach to crop protection that uses increased information and improved decision-making paradigms to reduce purchased inputs and improve economic, social and environment conditions on the farm and in society (Allen and Rajotte, 1990). IPM is a comprehensive approach to pest control that uses combined means to reduce the status of pests to tolerable levels while maintaining a quality environment (Pedigo, 1991). IPM is also defined as the intelligent selection and use of pest-control tactics that will ensure favourable economic, ecological and sociological consequences (Luckman and Metcalf, 1994).

IPM is a dynamic and constantly evolving approach to crop protection in which all the suitable management tactics and available surveillance and forecasting information are utilized to develop a holistic management programme as part of a sustainable crop production technology (Dhaliwal and Arora, 2001). Here it needs to be emphasized that the aim of future IPM programmes should not be restricted merely to the efficient use of pesticides and product substitution (biorationals and botanicals in place of conventional insecticides) within an agricultural system that remains essentially unchanged (Table 1.4). Rather, these programmes should aim at fundamental structural changes through a better understanding of ecological processes and synergy between crops (van Veldhuizen and Hiemstra, 1993).

Kogan (1998) carried out numerical analyses of various definitions spanning the last 35 years and found that most of the authors depended on the following issues to capture the essence of the IPM concept:

- The appropriate selection of pest-control methods, used singly or in combination.
- The economic benefits to growers and to society.
- The decision rules that guide the selection of the control action.
- The need to consider impacts of multiple pests.

Taking into consideration all the above points and the current thought, Kogan (1998) put forward his definition:

Table 1.4. Approaches to insect pest management: retrospect and prospect (from Dhaliwal and Arora, 1994b).

		Pest management system			
No.	Parameter	Traditional	Industrial	Present IPM	IPM in sustainable agriculture
1	Goal	Reduce losses due to pests	Eliminate or reduce pest species	Reduce costs of production	Multiple-ecological, economic and social
2	Diversity	High	Low	Low or medium	High
3	Ecosystem stability	Uncertain	Highly unstable	Unstable	Striving towards stability and equilibrium
4	Spatial scale	Single farm	Single farm	Single farm or small region defined by pests	Biogeographical regions
5	Time scale	Long term	Immediate	Single season	Long-term steady-state oscillatory dynamics
6	Target	Single pest or closely related groups of pests	Single pest	Several pests around a crop and their natural enemies	Fauna and flora of a cultivated area and linkages with non-cultivated ecosystem
7	Criteria for intervention	Past experience	Calendar date or presence of pest	Economic threshold	Multiple criteria
8	Principal method	Cultural and mechanical measures	Pesticides	Resistant varieties, cultural practices, monitoring, product substitution, insecticide resistance management and multiple interventions	Agroecosystem design to minimize pest outbreaks and mixed strategies, including group action on an area-wide basis to complement pest controls aimed at individual fields
9	Research goal	Nil due to absence of organized effort	Improved pesticides	More kinds of interventions	Minimize need for intervention
10	Extension technique	Nil	Transfer of technology (TOT)	TOT	Complementarity between TOT and farmer-first (FF) mode
11	Effect on environmental quality	Usually negligible	Highly detrimental	Moderately detrimental	Negligible

IPM is a decision support system for the selection and use of pest control tactics, singly or harmoniously coordinated into a management strategy, based on cost/benefit analyses that take into account the interests of and impacts on producers, society and the environment.

Decision Making Systems

Pest management is a combination of processes that include decision making, taking action against a pest and obtaining the information to be used in reaching these decisions (Ruesink and Onstad, 1994). In assessing, evaluating and choosing a particular pest-control option, farmers are likely to take three major factors into account (Fig. 1.2):

- Farmers' perception of the problem and of potential solutions is the most important factor. Here, the farmer's ability to identify pests, his/her assessment of likely and potential pest losses and his/her opinion regarding the efficacy of different control options will affect the decision process.
- The way in which control options are assessed will depend on the farmers' objectives. Subsistence farmers may opt for a guaranteed food supply, while commercial farmers are more concerned with profit.
- The number of options that a farmer can feasibly use will depend on the constraints set by the resources available.

Various alternative pest-control options could be evaluated for their cost-effectiveness (Reichelderfer *et al.*, 1984):

- Determine from experimental results both the per hectare cost and a measure of effect of each alternative practice. If effectiveness can be measured in terms of output (yield and/or crop quality), use partial budgeting or other analytical techniques to evaluate alternatives. If effectiveness cannot be measured in these terms, proceed with determination of cost-effectiveness.
- Using the same units in which effectiveness is measured (e.g. reduction in pest numbers or damage), specify an effectiveness target that is appropriate, given the experimental data and information at hand (e.g. a 50% reduction in pest population).
- Multiply the cost of each practice times the effectiveness target and then divide that product by the actual level of effectiveness achieved by the practice. This gives a set of relative cost-effectiveness figures.
- Compare the cost-effectiveness of alternative practices. The practice that has associated with it the lowest cost to achieve the effectiveness target is the most cost-effective practice.

Implementation

Although IPM has been accepted in principle as the most attractive option for the protection of agricultural crops from the ravages of insect and non-insect pests, implementation at the farmers' level has as yet been rather limited. Some of the important constraints to wider adoption of IPM and suggested measures to overcome them are discussed in this section.

Constraints in IPM implementation

The Consultant Group of the IPM Task Force has conducted an in-depth study of the constraints on the implementation of IPM in developing countries, which can be categorized into the following five main groups (NRI, 1992; Alam, 2000).

Institutional constraints

IPM requires an interdisciplinary, multifunctional approach to solving pest problems. Fragmentation between disciplines, between research, extension and implementation and between institutes leads to a lack of institutional integration. Secondly, both the national programmes of developing countries and the donor agencies have lacked a policy commitment to IPM in the context of national economic planning and

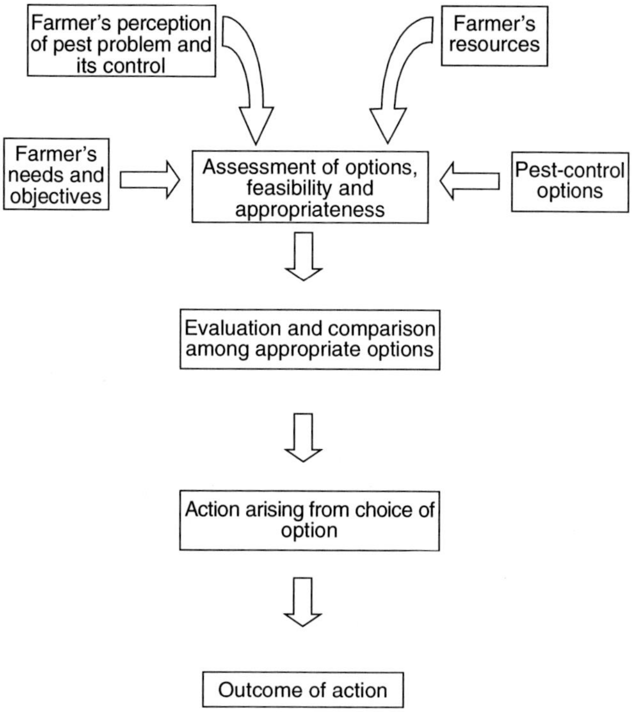

Fig. 1.2. The process of decision making in IPM (after Reichelderfer *et al.*, 1984).

agricultural development. This has resulted in a low priority for IPM from national programmes and donors alike. Thirdly, the traditional top-down research in many cases does not address the real needs of farmers, who eventually are the end-users and who elect to adopt or reject the technology based on its appropriateness. Institutional barriers to research scientists in national programmes conducting on-farm research in developing countries are real and need to be addressed.

Informational constraints

The lack of IPM information that could be used by farmers and extension workers is a major constraint in implementation. In a recent study regarding the implementation of IPM in Haryana, India, it was found that more than three-quarters of the farmers were not even aware of the concept of IPM. Even those who were aware of the concept reported that they lacked the skills necessary to practise IPM (Alam, 2000). While the individual control techniques are well known, little knowledge is available on using them in an integrated fashion under farm conditions. The lack of training materials, curricula and experienced teachers on the principles and practice of IPM is another major constraint. In many cases, the field level extension workers are not sufficiently trained in IPM to instil confidence in the farmers.

Sociological constraints

The conditioning of most farmers and farm-level extension workers by the pesticide industry has created a situation where chemicals are presented as highly effective and simple to apply. This acts as a major constraint in IPM implementation. There appears to be a direct conflict between the industry's objective of more sales and the IPM message of rational pesticide use in the eyes of farmers. There is a need for private industry and public-sector extension agencies to work in a more complementary manner. A majority of the farmers in a recent study in Haryana, India, expressed their lack of faith in IPM. They considered IPM practices to be risky as compared with the use of chemical pesticides (Alam, 2000).

Economic constraints

A major constraint, even if IPM is adopted in principle, is the funding for research, extension and farmer training needed for an accelerated programme. IPM must be viewed as an investment and, as with other forms of investment, requires an outlay. Starting with the Huffaker project, the US Environmental Protection Agency (USEPA) and USDA have financed a number of IPM programmes, resulting in substantial progress in the development and implementation of IPM. In the long run, IPM programmes may become self-generating due to savings on resource inputs for production. A majority of the farmers purchase pesticides on credit and depend on shopkeepers and pesticide dealers for information about pest-control methods.

Political constraints

The relatively low status of plant-protection workers in the administrative hierarchy is a constraint to general improvement in plant protection. Associated with the above is the moral and financial standing of these workers. The continuance of pesticide subsidies by the government for political reasons and their tie-up with the government-provided credit for crop production act as a major constraint to farmers' acceptance of IPM. Various vested interests associated with the pesticide trade also act as a political constraint on the implementation of IPM (Jiggins, 1996).

Strategies for IPM implementation

Acceleration of IPM implementation in developing countries requires farmers' participation, increased government support, an improved institutional infrastructure and a favourable environment.

Farmers' participation

It would not be an exaggeration to say that the dawn of civilization started with farmer innovation. Ever since that day, farmers have improved ways of growing crops through successive innovations. Prior to the emergence of crop-protection sciences and even before the broad outlines of the biology of pests were understood, farmers evolved many cultural, mechanical and physical control practices for the protection of their crops from insects and non-insect pests (Smith *et al.*, 1976). Farmers' innovations were the only source of improvements in crop production and protection technology until formal research by on-station scientists started complementing it during the late 18th and 19th centuries (Haverskort *et al.*, 1991).

Unfortunately, with the advent of modern high-tech agriculture, comprising of high-yielding varieties (HYVs), fertilizers and pesticides, farmers have been completely displaced from the research and development process. Instead, this role has been usurped by private industry and government agencies. The technology generated by farm scientists is being transferred through extension agencies to farmers. The new technology package has created a number of ecological and environmental problems. The alternative path of sustainable agriculture requires farmers' participation at every step of the research and development process in order to draw on their understanding of the local conditions and constraints, their innovativeness and their skills at making the best possible use of limited resources.

Placing the farmer at the centre of the development process is wholly consistent with the IPM goal of making the farmer a confident manager and decision maker, free from dependence on a constant stream of pest-control instructions from outside. The role of researchers, extension workers and non-governmental organizations (NGOs) is to act as consultants, facilitators and collaborators, stimulating and empowering the farmers to analyse their own situation, to experiment and to make constructive choices. A number of terms have been proposed for the new approach. These include:

'farmer first and last', 'farmer participatory research', 'farmer first', 'approach development', 'people-centred technology development' (PCTD) and 'participatory technology development' (PTD) as well as the old term 'sustainable agriculture' (Chambers *et al.*, 1991; Haverskort *et al.*, 1991). PTD serves to improve the experimental capacity of farmers and helps in the development of locally adapted improved technologies.

The approach has been used for implementation of IPM programmes in Indonesia (Matteson *et al.*, 1994). In this method, farmers are divided into small groups to monitor the crop and then each group analyses the field situation by identifying the key factors. Group members then decide whether any action is required. At a combined meeting, each group presents and defends its summary to the other trainees. The trainer facilitates by asking leading questions or adding technical information if necessary. This process allows farmers to integrate and practise their skills and knowledge and gives trainers an opportunity to evaluate the trainees' ability. Thousands of farmers have been trained utilizing this approach and it is being tried on a pilot scale. A survey among these farmers during the first post-training season revealed that they really decreased their frequency of pesticide sprays to a level consistently lower than that of non-IPM farmers. The percentage of farmers not applying pesticides was also significantly higher among the trained farmers. In spite of lower pest-control expenditures, these farmers obtained higher yields than the non-IPM farmers. This programme has been extended to several other Asian countries and the evident advantages of the approach are a marked reduction in the use of pesticides, with measurable benefits to the environment (APO, 1996; Heinrichs, 1998; Ooi, 2000).

Government support

Both the national programmes of developing countries and the donor agencies must have a policy commitment to IPM in the context of national economic planning and agricultural development. The costs to developing countries of not bringing their policies in line with the objectives of IPM are relatively greater than the costs to developed countries. National policies to promote IPM require close regulation at all stages related to the importation and/or manufacture, distribution, use and disposal of pesticides. In the case of pesticides that do not meet prescribed standards for safety, persistence, etc., import and manufacturing bans should be enacted. At a minimum, the conditions laid out by the FAO *Code of Conduct on the Regulation, Distribution and Use of Pesticides* should be adopted. Pesticide subsidies need to be eliminated in order to make IPM an attractive alternative. The funds thus saved may be utilized for the implementation of IPM. Funds may also be diverted from some of the current research programmes to IPM-oriented plant-protection programmes. Additional monetary resources may be generated through cooperation with bilateral/multilateral agencies willing to support such programmes (NRI, 1992).

Legislative measures

IPM is an information system and its adoption reduces pest-control costs. The alternative to IPM is the indiscriminate use of broad-spectrum synthetic organic pesticides. Unfortunately, while pesticide manufacturers and users (farmers) derive the full benefits from the use of these chemicals, they pass on the environmental and ecological costs of their use to the society as a whole. If they are made to bear the full cost of the use of these toxicants, they may find IPM a more economical and attractive alternative. This could be achieved by enforcing suitable legislative measures.

Secondly, the success of an IPM programme in any geographical region depends upon its implementation by all the farmers in the area. Ideally, farmers may voluntarily adopt an IPM programme but some farmers may hold out. Such farmers, called 'spoiler holdouts', may impair the success of a programme by failing to adopt a necessary practice, thus causing damage to adjacent areas. This is especially important in the case of mobile pests. In addition, some farmers may free-ride and thus shift the costs of implementing and managing a programme to a group of participating farmers. To overcome 'spoiler holdouts' and 'free riders', it may be necessary to impose a programme upon an unwilling minority through suitable legislative measures (Tarlock, 1980).

Improved institutional infrastructure

IPM cannot be implemented unless there is a basic infrastructure for plant protection in a country. There is a need to develop and support national programme capabilities for on-farm testing and technology extrapolation. At the international level, the establishment of an IPM working group to coordinate and monitor the funding of IPM projects is bound to provide an impetus to the implementation of IPM.

IPM is predominantly a knowledge technology, the use of which requires training of the many groups involved. There is currently little training material for most of these groups, including farmers, extension personnel and researchers. If IPM is to become the major approach for pest management in the developing world, this deficiency must be remedied urgently (NRI, 1992). Another aspect requiring greater attention is coordination of effort within and between countries (which is especially important with mobile insect species, such as the brown planthopper), between national research, training and implementation institutes/programmes, and among international development agencies.

Lack of a reliable database has also hampered the progress of IPM programmes. A reliable source of accurate information on the status of crops and pests in farmers' fields is necessary for many IPM activities. Most of the successful IPM programmes in both developed and developing countries have a reasonably accurate system of monitoring and evaluating various biological and environmental parameters in the agroecosystem. A reliable database on crop yield and pest losses is required for planning and resource allocation at the national and international level. Systems analysis has been used as a problem-diagnosis tool for IPM in developed-country cropping systems and may be used in developing countries as well.

Improved awareness

Increased education and awareness regarding the objectives, techniques and impact of IPM programmes are required at all levels including policy makers, planners, farmers, consumers and the general public. The importance and benefits of pesticides are being overemphasized by a multibillion-dollar industry utilizing the services of not only their salesmen but also agricultural scientists, administrators and planners. There is not yet a strong market in IPM information. Policy makers and planners need to be convinced that without IPM current agricultural production systems are not sustainable. Similarly, much important information that might induce a farmer to adopt IPM is not immediately observable and is therefore not sought by the farmer. A manufacturer has no incentive to recommend a programme that uses less pesticide or even selective pesticides that kill a limited range of pests (Tarlock, 1980).

Consumer groups and the general public may also be able to support the implementation of IPM programmes by demanding residue-free commodities. There is now a distinct market for organically produced food and other products. For example, there is a major programme at Cornell University, USA, for pesticide-reduced produce. NGOs and consumer groups need to be strengthened in developing countries, so that there is a public-oriented movement for implementation of IPM.

Future Outlook

Initially, IPM programmes evolved as a result of the pest problems caused by repeated and excessive use of pesticides and increasing cases of pest resistance to these chemicals. It is only during the past few years that the economic and social aspects of IPM have also received increasing attention (Fig. 1.3). It is now being increasingly recognized that modern agriculture cannot sustain the present productivity levels with the exclusive use of pesticides. Increasing pest problems and disruptions in agroecosystems can only be cor-

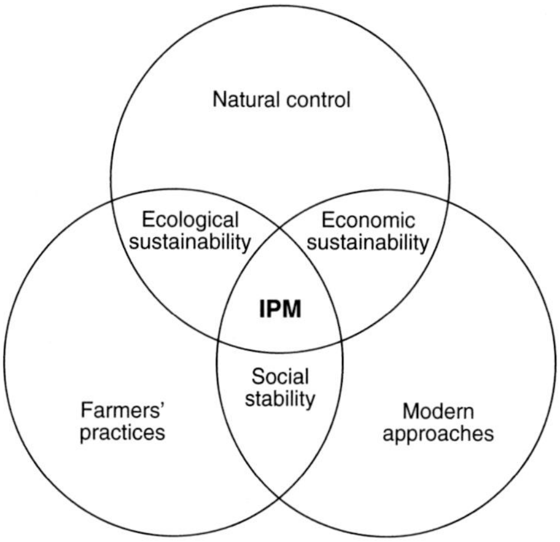

Fig. 1.3. Role of IPM in ecosystem stability (after Dhaliwal and Heinrichs,1998)

rected by use the of holistic pest-management programmes. If the environmental and social costs of pesticide use are taken into account, IPM appears to be a more attractive alternative with lower economic costs. The IPM programmes do not endanger non-target organisms, nor do they pollute the soil, water and air. IPM builds upon indigenous farming knowledge, treating traditional cultivation practices as components of location-specific IPM practices. This is especially important for farmers in developing countries, where traditional agricultural systems are based on indigenous farming practices. The incorporation of IPM into these practices helps the farmers to modernize while maintaining their cultural roots.

There needs to be a paradigm shift in the pest-management approach from managing components or individual organisms to an approach that examines processes, flows and interactions between populations of plants and populations of pest, beneficial and innocuous organisms (Dent, 1995; Hall, 1995; NRC, 1996; Kogan, 1998). There is a need for a better understanding of factors that affect ecosystem stability, the population dynamics of pests and the ability of populations to recover from stress. Technologies generated for the control of one group of pests, such as insects, diseases, weeds or nematodes, must be integrated in a crop-production system. There is a need to integrate the traditional practices of farmers with the advanced biological technologies now available in order to develop a profitable, safe and sustainable approach to pest management. It must be remembered that the success of any IPM programme will depend upon the effective transfer of technology to farmers. Therefore, the quick transfer, synthesis and simplification of information among farmers, suppliers of agricultural inputs and public agencies would speed up the successful development and implementation of IPM. Finally, the challenge to develop new technology, new management skills and new concepts of integration for pest management must be met so that we may protect our environment and provide a continuous supply of safe and nutritious food for the rapidly expanding world population.

References

Alam, G. (2000) *A Study of Biopesticides and Biofertilizers in Haryana, India*. Gatekeeper Series No. 93, International Institute of Environment and Development, London.

Allen, W.A. and Rajotte, E.G. (1990) The changing role of extension entomology in the IPM era. *Annual Review of Entomology* 25, 379–397.

Altieri, M.A. (1995) Escaping the treadmill. *Ceres* 27, 15–23.

APO (1996) *Integrated Pest Management in Asia and the Pacific*. Asian Productivity Organization, Tokyo, 170 pp.

Baron, J.J. (2002) The role of reduced risk crop protection chemicals on the success of IPM in the US. In: *International IPM Conference Exploring New Frontiers in Integrated Pest Management, 24–26 March, 2002, Toronto, Ontario, Canada*, pp. 60–62.

Benbrook, C.M. (2002) Measuring IPM adoption and the IPM continuum. In: *International IPM Conference Exploring New Frontiers in Integrated Pest Management, 24–26 March, 2002, Toronto, Ontario, Canada*, pp. 48–50.

CEQ (1972) *Integrated Pest Management*. Council on Environmental Quality, Washington, DC, 41 pp.

Chambers, R., Pacey, A. and Thrupp, L.A. (eds) (1991) *Farmer First: Farmer Innovation and Agricultural Research.* Intermediate Technology Publications, London.
DeBach, P. (1964) *Biological Control of Insect Pests and Weeds.* Chapman & Hall, London.
Dent, D. (1995) *Integrated Pest Management.* Chapman & Hall, London, 356 pp.
Dhaliwal, G.S. and Arora, R. (eds) (1994a) *Trends in Agricultural Insect Pest Management.* Commonwealth Publishers, New Delhi, 547 pp.
Dhaliwal, G.S. and Arora, R. (1994b) Components of insect pest management: a critique. In: Dhaliwal, G.S. and Arora, R. (eds) *Trends in Agricultural Insect Pest Management.* Commonwealth Publishers, New Delhi, pp. 1–55.
Dhaliwal, G.S. and Arora, R. (2001) *Integrated Pest Management: Concepts and Approaches.* Kalyani Publishers, New Delhi, 427 pp.
Dhaliwal, G.S. and Heinrichs, E.A. (eds) (1998) *Critical Issues in Insect Pest Management.* Commonwealth Publishers, New Delhi, 287 pp.
Dhaliwal, G.S. and Singh, B. (2000) *Pesticides and Environment.* Commonwealth Publishers, New Delhi, 439 pp.
Dhaliwal, G.S., Arora, R. and Heinrichs, E.A. (1998) Insect pest management: from traditional to sustainable approach. In: Dhaliwal, G.S. and Heinrichs, E.A. (eds) *Critical Issues in Insect Pest Management.* Commonwealth Publishers, New Delhi, pp. 1–25.
Eveleens, K.G. (2002) International IPM implementation and adoption: the Asian experience. In: *International IPM Conference Exploring New Frontiers in integrated Pest Management, 24–26 March, 2002, Toronto, Canada*, pp. 75–77.
FAO (1967) *Report of the First Session of the FAO Panel of Experts on Integrated Pest Control.* Food and Agriculture Organization of the United Nations, Rome.
FAO (1990) *Mid-term Review of FAO Intercountry Program for the Development and Application of Integrated Pest Control in Rice in South and South-East Asia.* Mission Report, Food and Agriculture Organization, Rome.
FAO (1995) *Intercountry Programme for the Development and Application of Integrated Pest Control in Rice in South and South-East Asia.* FAO Plant Protection Service, Rome, Italy.
FAO (2001) *World Review of the State of Food and Agriculture.* Food and Agriculture Organization of the United Nations, Rome, Italy, 295 pp.
Frisbie, R.E. and Adkisson, P.C. (1985a) *Integrated Pest Management on Major Agricultural Systems.* Texas A&M University, College Station, Texas.
Frisbie, R.E. and Adkisson, P.L. (1985b) IPM: definitions and current status in US agriculture. In: Hoy, M.A. and Herzog, D.C. (eds) *Biological Control in Agricultural IPM Systems.* Academic Press, Orlando, Florida, pp. 41–51.
Frisbie, R.E. and Smith, J.W. Jr (1991) Biologically intensive integrated pest management: the future. In: Menn, J.J. and Steinhauer, A.L. (eds) *Progress and Perspectives for the 21st Century.* Entomological Society of America, Lanham, Maryland, pp. 151–164.
Geier, P.W. (1966) Management of insect pests. *Annual Review of Entomology* 11, 471–490.
Geier, P.W. and Clark, L.R. (1961) An ecological approach to pest control. In: *Proceedings of the Eighth Technical Meeting, 1960, Warsaw, Poland.* International Union for Conservation of Nature and Natural Resources, Warsaw, pp. 10–18.
Hall, R. (1995) Challenges and prospects of integrated pest management. In: Reuveni, R. (ed.) *Novel Approaches to Integrated Pest Management.* Lewis Publishers, Boca Raton, Florida, pp. 1–19.
Haverskort, B., van der Kamp, J. and Waters-Bayer, A. (1991) *Joining Farmers Experiments – Experiences in Participatory Technology Development.* Intermediate Technology Publications, London.
Heinrichs, E.A. (1998) IPM in the 21st century: challenges and opportunities. In: Dhaliwal, G.S. and Heinrichs, E.A. (eds) *Critical Issues in Insect Pest Management.* Commonwealth Publishers, New Delhi, pp. 267–276.
Hinrichsen, D. and Robey, B. (2000) *Population and the Environment: the Global Challenge.* Population Reports, Series M, No. 5, Johns Hopkins University School of Public Health, Baltimore, Maryland, 31 pp.
Hoskins, W.M., Borden, A.D. and Michellbacher, A.E. (1939) Recommendations for a more discriminating use of insecticides. In: *Proceedings of the 6th Pacific Science Congress*, Vol. 5, pp. 119–123.
Jiggins, J. (1996) Women and the re-making of civil society. *Forest Trees and People Newsletter* 30, 18–22.
Kamp, K., Gregory, R. and Chowhan, G. (1993) Fish cutting pesticide use. *LEISA* 9, 22–23.
Kenmore, P. (1997) A perspective on IPM. *LEISA* 13, 8–9.

Kogan, M. (1998) Integrated pest management: historical perspectives and contemporary developments. *Annual Review of Entomology* 43, 243–270.
Luckman, W.H. and Metcalf, R.L. (1994) The pest management concept. In: Metcalf, R.L. and Luckman, W.H. (eds) *Introduction to Insect Pest Management*. John Wiley & Sons, New York, pp. 1–34.
McNeal, C.D. Jr (1988) Integrated pest management. In: *Pesticides: Risks, Management, Alternatives*. Virginia Water Resources Research Center, Blacksburg, Virginia, pp. 9–12.
Matteson, P.C., Gallagher, K.D. and Kenmore, P.E. (1994) Extension of integrated pest management for planthoppers in Asian irrigated rice: empowering the user. In: Denno, R.F. and Perfect, T.J. (eds) *Planthoppers: Their Ecology and Management*. Chapman & Hall, London, pp. 656–685.
Meerman, F., Bruinsma, W., Vanhuis, A. and Terweel, P. (1997) Integrated pest management: smallholders fight back with IPM. *LEISA* 13, 4–5.
Metcalf, R.L. (1980) Changing role of insecticides in crop protection. *Annual Review of Entomology* 25, 215–226.
Michelbacher, A.E. and Bacon, O.G. (1952) Walnut insect and spider mite control in Northern California. *Journal of Economic Entomology* 45, 1020–1027.
NAS (1969) *Principles of Plant and Animal Pest Control*, Vol. 3. *Insect Management and Control*. National Academy of Sciences, Washington, DC.
NDRC (1989) *Intolerable Risk: Pesticides in Our Children's Food*. Natural Resource Defense Council, Washington, DC.
NRC (1996) *Ecologically Based Pest Management: New Solutions for a New Century*. National Academy Press, Washington, DC, 144 pp.
NRI (1992) *Integrated Pest Management in Developing Countries: Experience and Prospects*. National Resources Institute, Chatham, UK.
Oerke, E.-C., Dehne, H.-W., Schonbeack, F. and Weber, A. (1994) *Crop Production and Crop Protection*. Elsevier Science, Amsterdam, 808 pp.
Ooi, P.A.C. (2000) Present status of IPM in the Asian region. In: *Farmer-led Integrated Pest Management*. Asian Productivity Organization, Tokyo, pp. 21–29.
Pedigo, L.P. (1991) *Entomology and Pest Management*. Macmillan, New York.
Perkins, J.H. (1980) The quest for innovation in agricultural entomology. In: Pimentel, D. and Perkins, J.H. (eds) *Pest Control: Cultural and Environmental Aspects*. AAAS Selected Symposium 43, Westview Press, Boulder, Colorado, pp. 23–80.
Pradhan, S. (1983) *Agricultural Entomology and Pest Control*. Indian Council of Agricultural Research, New Delhi.
Quinghua, Z. (1995) Careful control: IPM in China. *Ceres* 27, 12–13.
Rajak, R.L., Diurakar, M.C. and Mishra, M.P. (1997) National IPM programme in India. *Pesticides Information* 23, 23–26.
Reichelderfer, K.H., Carlson, G.A. and Norton, G.A. (1984) *Economic Guidelines for Crop Pest Control*. FAO Plant Production and Protection Paper 58, Food and Agriculture Organization of the United Nations, Rome.
Ruesink, W.G. and Onstad, D.W. (1994) Systems analysis and modelling in pest management. In: Metcalf, R.L. and Luckman, W.H. (eds) *Introduction to Insect Pest Management*. John Wiley & Sons, New York, pp. 393–420.
Smith, R.F. (1978) History and complexity of integrated pest management. In: Smith, E.H. and Pimetel, D. (eds) *Pest Control Strategies*. Academic Press, New York, pp. 41–53.
Smith, R.F. and Allen, W.W. (1954) Insect control and the balance of nature. *Scientific American* 190, 38–92.
Smith, R.F. and van den Bosch, R. (1967) Integrated control. In: Kilgore, W.W. and Doutt, R.L. (eds) *Pest Control: Biological, Physical and Selected Chemical Methods*. Academic Press, New York, pp. 295–340.
Smith, R.F., Apple, J.L. and Bottrell, D.G. (1976) The origins of integrated pest management concepts for agricultural crops. In: Apple, J.L. and Smith, R.F. (eds) *Integrated Pest Management*. Plenum Press, New York, pp. 1–16.
Sorenson, A.A. (1994) *Proceedings Integrated Pest Management Forum, Arlington, Virginia*. American Farmland Trust, Dekalb, Illinois.
Stern, V.M., Smith, R.F., van den Bosch, R. and Hagen, K.S. (1959) The integration of chemical and biological control of the spotted lucerne aphid. 1. The integrated control concept. *Hilgardia* 29, 81–101.
Tarlock, A.D. (1980) Legal aspects of integrated pest management. In: Pimentel, D. and Perkins, J.H. (eds) *Pest Control: Cultural and Environmental Aspects*. AAAS Selected Symposium 43, Westview Press, Boulder, Colorado, pp. 217–236.

Van de Fliert, E. (1993) *Integrated Pest Mangement: Farmers' Field Schools Generate Sustainable Practices.* Wageningen Agricultural University, Wageningen, The Netherlands.
Van de Fliert, E. (1998) Integrated pest management: springboard to sustainable agriculture. In: Dhaliwal, G.S. and Heinrichs, E.A. (eds) *Critical Issues in Insect Pest Management.* Commonwealth Publishers, New Delhi, pp. 250–266.
Van Veldhuizen, L. and Hiemstra, W. (1993) Cutting back: cure or prevent. *ILEIA Newsletter* 9, 3–4.

2 Cultural Practices: Springboard to IPM

Waheed I. Bajwa and Marcos Kogan
Department of Entomology and Integrated Plant Protection Center, Oregon State University, Corvallis, OR 97331, USA
E-mail: bajwaw@science.oregonstate.edu; koganm@science.oregonstate.edu

Introduction

Each agricultural region of the world grows a set of crops, the basic elements of agroecosystems, and uses unique technological procedures that in combination constitute the typical cropping system of the region. Cropping systems evolved under the pressure of the region's prevalent agroecological conditions and the sociocultural and economic characteristics of its human population. Cropping systems range from extremely monotonous, such as the huge small-grain monocropping systems in the traditional bread baskets of the world, to the highly diversified mixed or intercropping systems that prevail in many subsistence agricultural production systems in the less developed countries of Central and South America, Africa and Asia. In the hierarchy of environmental forces that shape a regional cropping system, climate and soil must be at the top. Soil fertility and texture have shaped the nature of cropping systems throughout history, but current knowledge and availability of adequate amendments have allowed farmers to correct soil deficiencies in many parts of the world. Climate, however, is not amenable to large-scale correction; therefore, crop mixes available for local production systems are usually limited by climatic factors. For instance, apples, pears, cherries and other temperate-zone fruits cannot be economically grown under tropical or subtropical conditions, and the reverse is true for tropical tree crops, such as citrus, coffee, pawpaw, cocoa and coconuts. Average annual temperatures and photoperiods are the key climatic factors. Where irrigation is not feasible, precipitation is another climatic determinant of cropping-systems evolution. Not far behind the climatic and edaphic conditions in shaping the nature of regional cropping systems are the pest complexes that attack the otherwise regionally adapted crops. Throughout the evolution of cropping systems, the presence of weeds, plant pathogens and arthropod pests has led to the adoption of procedures that in the aggregate are defined as cultural controls. Cultural controls, therefore, represent a fundamental integrated pest management (IPM) tactic even if the adoption of the practices, as driven by biotic factors of the environment, may remain unrecognized by growers. One of the dilemmas in maximizing the efficiency of cultural controls in IPM, however, is that seldom, if ever, are cropping systems optimized to cope with pest problems. Usually economic determinants take absolute precedence over all other factors. For a long time, maize-growers in the Midwestern USA knew that they could reduce the impact of rootworm, *Diabrotica* spp., attacks if they rotated maize with soybean. Continuous planting of maize, however, was more profitable for a

 Integrated Pest Management: Potential, Constraints and Challenges (eds O. Koul, G.S. Dhaliwal and G.W. Cuperus)

large segment of the grower population that raised pigs or dairy cattle. The economic consideration often prevailed over the IPM consideration.

The main focus of cultural controls in IPM is to explore and enhance synergies of ecological processes that limit pest invasion and population growth in an agroecosystem. IPM systems are generally developed and implemented in a stepwise manner and the first step requires the full understanding of the cropping system that is targeted for management. Cultural-control approaches that best fit the nature of the cropping system form the foundation upon which the other IPM tactics will be implemented. Use of cultural practices for pest control is one of the oldest and most effective pest-management tactics. Cultural-control tactics are agronomic practices primarily aimed at the prevention and reduction of pest outbreaks by increasing pest mortality or reducing its rates of increase, dispersal and overall damage potential. At present, most successful IPM programmes for major annual and perennial crops are based on a combination of cultural control and biological control, coupled with moderate use of chemical pesticides.

Under the label 'cultural control' are included cropping-systems practices related to the crop ecology with an impact on crop/pest interactions and practices that have a physical or mechanical nature. Practices related to crop ecology include crop rotations, row spacing (for row crops), planting dates (for annual crops), inter- and mixed cropping, strip cropping and cover cropping, among others. Practices that have an effect on the pests because of direct physical or mechanical impact include ploughing, disc-harrowing, cultivating, burning, flooding and pruning, among others. So cultural controls operate either through physical forces that suppress or limit pest-population growth or by promoting conditions within the agroecosystem that are detrimental to the pest but favourable to its natural enemies. Therefore, conserving pests' natural enemies and promoting their effectiveness are essential parts of cultural-control programmes. The appropriate use of cultural-control practices can reduce the damage potential to crops of essentially all pests. Cultural control, when well planned and integrated into an IPM system, can provide economic control of many insect pests, plant pathogens and weeds and greatly reduce reliance on pesticides.

The ecology of the crop, including the surrounding vegetation, determines the potential for the proliferation of pests and the complement of the pests' natural enemies. In an optimal cropping system, producers strive to provide the best conditions for crop plants to express their yield potential and to create adverse conditions for the potential pests and the most favourable conditions for the pest's natural enemies to flourish. Many stages in crop husbandry may provide such enhancements, and each of these stages should be considered in designing and implementing an IPM programme. It is possible often to identify underlying weaknesses in the ecosystem and in the prevailing agronomic practices that have allowed organisms to reach pest status. Identification of these weaknesses requires a thorough knowledge of the bionomics, behaviour and ecology of the pest in relation to the crop and its surroundings. Such knowledge is essential for developing alternative cultural practices that limit the potential of an organism to reach pest status. If these practices are cost-effective and can be integrated easily with other production practices, they are usually readily adopted by growers. In many parts of the world, good cultural-control methods have become stable farming practices that serve multiple purposes and help maintain agroecosystem sustainability.

Agronomic practices may have a positive, negative or neutral impact on a pest and its natural enemies. Analysis of these impacts provides a basis for the development of cultural-control programmes. Cultural-control practices fall into three main categories: prevention, avoidance and suppression (Table 2.1). Individually or together, these practices improve the ability of a crop to withstand pest attack, make the crop less suitable for the pest or make it more suitable for natural enemies. Some of the practices involve many aspects of crop management, such as the use of pest-free seed, good sanitation and the

Table 2.1. Use of various cultural control practices for different categories of pests.

Cultural control strategy/tactics	Insects	Weeds	Diseases	Nematodes
Prevention/avoidance				
Clean seed and planting material	X	X	X	X
Selection of well-adapted cultivars (including pest-resistant cultivars)	X		X	X
Optimal crop nutrition	X	X	X	X
Optimal water management (timing and proper amount of irrigation)	X	X	X	X
General sanitation of farm equipment (cleaning of cultivating and harvesting equipment)	X	X	X	X
Soil tillage	X	X	X	X
Crop rotation	X	X	X	X
Selection of planting and harvesting dates	X	X	X	X
Trap crops	X			
Suppression				
Adjustment of seeding rates and optimal row spacing (optimal plant density and fast canopy closure)	X	X	X	
Cultivation or hand hoeing (mainly for weed suppression)		X		
Habitat management for natural-enemy enhancement (hedgerows, alternative crops, cover crops, mulches)	X	X	X	
Crop diversification/mixed cropping	X			X
Soil tillage (reduction of seasonal carry-over)	X		X	
Destruction of alternative hosts and volunteer-crop plants	X		X	X

destruction of plant residue to limit the spread of pests; provision of optimum growing conditions to minimize stress on the crop; tillage practices that disrupt the insect's life cycle and destroy crop residue; early or late planting and harvest dates to promote phenological asynchronies between the crop and the pests; and crop rotations that include non-susceptible crops. In addition, several specific practices fall under the concept of habitat management, where practices are implemented to render the crop environment less favourable for the pests or more favourable for the pests' natural enemies. Intercropping, mixed cropping, hedge-vegetation management, trap cropping, cover cropping and certain other methods help divert pests from the main crop and promote the activities of beneficials (Landis *et al.*, 2000). These practices are generally compatible among themselves and with other pest-control tactics within a comprehensive IPM strategy and thus are highly desirable for the management of all pests. Variety selection is a key cultural practice and selection of resistant varieties is a major IPM tactic. Most texts on IPM, however, deal with host-plant resistance as a separate tactic, not under cultural controls.

Characteristics of Cultural-control Practices

The following features characterize cultural-control practices within an IPM context:

- Cultural-control methods are simple modifications or adaptations of regular farm operations. The added cost of their incorporation into pest-management systems is minimal in most cases. They are often the only control measures economically feasible for low-value crops.

- Cultural-control practices generally produce no, or negligible, undesirable ecological consequences.
- Cultural control is primarily aimed at the prevention and reduction of pest outbreaks. Its effectiveness is dependent on long-term, careful planning. The objective is either to reduce pest numbers below economic injury levels or to keep pests at a level that allows natural or biological control to take effect.
- Results of cultural practices as applied to arthropods are often difficult to quantify, primarily because ecological relationships within crop systems are poorly understood and because baseline data on pest abundance in the absence of cultural controls are not available. The effects of physical methods for weed control and crop rotations for disease control are often dramatic and very well documented.
- Cultural-control tactics are an effective means of pest control. They do not result in total elimination of the pest, thus allowing for conservation of beneficial insects. In the USA biological control of musk thistle is an excellent example (Kok, 2001).
- Most cultural practices indirectly affect arthropod pests. They are relatively slow acting and thus cannot resolve a pest outbreak. They are important, however, in minimizing pest damage by preventing pest build-up, rather than relieving an already existing pest problem.
- Cultural-control methods make cropping systems less friendly to the establishment and proliferation of pest populations. While they are designed to have positive effects on farm ecology and pest management, negative impacts may also result due to variations in weather, changes in crop management or perturbations in agroecosystems.
- Timing is critical to the success of many cultural controls; accordingly, the implementation of cultural-control tactics requires thorough knowledge of pest ecology and its interaction with the cropping system. In designing cultural control, special attention is given to recognizing the 'weak link' in the pest's life cycle, pest–crop interactions or the 'flaws' in agronomic practice that have allowed organisms to reach pest status.
- An area-wide deployment of cultural-control practices is essential for the effectiveness of the practice in IPM. If deployed in only a small area, a mobile arthropod pest would merely move from surrounding areas; effective use of some of these practices requires cooperation among farmers within communities.
- Cultural-control methods are often pest-, crop- and region-specific. Care should be exercised in transferring tactics to a region with markedly different agroecological conditions.

Cultural Practices and Sustainable IPM Systems

Optimizing plant health and the biological diversity of the farm system is the foundation of effective cultural controls. Healthy and vigorous crops are inherently tolerant of pests or their effects. Decreased biodiversity tends to result in unstable agroecosystems prone to recurrent pest outbreaks, reduced soil health and gradually declining productivity. Systems high in biodiversity tend to be more dynamically stable. Such systems provide more checks and balances to component species, thus helping prevent any given species from overwhelming the system. Most cultural-control tactics lower pest numbers in an ecosystem, and some may affect the carrying capacity of the ecosystem. These impacts tend to accumulate over time to produce sustained suppression of pest populations, keeping them below economic injury levels (Fig. 2.1). Cultural-control practices, therefore, must be a continuous component of an IPM programme so as to have a positive impact on all other practices by reducing overall levels of pest populations (MacHardy, 2000). Cultural controls tend to favour natural biological control directly by providing shelter and nectar for predators and parasitoids or indirectly by reducing insecticide use. It is the level and integration of both cultural and biological control that determine the

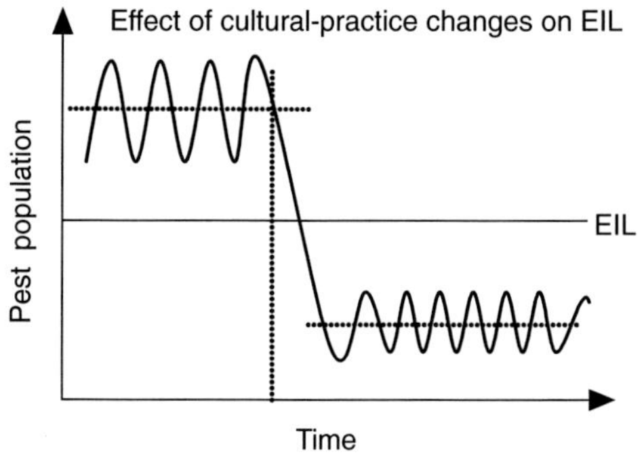

Fig. 2.1. Hypothetical effect of a change in cultural practices that lowers the general equilibrium position of a pest population, thus maintaining the population below the economic injury level (EIL).

nature and amount of other control actions to be taken at the farm level.

Cultural-control Practices and Contemporary Production Systems

The introduction of the concept of IPM in the late 1960s shifted the emphasis in pest control from a single-tactic, chemically based to a multitactic, ecologically based or biointensive system (Frisbie and Smith, 1991). Since the early days of IPM, cultural controls have been considered a first line of defence in many pest-management systems. Cultural practices alone may not give completely satisfactory pest control but, within an IPM system, they provide the matrix upon which other IPM tactics are deployed and often help reduce dependency on chemical pesticides (MacHardy, 2000). Ecologically based production systems, including organic farming (Brumfield and Ogier, 2000), total-habitat management (Prokopy, 1994; Kogan and Bajwa, 2001) and integrated fruit and crop production (Sansavini, 1997), take maximum advantage of farming practices that promote plant health and pests' natural controls and allow crops to escape or tolerate pest injury.

Cultural control is a cornerstone for most biointensive IPM programmes, where each component complements and often augments the effects of others (ecological synergism) (Fig. 2.2).

Cultural-control Practices

Table 2.1 lists the major cultural practices most commonly recommended for the control of weeds, plant pathogens and arthropods. These are discussed in some detail in the following sections. Comprehensive reviews of the major cultural-control methods are available in the literature. Variety selection and preference for adapted cultivars that incorporate resistance to arthropods and diseases are key components of the decision-making process in crop management. Host-plant resistance, however, is such an important component of IPM systems that it deserves special treatment in most texts. Detailed discussions of both traditional and genetically engineered plant resistance are offered in Chapters 6 and 7.

Sanitation

Sanitation involves removing and destroying overwintering sites, breeding refuges of arthropod pests and substrates for pathogen inoculum. Sanitation also prevents new pests from becoming established on the farm. This cultural-control method has been particularly useful for horticultural and tree-fruit crops. Fruit, twig and branch as well as root-crop pests can be affected by carefully conceived sanitation procedures. The most common means of field sanitation is destruction of crop residues by shredding and ploughing, separately or in combination. This process not only kills some pests directly but also speeds up natural rotting of the residues thus removing them as food or shelter source.

Removing crop residues can reduce the carry-over of pests from one season to the next. After harvest, destroying the stubble of cotton, maize and sugarcane is an important measure in the control of cotton bollworm, boll-weevil and pink bollworm in cotton, various corn borers and sugarcane borers (All, 1999). The practice of grazing by livestock of cotton fields after last picking is effective in reducing hibernating loads of pink bollworm and American bollworm in many regions (Bajwa, 1988). In the USA,

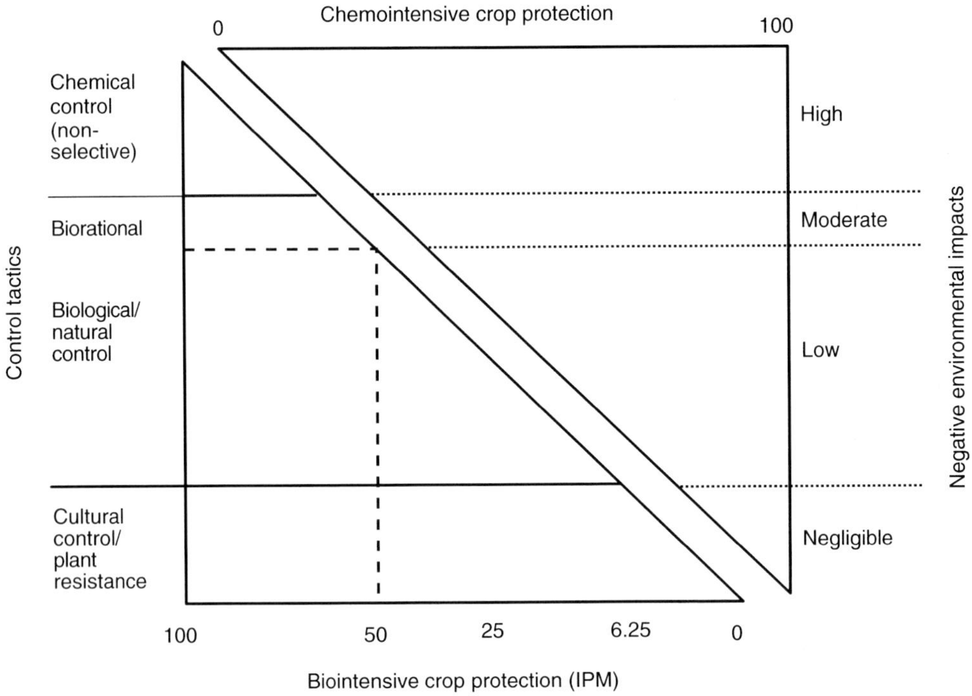

Fig. 2.2. Graphic representation of two opposing strategies of pest control. Triangle at left: biointensive pest control system (= IPM). These systems are integrative, stable and environmentally benign, with little reliance on broad-spectrum pesticides. The contribution of each control tactic to the stability of the system is represented as a proportion of the area of the triangle sector (or trapezoid) corresponding to that tactic. The various tactics complement and often potentiate each other (ecological synergism). Triangle at the right: chemointensive pest-control systems, on the other hand, are unstable (the ecological base is weak); control methods other than pesticides are not emphasized or their effects are masked by the antagonistic effects of chemical pesticides.

sheep grazing wheat stubble during the autumn and autumn/spring gave an effective control of wheat-stem sawfly, *Cephus cinctus* Norton (Hatfield *et al.*, 1999). In rice, destroying stubble and off-season sprouts reduces populations of the leafhopper, *Nephotettix impicticeps* Ishihara, and the whitebacked planthopper, *Sogatella furcifera* (Horvath) (Bajwa, 1988). In sugarcane and maize, destruction of cane trash and maize stalks in the winter significantly reduces the hibernating loads of several stem borers (Capinera, 2001).

Removal of fallen fruit from orchards and destruction of tree prunings are useful in reducing insect pests and plant-disease agents that overwinter in these materials. Destruction of prunings can control the peach borer, *Synanthedon exitiosa* (Say), and the lesser peach borer, *Synanthedon pictipes* (Grote and Robinson) (Cox and Atkins, 1964). Collecting and using dropped fruit or else destroying them reduces the populations of some important direct pests, such as plum curculio, *Conotrachelus nenuphar* (Herbst) (All, 1999), the codling moth, *Cydia pomonella* (Linnaeus) (Prokopy, 2001), the false codling moth, *Cryptophlebia leucotreta* (Meyrick), and many species of fruit flies (Bajwa, 1989; Stoll, 2000).

Other sanitation techniques include using pest-free seeds or transplants and decontaminating equipment, animals and other sources of food and shelter. Insects in cuttings or roots used in the vegetative propagation of crops can initiate infestations.

Seeding equipment should be clean and free of pests. Ideally, all farm machines should be cleaned before going from one field to another. Such procedures are essential to prevent the spread of soil-inhabiting pests, such as the grape phylloxera in vineyards of California and Oregon, in the USA (Hellman and Watson, 2000). In the tropics, the use of clean planting material reduces banana weevil, *Cosmopolites sordidus* (Germar), infestation by up to 80% (Stoll, 2000).

Soil tillage

While tillage may be a part of field sanitation, it can be an effective direct means of pest control by itself. Tilling the soil destroys life-cycle stages that occur in the soil or in crop residues. It destroys pests by mechanical action, starvation through debris destruction, desiccation and exposure to predators or adverse environmental conditions. Tillage may modify the soil microclimate, which will influence pest behaviour and plant growth. Often tillage timing and depth are the major considerations for the management of soil-inhabiting animal pests and critical factors in weed management. Timing is usually determined when pests are in an immobile stage (pupation or dormancy), and depth is recommended by the location of this stage in soil. Generally, tillage may be conducted in the autumn or early winter and in the spring before planting.

Soil-inhabiting pests such as rootworms, white grubs, wireworms and the overwintering larvae and pupae of Lepidoptera and Coleoptera may be exposed to desiccation or bird predation by ploughing. The pests that feed on stubble after harvest may starve if the ground is tilled (Speight *et al.*, 1999). Deep ploughing after harvest buries infested plant parts and stubble and destroys the larvae of pests such as army worm, *Pseudaletia unipunctata* (Harworth) (Capinera, 2001); wheat-stem sawfly, *C. cinctus*; maize earworm, *Helicoverpa zea* (Boddie); European corn borer, *Ostrinia nubilalis* (Hübner); soybean stem borer, *Dectes texanus* LeConte; grape berry moth, *Endopiza viteana* Clemens (Herzog and Funderburk, 1986); rice stem borers, *Tryporyza incertulas* (Walker) and *Chilo suppressalis* (Walker); and cotton bollworms, *Pectinophora gossypiella* (Saunders) and *Earias insulana* (Boisduval) (Bajwa, 1988). In the case of the European corn borer, the ploughing of stubble may result in a 90% reduction of hibernating larvae (Horn, 1988). Shallow autumn tillage may provide up to 90% sawfly control (Steffey *et al.*, 1992). If only spring tillage operations are performed, approximately 25% of larvae may be destroyed, depending upon the tillage implements used (Steffey *et al.*, 1992).

Reduced- or conservation-tillage practices may increase soil surface residues. These residues may have an impact on populations of certain pests. The presence of such residues repels the colonizing of a field by greenbugs, *Schizaphis graminum* (Rondani) in wheat and sorghum (Burton *et al.*, 1987), but attracts black cutworms in maize (Steffey *et al.*, 1992). Greenbugs prefer fields with more bare ground visible, while black cutworm prefers crop residue for oviposition. Reduced-tillage systems may have higher soil moisture and be slower to warm up in the spring, thus reducing crop growth. This may add to damage from soil pests (wireworms, white grubs and other seed and seedling pests) by increasing their feeding time on young plants (Steffey *et al.*, 1992).

Biological control agents are often affected by tillage practices. Discing or harrowing has fewer negative impacts on the parasitoid population than does ploughing (Herzog and Funderburk, 1986). Parasitoids of the cereal leaf beetle, *Oulema melanopus* (Linnaeus), can be severely affected by tillage operations, which has little effect on the pest (Pedigo, 2002). Reduced-tillage systems may increase populations of various predatory arthropods by increasing populations of their prey, such as other insects, mites and organisms that feed on decaying organic matter. Increased levels of predatory insects and predation on black cutworms, *Agrotis ipsilon* (Hufnagel), and maize earworm, *H. zea*, have been observed in reduced-tillage systems (Stinner and House, 1990).

Tillage is not always advantageous and can actually aggravate some pest problems.

For example, in some areas the soil surface tends to form a crust; keeping this crust intact can inhibit weed germination and/or prevent the penetration of soil-inhabiting pests (Norris *et al.*, 2003). Serious side effects of tillage are loss of organic matter, especially in warm soils, and accelerated loss of soil to wind and water erosion if the soil is left bare for an extended period.

Management of alternative hosts

Eliminating plants that can serve as alternative hosts when the crop is not present can suppress overwintering populations and reduce the growth rate of many pest populations. Vegetation serving as the alternative host may occur within the crop field and/or in surrounding areas. Whiteflies use many broad-leaved weeds as alternative hosts when suitable crops are not present (Norris *et al.*, 2003). Johnson grass is an excellent host of sorghum midge (Pedigo, 2002). Significant reductions of these pests have been reported where weed destruction by burning and other means was used. Many vegetable pests, such as squash- and stinkbugs, overwinter in crop debris and plant cover at the edge of plantings. Elimination of these hibernating habitats can significantly reduce infestations in squash, beans, cabbage and other vegetables (Capinera, 2001). Destruction of alternative hosts may also require careful scrutiny as it may eliminate an important habitat for beneficial insects.

Volunteer plants of a crop that remain in parts of a field after harvest may be a potential source of pest carry-over. These plants may harbour a large number of insect pests at times when pest presence would otherwise be impossible. The destruction of these volunteer plants is particularly important when crop rotation is practised to eliminate pests. For example, volunteer maize in soybean fields should be removed to prevent maize rootworm adults, *Diabrotica* spp., from laying eggs and producing larvae that would colonize maize the following season (Goodwin, 1985). Destruction of volunteer plants is also recommended for the suppression of other pests, such as Hessian fly (Lidell and Schuster, 1990); potato tuberworm, *Phthorimaea operculella* (Zeller); potato aphid, *Macrosiphum euphorbiae* (Thomas) (Capinera, 2001); cutworms, *Agrotis* spp. (van den Berg *et al.*, 1998); and the wheat curl mite, *Eriophyes tosichella* (Keifer) (Buntin *et al.*, 1991).

Maintaining and improving plant health

The primary goal of agricultural production is to maximize the yield of high-quality produce. Healthy plants are essential for optimal agricultural production. Stressed plants, including those with nutritionally induced low vigour, are often more susceptible to pest attack (Stoll, 1988). Vigorous plants are better able to tolerate pests, as well as producing high yields. In most crops, yield losses to a given degree of pest injury vary considerably depending on the vigour of the plants. However, there are documented instances of modern fertilizers and irrigation increasing vulnerability to certain pests (discussed below). Excessive nutrients (particularly nitrogen) and relative nutrient balance (ratios of nutrients) in soils affect the pest response to plants. Imbalances in the soil can make a plant more attractive to insect pests, less able to recover from pest damage or more susceptible to secondary infections by plant pathogens (Daane *et al.*, 1995; Phelan, 1997).

Cultural practices, such as proper irrigation and drainage, fertilization, row spacing and weed control, significantly influence the vigour of the crop and consequently the amount of pest damage. Undernourished plants, because of their pale or yellowish appearance, are often more attractive to colonizing aphids (Ferro, 1996). In some instances, however, the vigorous growth of plants may attract or enhance the development of many pest species. For example, leafhoppers and *Spodoptera littoralis* (Boisduval) have been most abundant on rice fertilized with high rates of nitrogen (Stoll, 2000). Excessive nitrogen may increase the incidence of *Tetranychus* mites (Stoll, 2000), fungal diseases and sucking insects (Flint and Gouveia, 2001) on an array of crops.

Succulent cotton growth attracts higher populations of the cotton aphid, *Aphis gossypii* (Glover), cotton fleahopper, *Pseudatamoscelis seriatus* (Reuter), and cotton bollworm, *H. zea* (Herzog and Funderburk, 1986). Increasing levels of soil fertility delay crop maturity in cotton, thereby reducing the potential for escape from pest injury (Anon., 1996). Increases in soil fertility of wheat result in increased wheat-stem sawfly injury due to the preference of ovipositing females for large succulent wheat plants (Morrill and Kushnak, 1996).

Good soils can improve yields and produce robust crops that are less vulnerable to pests. Soils used for years of continuous farming often require heavy fertilizer application to produce high yields. Soil quality can be maintained and enhanced in many ways, including incorporation of animal waste (manure), living plants or plant debris (compost). The addition of organic matter to soil is known to result in the suppression of a wide range of soil-borne plant pathogens (Cook and Baker, 1983). Application of manure to maize fields increases the predatory efficiency of mesostigmatid mites on maize rootworm larvae (Allee and Davis, 1996).

Water management can be used to grow more vigorous plants and thereby reduce losses. Excessive irrigation or frequent irrigations may favour the spread and development of many diseases and should be avoided. Cotton is severely stressed by inadequate irrigation, but excessive water may result in overly lush plants with higher insect densities and increased vulnerability to pest damage (Horn, 1988). Winter irrigation may reduce populations of overwintering pink bollworms by up to 50–70% (Bariola, 1983; Beasley, 1992). Cotton bollworms are attracted to succulent, rank-growing cotton plants; therefore, keeping water, fertilizer and plant density at recommended levels is important in order to avoid rank growth (Anon., 1999). Sprinkler irrigation has been effective in suppressing certain foliage-feeding insects by a washing and drowning action. The diamondback moth, *Plutella xylostella* (Linnaeus), and codling moth, *C. pomonella*, are effectively suppressed by frequent overhead irrigation of potatoes, head cabbage (McHugh and Foster, 1995) and apple trees (Knight, 1998), respectively. In California, sprinkler irrigation has been observed to suppress spider mites (Flint and Gouveia, 2001). Flooding is frequently used to reduce populations of sugarcane (Cherry, 1987; Deren *et al.*, 1993) and vegetable pests (Capinera, 2001). Paddy rice in the Orient has a complex biota (Kiritani, 2000). The biological impact of flooding in the rice paddy is a critical factor in the economic production of rice in vast regions of the world. Nevertheless, effective use of this technique depends on factors such as flood susceptibility and the stage of the pest species, the duration of flooding, water temperature and, perhaps more importantly, the availability and cost of water.

Timing of planting and harvest (disrupting crop–pest phenological synchrony)

Plant phenology can be manipulated to disrupt synchronization with the phenology of the major pests. It is sometimes possible to alter the timing of crop development by modifying regular cultural practices and thus to effect a substantial reduction in damage. This can be achieved by modifying planting time, by either delaying or advancing planting dates. In some cases, early-planted crops are less likely to suffer from pest outbreaks as they become well established before pests appear. They are either less palatable to herbivores or tolerant of higher pest densities without suffering much effect on yield. Early harvest often produces phenological asynchronies capable of disrupting a pest's life cycle, allowing harvest of the crop before the damaging state occurs. Planting early-maturing varieties often allows fields to escape infestations by late-season pests.

Early-planted maize is far less susceptible to maize earworm and stem borer, *Diatraea grandiosella* Dyar, damage than late-planted crops. The female *D. grandiosella* tends to lay fewer eggs on more mature plants and the plants have already passed their critical growth stage before significant numbers of larvae begin to feed (Herzog and Funderburk, 1986). In addition, early-planted maize can be harvested before many fully grown pre-diapause larvae have girdled the mature plants and caused yield losses through lodging of the plants (Roth *et al.*, 1995). The early planting of maize in East Africa is known to reduce problems with both maize leafhopper, *Cicadulina mbila* (Naude), and stalk borer, *Papaipema nebris* (Guenee) (Bajwa and Schaefers, 1998). The practice is so effective that no additional control measures are generally needed. Earlier-planted tomatoes in the western USA are far less likely to be infested by the tomato fruitworm, *H. zea*, than those planted later in the season (Anon., 1998). Early-season varieties of cotton avoid most boll-weevil and bollworm populations, and early-maturing soybean cultivars sufficiently evade bean-leaf beetles compared with standard varieties (Horn, 1988). Early-planted groundnuts avoid aphid damage in tropical Africa (Bajwa and Schaefers, 1998). Early planting of rice reduces or eliminates many insect-related problems (Speight *et al.*, 1999). This effect may not be a general rule for all pests and ecological situations. Early rice transplanting in South Korea may increase populations of striped rice borer, *C. suppressalis* (Ma and Lee, 1996).

Late planting of soybean interferes with the colonization patterns of the soybean thrips that are vectors of the bud-blight virus (Kogan *et al.*, 1999). Late planting of wheat has been used for a long time to manage Hessian flies. Adult Hessian flies have a very short lifespan (3–4 days) and oviposition occurs over a limited span of time during early autumn. If planting is delayed so that most of the flies have died before the wheat emerges, damaging infestations may be avoided. In regions where Hessian flies are a problem, fly-free dates have been established to guide autumn planting of wheat, based on the seasonal occurrence of the adults (Dufour, 2001). In situations in which migrant rather than resident populations are the major source of infestation, crop planting should be delayed until any major pest migration is over. Damage to sugarbeets by curly-top virus can be avoided or reduced by planting the crop after the spring migration of the beet leafhopper, a vector for the disease (Norris *et al.*, 2003).

Sometimes, it is possible to reduce pest populations or their damage by adjusting harvest time. As a general rule, crops should be harvested at the earliest possible date. Early harvesting of sorghum removes a large proportion of stem-borer populations; therefore, the crop should generally be harvested immediately after it attains physiological maturity (Omolo and Reddy, 1983). In lucerne, damage from the potato leafhopper, *Empoasca fabae* (Harris) and lucerne weevil, *Hypera postica* (Gyllenhal), can be minimized by early clipping at the early bloom or late bud stage (Steffy *et al.*, 1994). The lack of food, and hot, dry conditions after harvest can cause considerable mortality to the leafhopper and weevil larvae. Early planting reduces the loss of yield from maize ears that drop early because of European corn-borer tunnelling. However, early-harvested maize usually has a higher moisture content and must be dried before it is stored. Early sweet potato planting and harvesting is useful in white-fringed beetle, *Naupactus* spp., management programmes. These pests cause damage to roots late in the season, therefore, harvesting the crop before larvae reach sufficient size to cause serious feeding damage reduces the proportion of damaged and unmarketable roots at harvest (Zehnder *et al.*, 1998).

Crop rotation (increasing and maintaining temporal diversity)

Crop rotation means growing different crops in succession in the same field. It is especially effective against host-specific pests. Crop rotation drastically changes the environment, both above and below ground, usually to the disadvantage of pests of the previous crop. The same crop grown year after year on the same field will inevitably build up populations of organisms that feed on that plant or have a life cycle similar to that of the crop. It is important that the crops in a rotation system are genetically distant (belonging to different plant families) so that they do not have common pests. The focus is on either selecting rotations detrimental to certain pest species or avoiding rotations known to favour the pests. Crop rotation works by disrupting normal life cycles of pests by placing them in a non-host habitat (crop). It reduces pest pressure on all crops in the rotation by breaking the pest reproductive cycles. This practice seldom has any economic or ecological disadvantage; therefore, it is widely used even when crop damage is anticipated to be minimal (Herzog and Funderburk, 1986). Crop rotation is generally compatible with biological controls and forms the basis for IPM systems for many crops. Most common rotations include grass, legume and root crops. A leguminous crop in rotation generally replenishes plant nutrients, particularly nitrogen, thereby reducing the rates of needed chemical fertilizers. Also, rotation reduces the chance of pesticide build-up in the environment, thus decreasing the threat of pest resistance to pesticides (Reeves, 1994). Rotations that increase organic matter improve the environment for biological activity, which will increase the breakdown of pesticides.

Crop rotation is one of the oldest and most important measures for the control of pests that overwinter in the soil as eggs or partially grown larvae. It has been successfully used against many soil pests, including arthropods, plant-parasitic nematodes, fungal pathogens and bacterial pathogens. It is most effective against arthropod pests with a restricted plant-host range, long generation cycle (1 year or longer) and limited dispersal capability. Host selectivity may occur through either ovipositional or feeding behaviour.

Numerous species of major soil pests are successfully controlled by crop rotation. For example, the white-fringed weevil complex has limited dispersal capacity as the adult is unable to fly (Zehnder, 1997). These species are highly prolific on legumes; however, grasses, including maize, are in some way nutritionally deficient for supporting their feeding (Ferro, 1996). These pests cause no or low damage to grasses, but leguminous crops, soybean and groundnuts may suffer heavy losses. A soybean/maize rotation can therefore provide an excellent control, both effective and economical. Rotating potatoes with lucerne reduces wireworm damage, and rotating oats and maize reduces maize rootworm damage. Maize rootworms, *Diabrotica longicornis* Say and *Diabrotica virgifera* Le Conte, in the Midwestern USA have been effectively controlled by a 2-year rotation of maize with soybeans. Unfortunately, this tactic has been compromised in some areas where the rootworms have developed strains that can diapause for more than 1 year (Levine and Oloumi-Sadeghi, 1991).

Crop rotation may impose some limitations. Some crops used in rotation are of such low value that they contribute little to farm income. Also, an incorrect choice of crop sequence in a rotation can result in an elevated insect problem. For example, wireworms are more severe in potatoes following red clover or sweet clover (Norris *et al.*, 2003).

Interplanting or multiple cropping systems (maintaining and improving spatial diversity)

Multiple cropping or polyculture is typical of traditional farming systems in most developing countries. At present, there is insufficient experimental evidence that multiple cropping has a positive effect for pest management, although Altieri (1987, 1991, 1994), Wratten and van Emden (1995), Landis *et al.* (2000) and others provide abundant observational evidence that the inherent increase in biodiversity of multiple cropping systems increases the quality and quantity of the natural enemy fauna. The advantage of multiple cropping systems for IPM is postulated on the principle of habitat diversification. Monocultures inherently lack biodiversity as they are simplified and unstable agroecosystems, frequently prone to recurrent pest outbreaks that demand constant human intervention. Systems high in biodiversity tend to be more 'dynamically stable' because the variety of organisms provides more checks and balances on each other, thus helping prevent one species (i.e. pest species) from overwhelming the system. In IPM, biodiversity may create stability (but not always) within a crop season if employed as an area-wide approach. When applied to single fields, the approach may fail due to movement of pest organisms from adjacent fields. Monocultures open the way to pest infestations by providing concentrated resources and uniform physical conditions that promote pest invasions (Altieri, 1987). In these environments, the abundance and effectiveness of natural enemies are reduced because of inadequate alternative sources of food, shelter, breeding sites and other environmental factors. Increasing crop diversity, on the other hand, may be used to augment predator and parasitoid populations or to impair herbivores' ability to find and utilize their host plants. In multiple cropping systems certain plants may deter pests and reduce food supply for pests, while attracting and increasing an abundance of natural enemies. Pest levels are thus expected to be lower in polycultures.

Spatial arrangements used in multiple cropping are variations of row intercropping and strip cropping. Row intercropping is a system in which two or more crops are simultaneously planted in rows across a single field. The use of this practice as a strategy for weed control should be approached carefully. Intercropping may result in reduced yields of the main crop if competition for water or nutrients occurs. On the positive side, infestations of armyworm, *Spodoptera frugipereda* (J.E. Smith), in maize and *Empoasca* spp. (leafhoppers) and *Diabrotica* spp. (leaf beetles) in beans can be greatly reduced by interplanting the two crops (Altieri, 1987). Intercropping of soybean and maize increases the rate of parasitism by *Trichogramma* spp. (Altieri *et al.*, 1981). In Africa, intercropping of cereal crops (mainly maize and sorghum) with the non-host molasses grass, *Melinis minutiflora* (Beauv.), reduces infestation by stem borers, *Busseola fusca* Fuller and *Chilo partellus* (Swinhoe), in the main crop and also increases larval parasitism by *Cotesia sesamiae* (Cameron) (Khan *et al.*, 1997).

Strip cropping is the practice of growing two or more crops in different strips (usually four or more rows per strip) across a field wide enough for independent cultivation. For example, a scheme of alternating six-row blocks of soybean and maize or alternating strips of lucerne and cotton or lucerne and maize may reduce pest problems. Strip cropping may result in a more balanced insect population with an increase in beneficial insects.

An example is interplanting strips of lucerne in cotton for the control of lygus bugs, which prefer and will concentrate in the strips of uncut lucerne, leaving the cotton undamaged (Godfrey and Leigh, 1994). The lucerne may be harvested later as a forage crop.

In mixed-crop stands, it may be more difficult for pests to locate their host by either physical (visual clues) or chemical means (plant odours from non-host plants confuse feeding stimuli) (Bajwa and Shaefers, 1998). For example, thrips and whiteflies are attracted to green plants with a brown (soil) background, avoiding areas with full vegetation cover, such as a main crop and a cover crop between rows (Sullivan, 2001). Some intercrops have a spatial arrangement that produces the full vegetation cover that would be unfavourable for thrips and whiteflies. Other insects recognize their host plant by smell. Onions planted with carrots mask the smell of carrots from carrot flies (Sullivan, 2001).

Besides the potential IPM benefits, multiple cropping may also protect farmers against the risks of crop failure; if one crop within the system fails, the other may survive and compensate in yield to some extent, allowing the farmer an acceptable harvest. Despite all its potential benefits, much more research is needed on the complex interactions between various paired crops and their pest/predator complexes before the method will be widely accepted to replace large-scale monocultures. A major drawback of multiple cropping is the difficulty in mechanized planting, cultivating and harvesting.

Trap crops

Trap cropping is the practice of attracting pests to small plantings in or around a main crop or to an early planting of a crop on a small area. Trap crops are generally more favourable hosts for the target pest than the main crop. If a trap crop is maintained in a vigorous state, the pest may never leave the trap crop. If the pest population builds up and begins to leave, the trap crop can be mowed or sprayed to prevent damage to the main crop. This action does not affect the activities of beneficial species in the main crop. In many instances, trap crops can also serve as refugia or additional reservoirs for beneficial predators and parasitoids in the event that the adjacent crop field is treated.

Trap crops or trap plants have been in use against many insect pests, nematodes and plant pathogens. In beans, trap cropping can considerably reduce damage to Mexican bean beetle, *Epilachna varivestis* (Mulsant), and the bean leaf beetle, *Cerotoma trifurcata* (Forster). Early-maturing varieties can be planted 2 weeks prior to the main soybean crop. The adult beetles are attracted to these early-maturing trap crops and are then destroyed by cultivation or sprayed with an insecticide (Newsom and Herzog, 1977). Early-planted potatoes may act as a trap crop for Colorado potato beetles emerging in the spring (Hokkanen, 1991). Since the early potatoes are the only food source available, the beetles will assemble on these plants, where they can be controlled more easily. In Finland, mixed stands of trap plants (Chinese cabbage, oilseed and turnip rape, sunflower and marigold) near the main cauliflower plantings have been used for trapping the rape-blossom beetle, *Meligethes viridescens* (Fabricius). This beetle often ravages up to one-third of the whole harvest. As the beetle is highly mobile, several strips of trap plants are grown in the anticipated direction of infestation. Appropriately timed insecticide applications for trap cropping control the beetle and prevent its spread to the cauliflower plants. The technique has proved to increase by approximately 20% the marketable yield of the crop (Hokkanen, 1991). While some of these techniques still rely on insecticidal control, the area treated is greatly reduced (Hokkanen, 1991).

A possible limitation of trap cropping is the expense of producing and destroying a crop that brings no income. Nevertheless, in the southern USA and Hawaii, melon fields with small plantings of squash on the perimeter typically do not require insecticides. The practice reduces production costs for the main crop, conserves natural enemies and decreases the risk of secondary pest outbreaks. In this case, the squash trap crop enhances sustainability for the producer as a value-added crop. Sales of squash offset costs of the seed and pesticide and still provide additional income (Suszkiw, 1997).

Non-crop vegetation manipulation

Vegetation manipulation in agroecosystems and their surroundings is an important practice used to enhance beneficial arthropods in agricultural crops. For example, an orchard ground cover, if properly maintained, promotes the build-up of natural enemies of certain pests (USDA, 1998). Recently, several studies have demonstrated the potential for establishing flowering plants in or around farm fields to attract natural enemies and enhance biological control in the adjacent field (Altieri and Nicholls, 2000). In Europe, windbreaks and hedgerows have been used to encourage the build-up of natural enemies. Flowering strips on uncultivated field margins can encourage the build-up of syrphid fly and parasitoid populations, but plant age and composition appear to be important (Altieri and Nicholls, 2000). Unfortunately, the effectiveness of this practice is generally limited to areas of the crop close to the flowering strips (Alford, 2000). Modifying the wild vegetation surrounding crop fields and orchards may favour a natural balance between pest arthropods and their enemies (Rieux *et al.*, 1999). The technique is still in its early stages of development, but research in progress should help ascertain the role of the local flora and promising new plant introductions in and around agricultural fields. This research should help to clarify both the potential of the non-crop vegetation benefit as sources of natural enemies and the risk of harbouring phytophagous arthropods shared with the crops.

Cover crops are non-crop plant species grown either concurrently with the host crop (usually perennial plants) or in rotation with annual crops; they are generally not harvested. Examples include the establishment of pure or mixed stands of legumes and cereals to protect the soil against erosion. This technique ameliorates soil structure, enhances soil fertility and may help suppress certain weeds, arthropod pests and pathogens. Cover crops affect the ecology of orchards and vineyards by improving soil biology and fertility and by increasing biological control of insect pests by harbouring predators and parasitoids (Altieri and Nicholls, 2000). Cover crops attract and provide a nectar source for beneficial insects, spiders and mites.

Miscellaneous cultural-control practices

Increased plant density may sometimes be useful, but can add to production costs. Damage to seedlings by soil pests, such as cutworms, can sometimes be compensated for by higher seeding rates. Reducing row spacing causes the canopy to close early and improves conditions for predator colonization in many crops. Narrow row spacing in soybean decreases ovipositional preference of maize earworm moths as they prefer to oviposit in open-canopy fields. In contrast, damage to maize by larvae of the *Diabrotica undecimpunctata* Mannerheim borer decreases as plant density increases or when broad-leaved weeds are present rather than bare soil (Speight *et al.*, 1999). Planting of wheat at high densities and in narrow rows decreases moisture in stems, stem diameter and plant height. The wheat sawfly, *C. cinctus*, prefers larger, more succulent plants for oviposition, and damage to wheat decreases as seeding density increases and row spacing decreases (Herzog and Funderburk, 1986).

Mulches – natural or synthetic soil coverings – are useful for the suppression of weeds, insect pests and some plant diseases. A mulch can reduce the spread of soil-borne plant pathogens by preventing their transmission through soil splash. Winged aphids are repelled by reflective mulches (silver- or aluminium-coloured). Hay and straw mulches are more habitable than bare ground to some spider species and can reduce considerable damage to vegetable crops by insect pests (Reichert and Bishop, 1990). Living mulches of various clovers reduce insect-pest damage to vegetables and orchard crops (Bugg *et al.*, 1990) by providing essential resources for natural enemies. In some cases, mulching may provide a favourable environment for slugs and snails, which can be particularly damaging at the seedling stage.

Conclusions

Pest managers, in general, must learn to adjust the use of cultural controls to the features and properties of extant cropping systems. Powerful ecological, economic, cultural and social pressures have shaped the predominant cropping systems in most parts of the world. For example, as a consequence of the 1973 oil crisis, Brazil launched a vigorous campaign to promote the use of sugarcane-derived ethanol as a petrol substitute. Government subsidies and other incentives led growers in the state of São Paulo to replace a diverse agriculture that included some of the staples for low-income populations, such as rice and beans, with sugarcane plantations. In some areas, huge new monocrops became established almost overnight, covering millions of hectares. As expected, pest problems were aggravated and treated with an array of broad-spectrum pesticides. Industry and government gave little or no consideration to the resulting pest impacts of this shift of cropping systems until after the fact.

Thus, pest pressure is but one of the ecological forces that have influenced the evolution of cropping systems as they exist at present. Among the agricultural pests, weeds, more than either arthropod pests or pathogens, have influenced the development of cropping systems. Row spacing is adjusted for ease of cultivation. Vegetation management to enhance the natural enemies of arthropod pests and the use of cover crops are often disregarded by growers for fear that these techniques may make weed control more difficult. Certain rotations and better selection of regional crops could potentially improve overall regional crop health, but economic and cultural constraints often take these options away from the pest manager. As a result, successful use of cultural-control methods must take into account the nature of the dominant cropping system. Experiments conducted in isolated soybean fields in Louisiana showed that early planting of a border row of soybean, prior to planting of the bulk of the field, could be beneficial for the control of the bean-leaf beetle. A strip of about 10% of the total field area was planted. It attracted the colonizing beetles in large numbers. The strip was sprayed with an insecticide, thus eliminating most of the colonizing population. When the rest of the field was planted, it remained uninfested for most of the remainder of the season. The technique resulted in a 90% reduction in insecticide use. Despite the positive result of this trap-cropping experiment, the technique was not adopted by growers, who objected to moving twice to the same field the huge planters used in the industry. In addition, growers within a region tend to plant at different times; thus the early-planted fields in a region already acted as trap crops (Newsom *et al.*, 1980).

We do not presume to have offered an exhaustive discussion of cultural pest-control practices in this chapter. There is a vast literature on the subject (Herzog and Funderburk, 1986; Speight *et al.*, 1999; Landis *et al.*, 2000; Norris *et al.*, 2003) and the crop-specific literature offers the pest manager the best key to the prevalent practices in each region. This literature should be the first to be studied if one is to develop an IPM system with a strong foundation on cultural-control tactics. The effectiveness of cultural controls often rests on the complex interrelationships among many of these practices and with other IPM tactics, particularly biological control. Nearly every operation carried out in the field will have some effect, either good or bad, on current or potential pest problems. Understanding these relationships and acting to promote the positive ones are essential steps in the success of IPM programmes that aim at optimizing the role of cultural-control methods.

References

Alford, D.V. (2000) *Pest and Disease Management Handbook*. Blackwell Science, Oxford, 615 pp.

All, J.N. (1999) Cultural approaches to management arthropod pests. In: Ruberson, J.R. (ed.) *Handbook of Pest Management*. Marcel Dekker, New York, pp. 395–416.

Allee, L.L. and Davis, P.M. (1996) Effect of manure and corn hybrid on survival of western corn rootworm (Coleoptera: Chrysomelidae). *Environmental Entomology* 25, 801–809.

Altieri, M.A. (1987) *Agroecology: the Scientific Basis of Alternative Agriculture*. Westview Press, Boulder, Colorado, 227 pp.

Altieri, M.A. (1991) How best can we use biodiversity in agroecosystems. *Outlook on Agriculture* 20, 15–23.

Altieri, M.A. (1994) *Biodiversity and Pest Management in Agroecosystems*. Food Products Press, New York, 185 pp.

Altieri, M. and Nicholls, C.I. (2000) Applying agroecological concepts to the development of ecologically based pest management strategies. In: National Research Council (ed.) *Professional Societies and Ecologically Based Pest Management*. National Academy Press, Washington, DC, pp. 14–19.

Altieri, M.A., Lewis, W.J., Nordlund, D.A., Gueldner, R.C. and Todd, J.W. (1981) Chemical interactions between plants and *Trichogramma* wasps in Georgia soybean fields. *Protection Ecology* 3, 259–263.

Anon. (1996) *Integrated Pest Management for Cotton in the Western Region of the United States*. Publication 3305, University of California, Oakland, California, 164 pp.

Anon. (1998) *Integrated Pest Management for Tomatoes*, 4th edn. Publication 3274, University of California, Oakland, California, 118 pp.

Anon. (1999) *Integrated Pest Management for Apples and Pears*. Publication 3340, University of California, Oakland, California, 232 pp.

Bajwa, W.I. (1988) *Pest Management of Major Field Crops*. Agricultural Development Bank Pakistan, Islamabad, 60 pp.

Bajwa, W.I. (1989) *Citrus Pest Management*. Agricultural Development Bank Pakistan, Islamabad, 10 pp.

Bajwa, W.I. and Schaefers, G. (1998) *Indigenous Crop Protection Practices in Sub-Saharan East Africa, their Status and Significance Relative to Small Farmer IPM Programs in Developing Countries*. Available at: http://ippc.orst.edu/ipmafrica/

Bariola, L.A. (1983) Survival and emergence of overwintered pink bollworm moths (Lepidoptera: Gelechiidae). *Environmental Entomology* 12, 1877–1881.

Beasley, C.A. (1992) Winter irrigation reduces spring emergence of pink bollworm moths. In: *Proceedings Beltwide Cotton Production Research Conference*, Vol. 2. National Cotton Council of America, Memphis, Tennessee, pp. 943–944.

Brumfield, R.G. and Ogier, J.P. (2000) A review of organic horticulture and agriculture in the US. In: *Proceedings of the XIVth International Symposium on Horticultural Economics*. Acta Horticulturae, No. 536, St Peter Port, Guernsey, UK, pp. 21–28.

Bugg, R.L., Phatak, S.C. and Dutcher, J.D. (1990) Insects associated with cool-season cover crops in southern Georgia: implications for pest control in truck-farm and pecan agroecosystems. *Biological Agriculture and Horticulture* 7, 17–45.

Buntin, G.D., Cunfer, B.M. and Bridges, D.C. (1991) Impact of volunteer wheat on wheat insects in a wheat soybean double crop system. *Journal of Entomological Science* 26, 401–407.

Burton, R.L., Jones, O.R., Burd, J.D., Wicks, G.A. and Krenzer, E.G. Jr (1987) Damage by greenbug (Homoptera: Aphididae) to grain sorghum as affected by tillage, surface residues, and canopy. *Journal of Economic Entomology* 80, 792–798.

Capinera, J.L. (2001) *Handbook of Vegetable Pests*. Academic Press, New York, 728 pp.

Cherry, R.H. (1987) The effect of flooding on insect populations. *Bulletin of Agricultural Experiment Stations (University of Florida)* 870, 27–34.

Cook, R.J. and Baker, K.F. (1983) *The Nature of Practice of Biological Control of Plant Pathogens*. American Phytopathological Society, St Paul, Minnesota, 539 pp.

Cox, G.W. and Atkins, M.D. (1964) *Agricultural Ecology: an Analysis of World Food Production Systems*. W.H. Freeman, San Francisco, California, 721 pp.

Daane, K.M., Dlott, J.W., Johnson, R.S., Ramirez, H.T., Michailides, T.J., Yokota, G.Y., Crisosto, C.H. and Morgan, D.P. (1995) Excess nitrogen raises nectarine susceptibility to diseases and insects. *California Agriculture* 49, 13–18.

Deren, C.W., Cherry, R.H. and Snyder, G.H. (1993) Effect of flooding on selected sugarcane clones and soil insect pests. *Journal American Society of Sugar Cane Technologists* 13, 22–27.

Dufour, R. (2001) *Biointensive Integrated Pest Management*. Appropriate Technology Transfer for Rural Areas (ATTRA), Fayetteville, Arkansas, 52 pp.

Ferro, D.N. (1996) Cultural control. In: Radcliffe, E.B. and Hutchison, W.D. (eds) *Radcliffe's IPM World Textbook*. University of Minnesota, St Paul, Minnesota. Available at: http://ipmworld.umn.edu

Flint, M.L. and Gouveia, P. (2001) *IPM in Practice – Principles and Methods of Integrated Pest Management*. Publication 3418, University of California, Oakland, California, 296 pp.

Frisbie, R.E. and Smith, J.W. Jr (1991) Biologically intensive integrated pest management: the future. In: Menn, J.J. and Steinhauer, A.L. (eds) *Progress and Perspectives for the 21st Century*. Entomological Society of America, Lanham, Maryland, pp. 151–164.

Godfrey, L.D. and Leigh, T.F. (1994) Alfalfa harvest strategy effect on lygus bug (Hemiptera: Miridae) and insect predator population density: implications for use as trap crop in cotton. *Environmental Entomology* 23, 1106–1118.

Goodwin, D. (1985) The ecology of two species of corn rootworm (Coleoptera: Chrysomelidae) on volunteer corn in soybean fields in northern Illinois. *Bulletin of the Ecological Society of America* 66, 179–183.

Hatfield, P.G., Blodgett, S.L., Johnson, G.D., Denke, P.M., Kott, R.W. and Carroll, M.W. (1999) Sheep grazing to control wheat stem sawfly, a preliminary study. *Sheep and Goat Research Journal* 15, 159–160.

Hellman, E. and Watson, B. (2000) Reducing the risk of *Phylloxera* infestation. Northwest Berry and Grape Information Network. Electronic publication. Available at: http://berrygrape.orst.edu/fruitgrowing/grapes/phylrisk.htm

Herzog, D.C. and Funderburk, J.E. (1986) Ecological bases for habitat management and pest cultural control. In: Kogan, M. (ed.) *Ecological Theory and Integrated Pest Management Practice*. John Wiley & Sons, New York, pp. 217–250.

Hokkanen, H.M.T. (1991) Trap cropping in pest management. *Annual Review of Entomology* 36, 119–138.

Horn, D.J. (1988) *Ecological Approach to Pest Management*. Guilford Press, New York, 285 pp.

Khan, Z.R., Ampong-Nyarko, K., Chiliswa, P., Hassanali, A., Kimani, S., Lwande, W. and Overholt, C.M. (1997) Intercropping increases parasitism of pests. *Nature* 388, 631–632.
Kiritani, K. (2000) Integrated biodiversity management in paddy fields: shift of paradigm from IPM toward IBP. *IPM Reviews* 5, 175–183.
Knight, A.L. (1998) Management of codling moth (Lepidoptera: Tortricidae) in apple with overhead watering. *Journal of Economic Entomology* 91, 209–216.
Kogan, M. and Bajwa, W.I. (2001) IPM in the next century: how might things change? *Proceedings of Oregon Horticultural Society* 92, 51–70.
Kogan, M., Croft, B.A. and Sutherst, R.F. (1999) Applications of ecology for integrated pest management. In: Huffaker, C.B. and Gutierrez, A.P. (eds) *Ecological Entomology*. John Wiley & Sons, New York, pp. 681–736.
Kok, L.T. (2001) Classical biological control of nodding and plumeless thistles. *Biological Control: Theory and Applications in Pest Management* 21, 206–213.
Landis, D.A., Wratten, S.D. and Gurr, G.M. (2000) Habitat management to conserve natural enemies of arthropod pests in agriculture. *Annual Review of Entomology* 45, 175–201.
Levine, E. and Oloumi-Sadeghi, H. (1991) Management of diabroticite rootworms in corn. *Annual Review of Entomology* 36, 229–255.
Lidell, M.C. and Schuster, M.F. (1990) Distribution of the Hessian fly and its control in Texas. *Southwestern Entomologist* 15, 133–145.
Ma, K.C. and Lee, S.C. (1996) Occurrence of major rice insect pests at different transplanting times and fertilizer levels in paddy field. *Korean Journal of Applied Entomology* 35, 132–136.
MacHardy, W.E. (2000) Current status of IPM in apple orchards. *Crop Protection* 19, 801–806.
McHugh, J.J. Jr and Foster, R.E. (1995) Reduction of diamondback moth (Lepidoptera: Plutellidae) infestation in head cabbage by overhead irrigation. *Journal of Economic Entomology* 88, 162–168.
Morrill, W.L. and Kushnak, G.D. (1996) Wheat stem sawfly (Hymenoptera: Cephidae) adaptation to winter wheat. *Environmental Entomology* 25, 1128–1132.
Newsom, L.D. and Herzog, D.C. (1977) Trap crops for control of soybean pests. *Louisiana Agriculture* 20, 14–15.
Newsom, L.D., Kogan, M., Miner, F.D., Rabb, R.L., Tunnipseed, S.G. and Whitecomb, W.H. (1980) General accomplishments toward better pest control in soybean. In: Huffaker, C.B. (ed.) *New Technology of Pest Control*. Wiley, New York, pp. 51–98.
Norris, R.F., Caswell-Chen, E.P. and Kogan, M. (2003) *Concepts in Integrated Pest Management*. Prentice Hall, Upper Saddle River, New Jersey, 586 pp.
Omolo, E.O. and Reddy, S. (1983) An overview of cultural component of an integrated pest management systems in sorghum. In: *Africa: Seminar on the Use and Handling of Agricultural and Other Pest Control Chemicals*. Environment Liaison Centre in support of African NGO's Environment Network/Pesticide Action Network, Duduville, Nairobi, Kenya, pp. 100–103.
Pedigo, L.P. (2002) *Entomology and Pest Management*. Prentice Hall, Upper Saddle River, New Jersey, 742 pp.
Phelan, L. (1997) Soil management history and the role of plant mineral balance as a determinant of maize susceptibility to the European corn borer. *Biological Agriculture and Horticulture* 15, 25–34.
Prokopy, R. (1994) Integration in orchard pest and habitat management: a review. *Agriculture, Ecosystems and Environment* 50, 1–10.
Prokopy, R. (2001) Twenty years of apple production under an ecological approach to pest management. *Fruit Notes* 66, 3–10.
Reeves, D.W. (1994) Cover crops and rotations. In: Hatfield, J.L. and Stewart, B.A. (eds) *Crops Residue Management*. Lewis Publishers, Ann Arbor, Michigan, pp. 127–172.
Reichert, S.E. and Bishop, L. (1990) Prey control by an assemblage of generalist predators: spiders in garden test systems. *Ecology* 71, 1441–1450.
Rieux, R., Simon, S. and Defrance, H. (1999) Role of hedgerows and ground cover management on arthropod populations in pear orchards. *Agriculture, Ecosystems and Environment* 73, 119–127.
Roth, G.W., Calvin, D.D. and Lueloff, S.M. (1995) Tillage, nitrogen timing, and planting date effects on western corn rootworm injury to corn. *Journal of Agronomy* 87, 189–193.
Sansavini, S. (1997) Integrated fruit production in Europe: research and strategies for a sustainable industry. *Scientia Horticulturae* 68, 25–36.
Speight, M.R., Hunter, M.D. and Watt, A.D. (1999) *Ecology of Insects – Concepts and Applications*. Blackwell Science, London, 350 pp.

Steffey, K., Gray, M. and Weinzierl, R. (1992) Insect management. In: *Conservation Tillage Systems and Management*. Midwest Plan Services, Iowa State University, Ames, Iowa, pp. 67–74.

Steffey, K.L., Armbrust, E.J. and Onstad, D.W. (1994) Management of insects in lucerne. In: Metcalf, R.L. and Luckmann, W.H. (eds) *Introduction to Insect Pest Management*, 3rd edn. John Wiley & Sons, New York, pp. 469–506.

Stinner, B.R. and House, G.J. (1990) Arthropods and other invertebrates in conservation-tillage agriculture. *Annual Review of Entomology* 35, 299–318.

Stoll, G. (1988) Principles of preventive crop protection. In: Stoll, G. (ed.) *Natural Crop Protection in the Tropics*. Margraf Verlag, Weikersheim, pp. 14–23.

Stoll, G. (2000) *Natural Crop Protection in the Tropics*. Margraf Verlag, Weikersheim, 376 pp.

Sullivan, P. (2001) *Intercropping Principles and Practices: Agronomy Systems Guide*. Appropriate Technology Transfer for Rural Areas (ATTRA), Fayetteville, Arkansas, 16 pp.

Suszkiw, J. (1997) Melon growers' next battle cry against insect pests could be squash 'em. *Agricultural Research* 45, 16–17.

USDA (1998) *Managing Cover Crops Profitably*. Publication SX1005, Sustainable Agriculture Network, Washington, DC, 212 pp.

van den Berg, J., Nur, A.F. and Polaszek, A. (1998) Cultural control. In: Polaszek, A. (ed.) *African Cereal Stem Borers: Economic Importance, Taxonomy, Natural Enemies and Contro*l. CAB International, Wallingford, UK, pp. 333–347.

Wratten, S.D. and van Emden, H.F. (1995) Habitat management for enhanced activity of natural enemies of insect pests. In: Glen, D.M., Greaves, M.P. and Anderson, H.M. (eds) *Proceedings of the 13th Long Ashton International Symposium on Arable Ecosystems for the 21st Century*. Bristol, UK, p. 329.

Zehnder, G.W. (1997) Population dynamics of whitefringed beetle (Coleoptera: Chrysomelidae) in sweet potato in Alabama. *Environmental Entomology* 26, 727–735.

Zehnder, G.W., Briggs, T.H. and Pittsi, J.A. (1998) Management of whitefringed beetle (Coleoptera: Curculionidae) grub damage to sweet potato with adulticide treatments. *Journal of Economic Entomology* 91, 708–714.

3 The Relevance of Modelling in Successful Implementation of IPM

David E. Legg
Department of Renewable Resources, University of Wyoming
Laramie, WY 82071, USA
E-mail: DLEGG@uwyo.edu

Introduction

Integrated pest management (IPM) has evolved from its beginning as a concept of integrated control (Stern *et al.*, 1959) to a complex study of the agroecosystem. This change was both natural and necessary because conventional agroecosystems are typically dominated by one crop or animal species (host) and thus are not diverse. Ecosystems with limited diversity are vulnerable to the rapid colonization of the host by some biological organisms, and sometimes it can result in host depredation. Consequently, the task of many IPM practitioners is to increase the diversity of an agroecosystem as well as to increase the diversity of the pest-management tools that are used in that system; this will make them more stable and less dependent on pesticide usage. To this end, IPM very much encompasses and makes use of the cultural-control tactics and practices that were identified in Chapter 2. It also embraces the practice of using biopesticides. Host-plant resistance has always been a key element in many IPM programmes and transgenic techniques present many exciting opportunities for increasing the diversity of the agroecosystem. Biological control, particularly when viewed from a perspective of tritrophic interactions, is also an important part of the agroecosystem (see Chapter 4, this volume). Behaviour-modifying chemicals, as well as non-behaviour-modifying pesticides, have also played important roles in IPM. In addition, the behaviours of some pests can sometimes be used against them in a carefully crafted IPM programme. Finally, we must never forget that humans are an integral part of the agroecosystem, and it can be argued that the 'consumer' is a powerful, driving force in determining the composition of IPM programmes (see Chapter 11, this volume).

It goes without saying that humans both craft agroecosystems and study the interactions between pests, their hosts, their natural enemies and the environment. From such studies, IPM specialists obtain the information upon which agroecosystem changes are based. Also from these studies, 'tools' are developed to describe, analyse and even mimic parts of those systems. When those tools have been developed to a sufficient degree, they may then be used to predict what will happen in the future; predictions of the future are referred to as forecasts.

The tools I am referring to are models. For decades, models have been an integral part of IPM. For instance, the use of models has helped pest managers decide how the agroecosystem should be changed to favour economy and conservation and not to favour pests. Moreover, the use of models has allowed scientists to conduct simulated experiments when the conduct of those experiments would not have been possible. Further, models have been used whenever scientists wanted to explore as well as understand the complexities of agroecosystems.

In this chapter, I will define and discuss two basic terms: models and systems. I will then discuss the systems approach to modelling, which will be followed by discussions on various classification schemes of systems, the modelling process, the implementation of models and the identification of some commonly used models in IPM. Then I will discuss the teaching of some aspects of modelling to students in the classroom and to IPM practioners in the field. Finally, I will address the future of modelling in IPM.

Two Basic Terms

There are two basic terms that need to be understood; the 'models' and the 'systems'. In some circumstances, a 'model' represents an idealized situation or person. For example, the clothing and garment industry has the fashion model, who demonstrates the most beautiful way in which clothing may be worn. Estate agents have the model home, which represents a well-designed and richly decorated dwelling. Society has the model citizen, who is a person that everyone should strive to emulate. Then there is the model athlete, who is a person that excels beyond all others at his/her sport while also being a good person.

Happily, not all usage of the term model is of this sort. Architects and engineers often construct likenesses of whatever they are trying to build, and these constructs are called models. In addition, we can purchase miniature pieces of cars, aeroplanes, ships and the like and then assemble them; these too are called models. If the dimensions of such models are proportionally smaller versions of what is to be constructed, then they are said to be models of scale (e.g. 1 mm = 1 m). Models of scale have been used for centuries, particularly when constructing buildings and bridges.

What differentiates the former use of the word model from the latter? In the former, the focus is on an idealized representation of reality. In the latter, the focus is on an actual representation of reality. In IPM, we use models as actual representations of reality.

Models represent or mimic reality in several ways and, accordingly, there are several definitions as to what constitutes a model. Smith (1974) indicates that a model is a description of general ideas that include as little detail as possible. Jeffers (1978) defines a model as any formal expression of the relationships between defined symbols. Teng (1981, 1987) defines a model as any representation of a system in some form other than the original. Manetsch and Park (1982) define a model as an abstract representation of a real-world system that behaves like the real-world system in certain respects. Further, these authors indicate that a 'good' model represents the important aspects of the system for problem solving and minimizes 'behaviour' that is insignificant to the problem. Clearly, most definitions of models indicate that they represent something called systems and depend, therefore, on the definition of those systems.

A 'system', as defined by Miller and Miller (1984), is something that has a set of characteristics common to all systems and lacking in things that are not systems. Further, these authors indicate that a system has parts called units or components, which are interdependent and interact with one another. Focusing on living organisms, Teng (1987) indicates that a system cannot be properly understood or managed based on knowledge of some of its components. He also states that the components of a system interact with each other and are influenced in that interaction by external factors. Further, Teng (1987) indicates that the whole of the system is more than the sum of its parts. Manetsch and Park (1982) define a system as a set of interconnected elements organized towards a goal or set of goals separate from the environment, and are determined by factors completely independent of or external to the system. Teng (1987) correctly points out that the systems approach to problem solving differs from systems analysis, which is the analysis of system structure and behaviour, as well as from system control, which is the manipulation of input, system design, which is either the structuring of non-existing systems or the restructuring of existing systems, and system synthesis, which is a major rebuilding of systems through modelling.

From these definitions, it can be deduced that any specified system is composed of components, which are relevant and necessary to the system's function. Components of a system necessarily depend on one another, and together they function to achieve the goal of the system. One example of a system would be a wheat (*Triticum aestivum* L.) production system, which has the following components: crop, pests, soil, economics, environment, humans and pest and soil management.

Each component of a system can be subdivided. For example, the crop component of the aforementioned system could be divided into low-, mid- and high-latitude varieties. Moreover, varieties within a latitude could be divided into plants, with each plant being divided into leaves, roots, stems and flowers. Each leaf, for example, could then be divided into cells, and each cell could be divided into molecules. Each molecule could be divided into atoms, and each atom could be divided into subatomic particles. These are examples of how a system could be specified at any number of increasing levels of resolution and decreasing levels of scale. However, I could also have specified that system at any number of decreasing levels of resolution and increasing scale, by mimicking the relations between fields within farms, farms within regions, regions within continents or continents within the planet. Clearly, then, the specification of a system is a matter of resolution.

As was mentioned earlier, models are developed to mimic systems. However, models can also be developed to mimic system subcomponents. These are often referred to as submodels (e.g. Gelovani, 1984). Here, I simply note that the use of the terms model and submodel is somewhat subjective as a model that mimics one system may be a submodel in another system. For example, if a model represented or 'mimicked' the relations between plants in a field, that model could also be a submodel of a more inclusive model that would mimic the relations between fields within a farm. So where does a system begin and where does it end, and how can it be specified? The use of something called the systems approach to modelling may facilitate the answering of these questions.

The Systems Approach to Modelling

The systems approach, as defined by Teng (1987), is actually a problem-solving philosophy and methodology that are useful for guiding the generation of knowledge to support pest management and for synthesizing information into useful forms for delivery. The systems approach occurs in steps. As outlined by Jeffers (1978, 1984), these begin with recognizing the 'problem'. Then the problem is rigorously defined. Next, the goals and objectives of the problem-solving effort are explicitly stated. Then two or more potential solutions are generated to solve the problem. These are then employed in the modelling process (more on that later). After modelling, the outcomes are carefully evaluated, with potential courses of action being assessed. Finally, the course of action that shows the greatest promise for achieving the stated goals is taken (taking action). Placed into the context of a wheat production system, let us use the example of the pest *Diuraphis noxia* (Mordvilko) being accidentally introduced into the western wheat-growing areas of the USA. For many years after its introduction, this pest caused hundreds of millions of dollars (US) of damage (Legg and Archer, 1998). Initially, there were few alternatives to the use of insecticides, so insecticides were relied upon to keep *D. noxia* from causing significant economic losses (Legg and Archer, 1998). Therefore, one problem of immediate concern was that *D. noxia* was destroying a significant portion of the US wheat crop and insecticides were being heavily used for its control. A step towards solving this problem was to develop methods whereby the severity of *D. noxia* infestations could be gauged, relative to the cost of insecticide application, so that producers could assess whether such applications would be economically viable. Accomplishing this goal would reduce much of the uncertainty surrounding *D. noxia* infestations and, in doing so, would reduce the number of insecticide applications.

To satisfy this goal, three objectives were identified: (i) to establish the relationship between wheat yield and level of *D. noxia* infestation; (ii) to establish a method whereby the population dynamics of *D. noxia* could be mimicked; and (iii) to establish methods whereby *D. noxia* infestations could be quickly and reliably estimated. For the sake of simplicity, I will refer to these as the wheat response–*D. noxia* infestation system, the *D. noxia* population-dynamics system and the *D. noxia* population-estimation system. Once these were identified, experiments were designed and conducted to provide the necessary data for creating the models that mimicked each system. These models were then evaluated, courses of action were formulated and those that provided the best promise for satisfying the stated goals were implemented (Legg *et al.*, 1993; Legg and Archer, 1998). To summarize the generalized concepts of models and systems, it is sufficient to say that models represent systems, whereas systems are composed of objects that are united by their interactions to perform identifiable functions (Teng, 1987).

Classification Schemes for Systems

Some scientists, it seems, cannot resist the temptation to group different kinds of systems into categories. This is natural and serves to qualitatively describe systems of interest. In the literature, three such classification schemes are documented and will be referred to as the Miller and Miller (1984), Bawden *et al.* (1984), and Teng (1987) schemes. The Miller and Miller classification scheme identifies three classes of systems. These are the concrete, the abstracted and the conceptual systems. Concrete systems represent non-random accumulation of matter and energy in a region in physical space-time, organized into interacting, interrelated subsystems and components. In other words, concrete systems are 'phenomena' of the physical world. Concrete systems can be subdivided into subsystems; Miller and Miller give 19, four of which are the non-living, living, ecological and earth subsystems. Abstracted systems are actions that are abstracted from the behaviour of organisms. Conceptual systems, on the other hand, are systems of ideas that are expressed in symbolic form.

The Bawden *et al.* classification scheme distinguishes the 'soft' from the 'hard' systems. Soft systems are those for which the goals are not clearly recognizable and the outcomes are ambiguous and uncertain. However, hard systems are those that have clear goals and for which their outcomes are predictable. Finally, the Teng classification system is even more fundamental than the Bawden *et al.* system in that Teng divides the world into systems and non-systems.

Classification Schemes for Models

Up to now, I have avoided using terms that serve to classify models. This was deliberate, as there are many such classification schemes being used. One scheme, put forward by Richardson (1984), identifies 12 generalized types of models. Other classification schemes can be found in Jeffers (1978) and elsewhere (Manetsch and Park, 1982; Logan, 1994; Hess, 1996; Gutierrez, 2002). An important addition to these was put forward by Peck (2000), when he distinguished statistical from process models, the former being used to give a 'probabilistic interpretation of the data' without describing the underlying processes that drive the system, while the latter attempt to describe the permanent underlying biological processes that drive the system. Yet another classification scheme, which originated with Smith (1974), distinguished the practical (tactical) from the theoretical (strategic) models. These serve to identify just two of the many purposes for which modelling is undertaken (Hess, 1996). Other such purposes include crop growth, pest-population dynamics, sampling and sequential sampling. Also, models that describe or predict the change of something over space may be referred to as landscape or spatial models. Finally, some models are simply referred to by the name of the person or persons who either developed or popularized them. Two examples are the Nachman (Nachman, 1981) and Taylor's power law models (Taylor, 1961). Here, I briefly describe the Richardson and Jeffers schemes for classifying models, as they include most model types identified in the other schemes that are currently in use.

Richardson classification system

The Richardson system of model classification includes the sample, symbolism, simplification, analogy, scale forecast, paradigm perfection, life, caricature, computer, holism and design models. Modelling by sample involves selecting a representative number of 'individuals' from a specified population, assessing those individuals and inferring that assessment to the population at large; modelling by sample is an essential part of statistics. Modelling by symbolism involves the use of symbols to express and represent the relations and states of variables in a system; this type of modelling is primarily found in mathematics, studies in logic and decision making. Modelling by simplification involves the development of schemas of systems that are difficult or impossible to otherwise envision; it serves to simplify systems to the extent that just the basics are modelled. Modelling by analogy is a tangible representation of something that can be seen. Modelling by analogy requires (and provides) more detail than does modelling by simplification. An example of modelling by analogy is a map of roads or streets. Modelling by scale is producing a representative of some real-world entity, which resembles that entity in detail such that each part of the model is proportionally the correct size in relation to every other part of the model; as mentioned before, architects and engineers have used these models. Forecasting models will predict events that happen in the future. Examples include the times of sun- and moonrise (and set), the high and low temperatures for future dates, and the longevity of one's life. Modelling by perfecting the paradigm mimics systems through the use of rules; such models are essential to many knowledge-based systems (Stone *et al.*, 1986; Plant and Stone, 1991). Life models are a collection of methods whose commonality resides in the fact that they mimic human life processes; life models are necessary because it would be unethical, immoral and illegal to conduct certain experiments on humans. Caricature models are models that represent systems through the use of metaphors, similes, effigies and the like. Computer models mimic systems through the incorporation of rules and mathematical equations that express relationships between variables within systems. Computer models greatly facilitate the conduct of repetitive, complicated tasks. Also, they swiftly locate pieces of information that may be 'buried' in a great deal of literature. Modelling by holism represents a 'turning away' from the standard, reduction-driven systems approach to modelling and making use of philosophies and methods that have heretofore not been used (more on that later). Modelling by design ensures that the processes and schemas are constructed so that they will serve their intended purpose (i.e. will address the stated goals and objectives of the problem-solving venture).

Jeffers classification systems

Jeffers (1978) actually put forward three systems for classifying models. The first is based on a dichotomy of simulation and analytic models. Simulation models are those that can be specified by a routine of arithmetic operations (Jeffers, 1978), while simulation modelling involves the operation over time of a mathematical model that represents the structure and dynamics of a system. Simulation modelling is often conducted for the purpose of observing the system's behaviour under controlled or experimental conditions (Berryman and Pienaar, 1974). Simulation models perform many tasks, some of which are to solve differential equations, repeatedly apply transition matrices or repeatedly use random or pseudorandom number algorithms (Jeffers, 1978). Simulation models are very important in that they are used to conduct 'experiments' under conditions for which experiments could otherwise not be done. To simulate experiments, however, modellers must make assumptions about how the system works because not all components are included in the model and therefore the model is incomplete. As simulations often involve the repetitive solution of mathematical equations, as well as the handling of many individuals in specified populations, a computer is often needed to conduct them (e.g. Dowd *et al.*, 1984; Nachapong *et al.*, 1989; Culin *et al.*, 1990; Berry *et al.*, 1991; Follett *et al.*, 1993; Flinn and Hagstrum, 1995; Legg, 2000; Legg *et al.*, 2000, 2002; Arthur *et al.*, 2001). There are, however, some tasks that can be simulated without a computer (e.g. Penman and Chapman, 1982). Analytic models are those for which explicit formulae are derived for predicted values or distributions (Jeffers, 1978). Also, analytic models allow modellers to conduct in-depth explorations of model 'behaviour' (Peck, 2000). Analytic models tend to be less complex than are simulation models, though not all are necessarily simple. Typical analytic models are those used to describe gravity (Peck, 2000) or those used in analysis of variance (ANOVA), regression and multivariate analyses, as well as those that make use of the theoretical probability distributions of statistical applications (Jeffers, 1978).

The second of Jeffers' systems for classifying models is based on a fundamental dichotomy between word models and mathematical models. Word models are purely verbal descriptions of events, processes and relations. Often they are used to define the scope and depth of a problem. If systems are extremely simple, word models will perform well in mimicking them. However, if systems are not very simple, word models fail to mimic them. This is so because it requires a great number of words to describe the complex relationships that occur in some systems, as well as their feedbacks. In addition, it is unfortunate that some words have more than one meaning; such ambiguity, when it occurs, makes word models untenable (Jeffers, 1978). Mathematical models, on the other hand, are models that mimic systems through the use of symbolic logic. Such models are capable of expressing ideas and relations of great complexity while simultaneously retaining simplicity (Jeffers, 1978). Also, mathematical models are unambiguous. However, mathematical models must continually be checked for the presence of contradictions (Jacobsen, 1984).

Mathematical models are often divided into two groups: deterministic and stochastic (Jeffers, 1978; Peck, 2000). Deterministic models make use of the branch of mathematics that was developed when mathematics were first applied to physical problems and then to engineering problems (Jeffers, 1978; Logan, 1994). Sometimes deterministic models are composed of differential and difference equations (Jeffers, 1978); however, deterministic models can be represented by almost any kind of equation (Peck, 2000). For example, if a 22% infestation of *D. noxia* occurred on winter wheat in Wyoming, USA, then application of the deterministic model:

$$\textit{Yield loss} = \mathrm{EY} \times (0.5 \times \textit{per cent infestation}/100) \quad (1)$$

would determine the yield to be reduced by 47.08 l/ha (prediction), where *Yield loss* is the predicted yield loss (expressed in the same units as EY), EY is the expected yield that would occur if *D. noxia* were not infesting that field (say, 428 l/ha), and *per cent infestation* is the per cent of wheat tillers that are infested with at least 1.0 *D. noxia* (Legg *et al.*, 1993). A graphical representation of this model may be seen in Fig. 3.1.

It has long been known that many measurements taken by scientists do not represent the actual or true values of, say, population means. Rather, they come from distributions of measures that are taken on each of those populations (Salsburg, 2001). For these measures, deterministic models cannot exactly determine or predict the true values because the predictions will deviate from the true values by random amounts. In such cases, the application of stochastic models may be more appropriate in the modelling process. Stochastic models resemble deterministic models to an extent, but

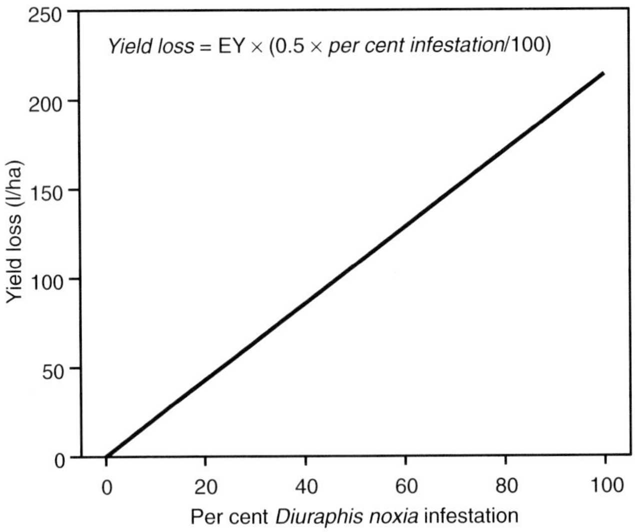

Fig. 3.1. Graphical representation of a deterministic model, which provides the exact yield losses that will occur at given levels of *Diuraphis noxia* infestation on winter wheat.

they also contain symbols, terms or algorithms that represent the deviations of predicted from measured values. Moreover, the behaviour of these deviations can be explained, to an extent, through probability distributions. For example, one stochastic model that is used when testing for equality of 'treatments' in a one-way ANOVA is as follows:

$$Y_{ij} = \mu + \tau_i + \xi_{ij} \qquad [2]$$

where Y_{ij} is a measure of the ith treatment and jth replicate, μ is the grand average of the experiment, τ_i is the ith treatment effect (i.e. the ith treatment average minus the grand average), and ξ_{ij} is the departure of each measured Y_{ij} from its predicted value (i.e. $\mu + \tau_i$). These types of models are referred to as linear statistical models (Cochran and Cox, 1957). Another stochastic model that is used when testing for relations between one dependent variable (Y) and one independent variable (X) is as follows:

$$Y_i = \beta_0 + \beta_1 X_i + \xi_i \qquad [3]$$

where β_0 and β_1 are the true y intercept and slope (Weisberg, 1980). These are sometimes referred to as regression models. Regression models can have more than one independent variable, as well as more than one slope, and these are referred to as multiple linear regression models.

The errors (ξ_{ij}) of linear statistical models and the errors (ξ_i) of regression models are known to have a mean of 0.0 and are used to calculate variance (σ^2), which describes the 'spread' of the errors about the mean. Also, the ξ_{ij} and ξ_i are assumed to conform to a specific probability distribution – in this case, the normal distribution (Snedecor and Cochran, 1967). Use of stochastic models in the form of linear statistical models and regression models have been very important in the successful implementation of IPM as their use has helped researchers to analyse the results of experiments (e.g. Legg *et al.*, 1987), describe ecological relationships (e.g. Legg and Chiang, 1984) and make predictions in IPM settings (Plant and Stone, 1991).

Stochastic functions can be added to almost any deterministic model. For example, if I wished to represent equation 1 as a stochastic model, I would first research which distribution describes the pattern of ξ (errors) about the predicted values, and then establish whether that distribution holds true for all values of expected yield in the absence of *D. noxia* (i.e. *EY*), as well as for all values of per cent infestation that would be encountered. Next I would establish whether

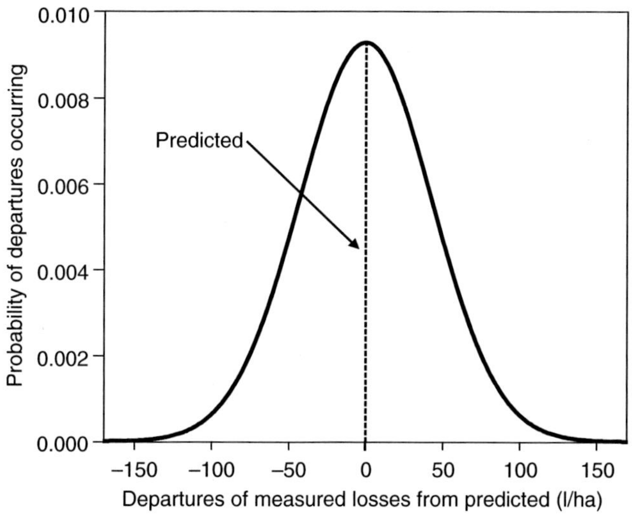

Fig. 3.2. Simulated probability distribution for the departures of measured yield losses (winter wheat) from predicted yield losses due to infestation of the insect *Diuraphis noxia*.

both the spread of the errors and the value for σ^2 were constant over all combinations of EY and per cent infestation that would be encountered. Once those steps were completed, I would then construct a stochastic representation of Equation 1:

$$\textit{Yield loss}_i = \mathrm{EY} \times (0.5 \times \textit{per cent infestation}/100) + \xi_i \quad [4]$$

where the 'spread' of the ξ_i is described by a specific distribution with a constant value for σ^2. If the spread of ξ_i is described by the normal distribution and the value for σ^2 is constant at, say, 1849, then, for any value of per cent infestation, the probabilities of obtaining certain values for yield losses would be as shown in Fig. 3.2. Note that the probabilities of yield losses are greatest for values of per cent infestation that are at or near the 'predicted value' and decline in a predictable manner as departures of measured yield losses (abscissa) increase (Fig. 3.2). The stochastic model, as represented by Equation 4, could be used for all kinds of purposes, one of which is to simulate a graphical representation of that model. This was done by calculating the predicted yield losses for each per cent infestation, beginning with 0.0 and ending with 100, in increments of 1.0, and then adding a random ξ_i, taken from a normal distribution with a mean of 0.0 and a σ^2 of 1849, to each predicted value (Press *et al.*, 1986). Using these simulated measures, I was then able to generate a graphical representation of model 4, where the predicted values (i.e. the 'line' from Equation 1) fails to equal each and every observed measure (dots) by some random amount (Fig. 3.3). As stochastic models are most useful when inadequate information exists for determining the outcome of each and every measure, the outcome of the modelling effort should be expressed as the chance of being within stated 'low' and 'high' values. For example, it is 90% certain that yield loss will be from 93 to 121 l/ha when per cent infestation is 50.

The third of Jeffers' systems for classifying models involves what he terms 'families of models.' These include the dynamic models, matrix models, multivariate models and optimization models. As pointed out by Jeffers (1978), these are not mutually exclusive as, for example, a dynamic model may also be a matrix model. The dynamic models are those that emphasize change in the variables of a system with respect to one another. They have the advantage of being very flexible and can make use of many mathematical equations to describe the change in specific

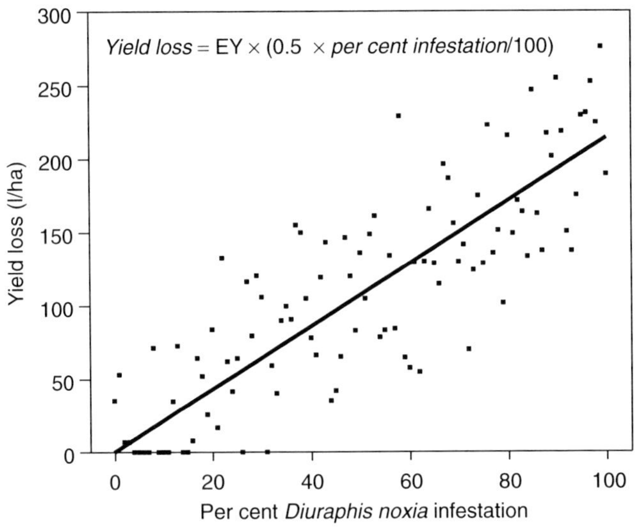

Fig. 3.3. Simulated distribution of observed yield losses (dots) about predicted levels of yield losses (line) due to infestation of the insect *Diuraphis noxia* on winter wheat.

parts of the system, relative to its other parts, at any point in time. The disadvantage of dynamic models is that they do not usually contain equations for all system components, as the description of most systems is incomplete. Dynamic models must therefore operate on the condition of making several assumptions. After a dynamic model is developed it can then be used to explore the system either for further understanding or for simulation. Examples of dynamic models include testing the effectiveness of pesticides (Schaalje, 1990), the growth of winter wheat in the Western Great Plains of the USA (McMaster and Smika, 1988) and the population dynamics of the Douglas-fir tussock moth, *Oryzia pseudotsugata* (McDunnough) (Berryman, 1991).

Matrix models use a mathematical technique that is often referred to as linear algebra. The use of matrix models involves the manipulation of values that are expressed in table-like entities, referred to as matrices, and row- and column-like entities referred to as vectors. These can be subjected to all sorts of mathematical operations and manipulations, and they most often mimic the population dynamics of biological organisms. The advantages of using matrix models lie in the fact that they are elegant in how they summarize information, they simplify procedures for solving complex problems and they can be placed into computer programs that conduct repetitive and sometimes complex mathematical procedures. One disadvantage of using matrix models is that they make use of a special notation commonly used with linear algebra, so they can be confusing to non-mathematically oriented IPM practitioners. Another disadvantage to the use of matrix models is that some computer programming skill may be needed to construct and maintain them. One type of matrix model is that of Lewis (1942) and Leslie (1945). Another is that of the Markov chain, which has been used to predict the probability of grasshopper outbreaks in Wyoming (Zimmerman, 1999). Multivariate models describe the pattern of relationships between several variables at the same time. These may be most useful for investigating ecological relationships in IPM. Some time-honoured multivariate techniques that have been used are principal component analysis, cluster analysis, discriminant analysis and canonical analysis. Optimization models are often used in operations research to search for a mathematical maximum or minimum, whichever is optimal. For IPM, practitioners often wish to maximize the return on a pest-management investment, so the application of optimization models may be found in the calculation of some economic thresholds. This is by no means a trivial task, as many variables are often used, with many different values being tried for each, to find the 'optimum'. Obviously, intensive computer programming is involved with the use of optimization models. One example of using an optimization model in IPM involved soybean pest management (Hutchins *et al.*, 1986). Another involved rangeland IPM with respect to grasshopper infestations (Davis *et al.*, 1992).

The Modelling Process

Modelling can be thought of as involving two phases. Phase I is the conception of the model, the construction of the model and the validation and verification of the model. Phase II is the implementation of the model.

Phase I

The first phase often progresses in a stepwise manner. To that end, both Jeffers (1978) and Teng (1987) have identified some of the steps in the process, which include: (i) defining and bounding the system that is to be modelled; (ii) evaluating the historical and current knowledge about the system; (iii) developing an initial conceptual (system) model; (iv) collecting data and constructing equations to describe the system; (v) structuring a detailed system model for computer modelling; (vi) translating the model into a selected language for computer performance; (vii) sensitivity analyses with verification and validation of the model; and (viii) model experimentation. Defining and bounding the system to be modelled is extremely important and, arguably, may be the most important step in the process. The objectives of this step should include the identification of how resolute the modelling process should be in order to address the stated problem, as well as the breadth of scale that should be considered for developing and applying the model. In other words, how complex or how simple does the model need to be to satisfy its stated purpose?

After these questions are answered, it is always good to conduct a thorough review of the literature and find out, from a historical perspective, what is known about the system of interest. Next, capture your ideas on how the model should be structured in a conceptual diagram. For this, I typically use flow charts, as I am comfortable with writing computer programs. However, just about any system of 'boxes' and arrows will be sufficient to show the variables and their relations to one another. After constructing a conceptual diagram, collect some data through either a series of designed experiments or observational studies, or both; then use these data to construct the 'equations' that mimic the system. Note that this step 'assumes' that mathematical models are to be constructed to mimic the system. This is not necessarily so, as for some problems, constructing an elegant word model or perhaps a knowledge-based system may prove satisfactory. Next, a detailed system is constructed for computer modelling. Here again, it is assumed that the identified system is sufficiently complicated for detailed analyses, as well as simulations, to be needed. Next, the constructed model is translated into a computer language for its use. Again, it is assumed that the model is mathematical in nature, is very complex, or both, and should be embedded in a computer program for ease of use. However, if the model is not of a mathematical nature, it may be represented by a knowledge-based system. Further, if the model is non-mathematical and is simple in its depth or breadth, a decision table or decision tree may be sufficient. Sensitivity analyses are then conducted on the model by varying the values for certain parameters in the equations, as well as the values for certain variables, to see if small changes in those parameters or variables induce small or large changes in the model's performance. In a parallel effort, the model must be tested in the 'real world' to see if it will perform at an acceptable level on independent sets of data. Finally, once the model has been so analysed, 'verified', and 'validated', it can be used to perform experiments to see how the system will behave under new sets of conditions.

This first phase to modelling, as challenging as it may seem, can be conducted if you have some training in and experience with modelling and if you possess a measure of 'modelling intuition.' In addition, having an open mind (as to which types of models to use), as well as using some creativity, can also help in the modelling process (Jeffers, 1978). However, it is the second phase of modelling that has turned out to be somewhat of a challenge.

Phase II

Model implementation involves three general steps: (i) introduction; (ii) adaptation; and (iii) incorporation by a specific agency, company or groups of individuals (user) (Kraemer, 1984). To date, much more attention has been given to the process of model development than has been given to the process of model implementation. As stated by Kraemer (1984), 'Some posit that this misplaced attention has resulted in a generally low level of model use and model success'. Model introduction refers to a period during which the model is considered for adoption. During this step, some early initial testing may be conducted and the results, along with the introductory information that was presented with the model, are used to make the decision on model adoption. Model adaptation refers to the period after model introduction during which broader support for the model is developed and plans are made for instructing and training practitioners on its use as well as the interpretation of its output. During model adaptation, the model begins to be widely used. Model incorporation is the step at which the model is no longer a new entity but, rather, becomes a routine part of the user's operation. Research has shown that the successful implementation of a model is influenced by at least three factors: (i) the inherent technical characteristics of the model itself; (ii) the social setting in which the model is used; and (iii) the uses and impacts of the model as experienced by the user (Kraemer, 1984). Interestingly, it has been hypothesized that a 'user' will not implement a model unless the model also serves some political interests (Kraemer, 1984).

Some Commonly Used Models in IPM

As was mentioned earlier, some of the most commonly used models in IPM are the linear statistical models of ANOVA as well as the regression models. Other commonly used models include crop-growth and crop-loss models, economic-threshold and injury-level models, sampling models and phenology models. Crop-growth and crop-loss models are commonly used in IPM. Many crop-growth models are process models that attempt to function on the physiological level (Pace and MacKenzie, 1987; Gutierrez, 2002). Most mimic plant growth by dividing the plant into its fundamental components, such as leaf, stem and root biomass accumulation, as well as photosynthesis and respiration. Some crop-growth models can be used for crop-loss assessment, through simulation. However, most crop-loss assessment is conducted through the use of regression models (e.g. Sah and MacKenzie, 1987; Shane and Teng, 1987; Walker, 1987; Mesbah *et al.*, 1994). Economic-threshold models are deterministic in nature and either contain or are linked to population-dynamics models (Chiang, 1979, 1983; Pedigo *et al.*, 1986); economic injury-level models are also deterministic in nature but do not contain the population-dynamics link. Economic-threshold models can also take the form of optimization models. Many examples exist in the literature of using the deterministic form of economic-threshold and economic injury-level models in IPM (e.g. Legg *et al.*, 1993). Examples also exist for using optimization models when calculating economic thresholds in IPM (e.g. Davis *et al.*, 1992).

Sampling models are particularly well used in IPM. The acts of sampling and sample inspection for assessing the abundance and presence of pests can take several forms (Legg and Archer, 1998). However, the sampling models serve to assist researchers and IPM practitioners in either the efficient estimation of pest-population abundance or in the rapid classification of a pest population relative to an economic (or action) threshold. The former are often referred to as precision-based sequential sampling models, as they serve to guide researchers (mostly) in determining how many samples are needed to estimate a population average with a certain predetermined level of precision (Hutchison, 1994); precision is a measure of the consistency of the estimates of the average (Legg and Moon, 1994). The latter are often referred to as classification-based sequential sampling models, as they serve to quickly classify infestations as being 'high' or 'low' relative to a threshold value, thus requiring less sampling effort to make a pest-management decision (Binns, 1994).

Finally, phenology models serve to predict, mostly in real time, the phenological development of pest populations. These are most useful for predicting the dates of first or peak emergence or the emergence of a second generation of a pest. Almost all phenology models are driven by ambient temperature, as plant-pathogenic organisms, weeds and pestiferous insects are poikilotherms (Legg *et al.*, 2002). However, some phenology models are driven by both temperature and moisture (e.g. Legg and Brewer, 1995). Phenology models can be developed either in the laboratory or in the field (Legg *et al.*, 1998b). If they are developed in the laboratory, the parameters may be meaningful in a biological sense. However, care must be taken to ensure that the model output, when used for IPM purposes, reflects pest development in the field. If they are developed in the field, however, the parameters may not be biologically meaningful but they may provide acceptable predictions (Legg *et al.*, 1998a).

Education and Modelling: Some Lessons Learned

Teaching people how to develop and use models is important for producing the next generation of modellers and ensuring that models will be used by IPM practitioners. Here, I distinguish between teaching students in the classroom and educating IPM practitioners outside the classroom. I will refer to the former as 'teaching' and the latter as 'extension' or 'outreach'. First, I will address teaching.

Teaching students about modelling actually has two perspectives: developing models and using models. Regarding the former, courses must be offered so that some students will learn how to develop models. In these courses, students learn how to use the modelling process and how to apply some modelling techniques. It has been my experience that most students who enrol in these courses are agriculturalists, biologists or natural resource scientists who also have some quantitative skills. Consequently, the teaching of 'modelling' must be done from a conceptual perspective using each of the following techniques: verbal, visual, activity-oriented and mathematically oriented methods. The use of verbal methods requires that I, as the instructor, use words to describe what is being done, why it is done and how it is done. For me, this requires careful thought and effort as such verbal descriptions are far longer and, in some cases, less elegant than using mathematical descriptions of the same. Nevertheless, verbal descriptions must be incorporated into each of the lectures as some students are very good listeners and learn primarily through verbal descriptions. Visual methods are extremely useful for some students as they are 'visually oriented' and will understand the concept or process only when a chart, graph or some other visual is used. Activity-oriented students are usually befuddled by either the verbal descriptions or the visual depictions I provide until I either work through an example or assign them a special project to be completed outside class (homework). Finally, there are some mathematically oriented students who are comfortable with and learn through the symbolic language of mathematics. For these, I provide explanations of processes and techniques using such tools as manipulated equations (all kinds), integrals and algebra.

Teaching students how to use models is necessarily different from teaching students how to develop models. This is so because these students are agriculturalists, biologists or natural resource scientists who have few quantitative skills. Instead, I teach these students a little about developing models but a lot about using models. Also, these students need to be taught to interpret the results or 'output' from models, as well as to 'experiment' with them by posing 'what if' scenarios and rerunning the models for each scenario.

Finally, extending models to IPM practitioners necessarily differs from teaching, as practitioners do not enrol in courses and therefore cannot be forced to use models for the conduct of pest management. Instead, I have learned that the development and implementation phases of modelling must be conducted with regular input from the people who will use those models. These people include representatives from producer groups, some university personnel and, perhaps, some government officials. Also, the administrator from the modeller's unit may be important as that individual provides resources for maintaining the models. Anyway, these individuals must work with the modellers because they have an interest in developing and maintaining the models as well as the computer programs in which they reside. This kind of 'partnership' is essential for developing products that are useful to and will be used by IPM practitioners.

The Future of Modelling in IPM

Given the importance of modelling and systems analysis in IPM, both should play important future roles. However, the types of models that may be used and the approach to designing and implementing those models may be different from the types of models that are currently being used and the approaches that are currently being employed by modellers. Plant and Stone (1991) have pointed out that traditional systems-level problems have been solved using systems analysis, along with mathematical models, which were central to the IPM projects of the 1970s and 1980s. More recently, however, some emphasis has been placed on the use of qualitative methods (or models) to solve agricultural problems because much of the present understanding of these systems is qualitative and is based on experience (Plant and Stone, 1991). Such knowledge is not easily quantifiable, which makes it difficult to construct mathematical models. In his 1994 article, Logan eloquently expressed how the 'infatuation' that IPM researchers and practitioners had with complex, mathematical simulation models waned during the 1980s, due in part to the unrealistic expectations that were put forward by the proponents of those models, the lack of the models' 'predictive power' and the environment in which some models were put together. In the general modelling literature, Richardson (1984) has summarized the essence of some papers by calling for a more inclusive approach to systems definitions and analysis, one that he termed 'holism'. Further, Richardson (1984) articulated the perception that dynamic models (and modellers) are in the process of a 'shift of paradigm' away from the classic application of reductionism, or Cartesian disassembly of systems, to a more inclusive approach to modelling those systems. In the IPM literature, this shift appears to be headed towards something called a 'whole-system', within which agriculture is viewed as an ecosystem that involves habitat management, crop attributes and multitrophic interactions as some of the principles that will guide decision making and promote agricultural sustainability (Rains *et al.*, 2002). Modelling changes in the landscape, using the science of geographical information systems, also appear likely. Access to models and the weather data that are needed to run the models will be increasingly made available on the Worldwide Web (e.g. http://okmesonet.ocs.ou.edu/) (Brock *et al.*, 1995). Finally, there are the knowledge-based systems. These are computer programs that solve complex problems within some defined area of knowledge (knowledge domain). Knowledge-based systems differ from traditional mathematical models in that they are designed to mimic the human reasoning processes, which rely on logic, beliefs, generalized rules, opinion, and experience; these are typically not quantifiable (Plant and Stone, 1991). Relatively new in their development, knowledge-based systems are extremely flexible and inexpensive to construct. They do, however, require constant attention by teams of experts, programmers, and 'knowledge engineers' to keep them current and valid.

References

Arthur, F.H., Throne, J.E., Maier, D.E. and Montross, M.D. (2001) Impact of aeration on maize weevil (Coleoptera: Curculionidae) populations in corn stored in the northern United States: simulation studies. *American Entomologist* 47, 104–110.

Bawden, R.J., Macadam, R.D., Packham, R.J. and Valentine, I. (1984) Systems thinking and practices in the education of agriculturalists. *Agricultural Systems* 13, 205–225.

Berry, J.S., Holtzer, T.O. and Norman, J.M. (1991) Experiments using a simulation model of the banks grass mite (Acari: Tetranychidae) and the predatory mite *Neoseiulus fallacies* (Acari: Phytoseiidae) in a corn microenvironment. *Environmental Entomology* 20, 1074–1078.

Berryman, A.A. (1991) Population theory: an essential ingredient in pest prediction, management, and policy-making. *American Entomologist* 37, 138–142.

Berryman, A.A. and Pienaar, L.V. (1974) Simulation: a powerful method of investigating the dynamics and management of insect populations. *Environmental Entomology* 3, 199–207.

Binns, M.R. (1994) Sequential sampling for classifying pest status. In: Pedigo, L.P. and Buntin, G.D. (eds) *Handbook of Sampling Methods for Arthropods in Agriculture*. CRC Press, Boca Raton, Florida, pp. 137–174.

Brock, F.V., Crawford, K.C., Elliott, R.L., Cuperus, G.W., Stadler, S.J., Johnson, H.L. and Eilts, M.D. (1995) The Oklahoma mesonet: a technical overview. *Journal of Atmospheric and Oceanic Technology* 12, 5–19.

Chiang, H.C. (1979) A general model of the economic threshold level of pest populations. *Food and Agricultural Organization of the United Nations Plant Protection Bulletin* 27, 71–73.

Chiang, H.C. (1983) Factors to be considered in refining a general model of economic threshold. *Entomophaga* 27, 99–103.

Cochran, W.G. and Cox, G.M. (1957) *Experimental Designs*, 2nd edn. John Wiley Press, New York, 611 pp.

Culin, J., Brown, S., Rogers, J., Scarborough, D., Swift, A., Cotterill, B. and Kovach, J. (1990) A simulation model examining boll weevil dispersal: historical and current situations. *Environmental Entomology* 19, 195–208.

Davis, R.M., Skold, M.D., Berry, J.S. and Kemp, W.P. (1992) The economic threshold for grasshopper control on public rangelands. *Journal of Agricultural and Resource Economics* 17, 56–65.

Dowd, P.F., Sparks, T.C. and Mitchell, F.L. (1984) A microcomputer simulation program for demonstrating the development of insecticide resistance. *American Entomologist* 30, 37–41.

Flinn, P.W. and Hagstrum, D.W. (1995) Simulation model of *Cephalonomia waterstoni* (Hymenoptera: Bethylidae) parasitizing the rusty grain beetle (Coleoptera: Cucujidae). *Environmental Entomology* 24, 1608–1615.

Follett, P.A., Kennedy, G.A. and Gould, F. (1993) REPO: a simulation model that explores the Colorado potato beetle (Coleoptera: Chrysomelidae) adaptation to insecticides. *Environmental Entomology* 22, 283–296.

Gelovani, V.A. (1984) An interactive modelling system as a tool for analyzing complex socio-economic problems. In: Richardson, J. (ed.) *Models of Reality Shaping Thought and Action*. Lomond Press, Mt Airy, Maryland, pp. 75–86.

Gutierrez, A.P. (2002) Modeling pest management. In: Pimentel, D. (ed.) *Encyclopedia of Pest Management*. Marcel Dekker, New York, pp. 500–503.

Hess, G.R. (1996) To analyse or to simulate, is that the question? *American Entomologist* 42, 14–16.

Hutchins, S.H., Higley, L.G., Pedigo, L.P. and Calkins, P.H. (1986) Linear programming model to optimize management decisions with multiple pests: an integrated soybean pest management example. *Bulletin of the Entomological Society of America* 32, 96–102.

Hutchison, W.D. (1994) Sequential sampling to determine population density. In: Pedigo, L.P. and Buntin, G.D. (eds) *Handbook of Sampling Methods for Arthropods in Agriculture*. CRC Press, Boca Raton, Florida, pp. 207–243.

Jacobsen, E. (1984) On logic, axioms, theorems, paradoxes and proofs. In: Richardson, J. (ed.) *Models of Reality Shaping Thought and Action*. Lomond Press, Mt Airy, Maryland, pp. 71–73.

Jeffers, J.N.R. (1978) *An Introduction to Systems Analysis, with Ecological Applications.* Edward Arnold Press, London, 198 pp.
Jeffers, J.N.R. (1984) The development of models in urban and regional planning. In: Richardson, J. (ed.) *Models of Reality Shaping Thought and Action.* Lomond Press, Mt Airy, Maryland, pp. 87–99.
Kraemer, K.L. (1984) The politics of model implementation. In: Richardson, J. (ed.) *Models of Reality Shaping Thought and Action.* Lomond Press, Mt Airy, Maryland, pp. 131–160.
Legg, D.E. (2000) Tables of computer-simulated errors for binomial sequential sampling plans. In: Pandalai, S.G. (ed.) *Recent Research Developments in Entomology,* Vol IV. Research Signpost Press, Trivandrum, pp. 95–106.
Legg, D.E. and Archer, T.L. (1998) Sampling methods, economic injury levels, and economic thresholds for the Russian wheat aphid (Homoptera: Aphididae). In: Quisenberry, S.S. and Peairs, F.B. (eds) *Response Model for an Introduced Pest – the Russian Wheat Aphid.* Thomas Say Publications in Entomology, Proceedings, Entomological Society of America, Lanham, Maryland, pp. 313–336.
Legg, D.E. and Brewer, M.J. (1995) Relating within-season Russian wheat aphid (Homoptera: Aphididae) population growth in dryland winter wheat to heat units and rainfall. *Journal of the Kansas Entomological Society* 68, 149–158.
Legg, D.E. and Chiang, H.C. (1984) European corn borer (Lepidoptera: Pyralidae) infestations: predicting second generation egg masses from blacklight trap captures and relating their abundance to several corn crop characters. *Journal of Economic Entomology* 77, 1432–1438.
Legg, D.E. and Moon, R.D. (1994) Bias and variability in statistical estimates. In: Pedigo, L.P. and Buntin, G.D. (eds) *Handbook of Sampling Methods for Arthropods in Agriculture.* CRC Press, Boca Raton, Florida, pp. 55–69.
Legg, D.E., Barney, R.J., Tipping, P.W. and Rodriguez, J.G. (1987) Preferred grain quantity and insect density for maize weevil (Coleoptera: Curculionidae) interaction studies. *Journal of Economic Entomology,* 80, 388–393.
Legg, D.E., Wangberg, J.K. and Kumar, R. (1993) Decision support software for implementation of Russian wheat aphid economic injury levels and thresholds. *Journal of Agricultural Entomology* 10, 205–213.
Legg, D.E., Van Vleet, S.M., Lloyd, J.E. and Zimmerman, K.M. (1998a) Calculating lower developmental thresholds of insects from field studies. *Recent Research Developments in Entomology* 2, 163–172.
Legg, D.E., Struttmann, J.M., Van Vleet, S.M. and Lloyd, J.E. (1998b) Bias and variability in lower developmental thresholds estimated from field studies. *Journal of Economic Entomology* 91, 891–898.
Legg, D.E., Van Vleet, S.M. and Lloyd, J.E. (2000) Simulated predictions of insect phenological events made by using mean and median functional lower developmental thresholds. *Journal of Economic Entomology* 93, 658–661.
Legg, D.E., Van Vleet, S.M., Ragsdale, D.W., Hansen, R.W., Chen, B.M., Skinner, L. and Lloyd, J.E. (2002) Required number of location-years for estimating functional lower developmental thresholds and required thermal summations of insects: first emergence of adult *Aphthona nigriscutis* Foudras as an example. *International Journal of Pest Management* 48, 147–154.
Leslie, P.H. (1945) On the use of matrices in certain population mathematics. *Biometrika* 33, 183–212.
Lewis, E.G. (1942) On the generation and growth of a population. *Sankhya* 6, 93–96.
Logan, J.A. (1994) In defence of big ugly models. *American Entomologist* 40, 202–207.
McMaster, G.S. and Smika, D.E. (1988) Estimation and evaluation of winter wheat phenology in the central great plains. *Agricultural and Forest Meteorology* 43, 1–18.
Manetsch, T.J. and Park, G.L. (1982) *System Analysis and Simulation with Application to Economic and Social Systems,* Vol. I, 4th edn. Engineering Library, Michigan State University, East Lansing, Michigan, 52 pp.
Mesbah, A., Miller, S.D., Fornstrom, K.G. and Legg, D.E. (1994) Kochia (*Kochia scoparia*) and green foxtail (*Setaria viridis*) interference in sugarbeets. *Weed Technology* 8, 754–759.
Miller, G.G. and Miller, J.L. (1984) The earth as a system. In: Richardson, J. (ed.) *Models of Reality Shaping Thought and Action.* Lomond Press, Mt Airy, Maryland, pp. 19–49.
Nachapong, M., Legg, D.E., Kittiboonya, S. and Wangboonkong, S. (1989) Validation of computer-simulated presence-absence sequential sampling plans for the cotton bollworm (*Heliothis armigera*) (Hübner) in cotton. *Thai Journal of Agricultural Science* 22, 293–302.
Nachman, G. (1981) A mathematical model of the functional relationship between density and spatial distribution of a population. *Journal of Animal Ecology* 50, 453–460.

Pace, M.E. and MacKenzie, D.R. (1987) Modeling of crop growth and yield for loss assessment. In: Teng, P.S. (ed.) *Crop Loss Assessment and Pest Management*. American Phytopathological Society Press, St Paul, Minnesota, pp. 30–36.

Peck, S.L. (2000) A tutorial for understanding ecological modelling papers for the nonmodeler. *American Entomologist* 46, 40–49.

Pedigo, L.P., Hutchins, S.H. and Higley, L.G. (1986) Economic injury levels in theory and practice. *Annual Review of Entomology* 31, 341–368.

Penman, D.R. and Chapman, R.B. (1982) Design of a field-sampling simulator. *Bulletin of the Entomological Society of America* 28, 143–145.

Plant, R.E. and Stone, N.D. (1991) *Knowledge-based Systems in Agriculture*. McGraw-Hill Press, New York, 364 pp.

Press, W.H., Flannery, B.P., Teukolsky, S.A. and Vetterling, W.T. (1986) *Numerical Recipes*. Cambridge University Press, New York, 203 pp.

Rains, G.C., Olson, D.M., Lewis, J.W. and Tumlinson, J.H. (2002) Systems management. In: Pimentel, D. (ed.) *Encyclopedia of Pest Management*. Marcel Dekker, New York, pp. 826–828.

Richardson, J.G. (1984) A primer of model systems. In: Richardson, J. (ed.) *Models of Reality Shaping Thought and Action*. Lomond Press, Mt Airy, Maryland, pp. 3–18.

Sah, D.N. and MacKenzie, D.R. (1987) Methods of generating different levels of disease epidemics in loss experiments. In: Teng, P.S. (ed.) *Crop Loss Assessment and Pest Management*. American Phytopathological Society Press, St Paul, Minnesota, pp. 90–96.

Salsburg, D. (2001) *The Lady Tasting Tea*. Freeman Press, New York, 24 pp.

Schaalje, G.B. (1990) Dynamic models of pesticide effectiveness. *Environmental Entomology* 19, 440–447.

Shane, W.W. and Teng, P.S. (1987) Generating the database for disease-loss modelling. In: Teng, P.S. (ed.) *Crop Loss Assessment and Pest Management*. American Phytopathological Society Press, St Paul, Minnesota, pp. 82–89.

Smith, M.J. (1974) *Models in Ecology*. Cambridge University Press, London, 145 pp.

Snedecor, G.W. and Cochran, W.G. (1967) *Statistical Methods*, 6th edn. Iowa State University Press, Ames, Iowa, 593 pp.

Stern, V.M., Smith, R.F., Van Den Bosch, R. and Hagen, K.S. (1959) The integrated control concept. *Hilgardia* 29, 81–101.

Stone, N.D., Coulson, R.N., Frisbie, R.E. and Loh, D.K. (1986) Expert systems in entomology: three approaches to problem solving. *American Entomologist* 32, 161–166.

Taylor, L.R. (1961) Aggregation, variance and the mean. *Nature* 189, 732–735.

Teng, P.S. (1981) Validation of computer models of plant disease epidemics: a review of philosophy and methodology. *Zeitschrift für Pflanzenkrankheiten und Pflanzenschutz* 88, 49–63.

Teng, P.S. (1987) The systems approach to pest management. In: Teng, P.S. (ed.) *Crop Loss Assessment and Pest Management*. American Phytopathological Society Press, St Paul, Minnesota, pp. 160–167.

Walker, P.T. (1987) Quantifying the relationship between insect populations, damage, yield, and economic thresholds. In: Teng, P.S. (ed.) *Crop Loss Assessment and Pest Management*. American Phytopathological Society Press, St Paul, Minnesota, pp.114–125.

Weisberg, S. (1980) *Applied Linear Regression*. John Wiley Press, New York, 283 pp.

Zimmerman, K.M. (1999) A spatial model for markov chain analysis of grasshopper population dynamics in Wyoming. MS thesis, University of Wyoming, Laramie, Wyoming, 91 pp.

4 Manipulation of Tritrophic Interactions for IPM

Robert H.J. Verkerk
Department of Biological Sciences, Imperial College London, Silwood Park, Ascot, Berkshire, SL5 7PY, UK
E-mail: RHJVerkerk@aol.com

Introduction

Until recently, there has been a tendency by those involved in integrated pest management (IPM) or integrated crop management (ICM) (Meerman *et al.*, 1996; Denyer, 2000) to be principally concerned with effects on herbivores or interactions between just two trophic levels. However, interest in the importance of interactions between the three or four trophic levels that characterize most natural systems and agroecosystems has been increasing rapidly during the last two decades.

Historically, the chemical-control exponent has rarely been concerned about host-plant effects, and has been interested primarily in chemical impacts on herbivores (the second trophic level; Fig. 4.1) and sometimes side effects on natural enemies (third trophic level; Fig. 4.1). Understanding of interactions between the levels has not been prioritized (Thomas, 1999). Most host-plant resistance specialists are concerned mainly with ways in which resistance factors within or on the plant (first trophic level; Fig. 4.1) affect the development of herbivores, while the biological control specialist tends to concentrate on ways in which predators, parasitoids or pathogens (third trophic level; Fig. 4.1) are able to limit the abundance of and damage caused by herbivorous pests

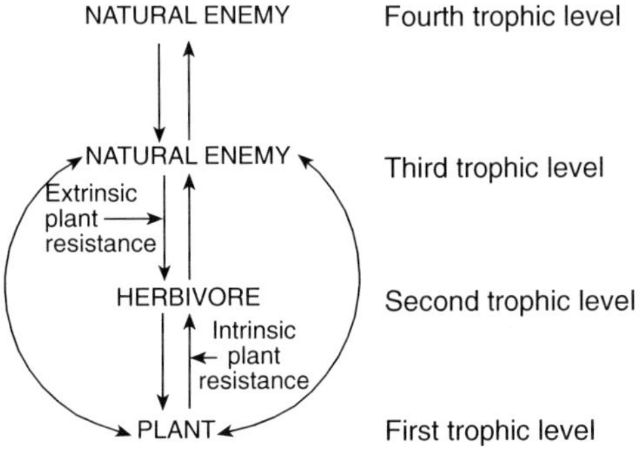

Fig. 4.1. Simplified multitrophic interactions in natural and agroecosystems (adapted from Price, 1986).

Following from this, the plant-resistance specialist has tended to have little interest in interactions between the first, second and third trophic levels, while the biological control specialist, in turn, has given only passing attention to interactions between the first and third trophic levels. In the case of some natural systems and many fewer crop-based systems, interactions between the third and fourth trophic levels have been well studied (e.g. parasitoid–hyperparasitoid interactions).

It is interesting that many traditional cultural practices – one of the less-studied aspects of pest-management science – exert their effects through complex multitrophic interactions, but it is exactly this complexity that makes such systems difficult to assess experimentally or validate conclusively across a broad range of environments.

The reality is that agroecosystems, especially perennial ones, which are not subject to intensive pesticide regimes, like natural systems, contain much more than simple tritrophic interactions. They support food-web interactions of varying complexity across multitrophic levels (Berryman *et al.*, 1995; Janssen *et al.*, 1998; Letourneau and Andow, 1999). Modelling approaches, using empirical data where these are available, are contributing substantially to the understanding of interactions in agroecosystems (Gutierrez, 1996).

Research on tritrophic interactions relating to arthropod herbivores has expanded rapidly since the late 1970s (e.g. Bergman and Tingey, 1979; Price *et al.*, 1980; Strong *et al.*, 1984; Boethel and Eikenbary, 1986; Price, 1986; Duffey and Bloem, 1987; Barbosa, 1988; Gutierrez *et al.*, 1988; Nordlund *et al.*, 1988; Whitman, 1988; Smith, 1989; Price *et al.*, 1990; Fritz, 1992; Hare, 1992; Vet and Dicke, 1992; Steinberg *et al.*, 1993; Godfray, 1994; Mattiacci *et al.*, 1994, 1995; Dicke *et al.*, 1998; Du *et al.*, 1998; Redman and Scriber, 2000; Walker and Jones, 2001). However, as implied above, there have been relatively few empirical studies on crop-based systems (van Emden, 1987, 1995; van Emden and Wratten, 1991; Hare, 1992; Shimoda *et al.*, 1997) and rigorous field studies have only been conducted in the last few years (Eigenbrode *et al.*, 1995; Camara, 1997; Fritz *et al.*, 1997; Karban and EnglishLoeb, 1997; Nwanze *et al.*, 1998; Hufbauer and Via, 1999; Theodoratus and Bowers, 1999; Lill and Marquis, 2001; Chen and Welter, 2002; Thaler, 2002).

Studies in the tritrophic area are taking on a new significance as less pesticide-dependent, more ecologically based approaches to the management of pests are being demanded (Poppy, 1997; Bottrell *et al.*, 1998; Verkerk *et al.*, 1998). As knowledge of interactions across multitrophic systems both in nature and in agroecosystems expands, researchers and pest-management practitioners are beginning to find ways of manipulating interactions across different trophic levels in order to develop more sustainable approaches to pest management. Some of the most relevant approaches are discussed in this chapter and are supported with examples. The chapter restricts itself to consideration of tritrophic aspects of arthropod pest management as arthropods have, to date been much better studied in this regard than pathogens.

Identifying Areas for Manipulation

There are a number of key areas where manipulation of crop–pest–natural-enemy interactions could provide substantial benefits in pest-management systems. The significant and growing evidence from fundamental research in certain areas, e.g. allelochemically mediated interactions, hold substantial promise with regard to the development of novel IPM or ICM techniques, but practical methods have yet to be developed for wide-scale application. In other areas, techniques involving tritrophic interactions are already practised. Although these areas are considered discretely under different headings, it is important to appreciate the considerable overlap between different techniques or approaches. Some of the conflicts or paradoxes that may be inherent in certain forms of manipulation, are also outlined.

Manipulation of host-plant quality

For many years, there was a widely held view that host-plant resistance should be seen as an integral component of IPM programmes, but it has been demonstrated that host-plant resistance is by no means always compatible with biological control (van Emden, 1991, 1995; Hare, 1992; Thomas and Waage, 1994). Hare (1992), for example, cited 16 studies where interactions between resistant crop varieties and natural enemies (parasitoids) were studied and the outcomes show a spectrum of interactions, ranging from synergistic, to additive, to none apparent through to disruptive or antagonistic. Negative interactions can occur because of the presence of secondary chemicals that are ingested or sequestered by natural enemies feeding on hosts present on resistant or partially resistant plants (Godfray, 1994). Specific toxic components in partially resistant soy bean plants can be particularly problematic in this regard (Orr and Boethel, 1985). van Emden (1995) suggests that negative interactions between potentially toxic secondary plant compounds and natural enemies are less likely in the case of natural enemies using phloem-feeding insects, such as aphids, as prey compared with chewing insects because, when present, these compounds are quite often at low concentrations or absent in the phloem (Ryan *et al.*, 1990).

van Emden (1995) cites a number of studies showing that partial plant resistance or environmental variables (e.g. reduced nitrogen fertilization of plants) can not only reduce aphid size and fecundity but may also substantially reduce (by *c.* 30%) the weight and fecundity of female parasitoids, *Aphidius rhopalosiphi* De Stefani Perez, emerging from the aphids. Also of potential concern to the biological control practitioner were the effects of plant resistance on sex ratio (10% fewer females on a partially resistant compared with a susceptible wheat variety) and parasitoid emergence success from aphid mummies (*c.* 20% reduction). A subsequent study has shown that linear furanocoumarins (secondary plant metabolites) in celery plants selectively and adversely affect the polyembryonic encyrtid parasitoid, *Copidosoma floridanum*, while not affecting its larval noctuid host, *Trichoplusia ni* (Hübner) (Reitz and Trumble, 1996). Negative interactions may also be caused by physical (or chemical) factors such leaf toughness, cuticle thickness, trichomes (glandular and non-glandular), galls and plant architecture (Price, 1986).

In terms of positive interactions that can be manipulated, van Emden and Wearing (1965) were the first to suggest that partial plant resistance, in combination with natural enemies, may give economic levels of control for some agricultural insect pests. However, since this time, a considerable body of work has accumulated regarding mechanisms that both support and oppose this notion. Bottrell *et al.* (1998) provide a useful review of plant-resistance–natural-enemy interactions. The authors emphasize that many of the interactions that occur in field systems are poorly understood and that combining natural enemies and plant resistance may slow the adaptation of some insect pests, while it may speed up adaptations of others.

Feeny (1976) and others (e.g. Moran and Hamilton, 1980; Price *et al.*, 1980; Leather, 1985) proposed that sublethal defences (e.g. 'digestibility reducers') in plants increase herbivore exposure to natural enemies: the slower the growth, the higher the mortality. This aside, several studies have shown that natural enemies foraging on leaf surfaces dislodge potential (aphid) prey to a greater extent on partially resistant compared with susceptible host plants. Gowling and van Emden (1994) showed this for *Metopolophium dirhodum* (Walker) and the parasitoid *A. rhopalosiphi* (on partially resistant and susceptible cultivars of wheat) in the glasshouse, as well as for *Brevicoryne brassicae* (Linnaeus) on Brussels sprouts in the field, where hover flies (Diptera: Syrphidae) were the main predatory group.

However, field studies on galling sawflies of willow showed the opposite phenomenon, i.e. that parasitism was greater in faster-growing larvae, implying that sublethal plant defences remain a paradox (Clancy and Price, 1987). Faster-growing larvae of the pyralid moth *Omphalocerca munroei* that fed on young leaves of their host plants (*Asimina* spp.) were also observed to be more vulnerable to predators than their slower-growing counterparts that fed on the preferred, older leaves of the same host plants (Damman, 1987). The latter studies support the view of Schultz (1983) that bigger individuals are a more profitable resource for predators. However, galling sawflies and the pyralid moth (a gregarious shelter-builder) both live in relatively protected but highly apparent environments (i.e. vulnerable to discovery by their enemies) and may not be ideal systems to test the slow growth/higher mortality hypothesis (Leather and Walsh, 1993). To examine this hypothesis, the latter authors infested 'poor'- and 'good'-quality hosts (*Pinus contorta* of two different origins) with the pine beauty moth, whose larvae are more exposed on foliage. They found that larvae reared on the 'good' host, in the absence of predators, were heavier and more numerous than those reared in a similar manner on the 'poor' host. However, with populations suffering predation, it was found that these heavier individuals had been the ones favoured by the predators. This result suggests that the bigger and healthier the host, the greater the reward for the predator (see Schultz, 1983), and that poor nutritional status may be a liability to the plant in these circumstances rather than an asset.

Host-plant quality may affect the influence of pathogens of insect pests. Differential susceptibility to a nuclear polyhedrosis virus (NPV) was found for two pests, beet armyworm, *Spodoptera exigua* (Hübner), and maize earworm, *Helicoverpa zea* (Boddie), fed on different host plants. Beet armyworm was most susceptible to the virus on tomato, least susceptible on cotton and intermediate on cole-wort, while maize earworm was most susceptible on maize, least on cotton and intermediate on bean (Farrar and Ridgway, 2000). The greenhouse whitefly, *Trialeurodes vaporariorum* (Westwood) was found to be considerably more susceptible to the fungal pathogens *Beauveria bassiana* and *Paecilomyces fumosoroseus* when reared on cucumber compared with tomato (Poprawski *et al.*, 2000).

Host-plant variation has been shown to contribute substantially to the toxicity of endotoxins of the bacterial pathogen *Bacillus thuringiensis* (*Bt*). More than fivefold differences in toxicity of *Bt* subsp. *aizawai* were found to diamondback moth, *Plutella xylostella* (Linnaeus), larvae when exposed to treated leaves of susceptible and partially resistant common cabbage (*Brassica oleracea* var. *capitata*) (Verkerk and Wright, 1996a), while a more than tenfold variation in toxicity of *Bt* subsp. *kurstaki* was shown to larvae of gypsy moth, *Lymantria dispar* (Linnaeus), when exposed on a range of different host plants (Farrar *et al.*, 1996).

There have been few studies looking at population-level effects of variable plant phenology. In one such study, the timing of egg hatch of the gypsy moth, *L. dispar* was manipulated in the field (Hunter and Elkinton, 2000). When larvae fed on old leaves, i.e. larvae released long before budburst, the adult fecundity of the moth declined owing to the low quality of the host plant. However, where the leaves were old, the dispersal rate (by 'ballooning') was greater than from young leaves and, since dispersal reduces herbivore density, natural-enemy-induced density-dependent mortality was also lower (Hunter and Elkinton, 2000). This shows that first and third trophic-level-mediated effects may counteract each other in relation to herbivore survival or abundance.

Finally, host-plant quality can have important implications for omnivores. This has been well studied in the case of western flower thrips, *Frankliniella occidentalis* (Pergande), which eats both animal prey (mite eggs) and host-plant foliage. According to foraging theory, omnivores balance their diet as a result of food quality, nutritional need and the availability of alternative foods. Although it has been shown that reducing plant quality can cause omnivores to shift towards relatively more predation than herbivory (Agrawal *et al.*, 1999), it was later shown that where plant resistance has been induced by herbivory (induced plant resistance), reduced prey density and quality may antagonize this shift towards increased predation (Agrawal and Klein, 2000).

Allelochemicals

Allelochemicals are those chemicals that mediate interspecific interactions and are distinguished from pheromones, which mediate intraspecific interactions. They have been shown to be among the most important factors controlling the sequence of natural-enemy host-searching and selection behaviours, and this has been particularly well studied in the case of parasitoids (Vinson, 1984; Lewis and Martin, 1990; Turlings and Tumlinson, 1991; Whitman and Eller, 1992; Tumlinson *et al.*, 1993).

The induction of plant defence through the release of allelochemicals as a response to herbivory has been amply demonstrated (e.g. Whitman and Eller, 1992; Steinberg *et al.*, 1993; Takabayashi *et al.*, 1994; Mattiacci *et al.*, 1995; Du *et al.*, 1996, 1998; Dicke *et al.*, 1998; van Poecke *et al.*, 2001). Such allelochemicals are often referred to as synomones, given that both the plant (emitter) and the natural enemy (receiver) benefit from the chemical communication. It has been shown that the allelochemicals released following herbivory not only influence specialist and/or generalist natural enemies but can also affect the behaviour of the herbivore and, through plant–plant communication, can make adjacent plants more attractive to natural enemies. The plant emissions can be herbivore-specific. De Moraes *et al.* (1998) showed that tobacco, cotton and maize plants each released distinct blends of volatiles when they were exposed to herbivory by two closely related Lepidoptera larvae, *Heliothis virescens* (Fabricius) and *H. zea*. In chemical/behavioural assays, these authors demonstrated that the specialist parasitoid *Cardiochiles nigriceps* Viereck was able to discriminate between these blends, being attracted specifically to its own host, *H. virescens*.

Despite recent progress and an abundance of research in the area of allelochemically mediated tritrophic interactions, there have yet to be significant advances in the applied area, so there are few examples of synomones being used practically as an IPM technique. Lewis and Nordlund (1984) demonstrated the potential of synomone-mediated attraction of natural enemies by applying synomone-containing extracts from maize and tomato. Parasitism by *Telenomus remus* Nixon of autumn armyworm, *Spodoptera frugiperda* (J.E. Smith) was increased twofold on plants treated with extracts, compared with the untreated plants. Limited field studies by Altieri *et al.* (1981) found that application of a water-based extract from an *Amaranthus* sp. significantly increased parasitism of *H. zea* eggs by naturally occurring *Trichogramma* spp. in various crops, including soybean, cowpea, tomato and cotton.

The prospects for manipulating predator–prey interactions via the first (plant) trophic level have yet to be fully realized. Dicke *et al.* (1990) provide a useful synthesis on specific areas of allelochemical manipulation, drawing heavily on research on herbivorous and predatory mite systems. These authors cite work relating to: enhancement of searching efficiency or triggering specific search modes in natural enemies; increasing host acceptability to natural enemies in mass rearing; using the response to an allelochemical as a criterion in the selection of natural enemies in control programmes; ensuring quality control of mass-reared natural enemies; and breeding of plant cultivars that have high emission rates of natural-enemy-attracting synomones.

Aphid parasitoids may be 'conditioned' during their larval stage to the specific host-plant variety on which the parasitoid's host is reared (Braimah and van Emden, 1994; van Emden, 1995). This is potentially of considerable relevance to biological control since the introduction of a crop variety with resistance based on an allelochemical may actually deter parasitoids and the existence of host-plant conditioning could offset the usefulness of weed-based reservoirs of polyphagous parasitoids (van Emden, 1995). This phenomenon may also cause suboptimal parasitism following introduction of an exotic parasitoid where mass rearing has occurred on a host-plant variety different from that constituting the target crop.

The rapidly growing body of work on allelochemicals demonstrates that the use of natural-enemy-attracting chemicals produced by plants may hold substantial promise for enhancement of biological control. This is most likely either through augmentation (field application of synomones) or by breeding of plants with elevated emission rates of synomones (see Genetic manipulation section below).

Crop diversification

The diversification of crop and neighbouring environments is widely regarded as benefiting biological control and sustainable agriculture systems. It is based on the premise that habitats that are structurally, biologically or temporally diverse provide greater levels of habitat diversity, which in turn diversifies the available prey, so increasing the abundance of natural enemies (Root, 1973).

There are many ways in which diversification of the crop environment can be undertaken; most practical attempts rely on polycultures (multiple cropping) or the use of refugia. Despite the theoretical benefits of crop diversification, results from studies attempting to verify the benefits or otherwise of such practices have been mixed (Sheehan, 1986) and demonstrate the importance of understanding the nature of multitrophic interactions. One of the best-known practical examples of the benefits of mixed cropping was the rediversification of the cotton agroecosystem in the Cañete Valley in Peru, after massive bollworm outbreaks followed the abandonment of mixed cropping in the 1950s (Doutt and Smith, 1971). However, it should be borne in mind that some of these benefits occurred as a result of the cessation of highly intensive insecticide application regimes.

Chen and Welter (2002) tested the hypothesis that herbivores are more abundant in agroecosystems compared with more diverse natural habitats, and studied the dynamics of the sunflower moth, *Homoeosoma electellum* (Hulst) and its natural enemies in agricultural and native sunflower habitats. The authors showed clearly that sunflower moths were consistently more abundant in the agricultural habitats (with domesticated sunflowers) compared with the native ones (containing wild ancestors of the domesticated varieties). Additionally, parasitism rates of the herbivore were between six and ten times higher in the native compared with the agricultural habitats.

One of the likely reasons for mixed results from studies investigating the effects of crop diversification on natural enemies is the presence of confounding factors (Barbosa and Wratten, 1998). Also, the results will inevitably be limited by the specific parameters measured.

Polycultures

Gold *et al.* (1989) showed that the density of plants in a polyculture may confound the influence of plant diversity. Apart from spatial and habitat differences, polycultures may give rise to changes in movement rate which can only be detected through season-long studies. For example, initial colonization by a eulophid parasitoid of the Mexican bean beetle, *Epilachna varivestis* Mulsant, was found to be greater in monoculture than in more diverse (intercropped) plots (Coll and Bottrell, 1996). But, because emigration from the monocultural plots occurred sooner, tenure time for the parasitoid in the intercropped plots was greater as the latter provided a more favourable habitat.

Intercropping may increase the effects of natural enemies because one of the intercropped plants provides allelochemical attraction or a nectar source for natural enemies or because the intercrop improves conditions (e.g. moisture, shelter) for ground-dwelling predators (van Emden, 1989). Read *et al.* (1970) suggested that planting cole-worts near beet might enhance biological control of beet pests because cole-worts attract braconid parasitoids, e.g. *Diaeretiella rapae* M'Intosh, to the general area.

Field studies in Mexico showed that parasitism of the pyralid *Diaphania hyalinata* was greater in tricultures (squash, maize, legumes) than in monocultures of squash (Letourneau, 1987), providing partial support for the 'enemies hypothesis' (Root, 1973), which dictates that natural enemies should be more abundant in diverse rather than in simple habitats. However, the author also found that parasitoid attack rates were elevated in maize monocultures and that predator abundance was not enhanced in the tricultures, some predator species being actually more abundant in the monocultures. The author concluded that particular vegetational associations can enhance natural-enemy activity, although the role of single-species effects can be obscured by a focus on diversity and these enhancement effects do not apply to all natural enemies (Letourneau, 1987).

Refugia and non-crop plants

Knowledge that uncultivated land can support a diverse range of natural enemies, which can help to control pests on agricultural crops, has long been known. However, in many industrialized countries, the mechanization of agriculture, as well as the availability of high-yielding crop varieties and synthetic pesticides, has meant that pest-management programmes have often failed to emphasize the importance of natural-enemy refugia. In Britain, for example, hedgerows and uncultivated field margins were destroyed as a result of these modern technologies and only recently have there been attempts to reverse this trend in the wake of increasing awareness of the problems associated with large-scale monocultures and over-reliance on pesticides (van Emden, 1990).

Within the crop, natural-enemy refugia can take the form of unsprayed crop areas, protected plant parts of the crop itself, non-crop plants that favour natural enemies or alternative hosts, which in turn encourage economically unimportant arthropods that provide an alternative food source for natural enemies. Outside the crop, uncultivated field margins, hedgerows, 'conservation headlands' and 'live fences' have all been used, at least in part, to provide refugia for natural enemies, which can subsequently move into the crop environment.

The effective use of refugia generally requires a thorough knowledge of crop–pest–natural-enemy interactions. Spatial scales are particularly important (Murdoch and Briggs, 1996). It is often not sufficient simply to leave uncultivated sections within or outside the crop. The floral composition of refuges, their location and their dimensions should be carefully considered in relation to the specific crop–pest–natural-enemy situation in question.

There are many examples in the literature indicating the effectiveness of natural-enemy refugia across a wide range of agroecosystems. Some examples are considered below.

The planting of lucerne (*Medicago sativa*) strips in cotton fields in Australia has been shown to increase the abundance of important predators (Mensah, 1999). Predatory beetles, bugs and lacewings were five- to sevenfold more abundant in the lucerne strips than in the cotton crop itself and these predators in turn migrated into the cotton and were on average twice as abundant in cotton with lucerne strips compared with cotton without the strips (Mensah, 1999).

Wild brassicas (e.g. *Barbarea vulgaris* and *Brassica kaber*) planted in the vicinity of cultivated brassicas provide floral nectar reserves for parasitoids of the diamondback moth, *P. xylostella*, so improving the potential for biological control (Idris and Grafius, 1996). Such non-crop plants can sometimes provide a dual function in pest management, acting both as refuge sites for natural enemies and as trap crops for the pest. However, there is also the possibility that some plants known to favour natural enemies will also attract pests to the vicinity.

Border-planting of *Phacelia tanacetifolia* (Hydrophyllaceae) as a floral (pollen) resource has also been shown to be effective in increasing the abundance of aphid-eating hover flies, the larvae of which are often key natural enemies of aphids (White *et al.*, 1995). Adult female hover flies generally need to consume pollen before they can lay fertile eggs so it is important that such floral reserves are present close to the crop.

The common farming practice of using plum trees as parasitoid refugia has been shown to improve biological control of the western grape leafhopper, *Erythroneura elegantula* Osborn in vineyards in the western USA (Murphy *et al.*, 1996). Almost twice the abundance of the egg parasite *Anagrus epos* Girault was found in vineyards with plum tree refugia compared with those without, the trees acting both as overwintering sites and as wind-breaks, which aid parasitoid flight and dispersal (Murphy *et al.*, 1996).

Protected parts of certain crops that are relatively free from the disturbances associated with agronomic practices can also provide important refugia for natural enemies. For example, the areas beneath the plucking surface of tea plants are well known as a refuge for a range of natural enemies of important tea pests. Research has shown that domatia, which are small, sometimes elaborate pits or shelters at the junctures of veins on the lower surface of leaves of some plants, are inhabited primarily by predatory rather than herbivorous arthropods. A group of researchers has shown that, by adding artificial domatia to cotton plants, fruit production could be increased by 30%, compared with control plants (Agrawal and Karban, 1997). This increase was caused by enhanced predation of herbivorous mites by predatory ones that were able to harbour in the artificial domatia (Agrawal and Karban, 1997).

Some natural enemies overwinter within plant parts so the removal of stubble, fruits or other crop litter may be counterproductive as these plant parts act as refugia for the natural enemies. For example, in Australia, several species of predatory mite, *Typhlodromus* spp., have been shown to overwinter in the calyx cavities of apple fruits, so that early-season phytophagous mite control can be improved if apples are left on the ground through winter (Gurr *et al.*, 1997).

Resource availability

Population ecologists are actively debating the relative importance of bottom-up (resource-driven) and top-down (natural-enemy-driven) processes in the regulation of herbivores (Walker and Jones, 2001; Denno *et al.*, 2002).

Walker and Jones (2001) argue that the absence of empirical data and the desire by researchers to favour one of these processes to the exclusion of the other have greatly hampered the understanding of the interactions between these processes. Very little work of this type has been conducted in relation to agricultural systems, but models using limited empirical data are increasingly being used to elucidate the importance of these processes (Gutierrez, 1996).

Denno *et al.* (2002) investigated the relative effects of bottom-up and top-down processes in both laboratory mesocosms and the field, using a non-agricultural Spartina grassland system. They found that changes in plant nutrition and structure had complex effects on natural-enemy and herbivore abundance and that these effects were not necessarily paralleled under laboratory and field conditions because of more complex interactions in the field. However, it was shown that in some situations resource availability can override the importance of natural enemies, and this has potential applications for pest management in agriculture.

The most common method of manipulating bottom-up processes in the field is by altering the fertilization regime. Differences in nutritional status of the host plant will have different effects on herbivores compared with natural enemies, and under some conditions the growth rate of natural enemy populations may not be able to 'keep up' with pest populations. Walde (1995) tested this hypothesis in a mite system and found that the growth rate of a phytophagous mite, *Panonychus ulmi* (Koch), in apple trees was non-linear in relation to increased addition of NKP fertilizer; the density increased and then decreased with the increasing N content of the leaves. Two key predatory mite species in this system were unable to compensate for the increased growth rate of the phytophagous mite, but the populations of the latter were none the less maintained beneath the economic threshold.

Apart from affecting growth rates, fertilization may affect sex ratios of natural enemies. In studies of interactions between cole-worts, diamondback moth and its ichneumonid parasitoid, *Diadegma insularis* (Cresson), sex ratios of parasitoids emerging from hosts on plants treated with high levels of nitrogen were found to be consistently female-biased compared with those of parasitoids emerging from unfertilized plants, where sex ratios were male-biased with the implication of lower potential parasitism (Fox *et al.*, 1990).

A comprehensive study of the effects of nitrogen treatment of cole-worts greens on the vulnerability of the cabbage white butterfly, *Pieris rapae* (Linnaeus), to natural enemies showed that its larvae grew more slowly under low-nitrogen conditions and were more susceptible to ground-dwelling predators (Loader and Damman, 1991). This supported the slow growth/high mortality hypotheses of Feeny (1976) and Moran and Hamilton (1980) (see above). Interestingly, however, Loader and Damman (1991) found that emergence of pupal parasites, e.g.

Pteromalus spp., was greater on high-nitrogen plants, larvae on the low-nitrogen plants frequently dying as a result of stress-induced bacterial infection, so preventing complete development of the parasitoids.

Crop background

Visual cues can influence host selection by both herbivores and natural enemies. Accordingly, crop background can have an influence on long-range searching. A number of insect pests and natural enemies have been shown to be differentially attracted to particular crop backgrounds. For example, cabbage aphid, *B. brassicae,* abundance was greater on Brussels sprout plants where they contrasted with a bare background compared with when the background consisted of a 'carpet' of weeds, cut regularly to avoid excessive competition with the crop (Smith, 1969). Smith (1969) and others (e.g. Kennedy *et al.*, 1961; van Emden, 1965) showed that aphids were more strongly attracted to the crop when it contrasted with the bare background, but that relative attraction between different groups of natural enemies (Coccinelidae, Cecidomyiidae, Chrysopidae, Anthocoridae and Syrphidae) varied significantly (Smith, 1969). Smith (1969) suggested that aphid abundance could have been affected by differential attraction of natural enemies, with anthocorid predators being most strikingly enhanced in the weedy habitats. Van Emden (1989) noted that certain species of hover fly lay more eggs on aphid-infested crops when the ground is covered, but warned that other syrphid species prefer crops surrounded by bare soil.

Exogenous interactions

The application of exogenous substances, such as botanical extracts, may have differential effects on pests and natural enemies that can be exploited as a biorational method in IPM. Although such substances do not necessarily affect interactions across three trophic levels, this possibility exists. The appreciation that partial host-plant resistance could enhance parasitism of the diamondback moth, *P. xylostella* (Verkerk, 1995), led to the testing of low (sublethal) concentrations of neem seed kernel extract (from *Azadirachta indica*), on a host plant highly susceptible to *P. xylostella.* This botanical treatment was shown to enhance parasitism success by *Diadegma semiclausum* Hellen in the laboratory, although a dose fivefold greater than this parasitism-enhancing dose led to a drastically reduced rate of pupal emergence so effectively inhibiting parasitism (Verkerk and Wright, 1994). Verkerk and Wright (1994) suggested that observed increases in the rates of parasitoid pupation when hosts were maintained either on partially resistant host plants or susceptible plants treated with neem could be caused by stress-induced impairment of the host's immune system. This phenomenon has previously been shown to occur in other host–parasitoid systems when the hosts were subjected to nutritional stress (see Godfray, 1994).

Low doses of a neem extract and azadirachtin (a primary active ingredient of neem) on plants have been shown to reduce the fecundity of several aphid species (Lowery and Isman, 1994; Koul, 2003). It has also been suggested that sublethal doses of azadirachtin may make a significant contribution to the control of the peach–potato aphid, *Myzus persicae* (Sulzer), while having little or no adverse impact on parasitism by *Aphidius matricariae* Haliday (Sugden, 1994; Mordue *et al.*, 1996). Laboratory studies showed that azadirachtin could enhance the toxicity of gypsy moth, *L. dispar*, NPV to its host, the combined effects of the insecticide and NPV being greater than either product applied on its own (Cook *et al.*, 1996).

In an IPM programme targeting the glasshouse whitefly, *T. vaporariorum*, conventional insecticides were used at one-third rates in conjunction with a mycoinsecticide (*B. bassiana*: Boveral®) and the encyrtid parasitoid, *Encarsia formosa* (Dirlbek *et al.*, 1992). The selection of appropriate insecticides (at reduced rates), careful timing and integration of all three control methods gave optimum whitefly control on certain lines (e.g. Transvaal daisy) where one method alone was found to be inadequate.

Although not tritrophic in nature, Chilcutt and Tabashnik (1997) investigated the effects of ditrophic (within-host) interactions between different phenotypes of *P. xylostella*, its endolarval braconid parasitoid *Cotesia plutellae* Kurdyumov and *Bt*. The outcome of the interaction was dependent on the host phenotype (*Bt*-resistant or *Bt*-susceptible). In susceptible hosts the parasitoid did not affect performance of the pathogen, in moderately resistant hosts the interaction was symmetrical and competitive, while highly resistant hosts were not susceptible to the pathogen, which created a refugium from competition for the parasitoid (Chilcutt and Tabashnik, 1997). These studies demonstrate the considerable scope for integration of chemical and biological control within a multitrophic context (see also Wright and Verkerk, 1995).

Genetic manipulation

To date, most of the work on genetic manipulation for pest-management purposes, through either plant breeding or genetic engineering, has involved host plants. This has been mainly through the development of transgenic crops containing endotoxins from *Bt* (Brar and Khush, 1993; Daly *et al.*, 1994; Tabashnik, 1994; Metz *et al.*, 1995). Virtually no work has been carried out on breeding programmes or genetic engineering that affects natural enemies, despite the recognized key importance of natural enemies in pest management (Groot and Dicke, 2002). Genetic manipulation of insects (pests and natural enemies) for pest management is in its infancy but it is thought that it will become increasingly important as non-pesticidal approaches gain increasing favour in IPM (DeVault *et al.*, 1996).

As transgenic crops become more widely adopted, concerns about their potential natural-enemy and non-target impacts, as well as the likelihood of their inducing resistance to toxic principles, are increasing. In a review by Groot and Dicke (2002), the possible effects of insect-resistant transgenic and non-transgenic crops on food webs are considered. Because *Bt* toxins maintain their toxicity when bound to the soil, there is a real possibility of non-target impacts, although the effects may be indirect and difficult to evaluate. Some fauna (e.g. earthworms and some non-target herbivores) are not directly affected by the toxins, but the toxins may be ingested subsequently by susceptible invertebrates (Groot and Dicke, 2002).

The potential risk of insects developing resistance to toxins in such transgenic crops has been recognized by many researchers (e.g. Strong *et al.*, 1990; Menken *et al.*, 1992; Daly *et al.*, 1994), while others have argued that the risk may be smaller than that of the use of the same toxic principles in sprays (e.g. Roush, 1994). Transgenic crops such as cotton have also been known to trigger secondary pest outbreaks (e.g. Heteroptera: Miridae: see Fitt *et al.*, 1994; Hardee and Bryan, 1997), which may need to be controlled with conventional insecticide sprays. These insecticides have the potential to harm natural enemies. Strong *et al.* (1990) argued that transgenic techniques need to be refined so that toxins are only expressed in a subset of crucial tissues and at specific developmental stages and they should also be integrated into an ecological framework if they are to be effective and contribute to biocontrol. Van Emden and Wratten (1991) warned that modern gene-transfer techniques aimed at creating resistant crop varieties are more likely than traditional plant-breeding methods to use an allelochemical mechanism of resistance (antibiosis), which might be damaging to natural enemies and other non-target organisms.

Genetic manipulation of crop plants to enhance nutrient uptake has been considered and could be compatible with IPM/ICM systems (Johansen *et al.*, 1995). The potential also exists for creating genetically engineered plants that emit increased amounts of natural-enemy-attracting volatiles (synomones), although such plants are yet to be available commercially.

Tritrophic Models

The intrinsic complexity of multitrophic interactions in agroecosystems and the difficulty of generating meaningful empirical data from the field have meant that modelling approaches have been used to facilitate the understanding of mechanisms and processes. Simulation models of crop–pest–natural-enemy systems used to date (e.g. Gutierrez, 1996) are based on detailed field and laboratory data. The models integrate the biology and physiology of consumers, which in turn affect resource acquisition and allocation and hence population birth and death rates. Resource availability (i.e. bottom-up effects) define the underlying base dynamics of each species, while natural enemies provide the top-down or regulatory effects (Gutierrez, 1996).

Simulation models have been used to assist in the tactical evaluation of various biological control and plant-resistance interactions (e.g. Gutierrez *et al.*, 1984, 1988, 1994) and simple analytical models (e.g. Hassell, 1978) have been used to make very general population-dynamics predictions. However, there have been very few studies that have examined tritrophic interactions using biologically realistic but quantitative population-dynamics models (see Gutierrez *et al.*, 1993). The relatively recent application of age-structured models written as a series of delay-differential equations (Nisbet and Gurney, 1983) has promoted a middle approach between tactical and strategic predictive modelling, which may lead to a better understanding of pest–natural-enemy interactions under field conditions. Such models have been termed 'models of intermediate complexity' (e.g. Godfray and Waage, 1991; Briggs and Godfray, 1995) and can be applied to tritrophic systems (Thomas and Waage, 1994), although to date these models have been largely concerned with second and third trophic-level interactions (Gutierrez *et al.*, 1994).

A complex, metapopulation tritrophic model of the African cassava system was developed recently by Gutierrez *et al.* (1999). The model is based on a single-patch age-structured population-dynamics model, taking into account key plants and animals in the system. It also accounts for movement rates of animals between patches, these being dependent on species-specific supply–demand relations. The model demonstrates the importance of plant-level effects on higher trophic levels. Such models can be very valuable tools in understanding agroecosystems, which can in turn facilitate better management of biological- or cultural-control programmes in IPM.

Conclusions

Improving the understanding of multitrophic systems is critical to the development of sustainable, less pesticide-dependent or pesticide-free pest-management systems (Bottrell *et al.*, 1998). This is because control techniques, or target pests, cannot be regarded in isolation – manipulation of any single factor will tend to have knock-on effects on different trophic levels. From a commercial point of view, the areas with greatest potential are likely to be allelochemical and genetic manipulations. However, it should be recognized that at present the biological control market is estimated to be in the region of only US$75 million (Waage, 1997), compared with the US$32 billion global agricultural-chemical market (Warrior, 2000).

Thomas (1999) reasons that increasing our basic understanding of how individual pest-control technologies act and interact will reveal new opportunities for improving pest control. There is a great

> need to break away from the existing single-technology, pesticide-dominated paradigm and to adopt a more ecological approach built around a fundamental understanding of population biology at the local farm level and the true integration of renewable technologies such as host plant resistance and natural biological control, which are available to even the most resource-poor farmers.
>
> (Thomas, 1999)

As the popularity of organic agriculture increases and pesticide-related problems (e.g. residues, resistance, side effects) continue to mount, there will be an ever greater requirement for effective and reasonably predictable natural control systems (see Verkerk and Wright, 1996b). A study of 18 commercial tomato farms in California, half certified organic production systems, half conventionally managed and using synthetic pesticides, showed clearly the viability of organic systems (Letourneau and Goldstein, 2001). Benefits of the organic systems included greater natural-enemy diversity and abundance of functional guilds.

In the interest of agricultural sustainability, tritrophic manipulation, as a distinct approach to biological and/or cultural control, is likely to be prioritized increasingly by both researchers and those responsible for the development and practical implementation of pest-management programmes. This process will be facilitated as improvements in the understanding of crop–pest–natural-enemy evolution and interactions are achieved (Bottrell *et al.*, 1998).

References

Agrawal, A.A. and Karban, R. (1997) Domatia mediate plant–arthropod mutualism. *Nature* 387, 562–563.

Agrawal, A.A. and Klein, C.N. (2000) What omnivores eat: direct effects of induced plant resistance on herbivores and indirect consequences for diet selection by omnivores. *Journal of Animal Ecology* 69, 525–535.

Agrawal, A.A., Kobayashi, C. and Thaler, J.S. (1999) Influence of prey availability and induced host-plant resistance on omnivory by western flower thrips. *Ecology* 80, 518–523.

Altieri, M.A., Lewis, W.J., Nordlund, D.A., Gueldner, R.C. and Todd, J.W. (1981) Chemical interactions between plants and *Trichogramma* wasps in Georgia soybean fields. *Protection Ecology* 3, 259–263.

Barbosa, P. (1988) Natural enemies and herbivore plant interactions: Influence of plant allelochemicals and host specificity. In: Barbosa, P. and Letourneau, D.K. (eds) *Novel Aspects of Insect–Plant Allelochemicals and Host Specificity.* John Wiley & Sons, New York, pp. 201–229.

Barbosa, P. and Wratten, S.D. (1998) Influence of plants on invertebrate predators: implications to conservation biological control. In: Barbosa, P. (ed.) *Conservation Biological Control.* Academic Press, San Diego, California, pp. 83–100.

Bergman, J.M. and Tingey, W.M. (1979) Aspects of interactions between plant genotypes and biological control. *Bulletin of the Entomological Society of America* 25, 275–279.

Berryman, A.A., Michalski, J., Gutierrez, A.P. and Arditi, R. (1995) Logistic theory of food web dynamics. *Ecology* 76, 336–343.

Boethel, D.J. and Eikenbary, R.D. (1986) *Interactions of Plant Resistance and Parasitoids and Predators of Insects.* Wiley & Sons, Chichester, UK, 224 pp.

Bottrell, D.G., Barbosa, P. and Gould, F. (1998) Manipulating natural enemies by plant variety selection and modification: a realistic strategy? *Annual Review of Entomology* 43, 347–367.

Braimah, H. and van Emden, H.F. (1994) The role of the plant in host acceptance by the parasitoid *Aphidius rhopalosiphi* (Hymenoptera: Braconidae). *Bulletin of Entomological Research* 84, 303–306.

Brar, D.S. and Khush, G.S. (1993) Application of biotechnology in integrated pest management. *Journal of Insect Science* 6, 7–14.

Briggs, C.J. and Godfray, H.C.J. (1995) The dynamics of insect–pathogen interactions in stage-structured populations. *American Naturalist* 145, 855–887.

Camara, M.D. (1997) Predator responses to sequestered plant toxins in buckeye caterpillars: are tritrophic interactions locally variable? *Journal of Chemical Ecology* 23, 2093–2106.

Chen, Y.H. and Welter, S.C. (2002) Abundance of a native moth *Homoeosoma electellum* (Lepidoptera: Pyralidae) and activity of indigenous parasitoids in native and agricultural sunflower habitats. *Environmental Entomology* 31, 626–636.

Chilcutt, C.F. and Tabashnik, B.E. (1997) Host-mediated competition between the pathogen *Bacillus thuringiensis* and the parasitoid *Cotesia plutellae* of the diamondback moth (Lepidoptera: Plutellidae). *Environmental Entomology* 26, 38–45.

Clancy, K.M. and Price, P.W. (1987) Rapid herbivore growth enhances enemy attack: sublethal plant defences remain a paradox. *Ecology* 68, 733–737.

Coll, M. and Bottrell, D.G. (1996) Movement of an insect parasitoid in simple and diverse plant assemblages. *Ecological Entomology* 21, 141–149.

Cook, S.P., Webb, R.E. and Thorpe, K.W. (1996) Potential enhancement of the gypsy moth (Lepidoptera: Lymantriidae) nuclear polyhedrosis virus with the triterpene azadirachtin. *Environmental Entomology* 25, 1209–1214.

Daly, J.C., Hokkanen, H.M.T. and Deacon, J. (1994) Ecology and resistance management for *Bacillus thuringiensis* transgenic plants. *Biocontrol Science and Technology* 4, 563–571.

Damman, H. (1987) Leaf quality and enemy avoidance by the larvae of a pyralid moth. *Ecology* 68, 88–97.

De Moraes, C.M., Lewis, W.J., Pare, P.W., Alborn, H.T. and Tumlinson, J.H. (1998) Herbivore-infested plants selectively attract parasitoids. *Nature* 393, 570–573.

Denno, R.F., Gratton, C., Peterson, M.A., Langellotto, G.A., Finke, D.L. and Huberty, A.F. (2002) Bottom-up forces mediate natural-enemy impact in a phytophagous insect community. *Ecology* 83, 1443–1458.

Denyer, R. (2000) Integrated crop management: introduction. *Pest Management Science* 56, 945–946.

DeVault, J.D., Hughes, K.J., Johnson, O.A. and Narang, S.K. (1996) Biotechnology and new integrated pest management approaches. *Bio-technology* 14, 46–49.

Dicke, M., Sabelis, M.W., Takabayashi, J., Bruin, J. and Posthumus, M.A. (1990) Plant strategies of manipulating predator-prey interactions through allelochemicals: prospects for application in pest control. *Journal of Chemical Ecology* 16, 3091–3118.

Dicke, M., Takabayashi, J., Posthumus, M.A., Schutte, C. and Krips, O.E. (1998) Plant-phytoseiid interactions mediated by herbivore-induced plant volatiles: variation in production of cues and in responses of predatory mites. *Experimental and Applied Acarology* 22, 311–333.

Dirlbek, J., Dirlbekova, O. and Jedlicka, M. (1992) The combined use of mycoinsecticide, parasitoid and chemical stressor in the control of greenhouse whitefly (*Trialeurodes vaporariorum* Westwood). *Ochrana Rostlin* 28, 71–77.

Doutt, R.L. and Smith, R.F. (1971) The pesticide syndrome – diagnosis and suggested prophylaxis. In: Huffaker, C.B. (ed.) *Biological Control*. Plenum Press, New York, pp. 3–15.

Du, Y.J., Poppy, G.M. and Powell, W. (1996) Relative importance of semiochemicals from first and second trophic levels in host foraging behaviour of *Aphidius ervi*. *Journal of Chemical Ecology* 22, 1591–1605.

Du, Y.J., Poppy, G.M., Powell, W., Pickett, J.A., Wadhams, L.J. and Woodcock, C.M. (1998) Identification of semiochemicals released during aphid feeding that attract parasitoid *Aphidius ervi*. *Journal of Chemical Ecology* 24, 1355–1368.

Duffey, S.S. and Bloem, K.A. (1987) Plant defence-herbivore-parasite interactions and biological control. In: Kogan, M. (ed.) *Ecological Theory and Integrated Pest Management Practice*. Wiley & Sons, New York, pp. 135–184.

Eigenbrode, S.D., Moodie, S. and Castagnola, T. (1995) Predators mediate host plant resistance to a phytophagous pest in cabbage with glossy leaf wax. *Entomologia Experimentalis et Applicata* 77, 335–342.

Farrar, R.R. and Ridgway, R.L. (2000) Host plant effects on the activity of selected nuclear polyhedrosis viruses against the corn earworm and beet armyworm (Lepidoptera : Noctuidae). *Environmental Entomology* 29, 108–119.

Farrar, R.R., Martin, P.A.W. and Ridgway, R.L. (1996) Host plant effects on activity of *Bacillus thuringiensis* against gypsy moth (Lepidoptera: Lymantriidae) larvae. *Environmental Entomology* 25, 1215–1223.

Feeny, P. (1976) Plant apparency and chemical defence. In: Wallace, J.W. and Mansell, R.L. (eds) *Biochemical Interaction between Plants and Insects*. Plenum Press, New York, pp. 1–40.

Fitt, G.P., Mares, C.L. and Llewellyn, D.J. (1994) Field-evaluation and potential ecological impact of transgenic cottons (*Gossypium hirsutum*) in Australia. *Biocontrol Science and Technology* 4, 535–548.

Fox, L.R., Letourneau, D.K., Eisenbach, J. and van Nouhuys, S. (1990) Parasitism rates and sex ratios of a parasitoid wasp: effects of herbivore and plant quality. *Oecologia* 83, 414–419.

Fritz, R.S. (1992) Community structure and species interactions of phytophagous insects on resistant and susceptible host plants. In: Fritz, R.S. and Simms, E.L. (eds) *Plant Resistance to Herbivores and Pathogens: Ecology, Evolution and Genetics*. University of Chicago Press, Chicago, Illinois, pp. 240–277.

Fritz, R.S., McDonough, S.E. and Rhoads, A.G. (1997) Effects of plant hybridization on herbivore-parasitoid interactions. *Oecologia* 110, 360–367.

Godfray, H.C.J. and Waage, J.K. (1991) Predictive modelling in biological control: the mango mealy bug (*Rastrococcus invadens*) and its parasitoids. *Journal of Applied Ecology* 28, 434–453.

Godfray, H.J.H. (1994) *Parasitoids – Behavioural and Evolutionary Ecology*. Princeton University Press, Princeton, New Jersey, 473 pp.

Gold, C.S., Altieri, M.A. and Bellotti, A.C. (1989) The effects of intercropping and mixed varieties of predators and parasitoids of cassava whiteflies (Hemiptera: Aleyrodidae). *Bulletin of Entomological Research* 79, 115–121.

Gowling, G.R. and van Emden, H.F. (1994) Falling aphids enhance impact of biological control by parasitoids on partially aphid-resistant plant varieties. *Annals of Applied Biology* 125, 233–242.
Groot, A.T. and Dicke, M. (2002) Insect-resistant transgenic plants in a multi-trophic context. *Plant Journal* 31, 387–406.
Gurr, G.M., Thwaite, W.G., Valentine, B.J. and Nicol, H.I. (1997) Factors affecting the presence of *Typhlodromus* spp. (Acarina: Phytoseiidae) in the calyx cavities of apple fruits and implications for integrated pest management. *Experimental and Applied Acarology* 21, 357–364.
Gutierrez, A.P. (1996) *Applied Population Ecology: a Supply–Demand Approach.* John Wiley & Sons, New York, 300 pp.
Gutierrez, A.P., Baumgaertner, J.V. and Summers, C.G. (1984) Multitrophic models of predator prey energetics. I. Age-specific energetics models – pea aphid *Acrythosiphon pisum* (Harris) (Homoptera: Aphidae) as an example. *Canadian Entomologist* 116, 924–932.
Gutierrez, A.P., Wermelinger, B., Schulthess, F., Baumaertner, J.V., Herren, H.R., Elliss, C.K. and Yaninek, J.S. (1988) Analysis of biological control of cassava pests in Africa. I. Simulation of carbon, nitrogen and water dynamics in cassava. *Journal of Applied Ecology* 25, 901–920.
Gutierrez, A.P., Neuenschwander, P. and van Alphen, J.J.M. (1993) Factors affecting biological control of cassava mealybug by exotic parasitoids: a ratio-dependent supply–demand driven model. *Journal of Applied Ecology* 30, 706–721.
Gutierrez, A.P., Mills, N.J., Schreiber, S.J. and Ellis, C.K. (1994) A physiologically based tritrophic perspective on bottom-up–top-down regulation of populations. *Ecology* 75, 2227–2242.
Gutierrez, A.P., Yaninek, J.S., Neuenschwander, P. and Ellis, C.K. (1999) A physiologically-based tritrophic metapopulation model of the African cassava food web. *Ecological Modelling* 123, 225–242.
Hardee, D.D. and Bryan, W.W. (1997) Influence of *Bacillus thuringiensis*-transgenic and nectarless cotton on insect populations with special emphasis on the tarnished plant bug (Heteroptera: Miridae). *Journal of Economic Entomology* 90, 663–668.
Hare, D.J. (1992) Effects of plant variation on herbivore–enemy interactions. In: Fritz, R.S. and Simms, E.L. (eds) *Plant Resistance to Herbivores and Pathogens.* University of Chicago Press, Chicago, Illinois, pp. 278–298.
Hassell, M.P. (1978) *The Dynamics of Arthropod Predator–Prey Systems.* Princeton University Press, Princeton, New Jersey, 237 pp.
Hufbauer, R.A. and Via, S. (1999) Evolution of an aphid–parasitoid interaction: variation in resistance to parasitism among aphid populations specialized on different plants. *Evolution* 53, 1435–1445.
Hunter, A.F. and Elkinton, J.S. (2000) Effects of synchrony with host plant on populations of a spring-feeding lepidopteran. *Ecology* 81, 1248–1261.
Idris, A.B. and Grafius, E. (1996) Effects of wild and cultivated host plants on oviposition, survival, and development of diamondback moth (Lepidoptera: Plutellidae) and its parasitoid *Diadegma insulare* (Hymenoptera: Ichneumonidae). *Environmental Entomology* 25, 825–833.
Janssen, A., Pallini, A., Venzon, M. and Sabelis, M.W. (1998) Behaviour and indirect interactions in food webs of plant-inhabiting arthropods. *Experimental and Applied Acarology* 22, 497–521.
Johansen, C., Lee, K.K., Sharma, K.K., Subbarao, G.V. and Kueneman, E.A. (1995) Genetic manipulation of crop plants to enhance integrated nutrient management in cropping systems. 1. Phosphorus. In: *Proceedings of an FAO–ICRISAT Expert Consultancy Workshop, Patancheru, India, 15–18 March 1994.* International Crops Research Institute for the Semi-Arid Tropics (ICRISAT), Patancheru, India, 177 pp.
Karban, R. and EnglishLoeb, G. (1997) Tachinid parasitoids affect host plant choice by caterpillars to increase caterpillar survival. *Ecology* 78, 603–611.
Kennedy, J.S., Booth, C.O. and Kershaw, W.J.S. (1961) Host finding by aphids in the field. III. Visual attraction. *Annals of Applied Biology* 49, 1.
Koul, O. (2003) Variable efficacy of neem-based formulations and azadirachtin to aphids and their natural enemies. In: Koul, O., Dhaliwal, G.S., Marwaha, S.S. and Arora, J.K. (eds) *Biopesticides and Pest Management,* Vol. 1. Campus Books International, New Delhi, pp. 64–74.
Leather, S.R. (1985) Oviposition preferences in relation to larval growth rates and survival in the pine beauty moth, *Panolis flammea. Ecological Entomology* 10, 213–217.
Leather, S.R. and Walsh, P.J. (1993) Sub-lethal plant defences: the paradox remains. *Oecologia* 93, 153–155.
Letourneau, D.K. (1987) The enemies hypothesis: tritrophic interactions and vegetational diversity in tropical agroecosystems. *Ecology* 68, 1616–1622.
Letourneau, D.K. and Andow, D.A. (1999) Natural-enemy food webs. *Ecological Applications* 9, 363–364.

Letourneau, D.K. and Goldstein, B. (2001) Pest damage and arthropod community structure in organic vs. conventional tomato production in California. *Journal of Applied Ecology* 38, 557–570.
Lewis, W.J. and Martin, W.R. (1990) Semiochemicals for use with parasitoids: status and future. *Journal of Chemical Ecology* 16, 3067–3089.
Lewis, W.J. and Nordlund, D.A. (1984) Semiochemicals influencing fall armyworm parasitoid behavior: implications for behavioral manipulation. *Florida Entomologist* 67, 343–349.
Lill, J.T. and Marquis, R.J. (2001) The effects of leaf quality on herbivore performance and attack from natural enemies. *Oecologia* 126, 418–428.
Loader, C. and Damman, H. (1991) Nitrogen content of food plants and vulnerability of *Pieris rapae* to natural enemies. *Ecology* 72, 1586–1590.
Lowery, D.T. and Isman, M.B. (1994) Insect growth-regulating effects of a neem extract and azadirachtin on aphids. *Entomologia Experimentalis et Applicata* 72, 77–84.
Mattiacci, L., Dicke, M. and Posthumus, M.A. (1994) Induction of parasitoid attracting synomone in Brussels sprouts plants by feeding of *Pieris brassicae* larvae – role of mechanical damage and herbivore elicitor. *Journal of Chemical Ecology* 20, 2229–2247.
Mattiacci, L., Dicke, M. and Posthumus, M.A. (1995) Beta-glucosidase – an elicitor of herbivore-induced plant odor that attracts host-searching parasitic wasps. *Proceedings of the National Academy of Sciences of the USA* 92, 2036–2040.
Meerman, F., van den Ven, G.W.J., van Keulen, H. and Breman, H. (1996) Integrated crop management: an approach to sustainable agricultural development. *International Journal of Pest Management* 42, 13–24.
Menken, S.B.J., Visser, J.H. and Harrewijn, P. (1992) *Proceedings of the 8th International Symposium on Insect–Plant Relationships*. Series Entomologica Vol. 49, Kluwer Academic Publishers, Dordrecht, The Netherlands, 424 pp.
Mensah, R.K. (1999) Habitat diversity: implications for the conservation and use of predatory insects of *Helicoverpa* spp. in cotton systems in Australia. *International Journal of Pest Management* 45, 91–100.
Metz, T.D., Roush, R.T., Tang, J.D., Shelton, A.M. and Earle, E.D. (1995) Transgenic broccoli expressing a *Bacillus thuringiensis* insecticidal crystal protein – implications for pest resistance management strategies. *Molecular Breeding* 1, 309–317.
Moran, N. and Hamilton, W.D. (1980) Low nutritive quality as defence against herbivores. *Journal of Theoretical Biology* 86, 247–254.
Mordue, A.J., Nisbet, A.J., Nasiruddin, M. and Walker, E. (1996) Differential thresholds of azadirachtin for feeding deterrence and toxicity in locusts and an aphid. *Entomologia Experimentalis et Applicata* 80, 69–72.
Murdoch, W.W. and Briggs, C.J. (1996) Theory for biological control: recent developments. *Ecology* 77, 2001–2013.
Murphy, B.C., Rosenheim, J.A. and Granett, J. (1996) Habitat diversification for improving biological control: abundance of *Anagrus epos* (Hymenoptera: Mymaridae) in grape vineyards. *Environmental Entomology* 25, 495–504.
Nisbet, R.M. and Gurney, W.S.C. (1983) The systematic formulation of population models for insects with dynamically varying instar duration. *Theoretical Population Biology* 23, 114–135.
Nordlund, D.A., Lewis, W.J. and Altieri, M.A. (1988) Influences of plant produced allelochemicals on the host–prey selection behaviour of entomophagous insects. In: Barbosa, P. and Letourneau, D.K. (eds) *Novel Aspects of Insect–Plant Interactions*. John Wiley & Sons, New York, pp. 65–95.
Nwanze, K.F., Reddy, Y.V.R., Nwilene, F.E., Kausalya, K.G. and Reddy, D.D.R. (1998) Tritrophic interactions in sorghum, midge (*Stenodiplosis sorghicola*) and its parasitoid (*Aprostocetus* spp.). *Crop Protection* 17, 165–169.
Orr, D.B. and Boethel, D.J. (1985) Comparative development of *Copidosoma truncatellum* (Hymenoptera: Encyrtidae) and its host, *Pseudoplusia includens* (Lepidoptera: Noctuidae), on resistant and susceptible soybean genotypes. *Environmental Entomology* 14, 612–616.
Poppy, G.M. (1997) Tritrophic interactions: improving ecological understanding and biological control? *Endeavour* 21, 61–65.
Poprawski, T.J., Greenberg, S.M. and Ciomperlik, M.A. (2000) Effect of host plant on *Beauveria bassiana* and *Paecilomyces fumosoroseus*-induced mortality of *Trialeurodes vaporariorum* (Homoptera : Aleyrodidae). *Environmental Entomology* 29, 1048–1053.
Price, P.W. (1986) Ecological aspects of host plant resistance and biological control: interactions among three trophic levels. In: Boethel, D.J. and Eikenbary, R.D. (eds) *Interactions of Plant Resistance and Parasitoids and Predators of Insects*. Ellis Horwood, Chichester, UK, pp. 11–30.

Price, P.W., Bouton, C.E., Gross, P., McPheron, B.A., Thompson, J.N. and Weis, A.E. (1980) Interactions among three trophic levels: influence of plants on interactions between insect herbivores and natural enemies. *Annual Review of Ecology and Systematics* 11, 41–65.

Price, P.W., Cobb, N., Craig, T.P, Fernandes, G.W., Tami, J.K., Mopper, S. and Preszler, R.W. (1990) Insect herbivore population dynamics on trees and shrubs: new approaches relevant to latent and eruptive species and life table development. In: Bernays, E.A. (ed.) *Insect–Plant Interactions*, Vol. 2. CRC Press, Boca Raton, Florida, pp. 2–38.

Read, D.P., Feeny, P.P. and Root, R.B. (1970) Habitat selection by the aphid parasite *Diaeretiella rapae* (Hymenoptera: Braconidae) and hyperparasite *Charips brassicae* (Hymenoptera: Cynipidae). *Canadian Entomologist* 102, 1567–1578.

Redman, A.M. and Scriber, J.M. (2000) Competition between the gypsy moth, *Lymantria dispar*, and the northern tiger swallowtail, *Papilio canadensis*: interactions mediated by host plant chemistry, pathogens, and parasitoids. *Oecologia* 125, 218–228.

Reitz, S.R. and Trumble, J.T. (1996) Tritrophic interactions among linear furanocoumarins, the herbivore *Trichoplusia ni* (Lepidoptera: Noctuidae), and the polyembryonic parasitoid *Copidosoma floridanum* (Hymenoptera: Enccyrtidae). *Environmental Entomology* 25, 1391–1397.

Root, R.B. (1973) Organization of a plant-arthropod association in simple and diverse habitats: the fauna of collards (*Brassica oleracea*). *Ecological Monographs* 43, 95–124.

Roush, R.T. (1994) Managing pests and their resistance to *Bacillus thuringiensis* – can transgenic crops be better than sprays? *Biocontrol Science and Technology* 4, 501–516.

Ryan, J.D., Morgham, P.E., Richardson, R.C., Johnson, R.C., Mort, A.J. and Eikenbary, R.D. (1990) Greenbug and wheat: a model system for the study of phytotoxic Homoptera. In: Campbell, R.K. and Eikenbary, R.D. (eds) *Aphid–Plant Genotype Interactions*. Elsevier, Amsterdam, pp. 171–186

Schultz, J.C. (1983) Impact of variable plant defensive chemistry on susceptibility of insects to natural enemies. *American Chemical Society Symposium Series* 208, 37–54.

Sheehan, W. (1986) Response by specialist and generalist natural enemies to agroecosystem diversification: a selective review. *Environmental Entomology* 15, 456–461.

Shimoda, T., Takabayashi, J., Ashihara, W. and Takafuji, A. (1997) Response of predatory insect *Scolothrips takahashii* toward herbivore-induced plant volatiles under laboratory and field conditions. *Journal of Chemical Ecology* 23, 2033–2048.

Smith, C.M. (1989) *Plant Resistance to Insects: a Fundamental Approach*. John Wiley & Sons, New York, 286 pp.

Smith, J.G. (1969) Some effects of crop background on populations of aphids and their natural enemies on brussels sprouts. *Annals of Applied Biology* 63, 326–330.

Steinberg, S., Dicke, M. and Vet, L.E.M. (1993) Relative importance of infochemicals from 1st and 2nd trophic level in long-range host location by the larval parasitoid *Cotesia glomerata*. *Journal of Chemical Ecology* 19, 47–59.

Strong, D.R., Lawton, J.H. and Southwood, T.R.E. (1984) *Insects on Plants*. Blackwell Scientific Publications, Oxford, 313 pp.

Strong, D.R., Baker, R.R. and Dunn, P.E. (1990) Interface of natural enemy and environment. *UCLA Symposia on Molecular and Cellular Biology* 112, 57–64.

Sugden, M.R. (1994) The effects of azadirachtin on *Myzus persicae* and its hymenopterous parasitoid *Aphidius matricariae*. MSc thesis, University of Aberdeen, UK.

Tabashnik, B.E. (1994) Evolution of resistance to *Bacillus thuringiensis*. *Annual Review of Entomology* 39, 47–79.

Takabayashi, J., Dicke, M. and Posthumus, M.A. (1994) Volatile herbivore-induced terpenoids in plant mite interactions – variation caused by biotic and abiotic factors. *Journal of Chemical Ecology* 20, 1329–1354.

Thaler, J.S. (2002) Effect of jasmonate-induced plant responses on the natural enemies of herbivores. *Journal of Animal Ecology* 71, 141–150.

Theodoratus, D.H. and Bowers, M.D. (1999) Effects of sequestered iridoid glycosides on prey choice of the prairie wolf spider, *Lycosa carolinensis*. *Journal of Chemical Ecology*, 25, 283–295.

Thomas, M.B. (1999) Ecological approaches and the development of 'truly integrated' pest management. *Proceedings of the National Academy of Sciences of the USA* 96, 5944–5951.

Thomas, M.B. and Waage, J.K. (1994) *Integration of Biological Control and Host Plant Resistance Breeding*. CAB International, Wallingford, UK, 139 pp.

Tumlinson, J.H., Turlings, T.C.J. and Lewis, W.J. (1993) Semiochemically mediated foraging behavior in beneficial parasitic insects. *Archives of Insect Biochemistry and Physiology* 22, 385–391.

Turlings, T.C.J. and Tumlinson, J.H. (1991) Do parasitoids use herbivore-induced plant chemical defenses to locate hosts? *Florida Entomologist* 74, 42–50.
van Emden, H.F. (1965) The effect of uncultivated land on the distribution of cabbage aphid (*Brevicoryne brassicae*) on an adjacent crop. *Journal of Applied Ecology* 2, 171.
van Emden, H.F. (1987) Cultural methods: the plant. In: Burn, A.J., Coaker, T.H. and Jepson, P.C. (eds) *Integrated Pest Management*. Academic Press, London, pp. 27–68.
van Emden, H.F. (1989) *Pest Control*, 2nd edn. Edward Arnold, London, 117 pp.
van Emden, H.F. (1990) Plant diversity and natural enemy efficiency in agroecosystems. In: Mackauer, M., Ehler, L.E. and Roland J. (eds) *Critical Issues in Biological Control*. Intercept, Andover, UK, pp. 63–80.
van Emden, H.F. (1991) The role of host plant resistance in insect pest mis-management. *Bulletin of Entomological Research* 81, 123–126.
van Emden, H.F. (1995) Host plant–Aphidophaga interactions. *Agriculture, Ecosystems and Environment* 52, 3–11.
van Emden, H.F. and Wearing, C.H. (1965) The role of the host plant in delaying economic damage levels in crops. *Annals of Applied Biology* 56, 323–324.
van Emden, H.F. and Wratten, S.D. (1991) Tri-trophic interactions involving plants in the biological control of aphids. In: Peters, D.C., Webster, J.A. and Chouber, C.S. (eds) *Aphid–Plant Interactions: Populations to Molecules*. Agricultural Research Service, USDA Oklahoma State University, Stillwater, Oklahoma, pp. 29–43.
van Poecke, R.M.P., Posthumus, M.A. and Dicke, M. (2001) Herbivore-induced volatile production by *Arabidopsis thaliana* leads to attraction of the parasitoid *Cotesia rubecula*: chemical, behavioral, and gene-expression analysis. *Journal of Chemical Ecology* 27, 1911–1928.
Verkerk, R.H.J. (1995) Studies on interactions between diamondback moth, host plants, endolarval parasitoids and selective toxicants. PhD thesis, University of London, UK, 244 pp.
Verkerk, R.H.J. and Wright, D.J. (1994) The potential for induced extrinsic host plant resistance in IRM strategies targeting the diamondback moth. In: *Proceedings – Brighton Crop Protection Conference: Pests and Diseases – 1994*. British Crop Protection Council, Farnham, UK, pp. 457–462.
Verkerk, R.H.J. and Wright, D.J. (1996a) The effects of host plant-selective insecticide interactions on larvae of *Plutella xylostella* (Lepidoptera: Yponomeutidae) in the laboratory. *Pesticide Science* 44, 171–181.
Verkerk, R.H.J. and Wright, D.J. (1996b) Multitrophic interactions and management of the diamondback moth. *Bulletin of Entomological Research* 86, 205–216.
Verkerk, R.H.J., Leather, S.R. and Wright, D.J. (1998) The potential for manipulating crop pest-natural enemy interactions for improved insect pest management. *Bulletin of Entomological Research* 88, 493–501.
Vet, L.E.M. and Dicke, M. (1992) Ecology of infochemical use by natural enemies in a tritrophic context. *Annual Review of Entomology* 37, 141–172.
Vinson, S.B. (1984) How parasitoid locate their hosts: a case of insect espionage. In: Lewis, T. (ed.) *Insect Communications*. Academic Press, London, pp. 325–348.
Waage, J. (1997) What does biotechnology bring to integrated pest management? *Biotechnology and Development Monitor* 32, 19–21.
Walde, S.J. (1995) How quality of host-plant affects a predator-prey interaction in biological control. *Ecology* 76, 1206–1219.
Walker, M. and Jones, T.H. (2001) Relative roles of top-down and bottom-up forces in terrestrial tritrophic plant-insect herbivore-natural enemy systems. *Oikos* 93, 177–187.
Warrior, P. (2000) Living systems as natural crop-protection agents. *Journal of Pest Management Science* 56, 681–687.
White, A.J., Wratten, S.D., Berry, N.A. and Weigmann, U. (1995) Habitat manipulation to enhance biological control of *Brassica* pests by hover flies (Diptera: Syrphidae). *Journal of Economic Entomology* 88, 1171–1176.
Whitman, D.W. (1988) Allelochemical interactions among plants, herbivores and predators. In: Barbosa, P. and Letourneau, D.K. (eds) *Novel Aspects of Insect-Plant Interactions*. John Wiley & Sons, New York, pp. 11–64.
Whitman, D.W. and Eller, F.J. (1992) Orientation of *Microplitis croceipes* (Hymenoptera: Braconidae) to green leaf volatiles: dose–response curves. *Journal of Chemical Ecology* 18, 1743–1753.
Wright, D.J. and Verkerk, R.H.J. (1995) Integration of chemical and biological control systems for arthropods: evaluation in a multitrophic context. *Pesticide Science* 44, 207–218.

5 Behaviour-modifying Chemicals: Prospects and Constraints in IPM

Larry J. Gut,[1] Lukasz L. Stelinski,[1] Donald R. Thomson[2] and James R. Miller[1]
[1]*Department of Entomology, Michigan State University, East Lansing, Michigan, USA;*
[2]*DJS Consulting, Seattle, Washington, USA*

Introduction

Many organisms rely on chemical messages to communicate with each other or to find suitable hosts. Insects and other arthropods appear to be especially dependent on chemical stimuli for survival and reproduction. Essential insect behaviours that may be stimulated or inhibited by olfactory information include mating, feeding and oviposition. Chemical messages that trigger various behavioural responses are collectively referred to as semiochemicals. The term pheromone is used to describe compounds that operate intraspecifically (Karlson and Lüscher, 1959), while allelochemical is the general term for an interspecific effector (Whittaker and Feeney, 1971). Pheromones are categorized according to function, including sex pheromones, aggregation pheromones, alarm pheromones, trail pheromones and host-marking pheromones. Allelochemicals are classified based on the advantage of the message to the receiver or sender. Alomones benefit the sender, kairomones give advantage to the receiver and synomones benefit both parties. Examples of these kinds of semiochemicals include floral compounds that are attractive to pollinators and plant-produced feeding or oviposition deterrents or stimulants.

Since the early 1970s, considerable progress in understanding insect behaviour and complementary advances in analytical chemistry have led to the identification of thousands of pheromones and other semiochemicals (Mayer and McLaughlin, 1991; Hardie and Minks, 1999). Not surprisingly, a wide array of uses for semiochemicals has been tested over the past few decades and many practical applications have become integral components in pest-management programmes (Metcalf and Metcalf, 1992; Jones, 1998; Knight and Weissling, 1999; Suckling and Karg, 2000).

Applied scientists have long recognized the potential of using behaviourally active compounds to trap insects (Lanier, 1990). Monitoring the activity of insects with pheromone- or kairomone-baited traps is now a standard practice in the majority of pest-management systems. The information gathered can be used to time insecticide applications or for determining the need to treat. Trapping is also widely used to detect the presence of pests in quarantine or survey programmes.

The realization that behaviours critical to insect survival were strongly influenced by semiochemicals rapidly led to proposals for using these agents as practical tools for pest suppression (Wood *et al.*, 1970; Shorey and McKelvey, 1977). For example, some early pioneers proposed the tactic of broadcasting insect sex-attractant pheromones for direct control through communication disruption (Gaston *et al.*, 1967; Mitchell *et al.*, 1974; Shorey *et al.*, 1974). In addition to recognizing the strengths offered by semiochemical-based controls, the early innovators also appreciated the weaknesses of behavioural controls (Knipling, 1979; Ritter, 1979; Mitchell, 1981). These inherent strengths and weaknesses are well worth reviewing.

As rendered by Miller and Cowles (1990), damage to a crop can be approximated by the expression: damage $\propto$ D $\times$ A $\times$ S $\times$ T, where D is the density of pest in a crop, A is the acceptability of the crop to the pest, S is the suitability of the crop to the pest and T is the time of the interaction. Generally, as more of these parameters are reduced by a control tactic, the outcome for crop damage reduction improves. For example, a great strength of insecticides lies in the dramatic reduction in pest density, crop suitability and time of the interaction. The lack of a negative effect on acceptability is actually advantageous in this case because ingestion of the toxic substance is desired. Behavioural controls often do not kill the target pest. In this case, they do not directly reduce pest density or the time of pest–crop interactions, unless the mode of action is repellence. Thus, while insecticides can prove effective during their short lifespan on the crop, behavioural controls must remain effective for longer durations, usually throughout the lifetime of the pest. Additionally, behavioural controls can sometimes lose their effect over time due to changes in physiological and behavioural thresholds resulting from, for example, increased hunger or build-up of eggs (Miller and Cowles, 1990). The effectiveness of controls involving antibiosis is generally independent of pest density. In contrast, the efficacy of behavioural controls can be density-dependent. Under such conditions, behavioural-control measures become more suited for management of pest populations that are below the outbreak threshold rather than for suppression (*sensu* Knipling, 1979) after an outbreak has occurred.

Despite some fundamental limitations, the development and implementation of pest-management strategies and control tactics based on semiochemicals have progressed considerably since the earliest attempts at direct control. New regulations governing the use of pesticides, increasing environmental concerns and the regular occurrence of resistance to chemical controls have provided a strong impetus for the adoption of novel technologies, including behavioural controls. In addition, growing numbers of producers have recognized that the weaknesses of behavioural controls are to some extent counterbalanced by the advantage of negligible environmental impact and increased compatibility with biological and cultural methods of control.

Although the majority of successful uses of semiochemicals are for monitoring pest activity, there is an increasing number of examples of direct control with pheromones and other behaviour-modifying compounds. The various approaches in which semiochemicals are used in pest management are listed in Table 5.1. Monitoring the activity of insects using pheromone- or kairomone-baited traps is an integral part of many pest-management programs. Trapping is often the most efficient method of detecting the presence of a species or measuring its sea-

Table 5.1. Practical uses of semiochemicals in pest management (adapted from Knight and Weissling, 1999).

Monitoring
1. Detect the presence of a species
2. Measure seasonal activity and provide decision support
3. Evaluate the effectiveness of mating disruption
4. Assess levels of insecticide resistance

Direct control
1. Mass deployment of attractant-baited traps
2. Application of attract-and-kill formulations or devices
3. Pheromone-mediated mating disruption
4. Manipulation of natural enemies using allelochemicals
5. Pheromone-based interference with host location or acceptance
6. Plant allomone-based deterrence of feeding or oviposition
7. Application of pheromones to enhance pollination

sonal activity. This kind of information is helpful in making economically sound management decisions. Attractant-baited lures also form the basis for two direct control measures: mass trapping and attract-and-kill. Both strategies require a high level of attraction to the lure. Another set of direct controls using pheromones or kairomones is based on interfering with an insect's natural behavioural response to these compounds. The most developed tactic is termed mating disruption. This approach entails releasing large amounts of synthetic sex pheromone into the atmosphere of a crop in an effort to interfere with mate-finding, thereby controlling the pest by curtailing the reproductive phase of its life cycle.

Our chapter is not intended to be a comprehensive review of the possible uses of behaviour-modifying chemicals in pest management, nor will it attempt to identify all of the constraints to successful deployment of the various approaches. Several treatises covering these subjects have been published over the past 25 years (Shorey and McKelvey, 1977; Mitchell, 1981; Nordlund *et al.*, 1981; Kydonieus and Beroza, 1982; Jutsum and Gordon, 1989; Ridgway *et al.*, 1990; Cardé and Minks, 1995, 1997; Jones, 1998). In keeping with the intent of this book, we present an overview of the potential role of semiochemicals in pest management and provide some examples of successes. The core of the chapter, however, is a discussion of the constraints and future prospects for mating disruption. We have emphasized this approach over others largely for practical reasons. A review of the literature quickly reveals that mating disruption is the most highly developed and widely adopted semiochemical-based control tactic. In addition, over the past 5–10 years our personal efforts have focused on the practical and commercial use of various mating disruption technologies in plant protection.

Monitoring

Applications of trapping systems

The production of synthetic copies of numerous kinds of semiochemicals has led to the development and widespread commercial use of attractant-baited traps for monitoring and trapping insect pests in many agricultural production systems, government detection and quarantine programmes, and consumer-protection efforts. Unlike other sampling methods that may be very time consuming and require technical expertise, semiochemical-based monitoring is efficient and easy to use. In addition, the approach is effective over a range of pest densities and often provides the most practical means of monitoring adult activity.

Detect the presence of a species

Semiochemical-baited traps provide a relatively simple and reliable means of detecting pest infestations. They are one of the primary tools employed in quarantine surveys to determine the presence of a species and prevent its establishment and spread (Johnson and Schall, 1989; Schwalbe and Mastro, 1990). Traps are routinely deployed around airports and harbours to detect potential introductions of exotic pests at these high-risk sites. In a similar manner, large numbers of traps are used in regional survey programmes to determine the distribution of specific pests and to provide the information needed for preventing spread to new areas (Schwalbe and Mastro, 1990). The use of attractant-baited traps to demonstrate pest-free production zones is of increasing importance. The process often requires implementing a defined monitoring programme to allow for the export of an agricultural commodity to a specific country (Riherd *et al.*, 1994). The monitoring protocols can be quite demanding, often specifying the kind of trapping system, the density of traps and a specific treatment regimen if a single individual is detected.

Measure seasonal activity and provide decision support

Monitoring with attractant-baited traps is an important component of pest-management programmes. The combination of pheromone traps and degree-day models can provide a reliable method for monitoring adult activity, predicting egg hatch and timing insecticide sprays (Welch *et al.*, 1981; Riedl *et al.*, 1986; Knight and Croft, 1991). The precision of this method of timing sprays has been well documented for codling moth, a key pest of pome fruits throughout the world (Beers *et al.*, 1993). The phenology model for this pest is based on accumulating degree-days (base 10°C) beginning on the day the first moth is captured in a pheromone trap, provided moths are captured on two successive trapping dates. This start of sustained moth capture is referred to as biofix. The first spray is applied at 121 degree-days (°C) post-biofix, which coincides with the start of egg hatch. In 6 of 10 years in the state of Washington, USA, this model predicted the start of egg hatch on the same day that it was first observed in the field (Table 5.2) and there was never a discrepancy of more than 2 days between the predicted and observed event. Prior to the development of the degree-day approach, the first spray for codling moth in Washington State was timed on a calendar basis. The emergence of moths from overwintering sites was anticipated at full bloom on 'Delicious' apples and the first treatment was applied 21 days after this event. Based on this approach, treatments were often applied before egg hatch. In 7 of 10 years the predicted start of egg hatch occurred at least 3 days prior to observed hatch in the field and, moreover, predictions were over 13 days early on three occasions (Table 5.2). The use of a degree-day model rather than a calendar approach to time insecticide applications will become increasingly crucial in the coming years because many new insecticide chemistries, such as insect growth regulators, require precise timing as they are primarily active against specific instars or life stages.

Table 5.2. Comparison of a degree-day model with a calendar approach for timing the first insecticide spray for control of hatching codling-moth larvae (based on Beers *et al.*, 1993).

	Accuracy in timing sprays (days)[a]	
Year	Model	Calendar
1979	0	3
1980	0	4
1981	−2	13
1982	0	2
1983	−1	8
1984	0	18
1985	−1	1
1986	0	13
1987	2	−2
1993	0	6

[a] Positive numbers indicate predicted timing was too early, negative numbers indicate predicted timing was too late.

Semiochemical-based monitoring systems can also be used to assess population trends and determine the need to treat. Quantitative relationships between adult captures and counts of larval stages or signs of larval feeding, such as faeces or damage, have been found for pests of tree fruits (McBrien *et al.*, 1994; Bradley *et al.*, 1998), annual crops (Van Steenwyk *et al.*, 1983) and forests (Sanders, 1988; Evenden *et al.*, 1995; Morewood *et al.*, 2000). In some cases, captures in one year can be used to predict events in a subsequent year. McBrien *et al.* (1994) demonstrated a correlation between catches of male mullein bug, *Campylomma verbasci* (Herrich-Schaffer), in the autumn and the density of nymphs the following spring in apple orchards. There has been developed for the eastern spruce budworm, *Choristoneura fumiferana* (Clemens), an early-warning system that uses male captures in pheromone traps to predict a severe outbreak of this pest several years in advance (Sanders, 1988). Basing management decisions on adult catches rather than taking a preventive or calendar-based approach is a key step in many efforts to reduce insecticide inputs. The approach typically entails intervening with a spray only if catches exceed a predetermined level. The decision can be based on a single weekly catch, consecutive catches or cumulative catches over an extended period time, such as a generation. Sticky-coated red spheres baited with synthetic apple volatiles can be used to monitor apple maggot activity and to alert growers as to the need for a spray (Stanley *et al.*, 1987; Agnello *et al.*, 1990). An action threshold of eight flies per trap allowed for a 70% reduction in sprays and acceptable levels of control. Reducing the threshold to five flies per trap resulted in 0.6 fewer sprays, but the conservative threshold was more likely to be adopted by apple growers. Treatment thresholds for codling moth based on moth captures in pheromone traps have been developed for most pome fruit producing regions of the world (Riedl *et al.*, 1986; Wall, 1989).

Evaluate the effectiveness of mating disruption

An increasingly important use of attractant-baited traps is to measure the efficacy of mating-disruption formulations. Capture of zero (complete shutdown) or very few moths in a pheromone-baited trap has been used to indicate successful disruption of the target pest. The rationale behind this measure of mating-disruption effectiveness is that, if male moths were incapable of finding a lure releasing synthetic pheromone, they were also unable to find a female moth releasing natural pheromone. Unfortunately, it is not uncommon to record low moth catches in traps and still have less than adequate pest control in pheromone-treated plots (refer to several chapters in Ridgway *et al.*, 1990). In some cases, it is possible to greatly inhibit catches in pheromone traps but still detect substantial numbers of mated females (Rice and Kirsch, 1990; Atanassov *et al.*, 2002) or actual mating (Suckling and Shaw, 1992). In the former situation where mated females are found in treated plots, it is possible that a portion of these females immigrated from adjacent plots not treated with pheromone.

There have been two approaches to improving the usefulness of pheromone-baited traps as monitoring tools in pheromone-treated plots. Cardé and Elkinton (1984) suggested that a lure with an emission rate closer to the natural rate would seem to be the most suitable for measuring the efficacy of a disruption treatment. Charmillot (1990) took a different approach, opting to use lures with very high release rates as a means of following changes in adult population densities in spite of air permeation with pheromone. In a series of experiments over the course of 2 years, he showed that, for codling moth, the sensitivity of pheromone traps in pheromone-treated orchards could be improved by increasing the amount of codlemone in the lure. Moth catch in disrupted orchards increased considerably when traps were baited with 10 or 20 mg of pheromone. Others have subsequently confirmed a significantly greater sensitivity of 10 mg- compared with 1 mg-baited traps in disrupted orchards (Barrett, 1995; Judd *et al.*, 1996). A further increase in moth captures in orchards treated with mating-disruption dispensers can be achieved by placing traps in the upper canopy (Barrett, 1995). Although the high-load pheromone trap is a useful tool for monitoring codling-moth activity in disrupted orchards, when used alone it is not a reliable method for assessing the effectiveness of the pheromone treatment. Trapping should be used in conjunction with visual inspection of fruit for codling-moth damage. The biggest concern with using pheromone traps to measure the effectiveness of mating disruption, regardless of the lure load, is the regular occurrence of 'false negatives' (Knight, 1995). This refers to a situation where low or no catches are recorded and yet fruit injury occurs in the block.

Assess levels of insecticide resistance

A dramatic rise in resistance to insecticides over the past decade (Norris *et al.*, 2003) has brought about an acute need to have simple and reliable methods for monitoring its severity and distribution. Pheromone-trap bioassays are one of the more widely adopted methods for assessing the susceptibility of lepidopteran pests to insecticides. The basic approach was developed by Riedl *et al.* (1985) to test codling-moth susceptibility to azinphosmethyl. It has since been modified and used to evaluate the tolerance of other pests to various insecticides (Suckling *et al.*, 1985; Haynes *et al.*, 1987; Knight and Hull, 1989; Varela *et al.*, 1997; Shearer and Usmani, 2001). The bioassay entails collecting large numbers of males in traps and testing for expression of resistance by topical application of insecticides or through incorporation of insecticide into the glue. The major advantage of pheromone-trap bioassays over most other methods of assessing resistance is that many individuals can be tested without incurring the costs in time and money associated with rearing large numbers of larvae. The simplicity of the trap bioassay does not come without trade-offs. Dose–mortality responses may be strongly affected by a number of factors, including whether the toxicant was delivered topically or through the adhesive (Knight and Hull, 1989; Shearer and Riedl, 1994). The technique may overestimate the impact of resistance in the field because only males are captured and assayed. Sex-related differences in the tolerance to organophosphorous and carbamate insecticides have been discovered for populations of Oriental fruit moth in the USA, with females more susceptible than males (deLame *et al.*, 2001; Shearer and Usmani, 2001). A major limitation of the approach is that it cannot be used to monitor for resistance to materials whose primary mode of action requires ingestion. Unfortunately, this includes most of the newer insecticide chemistries, such as insect growth regulators and neonicotinoids.

Performance of trapping systems

The development and commercialization of pheromone- or kairomone-based monitoring systems has greatly enhanced society's ability to monitor and manage insect pests. As a means of estimating population density or predicting crop damage, however, the strategy has proved to be of more limited utility. Morse and Kulman (1985) found no significant relationship between catches of the yellow-headed spruce sawfly, *Pikonema alaskensis*, and the degree of defoliation inflicted by this pest. A lack of correlation between moth catches and egg numbers has been reported for several vegetable pests; examples are *Helicoverpa zea* (Boddie) on tomato (Campbell *et al.*, 1992) and field maize (Latheef *et al.*, 1991).

A trap will sample only a portion of the pest population in an area. For a trap to accurately reflect the true population density and thus the potential for crop loss, the trap's attractiveness should remain constant, as should the proportion of the individuals from the population that are captured. The inadequacy of attractant-baited monitoring systems as quantitative decision tools is due in part to the substantial effect that trap design, attractiveness of delivery device, trap maintenance programme, and placement in the crop have on the performance of the system. Various trap designs and means of dispensing attractants have been developed for monitoring insects (Cardé and Elkinton, 1984; Wall, 1989; Jones, 1998). The majority of trap types employ a sticky surface to capture attracted insects. Other trap designs rely on a flight barrier, often combined with a knockdown insecticide or a liquid trapping medium to retain attracted insects. Attractants are commonly formulated in reservoirs made of various materials, including rubber, polyethylene, polyvinyl chloride and hollow fibres.

The probability of insects finding a trap is highly dependent on the attractiveness of the lure and the placement of the trap. Traps may be placed within or outside the crop, on the edge or interior of the plot, high or low in the canopy or on different sides of a tree. The location of the trap with respect to these parameters can have a substantial impact on moth catch. For example, differences in captures due to vertical positioning of traps within the crop canopy have been documented in both annual and perennial crops. Riedl *et al.* (1979) demonstrated that catches of male codling moth varied substantially depending on trap placement in the tree canopy (Fig. 5.1). Higher catches were recorded in the upper compared with lower canopy positions. Very few moths were captured above or below the canopy. Simandl and Anderbrant (1995) documented a similar outcome for the sawfly, *Neodiprion sertifer* (Geoffroy). Pheromone traps installed in the crowns of conifers (> 9 m) caught up to 15 times as many males as did traps suspended at 2.5 m. Derrick *et al.* (1992) found that traps placed within maize fields at ear level (1.5 m) captured greater numbers of European corn borer, *Ostrinia nubilalis* (Hübner), than traps hung in the field at canopy height (3.0 m).

Trap design and maintenance will influence the likelihood of capturing an insect after it finds the trap. The effectiveness of many trap designs depends on maintaining the integrity of trap shape and the quality of the adhesive surface throughout the season.

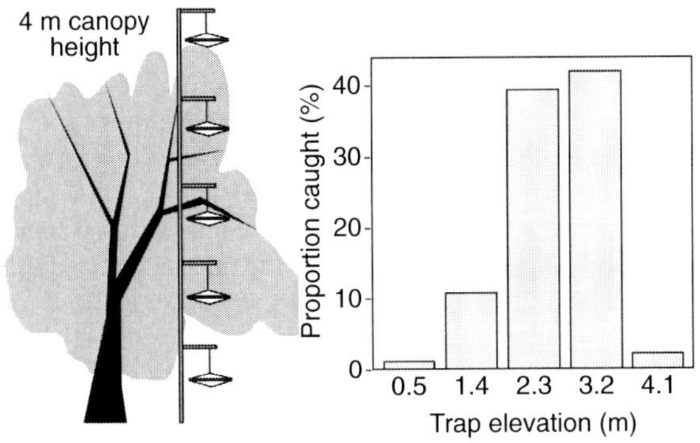

Fig. 5.1. The influence of canopy position on codling-moth captures in pheromone traps (from Riedl *et al.*, 1979).

Traps that depend on a sticky surface to capture insects can lose their ability to retain new arrivals if the surface becomes covered with debris or target and non-target insects. The efficiency of trapping for codling moth declined as catches exceeded 20–30 moths in traps with a 185 cm^2 sticky surface, while catches of 50–60 moths unacceptably reduced the performance of traps with a 360 cm^2 sticky surface (Riedl *et al.*, 1986). Under conditions of moderate or high pest pressure, traps can quickly become saturated with moths and diminish the likelihood of accurately measuring pest density by moth captures.

Future prospects

Drawing range of lures

A crucial problem with using pheromone-baited traps to measure pest density and predict damage is that insects may be attracted from distances well beyond the boundaries of the crop. Using the mark-and-recapture technique, Thayer (2002) demonstrated that male oblique-banded leaf-rollers could be captured in pheromone traps at distances of up to 440 m from the release point. The active space of a trap baited with 1 mg of pheromone was calculated to be 152,000 m^2 or nearly 15 ha. Baker and Roelofs (1981) recognized the problems associated with a large active space and proposed modifying the dose of the attractant to optimize the drawing range. Using this approach, they were able to restrict the sampling of *Cydia molesta* (Busck) to the immediate vicinity of the trap by lowering the dose. Faccioli *et al.* (1993) found that reducing the pheromone dosage in traps for the polyphagous pest *Argyrotaenia ljungiana* (Thunberg) enhanced the correlation of moth catch to larval infestation. The major reason for the improved precision was a decrease in the number of 'false positives' or cases of high catch with little or no concurring infestation. There are other possibilities for mitigating the effects of sampling large numbers of individuals from outside the crop on the usefulness of trapping information. The simplest approach for highly polyphagous pests would be to avoid placing traps close to the perimeter of a plot as a large portion of the area monitored by the trap would be outside its boundaries. However, placing traps in the interior still may not sufficiently limit the active space. A significant number of insects may move considerable distances prior to attraction, effectively increasing the area of influence of a trap (Wall and Perry, 1987). Knight and Hull (1988) found that a better correlation between pheromone-induced catches of tufted apple-bud moth, *Platynota idaeusalis* (Walker), and damage could be obtained if predictions were based on captures early in the flight, presumably before substantial movement by the moths, rather than over the entire flight period. Another promising approach is to bait traps with a lure that attracts both males and females.

Monitoring males and females

The use of a food bait or specific kairomone that attracts both sexes offers some distinct advantages over a sex-pheromone lure that attracts males only. Capture of females is more directly linked to oviposition and the potential for larval damage. Because females in many species do not disperse great distances, catch of females may more accurately reflect pest densities in the block that is being monitored.

The attraction to host volatiles should not be greatly impeded or suppressed by pheromone-based mating disruption, offering the possibility of a means of monitoring pest density in disrupted plots. Traps baited with a brown sugar and terpinyl acetate solution have been used to monitor Oriental fruit moth activity in pheromone-treated pome and stone-fruit orchards (Sexton and Il'ichev, 2000; Il'ichev and Sexton, 2002; Il'ichev *et al.*, 2002). These food traps are not specific to Oriental fruit moth, but they do capture females and males of this species and provide a more reliable measure of pest pressure than pheromone-baited traps. More recently, ethyl-(2*E*,4*Z*)-2,4-decadienoate (DA), a volatile present in the odour of ripe Bartlett pears, has been identified as a compound that is attractive to both sexes of codling moth (Light *et al.*, 2001). This pear ester is a stable compound that can be readily synthesized and loaded into a rubber septum to make a lure for monitoring both sexes of codling moth. Comparisons of DA and pheromone lures to monitor the activity of codling moth have been carried out for the past few years, most extensively in the western USA. Trials by a team of scientists (Light *et al.*, 2001) indicated that DA lure-baited traps provided good resolution of moth flight patterns for both sexes in conventional and especially mating-disrupted orchards.

A number of factors may influence the performance of traps baited with the DA lure and should be considered when using this trapping system. The crop or cultivar in which traps are placed influences moth catch in traps baited with the pear ester. Higher catches with this kairomone lure were recorded in walnut and in some late-season apples than in pear or in early-season apples (Light *et al.*, 2001; Alan Knight, Washington, USA, 2002, personal communication). Additional research on cultivar effects and the range of attraction is crucial to the development of a reliable pear ester-based monitoring system for codling moth.

Commercial implementation of monitoring programmes

Perhaps the greatest factor limiting the usefulness of attractant-baited traps is the incongruence between the monitoring programme that is developed and tested through experimentation and the system that is actually implemented in the field. Researchers focus on developing a tool and set of monitoring guidelines that are scientifically sound but often impractical. For example, treatment thresholds for codling moth in apple or pear were largely developed based on a trapping density of one trap per hectare because this was determined to be the active space of a trap. For an integrated pest management (IPM) practitioner, the economically 'practical space' is closer to one trap per 4 ha. A consultant is typically paid a set fee based on the size of the block monitored and must carefully consider the time involved in inspecting and maintaining traps in deciding how many to deploy. From the researcher's perspective, a consultant or other IPM practitioner often ignores critical aspects of trap placement or maintenance. In fruit orchards in the USA, codling-moth traps are routinely placed at 'truck-window' height, as well as on the very edge of a block. This certainly facilitates inspecting the trap but is counterproductive because very few moths will be caught. The typical justification for placing traps in less than optimum positions is that the best monitoring system is the one that provides the most reliable estimate of the population, not necessarily the catch of the greatest number of insects. However, it is essential to have at least adequate numbers upon which to base a decision. Moreover, there is likely to be no relationship between information garnered from a poorly run trapping programme and thresholds or other information that has been generated through more carefully conducted experiments.

The bottom line is that the best monitoring programme will develop through an awareness and consideration of the concerns and limitations expressed by the researcher and the practitioner, especially the cost of the programme. For example, effective monitoring of codling moth can be achieved by placing traps at mid-canopy height, and at a density of one trap every 4–6 ha on farms with large uniform plantings without a history of localized infestations. A tighter spacing of traps is required on farms with small plantings or a suspected uneven distribution of codling-moth pressure or in mating-disrupted blocks. Finally, in pheromone-treated orchards, consultants are urged to place traps high in the canopy.

Attraction–annihilation

Attraction–annihilation is probably the earliest use of attractants for pest control. Many early efforts were discouraging or, when effective control was demonstrated, practical deficiencies often inhibited the commercial adoption of the control system. The principles of attraction–annihilation and the potential of mass trapping, attract-and-kill and other practical applications of this pest-management approach have been reviewed comprehensively (Bakke and Lie, 1989; Lanier, 1990; Jones, 1998). Here we summarize the constraints, provide examples of some successes and offer some challenges for the future.

Mass trapping

The aim of mass trapping is to prevent crop damage by capturing a substantial proportion of a pest population prior to mating, oviposition or feeding. Success with this method requires the combination of a very attractive lure and a highly efficient trap. Although examples of mass trapping for pest control are fairly plentiful (see Table 5.3 for examples), most efforts have not been successful from the standpoint of commercial adoption.

Constraints

The commercial viability of mass trapping is limited by a number of practical and biological requirements. Of foremost importance

Table 5.3. Some promising applications of attraction–annihilation for insect pest management.

Approach	Pest	References
Mass trapping		
Attractants and water-based funnel traps	*Carpophilus* beetles	James *et al.*, 1996
Attractant-baited traps	Japanese beetle	Wawrzynski and Ascerno, 1998
Sex pheromone-baited traps	Chinese tortrix	Zhang *et al.*, 2002
Attractant-baited multisurface traps	Cigarette beetle	Buchelos and Levinson, 1993
Pheromone-based mass trapping	Ambrosia beetles	Borden, 1990
Inhibitor combined with mass trapping	Mountain pine beetle	Lindgren and Borden, 1993
Sex pheromone-based mass trapping	Beet armyworm	Park and Goh, 1992
Attract-and-kill		
Pesticide-treated spheres	Apple maggot fly	Prokopy *et al.*, 2000; Stelinski *et al.*, 2001
Pesticide-treated spheres	Blueberry maggot fly	Ayyappathe *et al.*, 2000; Stelinski and Liburd, 2001
Pheromone bait spray	Olive fly	Jones, 1998
Sex-pheromone-based attracticide	Codling moth	Charmillot *et al.*, 2000
Sex-pheromone-based attracticide	Light-brown apple moth	Suckling and Brockerhoff, 1999
Autodissemination		
Pheromone trap and fungus	Diamondback moth	Furlong *et al.*, 1995
Attractant trap and fungus	Japanese beetle	Klein and Lacey, 1999
Pheromone trap and baculovirus	Tobacco budworm	Jackson *et al.*, 1992

are the costs of substantial numbers of traps and the expenses for the material and labour needed to maintain them. The technique is more useful for the control of low-density than high-density populations. Traps can quickly become saturated with insects at high densities. If the attractant is a sex pheromone, mass trapping will require many traps to be effective against high-density populations because of competition with calling females (Knipling, 1979). The kind and potency of the attractant influence the outcome of a mass trapping programme. The chances for success are improved if both males and females are attracted to the trap. If only males are trapped, it is essential that they be captured before mating. In most insect species, males can mate more than once; thus a very high proportion of individuals needs to be removed from the population to obtain the high reduction in female fecundity required for control (Roelofs *et al.*, 1970). Therefore, optimizing the lure is critical to the success of mass trapping. Synthetic lures must compete well with natural attractive sources. The lure must be potent enough to draw in insects from a considerable distance without impeding progress once they are close to the source. Trap density is based on both economic considerations and the attractiveness of the lure. The spacing of traps should be such that competition between traps occurs, but just to a level that does not reduce the total kill (Lanier, 1990). Finally, as pointed out in the monitoring section, the ability to attract and retain very high numbers of individuals will be affected by trap design, placement and maintenance.

Prospects

The success of a mass-trapping programme is directly related to the desired outcome. A very high percentage of the individuals in the population must be removed if the tolerance for damage is very low. Pests that cause direct damage to a crop and have tolerances near zero are probably not the best choices for this control tactic. Better opportunities for mass trapping are in situations where there is some flexibility in the desired outcome, such as where the public is willing to tolerate some pest damage in exchange for a perceived reduced exposure to insecticides. In parks and city plantings, 80–90% rather than 100% control may be acceptable. Male removal using sex-pheromone traps was demonstrated to be an effective means of controlling Chinese tortrix, *Cydia trasias*, on street-planted Chinese scholar trees (Zhang *et al.*, 2002). Four years of mass trapping with a sex and floral lure reduced a small pocket of Japanese beetles in a city park by 97% (Wawrzynski and Ascerno, 1998).

The prospects for mass trapping are enhanced if population densities are low or if the technique is carried out in an area where immigration by the pest from outside the treated area is limited. The success of the above-mentioned efforts in urban or park settings was, in part, due to the isolation of the sites and the relatively low pest densities. Attempts to control Japanese beetle in other settings or where population densities were high have proved ineffective (Klein, 1981; Gordon and Potter, 1985, 1986). Food warehouses and other enclosed situations provide a high level of isolation, which should enhance the prospects for mass trapping (Suckling and Karg, 2000). Control of Mediterranean flour moths, *Ephestia kuehniella* Zeller, in flour mills was achieved through the mass deployment of pheromone-baited funnel traps combined with careful cleaning of the rooms and machinery (Trematerra and Battaini, 1987).

Perhaps the most successful use of this tactic has been for the control of some species of forest beetles. Semiochemicals play a major role in the process of host colonization by bark beetles. Intensive trapping of bark beetles for pest control is facilitated by their high dependence on aggregation pheromones that are attractive to both sexes as they mass attack a host. Mass trapping combined with other control measures, such as sanitation cutting, were used to control populations of the conifer bark beetle, *Ips typographus* (Linnaeus), in Norway and Sweden (Bakke and Lie, 1989). One of the most effective uses of mass trapping has been for control of ambrosia beetles in timber-processing facilities in British Columbia (Borden, 1990). In this case, the programme probably benefited from the trapping area being somewhat isolated from beetle populations in the forest (Schlyter and Birgersson, 1999). Control of some forest beetles may be enhanced by use of deterrents to 'push' the target beetles away from a host, combined with attractant-baited traps or trap trees to 'pull' them away (Borden, 1997). Aggregation and anti-aggregation pheromones were successfully used in this push–pull manner against the mountain pine beetle, *Dendroctonus ponderosae* Hopkins (Lindgren and Borden, 1993).

Attract-and-kill

As a control tactic, attract-and-kill is similar to mass trapping in that an attractant-based system is used to eliminate a substantial proportion of a pest population and thereby prevent unacceptable levels of crop damage. The major difference is that the attract-and-kill approach relies on a toxicant, rather than a trap, to remove individuals that respond to the synthetic attractant. In many ways this technique suffers from the same constraints as outlined for mass trapping, including population density, attractiveness of the lure and efficiency of the method of killing. A major advantage of the attract-and-kill approach is that the problem of trap saturation can be eliminated. This may improve the effectiveness of control in high-density situations. The issues of trap maintenance and the high cost of the control programme can also be mitigated to some extent, especially if the system relies on attracting the insect to a plant surface that has been treated with an insecticide rather than to some kind of target device (Jones, 1998).

Attract-and-kill formulations have been developed for control of various beetles, moths and especially flies (see examples in Table 5.3). Some of the earliest and most widely tested applications of attractants in combination with insecticides have been for control of tephritid fruit flies (Jones, 1998). They have largely evolved from attempts to mass trap these insects, which often failed because of the problem of trap saturation. Jones (1998) has summarized efforts to control the olive fly, *Bactrocera olea*e Gmelin, in Greece through various attract-and-kill strategies. His review illustrates well the process of developing this approach for fruit-fly control. Protein/insecticide-bait sprays have been used for control of this pest in most Mediterranean olive-growing areas for a number of years. A major concern with this tactic is that the bait is highly attractive and toxic to natural enemies. To overcome the detrimental effects on natural enemies, a system was developed based on the use of target traps baited with either a food-attractant or a sex-pheromone dispenser. This target-device method of controlling *B. oleae* was effective at reducing fruit infestation, especially when applied on an area-wide basis. In addition, the effectiveness of the device allowed for a reduction in the use of bait sprays and an accompanying increase in natural-enemy populations. The most recent development for fruit-fly control has been a microencapsulated sprayable formulation comprised of the sex pheromone of this species, 1,7-dioxaspiro, and an insecticide (either malathion or dimethoate). Interestingly, the pheromone-bait spray has provided significant reductions in fruit infestation, while attempts to use the pheromone as a mating disruptant only have failed because the approach produces substantial immigrations of male and female olive flies into the treated area.

Concerted efforts to develop lure-baited trapping systems for control of *Rhagoletis* fruit flies are ongoing in the eastern and Mid-western USA. Some early success was achieved using sticky-coated red spheres for direct control of *Rhagoletis pomonella* (Walsh) (Duan and Prokopy, 1995). A major impediment to commercial adoption of this control system was the high level of maintenance required to ensure trap effectiveness. Recent efforts, therefore, have focused on developing a system that relies on a small dose of toxicant, rather than a sticky material, to kill alighting flies. Biodegradable or wooden spheres laced with a low dose of imidacloprid show promise for control of *R. pomonella* in apple and *Rhagoletis mendax* Curran in blueberry (Hu *et al.*, 1998; Liburd *et al.*, 1999; Ayyappathe *et al.*, 2000; Prokopy *et al.*, 2000; Stelinski and Liburd, 2001; Stelinski *et al.* 2001). Pesticide-treated spheres rely on a combination of attractants, a feeding stimulant and a toxicant to lure and kill the target pest. Sphere shape and colour and fruit-volatile or food-based lures are the major attractants. The insecticide is incorporated into the latex paint used to colour the spheres, which aids in maintaining the residual activity of the toxicant. Sucrose is used as a feeding stimulant, which coerces flies to ingest lethal doses of toxicant. Placement of biodegradable spheres baited with an attractive component of host odour, butyl hexanoate, on perimeter trees of commercial apple blocks were nearly as effective as insecticide sprays at intercepting apple maggot flies and preventing fruit injury (Prokopy *et al.*, 2000).

Autodissemination

Autodissemination is an innovative and promising control technique that combines an attractive lure with an entomopathogen. Suckling and Karg (2000) recently proposed the term 'lure and infect' to describe this approach and provided a good summary of its limitations and unique advantages. Individuals that arrive at the source are not killed, but rather are inoculated with the pathogen and hopefully magnify the treatment by spreading the disease to other individuals. The host specificity of the pathogens means that the method will be highly compatible with biological control.

The approach has been attempted using a variety of disease organisms (Table 5.3). Shapas *et al.* (1977) substantially suppressed populations of a stored-product pest, *Trogoderma glabrum* (Herbst), using a combination of its pheromone and spores of a pathogenic protozoan, *Mattesia trogodermae*. Autodissemination of a baculovirus for management of tobacco bud-worm has been tested by Jackson *et al.* (1992). The effectiveness of a pheromone trap designed to deliver conidia of a fungal pathogen has been explored for control of the diamondback moth, *Plutella xylostella* (Linnaeus) (Furlong *et al.*, 1995). A fungus is also being developed for use against Japanese beetle. A Trece Catch Can Japanese-beetle trap modified to serve as an inoculation chamber has been tested for control of this pest, using the fungal pathogen *Metarhizium anisopliae* (Klein and Lacey, 1999). High levels of mortality were recorded for beetles emerging from the trapping device. In addition, it was demonstrated that contaminated beetles could pass the fungus to untreated beetles in quantities sufficient to kill a high proportion of the population.

A number of deficiencies will have to be overcome to make the autodissemination approach a commercially viable option. Some innovation will be required to design effective transfer stations. For example, Vega *et al.* (1995) invented an autoinoculating device that induces sap beetles to pick up whatever microorganism is loaded into it. The pathogens that are placed in delivery stations should also be readily transferred between individuals. Suckling and Karg (2000) pointed out that fungi might be the best candidates as they are transferred between adults and larvae, and do not require consumption or copulation to become pathogenic. Once an appropriate pathogen is selected, a formulation must be developed that protects the organism from environmental degradation. A major constraint with these systems, as with mass-trapping strategies, is likely to be the ability to make them cost-effective as many bait stations may need to be deployed for the approach to be effective. Finally, a general public concern over the production of pathogens and their release into the environment may limit the development and acceptance of this technique.

Mating Disruption

The most successful approach using semiochemicals for pest control over the last few decades has been the release of large amounts of synthetic pheromone into a crop in an effort to prevent or delay mating. The potential to control insects through mating disruption was first demonstrated for *Trichoplusia ni* (Hübner) over 30 years ago (Gaston *et al.*, 1967). Similar efforts with other moth species confirmed that dispensing large quantities of pheromone into a crop could disrupt mate location, thereby controlling the pest by interfering with the fertilization of eggs (Mitchell *et al.*, 1974; Shorey *et al.*, 1974; Taschenberg *et al.*, 1974; Rothschild, 1975).

There has been considerable progress in the application of formulated pheromone for direct pest control since the first promising trials. Through the combined efforts of researchers, private entrepreneurs, extension personnel and others, mating disruption has become an accepted control option for a number of lepidopteran pests of fruits, vegetables and forests (Ridgway *et al.*, 1990; Cardé and Minks, 1995, 1997). A listing of commercial formulations currently registered for use in North America and the estimated total area treated in 2002 (Table 5.4) provides strong testimony to the success of this approach.

A number of developments had to occur in order to make mating disruption an effective and economically viable control tactic. Continual advances in understanding the many biological characteristics, behavioural and otherwise, that influence the outcome of a mating-disruption programme were certainly instrumental in paving the way. Some advances were more technical in nature, such as the development of new techniques for identification and synthesis of pheromones and devices for releasing the pheromone over an extended period of time. Often overlooked was the strong involvement of government agencies, both in research and in technology transfer. Companies developing mating-disruption products typically had very small research budgets, with support provided primarily as donations of product to government researchers to conduct efficacy trials. Government agencies have played a particularly important educational role in area-wide projects and deserve a great deal of the credit for providing technical expertise and demonstrating the benefits of mating disruption (Staten *et al.*, 1997; Calkins *et al.*, 2000; Il'ichev *et al.*, 2002). Finally, changes in regulatory requirements were made that accelerated the registration process (Thomson *et al.*, 1998).

Although substantial inroads into commercial markets have been made since the early 1980s, disruptants and other semiochemical products continue to hold a rather small share of the total pest-control market. Jones and Casagrande (2000) placed the

Table 5.4. Commercial disruption formulations registered for use in North America and estimated area treated in 2002.

		Number of formulations				
Crop	Pest	Reservoir-type	Other	Hectares treated	Per cent of total hectares planted	Current market value ($US)
Almond	Peach-twig borer	2	1	200	< 0.1	30,000
Apple	Codling moth	6	4	45,000	18.0	13,750,000
	Leaf-roller	2	1	1,600	0.6	200,000
	Oriental fruit moth	2	2	400	0.1	40,000
Cotton	Pink bollworm	2	2	8,900	0.2	562,000
Cranberry	Black-headed fireworm	1	2	200	1.5	25,000
	Sparganothis fruitworm		1	100	0.7	12,500
Grape	Grape berry moth	2	1	300	0.1	18,750
	Omnivorous leaf-roller	1	1	500	0.1	31,250
Peach	Oriental fruit moth	4	3	17,800	26.2	1,780,000
	Peach-twig borer	2	1	600	0.9	90,000
Pear	Codling moth	6	4	8,000	28.6	2,200,000
Tomato	Tomato pinworm	2	2	10,000	5.6	72,000
Walnut	Codling moth	1	3	1,200	1.5	330,000

worldwide sales of semiochemical products at about US$70–80 million or about 1% of the agrochemical market. More recently, the world pheromone market for mating-disruption products was estimated to be $80 million (K. Ogawa, Japan, 2002, personal communication). A limited understanding of what is required to achieve success, technical deficiencies and tightening profit margins in agriculture are the principal factors that have slowed the rate of adoption of mating disruption.

Even some of the failures have been useful for identifying patterns that will improve our ability to select pest species, production systems and specific sites that will maximize the chance for the success of mating disruption from the outset. The discussion that follows is an attempt to identify and synthesize the patterns that have emerged since the first disruption field trials were conducted some 40 years ago.

To a large extent, the potential for the success of mating disruption is determined by the following set of biological parameters:

- Biology/ecology of the target species.
- Male sensitivity to the pheromone (physiological).
- Chemical characteristics of the pheromone.
- Influence of the physical environment.

It is how this suite of features plays out for a specific pest, commodity or site that determines the suitability of mating disruption for that particular situation. Once the decision is made to implement a mating-disruption programme, its commercial success depends on a set of operational parameters that provide for effective delivery of the active ingredient. Practical considerations, such as cost, ease of application and the reliability of a scouting programme will determine the extent to which users will embrace this technique.

Pest biology and ecology

A number of biological and ecological attributes are likely to influence the suitability of a pest species as a candidate for control by mating disruption (Rothschild, 1981; Cardé and Minks, 1995; Sanders, 1997). The success of mating disruption is dependent upon population density, the ecological setting of the treated area, and specific traits of the target species. Key biological traits include host specificity, dispersal capacity, number of generations and adult lifespan, fecundity, characteristics of pheromone emission by the female and other aspects of mating behaviour.

Experience teaches us that the effectiveness of mating disruption is compromised in situations where mated females invade from outside the treated area. Treatment with pheromone alone provides no protection against immigrating females, which can readily deposit fertile eggs. Immigration of gravid females is believed to be a key process that contributes to the development of border infestations (Tatsuki, 1990; Il'ichev *et al.*, 2002). It follows that a high dispersal capacity and a wide host range are life-history traits that tend to reduce the suitability of a species for mating disruption. Furthermore, the impact of these traits should be most pronounced where the area treated with disruptant is adjacent to untreated areas that harbour hosts for the target pest.

As with other semiochemical-based control tactics, there is often a strong interaction between population density and the effectiveness of mating disruption. In many instances, high-density populations are more difficult to control with this technique than less dense populations. For example, the best disruption of codling moth has been achieved where pest pressure is low, while attempts to control high-pressure populations have been problematic at best (Trimble, 1995; Gut and Brunner, 1998; Vickers *et al.*, 1998). In an effort to mitigate the effects of population density, growers typically apply one or more companion insecticide sprays to reduce pest pressure.

It is worth noting that, for some pest species, mating disruption is equally effective over a range of population densities. Field trials to determine the effectiveness of disruption for control of peach-tree borer, *Synanthedon exitosa* (Say), were conducted in peach orchards in Georgia, USA, using Hercon vinyl laminated dispensers loaded with 43 mg of the main component of the pheromone for this species, *(Z),(Z)*-3,13-octadecadienyl acetate (Snow, 1990). Excellent control was achieved under either moderate or heavy population pressure. Damage was recorded in the high-density area in the first year of the trial, but no mating was recorded on mating tables, and larval infestation was attributed to infiltration of mated females from outside the treated area. After 2 years, it was impossible to conduct further trials in these peach blocks because the population had essentially been eliminated. We will return to this apparent disparity in the effect of density on mating disruption, as we believe it is of critical importance in identifying promising targets for this technique.

Male response to pheromone

The male's response to pheromone is perhaps the most important biological characteristic determining the outcome of a mating-disruption programme.

Perception

The most important and specialized mate-detection organs in moths are the antennae. These comb-like or hairy, rod-shaped structures are adapted to sift odorant molecules from the air, which are then perceived by specifically tuned receptor cells. These olfactory receptor neurones elicit receptor potentials in response to species-specific pheromone components which manifest in specific patterns of action potentials that convey information about both odorant quantity and quality to the moth's brain (Kaissling, 1986). Usually, just a few receptor neurones are situated within hair-like, odour-perceiving structures, called sensilla, covering the external surfaces of moth antennae. Pheromone molecules adsorb on to the surface of sensilla and diffuse into the interior sensillum through minute pores sprinkled over the exterior sensillar shaft. Once inside the sensillum lymph, the hydrophobic pheromone molecules are dissolved through the association with pheromone-binding proteins (Klein, 1987). Finally, the binding protein complexes travel from the pores to receptors on the dendritic membranes of the odour neurons. After activation of the dendritic receptor, pheromone molecules must be rapidly removed in order for the moth to detect further stimuli. This is thought to be achieved by enzymatic degradation of the pheromone molecules (Vogt and Riddiford, 1986; Rybczynski *et al.*, 1989). The rapid termination of the pheromone signal by degrading enzymes is required for the high quantitative and temporal resolution of the odour signal (Stengl *et al.*, 1992).

The electric potential generated across moth antennae after stimulation with their pheromone was successfully measured by Schneider (1956, 1962) using the electroantennogram (EAG) technique. An EAG measures the depolarization of receptor potentials summed across the antennal olfactory neurones over the length of the antenna. EAGs have been an effective means of identifying the sensitivity of moths to odorants, quantifying dose–response relationships and measuring adaptation effects at the peripheral level of odour detection. Other researchers began to perform extracellular recordings from single sensilla and thus demonstrated the specificity of populations of sensilla to odorants in various insect taxa (Den Otter, 1977; Dickens, 1979; Fonta and Masson, 1987; Almaas and Mustaparta, 1990).

Moth pheromones are commonly comprised of complex blends of components (Tamaki, 1979; O'Connell, 1981). Although the major components of such blends often elicit some behavioural responses typical of males responding to calling females, usually the full complement and correct ratio of components are required to induce the complex sequence of male sexual behaviours (Linn *et al.*, 1984). Such sensitivity to specific blend ratios is believed to function as a mechanism for maintaining species isolation (Linn and Roelofs, 1983). It is thought that numerous types of narrowly tuned receptor neurones are specialized for detecting each separate pheromone component; this is known as the component hypothesis (O'Connell, 1972; Den Otter, 1977; Akers and O'Connell, 1988; Almaas *et al.*, 1991). However, there is also evidence that separate components of a multi-component pheromone blend may interact with common receptor binding sites; the blend hypothesis (O'Connell, 1985; Christensen *et al.*, 1989). Receptors having a very high affinity for pheromonal compounds are said to be 'tuned' to that odour stimulus. Such 'tuning' is based on molecular shape, length of the carbon chain and position of double bonds and functional groups (Todd and Baker, 1997).

Orientation

A female-produced pheromone plume is a filamentous structure of varying internal concentration detected by males as a series of stimulus pulses of varying duration and concentration (Murlis *et al.*, 1992). The intensity at which pheromone molecules are detected by the antennae determines the rate of action potentials generated. This information is passed to higher processing centres in the brain that control the rate of casting and counterturning behaviour, flight speed and orientation up the pheromone plume by the moth (Baker *et al.*, 1985). This plume-following behaviour brings males within close proximity of the calling female. At this point, the high pheromone concentration, along with visual cues, arrests flight. The male and female may then undergo courtship behaviours (Baker and Cardé, 1979) and then mate. The disruption of any or the entire above-described plume following and courtship behaviours is referred to as mating disruption or pheromone confusion.

Mechanisms for disruption

The most popular explanatory models for disruption of pheromone-based communication are: (i) sensory adaptation at the peripheral level affecting olfactory receptors; (ii) habituation affecting processing of and normal responsiveness to olfactory information reaching the central nervous system (CNS); (iii) camouflage of female-produced plumes; and (iv) false-trail-following of synthetic pheromone plumes by male moths (Rothschild, 1981; Bartell, 1982; Cardé, 1990). The first two mechanisms result from prolonged exposure to high and/or constant concentrations of pheromone delivered by synthetic dispensing systems. Under the third mechanism, it is assumed that male moths are unable to distinguish between female-produced pheromone plumes and the background concentration of pheromone emanating from dispensers. Finally, the fourth model postulates that males may follow the pheromone plumes generated by point sources of synthetic dispensers. Real females are thought to be out-competed by false plumes. Different combinations of the above mechanisms may be important in practical mating-disruption programmes based on crop canopy structure, wind speed, pheromone chemistry, the pheromone delivery system and the insect species.

Permeation of agricultural habitats with synthetic pheromones presumably exposes the target pest moths within those localities to unnaturally high and/or constant doses of pheromone. Given this presumption, many investigations have examined the effects of short and prolonged exposures of moths to their species-specific synthetic pheromones and geometric isomers (Bartell and Roelofs, 1973; Bartell and Lawrence, 1976a,b; Linn and Roelofs, 1981; Sanders, 1985). Bartell and Lawrence (1976a,b) differentiated between two possible effects of prolonged pheromone exposure: they called the effect operating at the peripheral level 'adaptation' and the effect operating at the level of the CNS 'habituation'. Many experiments directed at establishing whether adaptation or habituation validly explains mating disruption have not adequately differentiated between these two mechanisms, effectively lumping peripheral and central effects together (Rumbo and Vickers, 1997; Sanders, 1997).

Numerous basic investigations have shown that exposing male moths to their pheromone decreases subsequent stereotyped behavioural responses, such as wing fanning and rapid walking. In addition, pre-exposed male moths were less successful in orienting towards pheromone point sources, as measured by mark–release–recapture and wind-tunnel studies (Rumbo and Vickers, 1997; Daly and Figueredo, 2000). Such studies provide evidence that either habituation in the CNS or adaptation of the peripheral sensory apparatus had affected subsequent male behaviours. At least one study (Kuenen and Baker, 1981) demonstrated that pheromone exposure of male *T. ni* disrupted normal orientation responses of males with no corresponding effect on the olfactory receptor neurones as measured by EAGs. These results implicate habituation of the CNS as opposed to adaptation of the peripheral receptors on the antennae as the cause of the subsequent aberrant male behaviours. Furthermore, a later study demonstrated that male *C. molesta* exhibited days-long habituation after exposure to their pheromone (Figueredo and Baker, 1992). Similarly, wind-tunnel and field experiments on *Heliothis virescens* (Fabricius) implicated CNS habituation lasting up to 96 h as the major mechanism for decreasing male moth response to female pheromone and as the underlying means for pheromone-based mating disruption of this species (Daly and Figueredo, 2000).

Bartell and Lawrence (1976b) suggested that male moth exposures to pulsed pheromonal stimuli would more effectively reduce normal sexual responses compared with constant stimulation, because peripheral adaptation would be circumvented, allowing for greater central habituation. Kuenen and Baker (1981) obtained data supporting this hypothesis for *T. ni* by showing that pulsed rather than constant pre-exposure resulted in greater disorientation of subsequent sexual responses. Also, they demonstrated decreased EAG amplitudes with concurrent exposure, indicating that receptor adaptation was occurring. Therefore, this study implicated receptor adaptation in the antennae as an impediment for CNS habituation. In other words, a pheromone-exposed antenna becomes adapted and thus fires fewer action potentials in response to later pheromonal stimulation. In this sense, an adapted antenna can be considered as a filtering mechanism of the sensory information that would otherwise flood the CNS, perhaps preventing habituation and the associated reduction of normal sexual behavioural responses.

Recent studies with vertebrate olfactory receptor neurones distinguish three different forms of adaptation based on their different onset and recovery time courses and their pharmacological properties (Zufall and Leinders-Zufall, 2000). The two short-lived variants have onset times on the order of 100 ms and 4 s and corresponding recovery times of 10 s and 1.5 min, respectively. The third type of adaptation is characterized as 'long-lasting'; onset occurs after an exposure of 25 s and subsequent recovery takes place after 6 min. In addition, research has shown that these three types of adaptation are mediated by separate molecular mechanisms (Zufall and Leinders-Zufall, 2000).

There is also evidence for a distinction between long-lasting and short-lived variants of peripheral adaptation in insects. Kuenen and Baker (1981) documented a short-lived form of pheromonal adaptation in *T. ni*, using EAGs. In this case full receptor-cell recovery occurred within 1 min of exposure. In contrast, Schmitz *et al.* (1997) recorded a longer-lasting antennal adaptation in *Lobesia botrana* Denis and Schiffermüller, using EAGs, and observed that receptor recovery took place after 5 min. Stelinski *et al.* (2003a) have recently documented a 'long-lasting' form of peripheral adaptation in *Choristoneura rosaceana* (Harris), along with no such corresponding adaptation in *Argyrotaenia velutinana* (Walker). Exposure to high doses of pheromone for 60 min reduced sensory responsiveness of *C. rosaceana* by up to 60%, while identical exposure of *A. velutinana* yielded no long-lasting peripheral sensory adaptation (Fig. 5.2). Neither species adapted after only 5 min of exposure to pheromone. The EAG responses of *C. rosaceana* were lowered by 55–58% following exposure to pheromone for 15 min and made a linear recovery to 70–100% of the pre-exposure response within 12.5 min at a rate of 3–4%/min (Stelinski *et al.*, 2003a). By performing recordings from single antennal neurones, Baker *et al.* (1989) showed that male *Agrotis segetum* (Denis and Schiffermüller) olfactory receptor neurones adapted when they were exposed to high pheromone concentrations known to cause in-flight arrestment of progress toward the source. Using

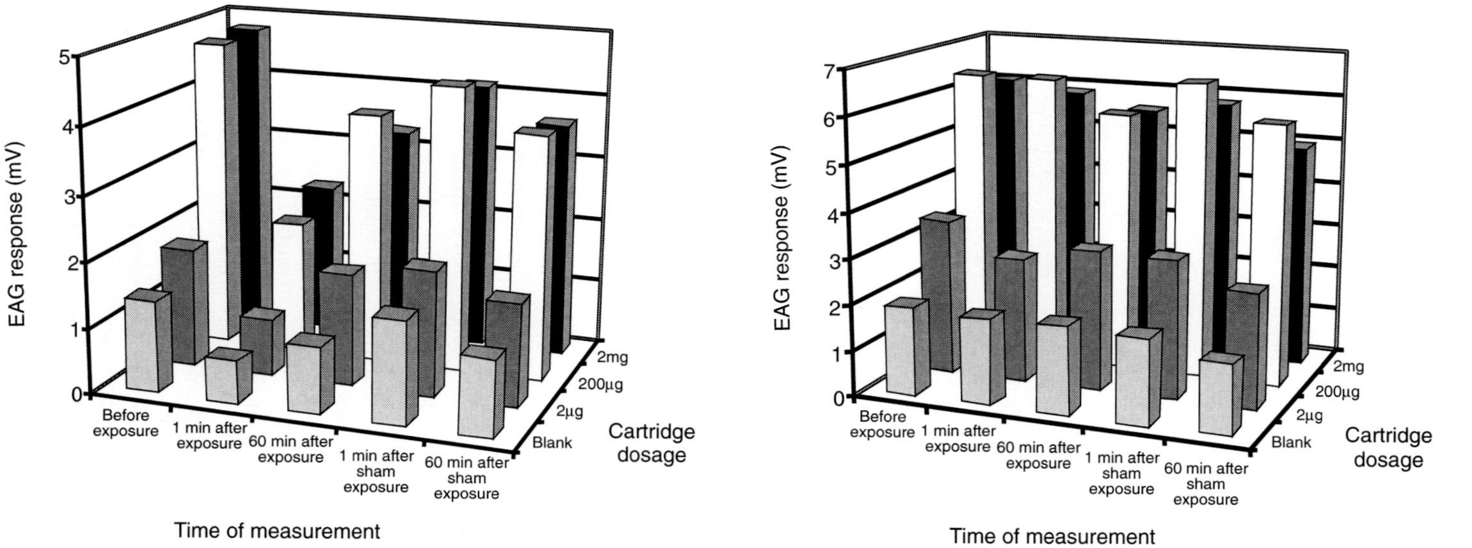

Fig. 5.2. Peripheral response of *Choristoneura rosaceana* (left) and *Argyrotaenia velutinana* (right) to the main component of their pheromonal blend and traces of its geometric isomer (*Z*11–14Ac and *E*11–14Ac, respectively) as measured by EAGs following 60 min exposure to high pheromone concentrations (adapted from Stelinski *et al.*, 2003a).

the same technique, they also showed that antennal neurones from *H. virescens* failed to adapt regardless of concentration. Baker *et al.* (1989), proposed that, given the low emission rate of (Z)-11–16:Ald from the rubber septa employed in their study, it was unlikely that *H. virescens* neurones were challenged to the same degree as those of *A. segetum* had been by the more volatile pheromone of that species. Alternatively, we suggest that *A. segetum* and *H. virescens* may differ in their susceptibility to peripheral sensory adaptation, as was observed with *C. rosaceana* and *A. velutinana*. Given these data and the fact that chemical signalling pathways are conserved among insects and vertebrates (Fein, 1986), it is possible that different molecular mechanisms and signal-transduction pathways may be involved in different forms of insect odour adaptation; these mechanisms may differ across insect taxa.

Chemical characteristics of the pheromone

Moth sex-attractant pheromones are comprised of blends of straight-chain hydrocarbons, alcohols, aldehydes and acetates, varying in chain length from *c.* C10 to C20 and in number and isomeric configuration of internal double bonds (Tamaki, 1979; O'Connell, 1981; Chapman, 1998). All are lipids – soluble in organic solvents rather than water. In pure form, most are oils at room temperature. Moreover, it is correctly understood that these agents of long-distance sexual communication operate as volatiles. As such, sex pheromones are invisible when they travel through air; yet their powerful effects on moth behaviours provide convincing evidence that atmospheric transfer of information is occurring.

Given their many similarities in the visible liquid state, it is understandable that persons working with pheromones can get lulled into envisioning that the behaviours of all these chemicals in the environment are similar. However, as the next sections will document, such a conclusion is a risky overgeneralization. Differences in molecular weight and functional group can profoundly influence rates of evaporation, dispersion in air and adsorption on to solid surfaces. Collectively, differences in these physicochemical properties can translate into highly disparate residence times of semiochemicals in the environment and the need for tailoring application technologies to suit the particular molecular specimen being used to manipulate a pest insect.

Vapour pressures

The pressure attained at equilibrium by the volume of pure vapour that builds up over a pure liquid or solid in a closed vessel is known as vapour pressure (VP). This property is expressed in pressure units, e.g. mmHg. VP is an equilibrial measure – the net value of the propensities of molecules to evaporate, as well as to condense. Compounds with high VPs evaporate rapidly, while those with low VPs evaporate slowly. Moreover, VP varies exponentially with molecular weight. Therefore, plots of molecular weights of compounds in a particular class (e.g. alcohols) against $\log_{10}$ of VP yield straight lines (Fig. 5.3).

Several notable conclusions arise from Fig. 5.3. First, VPs vary dramatically with molecular weight. Small organic molecules with low polarity, e.g. methyl acetate ($C_3H_6O_2$, MW = 74) or hexane (C_6H_{14}, MW = 86) have VPs higher than 100 mmHg. In contrast, only modestly larger molecules of similar polarity, e.g. pentyl acetate ($C_7H_{14}O_2$, MW = 130) or decane ($C_{10}H_{22}$, MW = 142) have dramatically lower VPs, < 10 mmHg. Back-calculations from the regression equations given in Fig. 5.3 reveal a consistent threefold decrease in VP for each additional $-CH_2-$ unit added to any compound in a given molecular series. Secondly, adding a particular functional group to a straight-chain hydrocarbon base may strongly influence VP while adding another may not. For example, adding an –OH moiety to the corresponding hydrocarbon decreases VP by nearly 100-fold, as derived from the difference in *y* intercepts of the regression equations (Fig. 5.3). Only *c.* 4% of this change can be explained by increasing the molecular weight by 16 (one oxygen atom), leaving some 96% of the effect attributable to altered chemical characteristics – most probably increased polarity due to the prevalent hydrogen bonding of alcohols (Morrison and Boyd, 1974). An even greater effect would be expected from organic acids. In contrast, nearly all of the effect of adding an acetate moiety to a hydrocarbon base is explained by increasing molecular weight, as evidenced by acetate data adherence to the hydrocarbon line (Fig. 5.3) when regressed against molecular weight. VP data for aldehydes fall between the lines for alcohols and acetates (Fig. 5.3), consistent with their known intermediate polarity.

The evaporation rate of pheromones measured in flowing open air at field-relevant temperatures is probably a parameter of greater direct relevance to applied chemical ecologists than is VP. Under these conditions, evaporation apart from condensation would be the main effect measured. Unfortunately, such data are not readily available even for standard reference compounds like those in Fig. 5.3. However, some predictions can be deduced from an understanding of physicochemical properties as to how VP measures would translate into open evaporation measures. At ordinary field temperatures, small compounds (MW < 100) have a much higher propensity to evaporate than to condense (see following section). The converse is true for large compounds (MW > 200) at similar temperatures; condensation is much more favoured than evaporation. It follows that removing condensation effects from equilibrial VP measures will raise the rate of evaporation for large compounds proportionately more than for small compounds. Thus, in a plot with the log of the evaporation rate in open air on the *y* axis against molecular weight on the *x* axis, the slope of the resultant line would be expected to be smaller than the threefold reduction for VP values seen with the addition of each $-CH_2-$ (Fig. 5.3). In other words, the effect of molecular weight would be less severe on

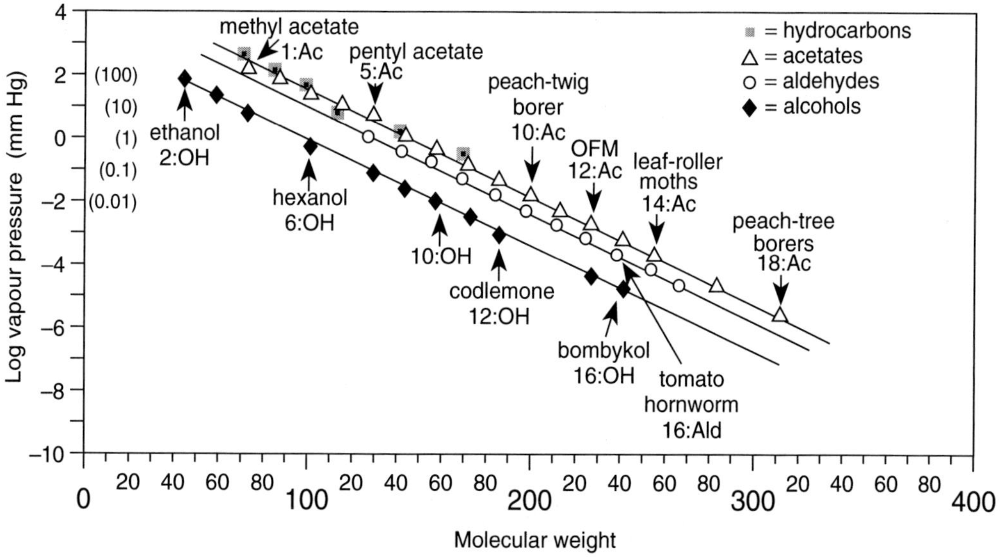

Fig. 5.3. Vapour pressures for straight-chain hydrocarbons, acetates, aldehydes and alcohols at 25°C as a function of molecular weight. Data points were obtained from manufacturers' MSDS sheets via website (http://hazard.com/msds/gn). Estimates of vapour pressures for the selection of moth sex pheromones shown by arrows were extrapolated using the dashed lines. For the solid lines, regression equations were: (alcohols) log VP = −0.034 × MW + 3.31; R^2 = 0.99. Not being statistically distinguishable, data for the hydrocarbons and acetates were combined: log VP = −0.030 × MW + 4.50; R^2 = 0.96.

evaporation rate in free moving air than it is for equilibrial VP.

Although not originally done for this purpose, we have produced some data on the relative rates of evaporation of pheromonal compounds in open moving air as a function of molecular weight. In one situation, the rates of evaporation of several milligrams of the pheromones of Oriental fruit moth (12:Ac) and peach-tree borer (18:Ac) loaded on to filter-paper discs were compared in a wind tunnel operating at *c.* 30°C and a wind speed of 2 m/s. In another situation, an ethanolic solution of long-chain primary alcohols was sprayed on to apple trees under typical Michigan midsummer field conditions and the disappearance of the compounds from leaves was measured over days by gas–liquid chromatography (GLC).

The resultant data are co-plotted in Fig. 5.4 as time for disappearance of half the sample of applied compound against molecular weight. Although these data should be considered preliminary, the outcome supports the above prediction of a less severe molecular weight effect for free evaporation than for equilibrial VP. For example, the difference in evaporation rates for 12:Ac vs. 18:Ac was less than two orders of magnitude, rather than the nearly three orders of magnitude predicted for VP. The 2.4-fold decrease in evaporation rate for each additional carbon compares well to the 2.7-fold decrease in evaporation rate per additional carbon recorded by McDonough *et al.* (1989) for acetates evaporating from rubber septa. Clearly, molecular weight still has a dramatic effect on pheromone longevity in the open air and explains the relationship across data points of Fig. 5.4 quite well. Half of the 12:Ac and 14:OH samples disappeared in less than 7 and 3 h, respectively. Over half of the 12:OH sample had evaporated in the first hour after application. Thus, the longevity at the site of release of pheromonal compounds smaller than this is fleeting under summertime field conditions.

On the other hand, the longevity of the larger moth sex pheromones was appreciable. It took 10 days for half of the sample of 18:Ac to evaporate at *c.* 30°C. Corresponding values for C18 and C16 alcohols were 4 days

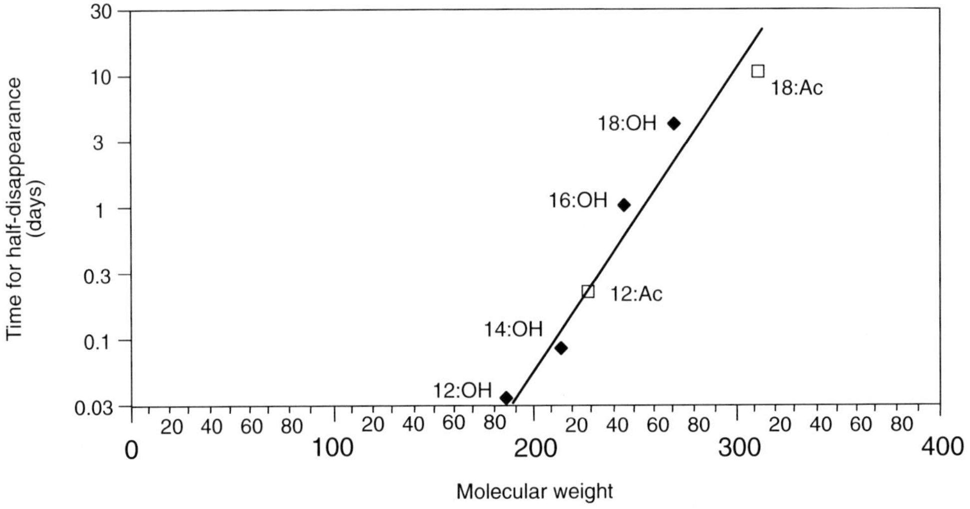

Fig. 5.4. Relative evaporation rates of differently sized straight-chain alcohols or acetates in moving air or under Michigan summertime conditions. Comparisons within a compound type (same functional group) are reliable; comparisons across compound types should be considered preliminary, as test conditions were not identical.

and 1 day, respectively, measures thought (by interpolation using Fig. 5.4) likely to correspond to those for C16 and C14 acetates.

These large differences in VP and evaporation rates across the spectrum of moth sex-pheromonal chemicals argues that the challenges in formulating them for use in pest control will likewise vary widely. For example, formulating pheromone to successfully meter it out for 1 month above a disruption threshold should be much easier for peach-tree borers than for codling moth. For peach-tree borer pheromone, the release profile (rate over time) without additives naturally lies much closer to the desired straight-line profile (representing constant release) than is true for the pheromone of codling moth or peach-twig borer, whose release profiles are naturally severely concave. Whether or not the underlying reasons were understood, this pattern of achieving greater success in formulating large vs. small pheromones is widely experienced by those manufacturing and testing pheromone formulations used as lures for traps and especially as devices for mating disruption.

Adsorption of pheromones on surfaces – condensation and partitioning

Another common misconception concerning the behaviour of volatile chemicals is that, once they do evaporate, it is difficult for them to be retrieved from the vapour state – i.e. achieve condensation. However, condensation is another physicochemical property that is highly influenced by the molecular weight and the functional group of given molecules (Miller, 2004). It turns out that it is remarkably easy for pheromonal compounds larger than molecular weight 200 to adsorb on to solid surfaces at ordinary environmental temperatures. The requirements are simply: (i) a short time (probably only seconds or less) to pass after evaporation and during which large molecules in the vapour state distribute some of their atypically high energy in collisions with the small gaseous molecules comprising air; and (ii) collision with a surface sufficiently large to allow multiple sites of contact between pheromone molecule and that surface. If the forces of cohesion to various sites along the molecule collectively exceed some energetic threshold, the molecule becomes 'adsorbed' and begins to skid about on the surface of the solid in two- rather than three-dimensional space, as was the case in air. The temperature of this surface establishes the probability that this 'captured' molecule will, by chance, at some point in time, receive a sufficient energetic boost to tear it from this surface and again thrust it into the vapour state (re-evaporation). The process of molecules travelling through space, sometimes in the vapour state while at other times adsorbed on to or permeating into surfaces, is known as 'partitioning'. Significant partitioning occurs for all moth sex-pheromone molecules at normal field temperatures, provided the pheromone molecules are released into an environment in which contacts with solid surfaces are probable.

We used the technique of gas–solid chromatography to quantify the degree to which moth sex pheromones of various molecular weights and functional groups partition between air and various types of solid surfaces at 25°C. Standard gas chromatography (GC) measures and equations were used to generate partitioning coefficients (C_ps), which express the ratio of the compound on a solid surface over that in the adjacent gaseous phase. For example, a C_p of 1.0 means there was one pheromone molecule on the solid for every molecule in the gas immediately over that solid. A C_p of 10 indicates ten times more pheromone molecules on the surface than in the vapour immediately over it, etc. Perhaps a more intuitive way to view this phenomenon is that C_p values equate to compound 'stickiness'; the higher the C_p value, the more sticky is the compound when contacting a solid.

Figure 5.5 reveals the C_p values for various straight-chain pheromonal compounds travelling through a 2 m long × 2 mm inside diameter (i.d.) glass tube at 25°C with a nitrogen flow rate of 20 ml/min. Hydrocarbons were the least sticky. Nevertheless, even hydrocarbons with carbon numbers of 13, 14 and 15 partitioned

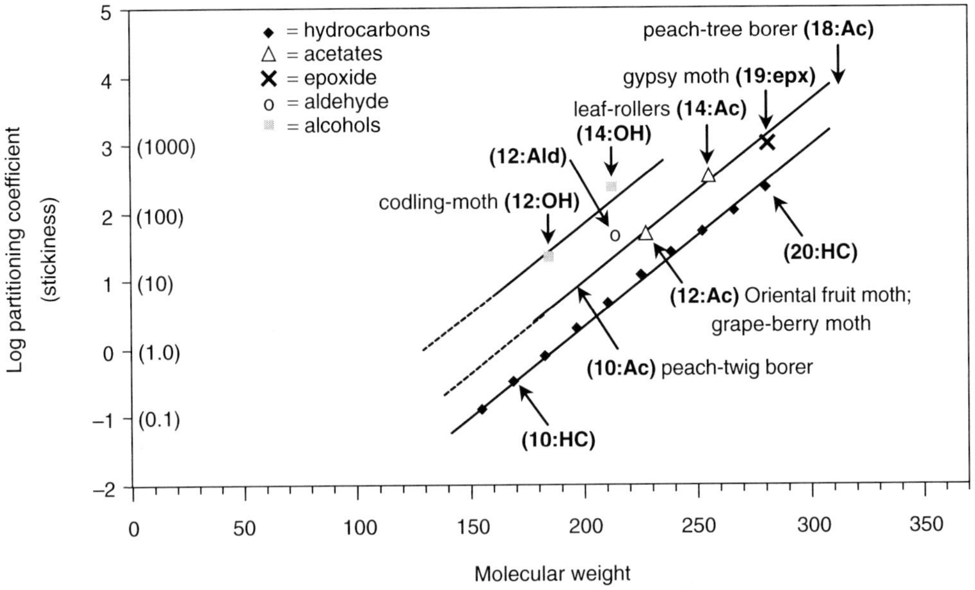

Fig. 5.5. Partitioning coefficients of straight-chain hydrocarbons and selected pheromonal compounds when moving through a narrow-bore glass tube as influenced by functional group. All data points were measured by Miller (2004); those indicated only by an arrow are extrapolated estimates.

sufficiently to be completely resolved from one another while traversing this column; they emerged in 5, 9 and 16 min, respectively. As was true for VP, stickiness increased by a consistent threefold for every additional $-CH_2-$ added to the hydrocarbons as well as their oxygenated derivatives. The preference of the C20 hydrocarbon was nearly 300-fold higher for the glass surface than for the gas immediately over it. Moreover, hydrocarbon stickiness was stable and independent of the tube composition – glass, silylanized glass, stainless steel or Teflon. This argues that the condensation effect is driven mainly by molecular energetics and is not explained only by specialized binding sites on the solid surface.

Oxygenation of the hydrocarbons further increased C_p as compared with just raising the molecular weight. Adding ester, epoxide, aldehyde and hydroxy oxygens increased C_p by 1.5-, 1.5-, 10-, and 30-fold, respectively, over the corresponding hydrocarbon for the glass-tube system (Fig. 5.5). Relative to one another, the C_ps for various familiar pheromone molecules vary immensely, as visually illustrated in Fig. 5.6. For example, the C_p of 18:Ac is 100 times larger than that of 12:Ac. If the font size for the 18:Ac were adjusted to fit on the page of Fig. 5.6, the printing for 12:Ac would be illegible. Hopefully, such contrasts drive home the conclusion that pheromones can differ greatly in their physicochemical behaviour when chain lengths vary by seemingly just a few carbons.

Compared with the glass tube, C_ps for the oxygenated pheromonal compounds were even higher for a stainless-steel tube but lower for a Teflon tube. It is well established that the lipid pheromones under consideration are strongly retained, even at elevated temperatures, when moving through GC columns with a thin film of wax used as the stationary phase. Thus, surfaces at normal environmental temperatures, such as plant leaves and insect cuticle, which are coated with waxes, would be expected to have a strong affinity for pheromonal compounds, leading to strong adsorption and high C_ps.

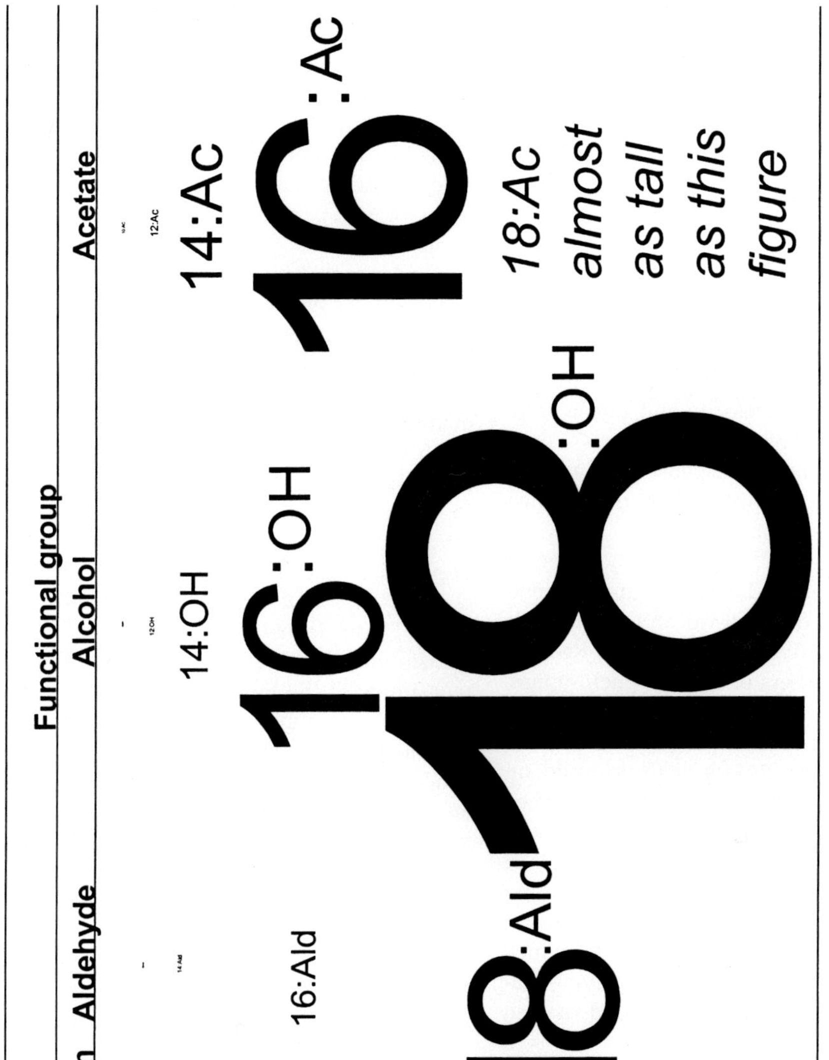

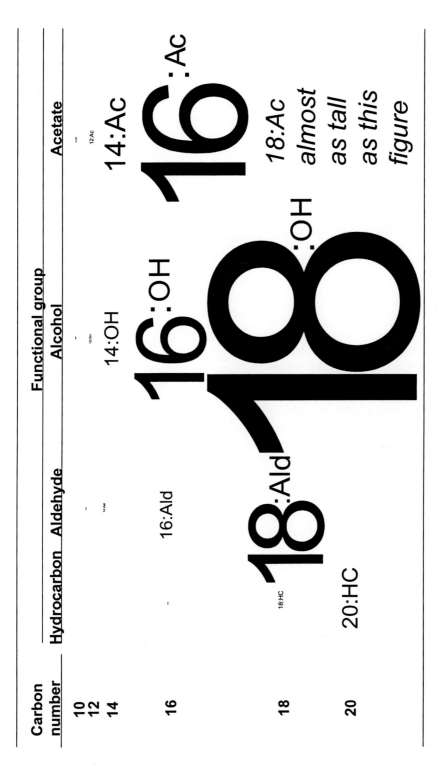

Fig. 5.6. Relative magnitude of partitioning coefficients of representative moth sex pheromone molecules as depicted by font size. Data derived from Fig. 5.5.

Possible consequences of pheromone partitioning effects in the field and evidence from the literature

Based on the above data, the fates of pheromones deployed in crops to achieve mating disruption can be expected to vary as a function of compound molecular weight and functional group. As covered in a previous section, pheromone evaporation rates from dispensers vary widely with molecular weight. After the given molecules have evaporated, the degree to which they partition on to foliage as they are dispersed by wind again depends on the molecular properties under discussion, as well as such factors as: temperature, wind velocity and foliage density. Substantial pheromone 'haloes' could build up around a dispenser of a large pheromone, such as the doubly unsaturated 18:Ac of peach-tree borers.

This effect will be especially large if foliage is dense, wind velocities are low, temperatures are not excessive and compounds do not quickly degrade chemically. Relative to 18:Ac, dispensers of smaller pheromones, such as codlemone ((*E*),(*E*)-8,10-12:OH), under the same conducive environmental conditions are much less likely to build up sizeable haloes, because of much reduced stickiness once evaporated and greatly increased evaporation rates after incidents of adsorption. An additional factor hindering codlemone build-up in the crop is the instability of its conjugated diene system, which is susceptible to oxidation and polymerization when exposed to sunlight and oxygen (Millar, 1995).

The picture we wish to paint here for pheromone-based disruption is that the movement of pheromones through crop foliage can be viewed as occurring roughly by a process of inefficient chromatography, where the wind is the mobile phase, the foliage is the column packing, the waxes on the leaves are the stationary phase and pheromone is constantly being injected into the system. In such a system, pheromone build-up over time will be positively correlated with the density of the packing, the molecular weight of the pheromone and its stickiness to the stationary phase. Pheromone build-up over time will be inversely correlated with wind velocity and temperature. Hopefully, an understanding of these fundamental chemical principles and the reference data provided above will assist pest managers in making implementation decisions for pheromone disruption in the face of limited data on the expected behaviours of the particular pheromone components in the field.

Evidence that pheromone partitions on to foliage under field conditions is documented in the literature. However, the range of chemical structures for which it is characterized is, as yet, severely limited. The early evidence of this partitioning phenomenon by moth sex pheromones arose from observations that foliage near calling female moths or pheromone-baited traps continued to be attractive after the calling female had departed or the trap was removed (Wall *et al.*, 1981; Wall and Perry, 1983; Noldus *et al.*, 1991). Upon development of the field EAG measurement technique, Karg *et al.* (1990), working with the European grapevine moth and its doubly unsaturated 12:Ac pheromone, demonstrated that pheromone concentration in the air of disruption plots and the structure of the pheromone cloud in treated vineyards were highly dependent on the density of vegetation in the target area. Dense foliage was found to be conducive to pheromone build-up. Two possible reasons were given for this effect. First, dense foliage reduces wind velocity, thus reducing dilution of the pheromone emanating at a roughly constant rate from the hand-applied polyethylene dispensers used in these tests. Secondly, Karg *et al.* (1990) concluded that pheromone was adsorbing on to the grape leaves and being re-released over time. Proposed positive effects for disruption were the wider and more homogeneous dispersal of pheromone throughout the crop.

Working with colleagues in New Zealand, Karg *et al.* (1994) found that leaves placed downwind of a rope releasing (*E*)-11-14:Ac became maximally loaded with the pheromone of the light-brown apple moth in just 3 min, as measured by EAG. Such leaves were then shown to release EAG-detectable levels of pheromone for at least 24 h. Moreover, leaves having ample opportunity to adsorb pheromone were slightly but significantly attractive to male moths for more than 24 h when used as lures in traps. Indeed, a halo of pheromone was found around rope dispensers that had aged in the field for some days. However, for this 14:Ac, its radius was measurable only within about 15 cm of the source.

This team (Suckling *et al.*, 1996) went on to find that shut-down of trap catches continued for a day or two after pheromone dispensers were removed from disruption plots. This disruption 'ghost effect' was rightly attributed to the build-up of a so-called 'buffer' of pheromone thought to have travelled away from the dispensers and yet remaining in the crop in concentrations that remained disruptive. This several-day 'ghost effect' for the C-14 acetates of *Epiphyas postvittana* (Walker) corresponds well with our estimate from Fig. 5.4 of approximately a 1-day interval for the half-disappearance of a 14:Ac.

Important questions remain to be answered with respect to these partitioning effects. Field studies should be extended to larger pheromones, which would probably produce larger haloes, as predicted by the data in Fig. 5.5. Attention should also be directed towards whether the pheromone adsorbed on or absorbed into leaves remains there or is metabolized. The unexplained rapid disappearance of 14:Ac topically applied to apple leaves (Suckling *et al.*, 1996) provides some contradictory evidence to the simple notion that foliage would be an inert buffer for pheromone. This outcome is consistent with the unpublished preliminary observation by J.R. Miller that more than 60% of the 30 μg of ^{14}C-labelled (Z)-11–14:Ac released into a closed vessel was taken up by apple leaves and was no longer extractable from the leaf surface after several days. Nevertheless, multiple lines of evidence now support the reality and potential importance of this partitioning phenomenon within crops.

The degree to which pheromone adsorbed some distance from a dispenser contributes to disruption of communication remains an open question. Of course, the plume of pheromone vapour coming directly from the dispenser is thought to play a key role in the success of mating disruption, consistent with conventional interpretations (Cardé and Minks, 1995; Sanders, 1997). The build-up of haloes around dispensers would definitely enlarge plumes and increase coverage by pheromones. It would probably also expand the area within the crop where resting moths are exposed to adsorbed pheromone, as suggested by Karg *et al.* (1994) and Karg and Sauer (1997). The possible long-lasting effects of such prolonged exposure deserve increased attention.

Influence of the physical environment

It is well understood that environmental factors have a major impact on the field stability and longevity of mating-disruption formulations (Weatherston, 1990). Exposure to ultraviolet (UV) radiation and heat can degrade pheromones through oxidative decomposition, isomerization of double bonds and other chemical processes. Mating disruption may be easier to accomplish in a full-canopied crop because the pheromone is less exposed to UV. Temperature also has a major influence on the rate of pheromone emission. High temperatures increase the pheromone release rate. This may be beneficial in the short term, but ultimately it may result in reduced longevity of a disruptant. Thus, disruption formulations may require a higher pheromone-loading rate in hot, compared with cool production regions. It is also possible that cool temperatures during critical periods can reduce pheromone emission rates below the levels needed for disruption (Howell, 1992). Rainfall can adversely affect sprayable formulations by washing off a portion of the capsules or beads (Waldstein and Gut, 2003b).

The distribution of pheromone is influenced by several physical factors, including wind, field or orchard topography and shape and canopy structure (Karg *et al.*, 1990; Färbert *et al.*, 1997). The best opportunity for disruption is where physical conditions allow for the uniform distribution of pheromone.

Thus, sites that are relatively calm and flat are better candidates for mating disruption than sites that experience frequent high winds or have steep slopes (Gut and Brunner, 1996). In addition, orchards with large numbers of missing trees or uneven canopies are considered poor candidates for mating disruption.

Operational factors that determine the level of success

In practice, the success of mating disruption depends on the cost-effective delivery of an appropriate blend, amount and spatial distribution of pheromone for an extended period of time. Suckling and Karg (2000) identified several operational factors that affect pheromone delivery and thus the efficacy of mating disruption. Of crucial importance is the type of formulation. To a great extent, the choice of formulation defines the other operational factors. Delivery systems vary in the rate and consistency of pheromone release. They also differ in their ability to limit the impacts of temperature and UV radiation on rates of emission and pheromone stability. The effectiveness of mating disruption further depends on application parameters, such as the timing and distribution of the pheromone treatment. Ultimately, commercial success hinges on providing economically viable mating-disruption formulations.

Types of formulations

At present, pheromone-based mating disruption is largely achieved through the manual application of reservoir-type release devices (Cardé and Minks, 1995; see also Table 5.4). Pheromone is enclosed in plastic or dispersed in synthetic polymers and slowly diffuses from these reservoirs over a period of several months. Pheromone is dispensed at rates of up to several micrograms per hour, which is at least 100-fold greater than the release rate of pheromone from a calling female (Sanders, 1997). Reservoir dispensers are hand-applied at a rate of at least 250 sources per hectare. Mainly for economical reasons, the highest recommended deployment rates are usually in the range of 500–1000 dispensers per hectare.

Two kinds of formulations have been developed that allow the pheromone to be sprayed on the crop either by ground or by air. Pheromone can be formulated into plastic flakes or chopped fibres designed to release pheromone at about the same rate as a calling virgin female (Brooks, 1980; Swenson and Weatherston, 1989). Female-equivalent formulations are sprayed on to a crop using custom-designed equipment, often with a sticker added so that the particles will adhere to foliage. A similar approach is to disperse pheromone in microcapsules or beads (Balken, 1980; Hall *et al.*, 1982). The individual particles in these formulations are small enough to be applied through conventional spray equipment, but as a consequence, pheromone is released at very low rates, below those of individual calling females (Sanders, 1997). Doane (1999) has recently provided a good overview of the current status of microencapsulated pheromone formulations. This technology has proved to be efficacious against several pests, especially the pink bollworm and tomato pinworm.

A recently developed approach to formulating and releasing insect sex attractants is through the use of aerosol-emitting devices, such as 'puffers' (Shorey *et al.*, 1996), 'misters' (Mafra-Neto and Baker, 1996) or 'microsprayers' (Isaacs *et al.*, 1999). Aerosol emitters are deployed at densities of only two to five per hectare, but each unit releases several milligrams of pheromone at least every 15–30 min. These super-low density devices control the release of pheromone mechanically to provide a constant predetermined release rate and a stable environment for the large volume of pheromone prior to its release. Aerosol emitters deployed at low densities have been tested for the disruption of insect pests in field crops (Shorey *et al.*, 1996; Baker *et al.*, 1997), tree crops (Shorey and Gerber, 1996a,b; Shorey *et al.*, 1996), stored products (Mafra-Neto and Baker, 1996) and cranberry bogs (Baker *et al.*, 1997). Success has varied widely; the consistently best outcomes were obtained with larger rather than smaller MW pheromones.

Application parameters

The success of mating disruption depends on achieving an even distribution of sufficient quantities of pheromone. The most widely used pheromone formulations, including polyethylene tubes, laminates and plastic membranes or ampullae, are typically deployed at rates of 250–1000 units per hectare. The total pheromone load varies by formulation and target species and can range from 20 to 2000 g/ha in season-long disruption programmes. The amount of pheromone delivered per unit area by sprayable formulations and aerosol devices has largely been set by successes achieved using given numbers of reservoir dispensers. Microencapsulated formulations are typically applied at rates of 20–100 g of pheromone per hectare and aim to release pheromone over a period of 3–4 weeks. Multiple applications are required to achieve season-long inhibition of mate location. Aerosol emitters may be programmed to release over 700 mg of pheromone per day and may emit this large volume of active ingredient for over 120 days. At this rate, the total pheromone load provided by five units per hectare would be over 400 g.

Crop damage in pheromone-treated plantings is often greater on the borders than in the interior (Knight, 1995; Gut and Brunner, 1998). Two processes are thought to contribute to the development of border infestations. Mated females emigrate from adjacent plantings that are not treated with pheromone. In addition, pheromone concentrations may be lower on the borders than in the interior (Ogawa, 1990; Karg and Sauer, 1995; Sauer and Karg, 1998), thus increasing the likelihood of males locating females and mating on the borders. Among the tactics used to protect borders is the application of extra pheromone near the edge of the planting or the extension of the pheromone treatment into adjacent plantings, if possible. Maximizing the amount of field or orchard interior relative to edge is the best protection against border damage (Fig. 5.7). Long, narrow plots are especially poor choices for mating disruption. The best approach for limiting edge effects is area-wide treatment with pheromone. Hot spots and edge infestations of Oriental fruit moth were eliminated after implementing an area-wide disruption programme in Australia (Il'ichev *et al.*, 2002).

Large-canopied crops present special challenges for achieving an effective distribution of pheromone. The positioning of hand-applied dispensers within the canopy of tree crops can dramatically affect the efficacy of mating disruption. Weisling and Knight (1995) demonstrated that significant mating of codling moth occurred when dispensers were placed in the middle, rather than in the upper portion of apple-tree canopies; the best control of codling moth with mating disruption has been achieved when dispensers are placed in the upper third of the tree canopy.

Determining the most effective pattern of pheromone distribution has been largely guided by trial and error, with two primary deployment strategies being adopted: (i) a uniform distribution; or (ii) a distribution that provides for higher concentrations along the edge rather than in the plot interior. Hand-application of reservoir dispensers in a predetermined pattern distributes the pheromone load uniformly within the treated area. The recommended application rate always specifies a minimum number of dispensers needed to achieve suppression. The conventional thinking has been that sufficient numbers of dispensers are needed to provide enough false trails to compete with females. Thus, higher application rates are needed when population pressure is higher. Alternatively, we suggest that enough dispensers are needed to 'charge' or impregnate the area around the dispenser with sufficiently high amounts of pheromone. In this view, applying too few dispensers increases the chances of creating areas with inadequate pheromone coverage.

In principle, sprayable formulations should provide a highly uniform distribu-

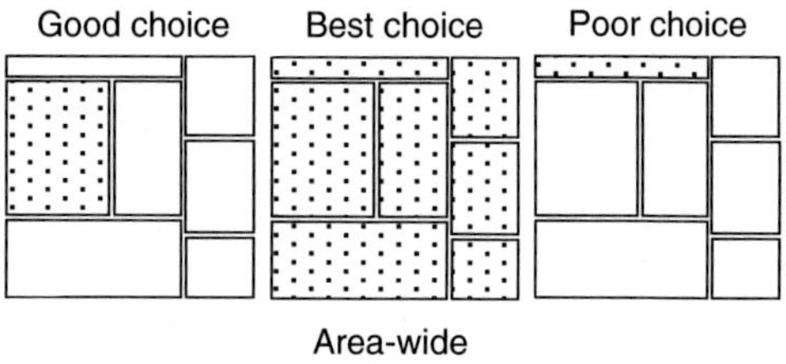

Fig. 5.7. Selection of appropriate plots for control with pheromones.

tion of pheromone. However, this assumes that capsules or fibres readily stick to plant surfaces. Waldstein and Gut (2003a) found considerable variation in the propensity of microcapsules to stick to various plant surfaces. Apple wood is a better substrate than fruit, which is superior to leaves. Sticking agents are often added to sprayable formulations to improve adhesion. Even with the addition of stickers, a high percentage of microcapsules may be lost following as little as 10 cm of rain (Waldstein and Gut, 2003b).

Wind is the primary means for dispersing pheromone that is emitted from aerosol devices. Pheromone distribution is also accomplished secondarily through a process of adsorption on to plant surfaces and subsequent release back into the environment (Suckling *et al.*, 1996; refer also to our earlier discussion of chemical characteristics of the pheromone). Each unit is required to disperse pheromone over an area of at least 0.5 ha. Whether this can be achieved is questionable. The rate and extent of dispersion may vary considerably depending on several factors, including wind conditions, plant canopy structure and chemical properties, such as relative molecular mass, inherent to the specific pheromone released. A grid of pheromone-baited traps has been used to map the apparent distribution of pheromone in 0.8 ha apple plots treated with an aerosol device, the Microsprayer (Gut and Garcia-Salazar, 2001, unpublished data). Examples of the patterns that were observed for *C. pomonella* and *A. velutinana* after the Microsprayers had been running for 2 days are presented in Fig. 5.8. Patterns of moth captures indicated that pheromone coverage was poor or absent in the areas upwind from the unit. Interestingly, a 'ghost' plume (*sensu* Suckling *et al.*, 1996) of pheromone was observed for several days after the units were turned off, supporting the role of secondary release of adsorbed pheromone from foliage as an important means of maintaining adequate pheromone levels throughout the treated area (Gut and Garcia-Salazar, 2001, unpublished data).

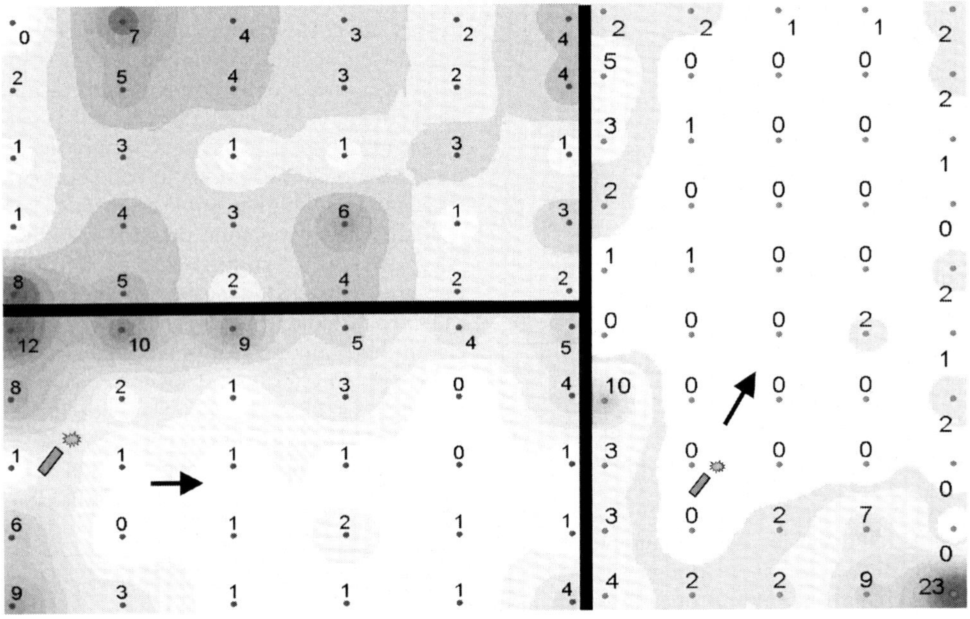

Fig. 5.8. Spatial mapping of moth captures using a grid of 30 pheromone-baited traps in 0.8-ha apple plots. Numbers are total catches per day in a single trap. Captures of *Cydia promonella* males in a plot not treated with pheromone (top left). Captures of *Cydia pomonella* males in a plot treated with a single aerosol device, the MSU Microsprayer, (rectangular symbol) emitting the major component of the pheromone of this species, (*E*) (*E*) 8, 10–12 : OH (bottom left). Captures of *Argyrotaenia velutinana* males in a plot treated with a single aerosol device, the MSU Microsprayer, emitting the major components of the pheromone of this species, *Z*11–14OAc and *E*11–14OAc in a 96 : 4 ratio (right panel). Arrows indicate the prevailing wind direction.

Comparison of delivery systems

Rather than predicting that any particular formulation of pheromones will out-compete and displace all others, we envision that each may find a niche within pest control. Microencapsulated pheromones may be the most convenient and quick to apply, but each application lasts for periods considerably shorter than hand- or aerosol-applied materials last. This weakness may be circumvented through the application of very low rates of pheromone at frequent intervals. Widely spaced devices appear to be particularly effective for pests that use large pheromone compounds, which have a high propensity for sticking to foliage upon wind dispersal. Moreover, such devices provide the option of releasing a mixture of pheromones of several pests simultaneously. Hand-applied pheromone formulations offer the advantages of longevity and a greater guarantee of uniform pheromone dispersal throughout the crop. However, they are more labour-intensive to apply than other techniques. This is especially true if dispensers must be applied in the upper canopy of tall trees, which is currently the recommended application protocol for disruption of some fruit pests (Gut and Brunner, 1996). In addition, the rate of release is greatly affected by environmental conditions, particularly temperature. There may be considerable variation in emission rate and field longevity among formulations or within formulations between years or in different growing regions (Thomson *et al.*, 1998).

Economic considerations

Many growers are apprehensive about using mating-disruption products because the input costs are high compared with most alternatives. These include the cost of the product and additional expenses to apply reservoir-type dispensers and intensively monitor pest populations, as well as other indirect costs associated with using an approach that only controls specifically targeted species.

In an attempt to address questions concerning the high input costs associated with mating disruption, Epstein *et al.* (2002) developed a model for evaluating the economics of pheromone-based management programmes. The model factored in not only input costs, but how those inputs affected fruit quality, the quantity of fruit harvested and the price received for that fruit. An approach known as ranging analysis, in the economic literature, was used. Ranging analysis takes into account variation in yield and price, with price varying depending on whether fruit was destined for the fresh or processing market. Direct comparisons of the economics of mating-disruption-based management programmes versus insecticide-only programmes demonstrated that in many cases the pheromone programmes were as inexpensive as or cheaper than the conventional programmes. The use of sprayable pheromone to enhance control of leaf-rollers was the most economical mating-disruption program, yielding increased profits of US$20/acre when the yield was average and 75% of the fruit was of fresh-market grade. The major impetus for conducting the economic analysis was to determine whether initial input costs could lead to future profits through decreased insecticide use and a larger harvest of high-quality fruit. The model demonstrated the economic benefit of eliminating some insecticide applications for secondary pests if fewer sprays were used to control primary pests. Presenting this information to growers was crucial to the recruitment process and enabled the project to expand from 500 to 8000 acres over a 3-year period.

Difficulties in measuring the efficacy of mating-disruption formulations in small plots has led to an overwhelming reliance on conducting such trials in large commercial plantings. The major drawbacks of on-farm experiments have been that untreated controls are usually not available and that a high level of replication cannot be accomplished. Furthermore, plots often vary substantially relative to initial pest densities and pesticide inputs throughout the season. Obtaining meaningful results from on-farm trials is indeed challenging. One approach, however, has provided a very effective means of demonstrating the economic viability of mating-disruption programmes. The general experimental design is to overlay a pheromone treatment on a portion of a large block that is being treated with insecticides. As long as the entire block receives the same sprays, differences in fruit injury can be credited to the added benefit of the disruption treatment. Gut *et al.* (1999) reported a three- to sixfold reduction in leaf-roller damage when mating disruption was overlaid on an insecticide programme. The growing reliance on mating disruption for control of codling moth can be attributed, in part, to the discovery by fruit growers that substantially better levels of control can be achieved by using pheromones in conjunction with a full-season insecticide programme than by relying on insecticides alone.

Towards increased success and adoption

The fact that producers rely on the deployment of pheromones for pest control on 150,000 to 200,000 ha worldwide (D. Thomson, unpublished data) justifies continued efforts to develop and implement management programmes based on the use of these behaviour-modifying chemicals. But what can be done to accelerate the successful use of mating disruption and minimize failures? There is general agreement among scientists working with this technique that a better understanding of the mechanisms underlying disruption, coupled with a good working knowledge of the biology, behaviour and mating system of target pests, is needed to improve success. Although we agree, in principle, with the above goal, we also believe that practical solutions are needed to facilitate the successful adoption of mating disruption. Despite examples of marginal efficacy in field trials and insufficient knowledge of how to improve performance, we believe that many new products will be introduced into the market-place over the next decade.

Target species selection

In reviewing the status of mating disruption of moth pests, Cardé and Minks (1995) concluded that not all species would prove to be susceptible to this technique. As a caveat to this statement, we would add that not all pest species susceptible to mating disruption may be equally controlled. For some species, such as peach-tree borer, red-banded leaf-roller and Oriental fruit moth, treatment with pheromone alone is often sufficient to mitigate crop damage. Other pests, such as the codling moth and certain leaf-rollers, appear to be more difficult to control using only mating disruption. Providing a framework that identifies certain pest species as more amenable to mating disruption than others should improve the likelihood of success. It would also minimize the considerable time and money spent on field trials that from the outset are unlikely to be successful – or are likely to fail. Taking a very practical approach, we have examined many of the successes and failures of mating disruption and propose that the following set of criteria can be used for rating species from easy to difficult to control using mating disruption:

- Extent to which moth captures in traps are inhibited.
- Number of dispensers necessary per unit area to achieve control.
- Pheromone dispensed per unit area to achieve control.
- Extent to which the outcome is density-dependent.

Complete or nearly complete trap shutdown is consistently achieved for some species, while for others it occurs rarely, if at all. The leaf-rollers *A. velutinana* and *C. rosaceana* are representative of the two extremes. The effectiveness of an aerosol delivery system, the MSU Microsprayer, for controlling these two species was evaluated in replicated trials in orchards in Michigan, USA. Microsprayers were placed in 0.8 ha apple blocks at densities of five units per hectare. Each unit was programmed to release a total of 410 mg of a 96 : 4 blend of (*Z*)-11–14:OAc : (*E*)-11–14:OAc per day. Moth catches in pheromone-baited traps were recorded weekly for each species in the Microsprayer-treated plots and in nearby plots not treated with pheromone. The pheromone treatment provided a very high level of inhibition of moth captures for *A. velutinana*, averaging 99.7% across four replicates (Fig. 5.9, top). In contrast, an average of 85% inhibition was recorded for *C. rosaceana* in the same plots (Fig. 5.9, bottom). Similar reductions in moth catches for these species have been recorded following the deployment of hand-applied devices or microencapsulated sprayable formulations releasing the same quantities and blend of active ingredients. Additional examples of species that are effectively inhibited from orienting to pheromone-baited traps through the deployment of a disruptant, include the tomato pinworm (Jenkins *et al.*, 1990), lesser peach-tree borer (Snow, 1990) and Oriental fruit moth (Rice and Kirsch, 1990).

Determining the amount of pheromone needed to interfere with mate location in the field has proved difficult. There have been a few examples where pheromone concentrations required for disruption in the field have been estimated (Rothschild, 1975; Koch *et al.*, 2002), but these have largely served as broad guidelines. The resultant limit in our understanding of the required airborne pheromone concentration needed to achieve communication disruption has led to the use of a trial-and-error approach in determining effective application rates. This has typically involved direct comparisons of various rates in small- and large-plot field trials. Based on this empirical approach, application rates ranging from 250 to 1000 dispensers per hectare have been established for polyethylene tubes, ampullae and other hand-applied dispensers. The number of devices deployed depends on the target species and may also vary in accordance with anticipated levels of pest pressure. The pheromone loading rate in the dispensers is often tuned to the requirements for communication disruption of a particular species. Other considerations include the cost of the active ingredient and the length of time a formulation must release pheromone. For example, a higher loading rate is required for season-long disruption of multiple-generation pests with extended periods of activity.

The peach-tree borer, *Synanthedon exitosa* (Say), is an example of a species that is 'easy' to disrupt using pheromone. A low application rate of *c.* 250 dispensers per hectare (one per tree), with each device releasing *c.* 0.3 mg of pheromone per day, provides levels of inhibition of moth catches in traps usually in excess of 97%. In addition, the level of disorientation to traps and decrease in larval densities appears to be independent of pest pressure (Snow, 1990). Similar levels of disorientation to traps and population suppression are readily achieved for the Oriental fruit moth, *C. molesta* (Rice and Kirsch, 1990; Vickers, 1990). In contrast, the oblique-banded leaf-roller, *C. rosaceana*, exemplifies a species that is 'difficult' to control using this technique. An application rate of 1000 dispensers per hectare, with each device loaded with 80–160 g of pheromone, provides levels of trap inhibition in the range of 85–92%. Doubling or tripling the rate of deployment does not significantly improve the level of inhibition of moth captures (Lawson *et al.*, 1996). Also, the level of disorientation to traps and suppression of the larval population appears to be highly related to population density, with commercially acceptable impacts only observed under low pest pressure (Novak *et al.*, 1978; Reissig *et al.*, 1978; Roelofs and Novak, 1981; Deland *et al.*, 1994; Agnello *et al.*, 1996; Lawson *et al.*, 1996; Knight, 1997). The codling moth, *C. pomonella*, is another example of a pest that appears to be relatively difficult to control through communication disruption. Deployment of 500 dispensers per hectare is only sufficient if pest densities are low. At higher densities, application rates up to 1000 devices per hectare and companion insecticide sprays are needed to achieve commercially acceptable control.

If indeed some species are highly amenable to disruption and others are not, then what are the factors that allow the more resilient species to operate in environments permeated with synthetic pheromone? We have documented in a previous section (Male response to pheromone) that there are fundamental differences in the capacities of species to become adapted and/or habituated when exposed to high doses of synthetic pheromone. In addition, we know that species differ in their dispersal and reproductive capabilities. Finally, we have provided evidence in an earlier section

(Chemical characteristics of the pheromone) that pheromones vary considerably with respect to rates of evaporation, dispersion in air and adsorption on to solid surfaces. Collectively, differences in these physico-chemical properties can have profound effects on the longevity and movement of pheromones in the environment. It is these basic differences in the properties of moths and their pheromones that make some species highly susceptible to mating disruption, while others are capable of averting the effects of this control technique (Fig. 5.9).

Certain lepidopteran species appear to be good 'adapters' in that they are physiologically capable of decreasing their sensitivity to pheromone for an extended period of time following pheromone exposure. We speculate that, under pheromone mating-disruption regimes, long-lasting adaptation may confer an advantage to moth species such as *C. rosaceana* relative to species such as *A. velutinana*, which do not appear to exhibit a physiological capacity for a long-lasting form of adaptation (Stelinski *et al.*, 2003a). For example, perhaps moths under long-lasting adaptation might sufficiently subdue their overt sexual responses so as to preclude exhaustion and cause them to depart from extraordinarily high-concentration pheromone sites where the likelihood of mate-finding is low. If they then happen to arrive in a location of low or no pheromone, disadaptation would occur within a short interval (10–15 min) and their ability to discriminate and orient to a natural pheromone plume would be restored, provided the possible effects of CNS habituation were shielded (Bartell and Lawrence, 1976b; Kuenen and Baker, 1981). Evenden *et al.* (2000) subjected male *C.*

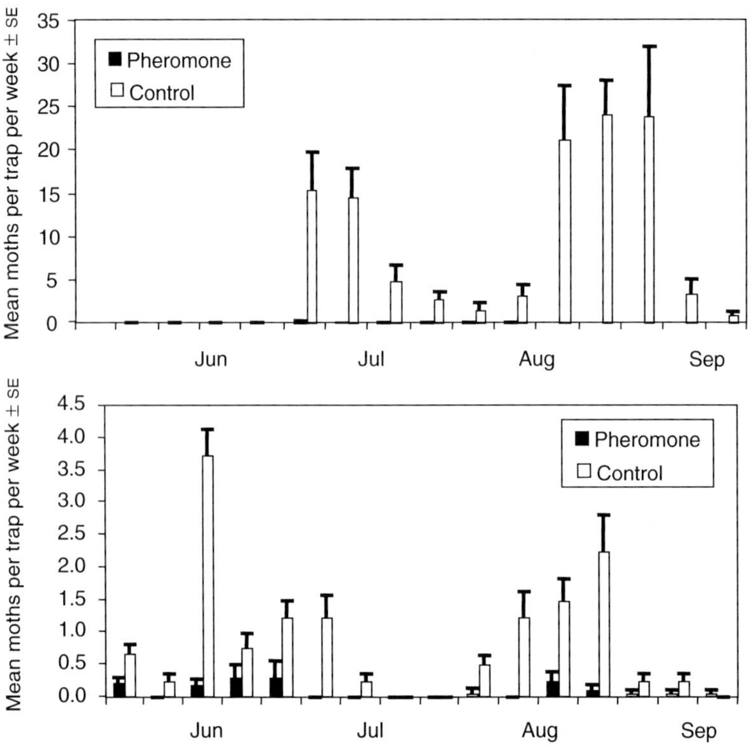

Fig. 5.9. Inhibition of *Argyrotaenia velutinana* (upper) and *Choristoneura rosaceana* (lower) male catches in pheromone-baited traps following deployment of a disruption treatment. An aerosol delivery system, the MSU Microsprayer, was placed in 0.8-ha apple plots at densities of five units per hectare. Each unit was programmed to release a total of 400 mg of a 96 : 4 blend of the main component of their pheromonal blend and traces of its geometric isomer (*Z*-11–14Ac and *E*11–14Ac). SE, standard error.

rosaceana to 1 h of constant pheromone exposure and then tested their behavioural response to a pheromone source in a wind-tunnel 10–30 min after the exposure treatments. Constant pre-exposure did not alter the proportion of males orienting upwind to pheromone plumes in the wind-tunnel. The conclusion was that habituation probably plays a minor role as a mechanism of mating disruption in this species. Evenden *et al.* (2000) chose to assay their pheromone-exposed *C. rosaceana* 10–30 min after treatment in order to avoid the effects of antennal adaptation. Our results showed that olfactory receptor neurones of *C. rosaceana* disadapt within 12.5 min after constant pre-exposure to pheromone. Therefore, the combined results of our study and that of Evenden *et al.* (2000) suggest that 'long-lasting' adaptation may shield *C. rosaceana* from CNS habituation, explaining why this species could successfully orient to pheromone point sources shortly after constant exposure to pheromone. Alternatively, both adaptation and habituation might occur, but recovery from both might be rapid.

Currently, hand-applied rope dispensers are the dominant method of dispensing pheromone for mating disruption of moth pests in orchards (Nagata, 1989; Agnello *et al.*, 1996; Knight *et al.*, 1998; Knight and Turner, 1999). The release rate for ropes marketed for leaf-roller moths averages *c.* 11 ng/s (Knight *et al.*, 1998; Knight and Turner, 1999). Moths within the treated crop could perceive the applied pheromone in several ways, including: (i) as a 'cloud' of pheromone resulting from a coalescence of plumes emanating from the many dispensers; (ii) as a localized plume downwind of a nearly dispenser; or (iii) a moth could even be attracted on to a dispenser. In our field tests, *C. rosaceana* did exhibit long-lasting adaptation upon exposure to pheromone ropes, but only when held within a few centimetres of the dispenser (Stelinski *et al.*, 2003b). Nevertheless, these results demonstrate that this phenomenon can occur under field conditions. Use of low-density, high-release dispensers, such as puffers (Shorey and Gerber, 1996a,b) or microsprayers (Isaacs *et al.*, 1999), offers an even greater opportunity for target moths to be exposed to extraordinarily high concentrations of pheromone as the pheromone solution emitted in an aerosol spray falls on to foliage and droplets of pure pheromone accumulate over time on the source tree. Moreover, large and highly concentrated plumes are thought to waft great distances downwind of the source trees. However, it remains to be determined whether moths exhibit the behavioural capacity for 'dosing' themselves with enough pheromone in the field for physiological phenomena such as long-lasting adaptation to be relevant under mating disruption.

Implementing a pheromone-based management programme for a pest that has been identified as easy to disrupt improves the chances for success, but it certainly does not ensure that control will be achieved or that it will be economical. The success of mating disruption in the field or orchard depends on cost-effective delivery of the active ingredient. Achieving this requires addressing the many factors or conditions, other than the moth and its pheromone, that have an impact on a mating disruption programme. We propose that, in practice, it is this set of conditions that determines the level of difficulty in meeting requirements for successful disruption of a particular pest species (Fig. 5.10). Operational requirements for successful mating disruption broadly include technical considerations, such as pheromone-delivery strategies, crop-management considerations and characteristics of the site, including initial pest density. Growers, consultants, extension personnel and others with a very applied viewpoint emphasize the need for disruption formulations that are economical relative to other control tactics, easy to use and compatible with current IPM programmes. From a technical standpoint, the delivery system that is selected must release the appropriate blend and amount of pheromone. It must also provide for an adequate distribution of the active ingredient over an extended period of time.

Perhaps the most crucial management decision is the selection of an appropriate area to be treated in terms of size and pest pressure. The likelihood of failure certainly

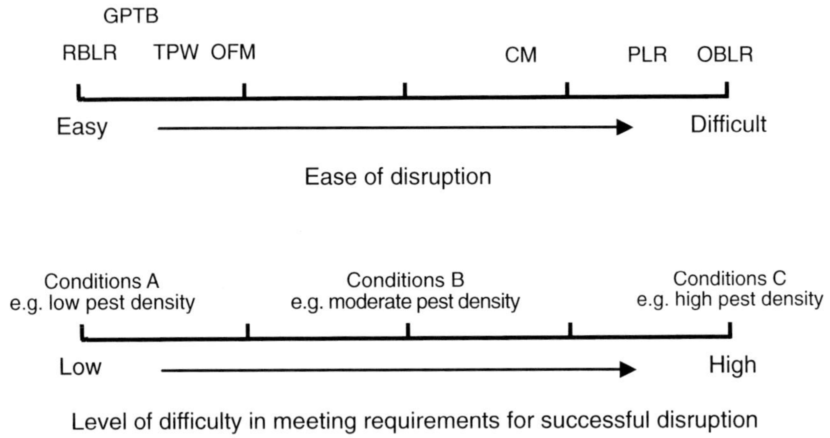

Fig. 5.10. Conceptualization of differences in the susceptibility of various species to mating disruption based on fundamental properties of the moths and their pheromones (upper). Species are red-banded leaf-roller (RBLR), greater peach-tree borer (GPTB), tomato pinworm (TPW), Oriental fruit moth (OFM), codling moth (CM), pandemis leaf-roller (PLR) and oblique-banded leaf-roller (OBLR). Conceptualization of differences in the level of difficulty of disrupting a particular species based on all of the factors other than the moth and its pheromone, such as physical characteristics of the site or starting pest density (lower).

increases if site selection does not minimize opportunities for immigration by mated females. In addition, the best successes for many species will be achieved where pest pressure is not too high. Operational requirements for successful disruption can vary depending on anticipated pest pressure. For example, a relatively low application rate of 500 dispensers per acre is sufficient for communication disruption and control of codling moth if pest pressure is low. The level of difficulty in meeting the requirements for successful disruption in this case is low (Fig. 5.10, conditions A). Control of codling moth can also be achieved under moderate pest pressure, but a full rate of 1000 dispensers per acre needs to be applied. Control of this pest is difficult to achieve where initial pest pressure is high, as even high application rates of at least 1000 dispensers per hectare cannot prevent mating. The difficulty in meeting the requirements for disruption of codling moth in these cases is moderate and high, respectively (Fig. 5.10, conditions B and C).

Traditionally, little attention has been focused on selecting appropriate targets prior to the development of a mating-disruption formulation. As pointed out by Doane (1999), there were suggestions over 20 years ago that habituation may be easier to elicit in species where fairly low doses impede the approach of males to a pheromone source (Cardé *et al.*, 1975; Baker and Roelofs, 1981). Yet, since these early observations, considerable attention has been focused on some species that have proved to be difficult candidates for disruption. The apparent driving force for these efforts is the importance of the pest economically. We propose that it may not be necessary to carry out detailed sets of experiments to determine the disruption capacity (high to low) among pest species. As a starting-point, the simplest measure may be, as was suggested over 25 years ago, the dose response of a species for orienting to various loading rates of pheromone lures. Some species, such as the Oriental fruit moth and tomato pinworm, are attracted to a narrow range of concentrations. A single hollow fibre attracted significantly more tomato pinworm moths than five or more fibre lures (Wyman, 1979). For Oriental fruit moth, high doses caused arrestment of upwind progress of males as they approached the source (Cardé *et al.*, 1975; Baker and Roelofs, 1981). A similar response was observed for spruce budworm, *C. fumiferana*, and Oriental fruit moth orienting to high-dose lures in wind tunnels (Sanders and Lucuik, 1996). All of these species, which are maximally attracted to low-dose lures, also appear to have a low capacity to avert communication disruption. In contrast, other species that exhibit high levels of attraction to a comparatively wider range of pheromone loadings appear to be more difficult to disrupt. Included in this second group is *C. rosaceana*, which is readily captured in traps baited with a wide range of dosages (Fig. 5.11).

Area-wide approach

The best successes with mating disruption have been achieved where large, contiguous areas have been treated with pheromone. Excellent control of Oriental fruit moth was obtained following implementation of area-wide disruption programmes in 1100 ha of apples and pears in Australia (Il'ichev *et al.*, 2002) and 1200 ha of mixed stone fruits in the Tulbagh valley in South Africa (Barnes and Blomefield, 1997). The US Department of Agriculture (USDA) sponsored the Codling Moth Area-wide Management Program (CAMP), adopting an IPM approach that relied on mating disruption technology and judicious and timely applications of insecticides for management of codling moth in pome fruit in the western USA (Calkins *et al.*, 2000; Brunner *et al.*, 2001). The number of CAMP sites and their size increased from an initial five sites totalling 1260 ha in 1995 to over a dozen sites totalling approximately 4000 ha by 1999. The results of this project were impressive, with pest densities, as measured by moth captures in pheromone traps and fruit injury at harvest, declining to very low levels following implementation of an area-wide approach (Fig. 5.12). In addition, direct comparisons with conventional programmes outside the project area revealed dramatic reductions in the number of insecticides applied for codling-moth control in CAMP orchards. Similar levels of success have been achieved through government-supported projects to control pink bollworm with pheromones in vast areas of cotton in the south-western USA (Staten *et al.*, 1997) and Egypt (Jones and Casagrande, 2000).

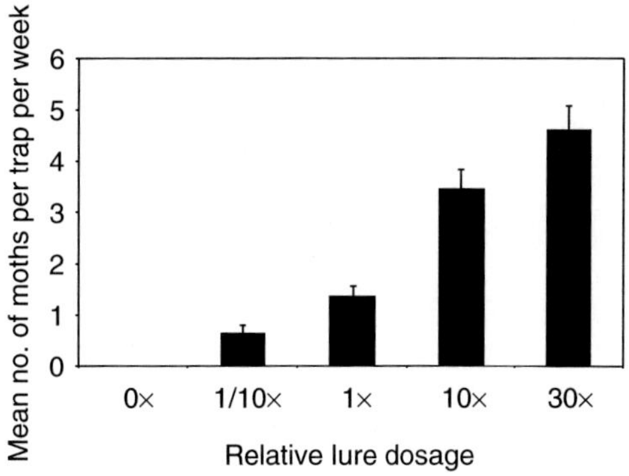

Fig. 5.11. Effect of lure dosage on captures of *Choristoneura rosaceana* in pheromone-baited traps (L. Stelinski, 2002, unpublished data).

Season-long versus targeted use of pheromones

Commercial development of disruption products has largely been geared towards

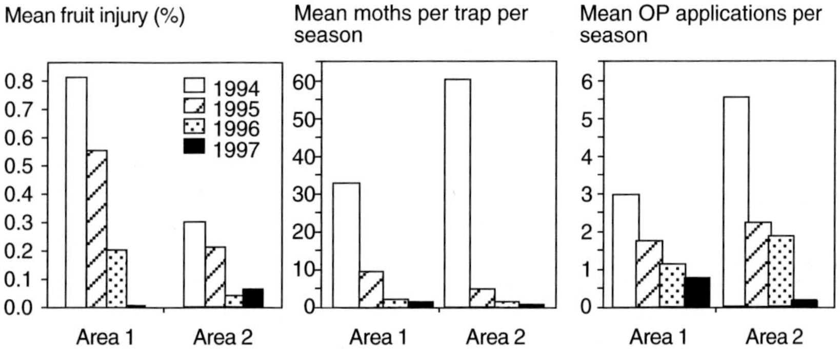

Fig. 5.12. Summary of results for two locations that were part of the Codling Moth Area-wide Management Program, USA, during the year prior to using mating disruption (1994) and the first 3 years of area-wide disruption (adapted from Calkins *et al.*, 2000; Brunner *et al.*, 2001). OP, organophosphorus insecticide.

providing season-long control of a pest population. The first application of pheromone is typically applied at or prior to the start of adult flight for the targeted pest. Additional treatments are made if the residual activity of the pheromone product does not cover the entire flight period. Although certain disruption products have achieved great success season-long, there are some critical limitations to this approach. Season-long control of some multivoltine species may require maintaining pheromone in the crop for over 180 days. This may be economically prohibitive or technically infeasible.

Targeted use of pheromones is an alternative approach that could provide new opportunities for some mating-disruption formulations. This strategy entails targeted use of a disruption formulation to affect a key period of adult activity. Sprayable pheromone formulations offer the greatest opportunity for incorporation into pest-management programmes in such a selective manner. We can envision microcapsules serving as a means for delivering a variety of insect pheromones. Perhaps they could be used in an 'off-the-shelf' approach, much like that adopted for lures used to attract pests to traps and monitor their activity. Suppliers of monitoring tools often rely on a single delivery device, such as a red septum, as the basis for a large product line. A key advantage of using pheromones in a targeted manner is that expensive mating-disruption products can be applied on an as-needed basis. This is in contrast to the preventive basis of using pheromones dictated by the season-long approach, with the disruption product typically applied prior to start of flight and often prior to any knowledge of pest density. In this situation, a grower must make an up-front investment in an expensive and, to some extent, risky technology. Microencapsulated formulations have recently been registered for suppression of Oriental fruit moth. These may be especially useful in crops such as apple, where this pest is generally only a problem late in the season.

Multispecies disruption

The extreme specificity of sex pheromones means there will be advantages and disadvantages to their use in pest management. Mating-disruption technologies are non-toxic to natural enemies. As a result, greater reliance on these highly selective tactics for pest control will increase the potential for biological control of secondary pests. On the other hand, the use of pheromone-based technology and the subsequent reduction in insecticides for a primary pest frequently result in outbreaks of other pests (Ridgway *et al*., 1990; Thomson *et al*., 1998). Growers are keenly aware of the potential risks and added costs associated with secondary-pest outbreaks.

The need to control several pest species is a major factor limiting the acceptance of disruption technologies in some crop-production systems. For example, over a dozen lepidopteran species can reach damaging levels in eastern apple-production systems (Epstein and Gut, 2000). Included in this mix is a complex of leaf-rollers that have overlapping activity periods and cause similar damage. Pfeiffer *et al*. (1993) tested the viability of mating disruption for a complex of four tortricids using various generic blends. All formulations proved ineffective for one or more species, presumably because they were not sufficiently similar to the natural blend. It may be worth revisiting this approach as new technologies are developed. For example, newly developed polyethylene twin-tube dispensers offer the possibility of emitting two generic blends, one from each tube, while applying only one device for both pheromone blends.

A few hand-applied dispensers containing the attractive blends for two or three species with very different pheromone components are either already commercially available or will soon be on the market. In large-plot field trials conducted over the past 2 years, multispecies formulations for codling moth and Oriental fruit moth or for these two species and oblique-banded leaf-roller have performed as well as the single-species products currently in commercial use (L.J. Gut, 2002, unpublished data). A concern with multiple-species reservoir-type dispensers is that the pheromones may interact in ways that reduce the efficacy of disruption against one or more target species. For example, Snow (1990) reported that release of only pure (*E*,*Z*)-3,13-octadecadenyl acetate (EZA) shut down captures of lesser peach-tree borer, while various blends, excluding pure EZA, significantly reduced captures of greater peach-tree borer. Successful disruption of lesser and greater peach-tree borer required that the specific blends most attractive to each species were provided in separate dispensers deployed on opposite sides of trees. Other problems that need to be addressed include the application timing and chemical interactions that affect the longevity of pheromone release. The spring emergence of Oriental fruit moth occurs *c.* 2 weeks before that of codling moth and *c.* 6 weeks prior to oblique-banded leaf-roller. A multispecies dispenser containing the pheromones of all three of these pests would have to be applied either prior to or after, the beginning of flight for at least one species. In addition, blending codlemone and (Z)-11–14:OAc in the same tube accelerates the release of codlemone, apparently because the acetate is a good solvent for the alcohol.

Sprayable formulations are now being tested or used commercially for a number of insect pests. It is possible that future pest-control programmes in tree-fruit crops may include sprayable products for codling moth, oblique-banded and red-banded leaf-rollers, tufted apple bud-moth and Oriental fruit moth. The advantage of this approach is that each spray can be timed to coincide with the start of flight for each pest. This may improve efficacy and be more economical, as expensive active ingredients are not wasted, contrary to when multispecies hand-applied dispensers are used.

Implementation and demonstration programmes: a key to success

A major determinant of the success of mating disruption for pest control is the extent to which growers will adopt these technologies, which are generally information-intensive, risky and expensive in comparison with conventional remedies. The best means of gaining acceptance for these sophisticated and less-certain pest-control methods is to demonstrate, on farm, that they are efficacious and economical. Investment in demonstration and implementation projects has been the key to widespread adoption of pheromone technologies in most cases where it has occurred. The rate of adoption of codling-moth mating disruption in Washington, USA, as estimated by Brunner *et al.* (2001) and J.F. Brunner (unpublished data), is illustrated in Fig. 5.13. Isomate-C® was registered in 1991 as the first mating-disruption product for codling-moth control in the USA. On-farm trials were immediately implemented by USDA and Washington State University researchers and comprised a large percentage of the acreage treated with pheromone in the first year. Parallel 3-year studies that followed the transition of apple-orchard management with conventional or organic pest-control programmes to a pheromone-based management programme

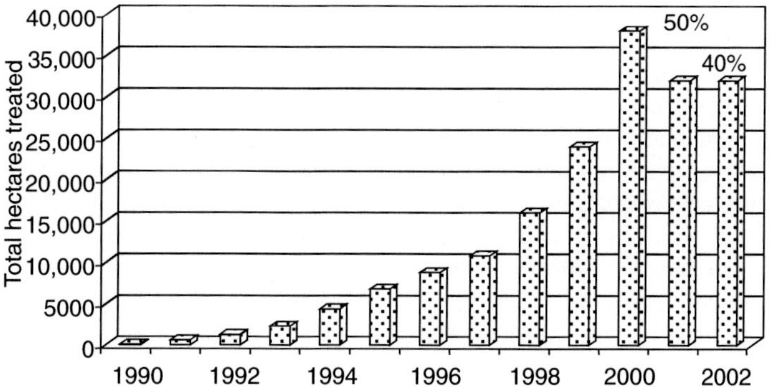

Fig. 5.13. Use of codling-moth mating-disruption products in Washington, USA (adapted from Brunner *et al.*, 2001).

provided the foundation for educating growers about the risks and benefits of this novel technology (Knight, 1995; Gut and Brunner, 1998). The number of growers trying the technique gradually increased and, in most cases, they achieved satisfactory control. Where failures did occur they were usually associated with smaller plot sizes, especially where neighbouring orchards served as a source of immigrating mated females. The USDA-sponsored CAMP was initiated in 1995 to address the issue of border effects and the need for improved monitoring and, in general, to facilitate greater adoption of mating disruption. The apple acreage under mating disruption in Washington increased dramatically over a period of 5 years, reaching a peak in 2000 at 37,000 ha, or 50% of state's total apple acreage. Perhaps the greatest achievement of this project was that the growers have continued coordinated efforts over large areas since the USDA terminated their role in the area-wide programme in 1999.

This 10-year transition to the large-scale, sustained use of mating disruption has been replicated in other production systems and in other parts of the world. The basic approach has been to establish a partnership between government or other public enterprises and producers. These demonstration sites serve as the catalyst and educational foundation for new sites in subsequent years, culminating in a large, coordinated effort between farmers and others involved in the production system. To be successful on an area-wide scale, the project must engage the infrastructure that is already in place in making management decisions. Critical participants include, among others, farmers, suppliers of control products, consultants and university extension and research personnel. A robust implementation programme involves educational programmes, including hands-on training, economic analysis, documentation of programme efficacy, regular communication between the participants and, often, some kind of financial incentive. To be successful, it must be an evolving process that develops and incorporates new control and monitoring technologies to overcome problems that are certain to arise. We believe this process is central to increased adoption of mating disruption.

Concluding Remarks

Substantial progress in developing semiochemical-based pest-management products that are commercially viable has certainly been made over the past 10 years, with mating-disruption formulations leading the way. We have no doubt that this trend will continue over the next decade. Two factors, one practical and the other biological, appear to provide the greatest impediment to the adoption of direct controls based on the use of behaviour-modifying chemicals. Mass trapping, attack-and-kill and mating disruption are expensive techniques compared with insecticidal control. It is hoped that the direct costs of the active ingredients will fall in the future. Meanwhile, the research community, farmers and others involved in commercial production systems must work together to develop cost-effective ways of using semiochemicals.

As recognized early on by Knipling (1979) and others, pest density is a very crucial limiting factor to pest control using semiochemicals. Two approaches that have successfully mitigated the problem of controlling high-density populations are area-wide programmes and the use of companion insecticide sprays. In the second approach, insecticides are used either to reduce initial population densities or as a supplement to the semiochemical-based tactic.

Wyatt (1997) proposed that, to make the task of developing mating disruption for various pests manageable, we should search for patterns and good predictors of success. We have proposed a set of criteria for predicting the success of mating disruption for a particular species or for the same species under varying conditions. Now, the challenge for us and hopefully also for others is to explore the usefulness of this scheme.

There is a heightened need for both good monitoring tools and alternatives to pesticides. As we mentioned at the outset, new regulations governing the use of pesticides and the regular occurrence of resistance to chemical controls provide a strong impetus for the adoption of behavioural controls. In addition, behavioural controls have the added advantages of negligible environmental impact and increased compatibility with biological and cultural methods of control. It is truly an exciting time to be working with semiochemicals.

References

Agnello, A.M., Spangler, S.M. and Reissig, W.H. (1990) Development and evaluation of a more efficient monitoring system for apple maggot (Diptera: Tephritidae). *Journal of Economic Entomology* 83, 539–546.

Agnello, A.M., Reissig, W.H., Spangler, S.M., Charlton, R.E. and Kain, D.P. (1996) Trap response and fruit damage by obliquebanded leafroller (Lepidoptera: Torticidae) in pheromone-treated apple orchards in New York. *Enviromental Entomology* 25, 268–282.

Akers, R.P. and O'Connell, R.J. (1988) The contribution of olfactory receptor neurones to the perception of pheromone component ratios in male redbanded leafroller moths. *Journal of Comparative Physiology* 163, 641–650.

Almaas, T.J. and Mustaparta, H. (1990) Pheromone reception in tobacco moth, *Heliothis virescens. Journal of Chemical Ecology* 16, 1331–1347.

Almaas, T.J., Christensen, T.A. and Mustaparta, H. (1991) Antennal receptor neurons encode several features of intra- and interspecific odorants in the male corn earworm moth *Helicoverpa zea. Journal of Comparative Physiology* 169, 249–258.

Atanassov, A., Shearer, P.W., Hamilton, G. and Polk, D. (2002) Development and implementation of a reduced risk peach arthropod management program in New Jersey. *Journal of Economic Entomology* 95, 803–812.

Ayyappathe, R., Poloavarapa, S. and McGuire, M. (2000) Effectiveness of thiomethoxam-coated spheres against blueberry maggot flies (Diptera: Tephritidae). *Journal of Economic Entomology* 93, 1473–1479.

Baker, T.C. and Cardé, R.T. (1979) Courtship behavior of the Oriental fruit moth (*Grapholitha molesta*): experimental analysis and consideration of the role of sexual selection in the evolution of courtship pheromones in the Lepidoptera. *Annals of the Entomological Society of America* 72, 173–188.

Baker, T.C. and Roelofs, W.L. (1981) Initiation and termination of Oriental fruit moth male response to pheromone concentrations in the field. *Environmental Entomology* 10, 211–218.

Baker, T.C., Willis, M.A., Haynes, K.F. and Phelan, P.L. (1985) A pulsed cloud of sex pheromone elicits upwind flight in male moths. *Physiological Entomology* 10, 257–265.

Baker, T.C., Hansson, B.S., Löfstedt, C. and Löfqvist, J. (1989) Adaptation of male moth antennal neurons in a pheromone plume is associated with cessation of pheromone-mediated flight. *Chemical Senses* 14, 439–448.

Baker, T.C., Dittl, T. and Mafra-Neto, A. (1997) Disruption of sex pheromone communication in the black-headed fireworm in Wisconsin cranberry marshes by using MSTRS devices. *Journal of Agricultural Entomology* 14, 449–457.

Bakke, A. and Lie, R. (1989) Mass trapping. In: Jutsum, A.R. and Gordon, R.F.S. (eds) *Insect Pheromones in Plant Protection*. Wiley, Chichester, UK, pp. 67–87.

Balken, J.A. (1980) Microencapsulation using coacervation/phase separation techniques. In: Kydonius, A.E. (ed.) *Controlled Release Technologies: Methods and Applications*, Vol. II. CRC Press, Boca Raton, Florida, pp. 83–105.

Barnes, B.N. and Blomefield, T.L. (1997) Goading growers towards mating disruption: the South African experience with *Grapholita molesta* and *Cydia pomonella* (Lepidoptera: Tortricidae). *IOBC WPRS Bulletin* 20, 45–56.

Barrett, B.A. (1995) Effect of synthetic pheromone permeation on captures of male codling moth (Lepidoptera: Tortricidae) in pheromone and virgin female moth-baited traps at different tree heights in small orchard blocks. *Environmental Entomology* 24, 1201–1206.

Bartell, R.J. (1982) Mechanisms of communication disruption by pheromone in the control of Lepidoptera: a review. *Physiological Entomology* 7, 353–364.

Bartell, R.J. and Lawrence, L.A. (1976a) Reduction in responsiveness of E*piphyyas postvittana* (Lepidoptera) to sex pheromone following pulsed pheromonal exposure. *Physiological Entomology* 2, 1–6.

Bartell, R.J. and Lawrence, L.A. (1976b) Reduction in responsiveness of male light-brown apple moths, *Epiphyas postvittana*, to sex pheromone following pulsed pre-exposure to pheromone components. *Physiological Entomology* 2, 89–95.

Bartell, R.J. and Roelofs, W.L. (1973) Inhibition of sexual response in males of the moth *Argyrotaenia velutinana* by brief exposures to synthetic pheromone or its geometrical isomer. *Journal of Insect Physiology* 19, 655–661.

Beers, E.H., Brunner, J.F., Willet, M.J. and Warner, G.M. (1993) *Orchard Pest Management: a Resource for the Pacific Northwest*. Good Fruit Grower, Yakima, Washington, 276 pp.

Borden, J.H. (1990) Use of semiochemicals to manage coniferous tree pests in Western Canada. In: Ridgway, R.L., Silverstein, R.M. and Inscoe, M.N. (eds) *Behavior-modifying Chemicals for Insect Management*. Marcel Dekker, New York, pp. 281–315.

Borden, J.H. (1997) Disruption of semiochemical-mediated aggregation in bark beetles. In: Cardé, R.T. and Minks, A.K. (eds) *Insect Pheromone Research, New Directions*. Chapman & Hall, New York, pp. 421–438.

Bradley, S.I., Walker, J.T.S., Wearing, C.H., Shaw, P.W. and Hodson, A.J. (1998) The use of pheromone traps for leafroller action thresholds in pipfruit. *Proceedings of the New Zealand Plant Protection Conference* 51, 173–178.

Brooks, T.W. (1980) Controlled vapor release from hollow fibers: theory and applications with insect pheromones. In: Kydonius, A.E. (ed.) *Controlled Release Technologies: Methods and Applications*, Vol II. CRC Press, Boca Raton, Florida, pp. 165–193.

Brunner, J., Welter, S., Calkins, C., Hilton, R., Beers, E., Dunley, J., Unruh, T., Knight, A., VanSteenwyk, R. and Van Buskirk, P. (2001) Mating disruption of codling moth: a perspective from the Western United States. *IOBC WPRS Bulletin* 25, 207–215.

Buchelos, C.T. and Levinson, A.R. (1993) Efficacy of multisurface traps and Lasiotraps with and without pheromone addition, for monitoring and mass-trapping of *Lasioderma serricorne* F. (Col., Anobiidae) in insecticide-free tobacco stores. *Journal of Applied Entomology* 116, 440–448.

Calkins, C.O., Knight, A.L., Richardson, G. and Bloem, K.A. (2000) Area-wide population suppression of codling moth. In: Tan, K.-H. (ed.) *Area-wide Control of Fruit Flies and Other Insect Pests*. Peneritg Universiti Sains Malaysia, Pulau Pinang, pp. 215–219.

Campbell, C.D., Walgenbach, J.F. and Kennedy, G.G. (1992) Comparison of black light and pheromone traps for monitoring *Helicoverpa zea* (Boddie) (Lepidoptera: Noctuidae) in tomato. *Journal of Agricultural Entomology* 9, 17–24.

Cardé, R.T. (1990) Principles of mating disruption. In: Ridgeway, R.L., Silverstein, R.M. and Inscoe, M.N. (eds) *Behavior-modifying Chemicals for Pest Management: Applications of Pheromones and Other Attractants*. Marcel Dekker, New York, pp. 47–71.

Cardé, R.T. and Elkinton, J.S. (1984) Field trapping with attractants: methods and interpretation. In: Hummel, H.E. and Miller, T.A. (eds) *Techniques in Pheromone Research*. Springer-Verlag, New York, pp. 111–129.

Cardé, R.T. and Minks, A.K. (1995) Control of moth pests by mating disruption: successes and constraints. *Annual Review of Entomology* 40, 559–585.

Cardé, R.T. and Minks, A.K. (eds) (1997) *Insect Pheromone Research, New Directions*. Chapman & Hall, New York, 684 pp.

Cardé, R.T., Baker, T.C. and Roelofs, W.L. (1975) Ethological function of components of a sex attractant system for Oriental fruit moth males, *Grapholita molesta* (Lepidoptera: Tortricidae). *Journal of Chemical Ecology* 1, 475–491.

Chapman, R.F. (1998) *The Insects: Structure and Function*. Cambridge University Press, Cambridge, UK, 770 pp.

Charmillot, P.-J. (1990) Mating disruption technique to control codling moth in western Switzerland. In: Ridgway, R.L., Silverstein, R.M. and Inscoe, M.N. (eds) *Behavior-modifying Chemicals for Insect Management*. Marcel Dekker, New York, pp. 165–182.

Charmillot, P.-J., Hofer, D. and Pasquier, D. (2000) Attract and kill: a new control of the codling moth *Cydia pomonella*. *Entomologia Experimentalis et Applicata* 94, 211–216.

Christensen, T.A., Mustaparta, H. and Hildebrand, J.G. (1989) Discrimination of sex pheromone blends in the olfactory system of the moth. *Chemical Senses* 14, 463–477.

Daly, K.C. and Figueredo, A.J. (2000) Habituation of sexual response in male *Heliothis* moths. *Physiological Entomology* 25, 180–191.

deLame, F.M., Hong, J.J., Shearer, P.W. and Brattsten, L.B. (2001) Sex-related differences in the tolerance of Oriental fruit moth, *Grapholita molesta*, to organophosphate insecticides. *Pest Management Science* 57, 827–832.

Deland, J.-P., Judd, G.J.R. and Roitberg, B.D. (1994) Disruption of pheromone communication in three sympatric leafroller (Lepidoptera: Tortricidae) pests of apple in British Columbia. *Environmental Entomology* 23, 1084–1090.

Den Otter, C.J. (1977) Single sensillum responses in the male moth *Adoxophyes orana* (F.v.R.) to female sex pheromone components and their geometrical isomers. *Journal of Comparative Physiology* 121, 205–222.

Derrick, M.E., Van Duyn, J.W., Sorenson, C.E. and Kennedy, G.C. (1992) Effect of pheromone trap placement on capture of male European corn borer (Lepidoptera: Pyralidae) in three North Carolina crops. *Environmental Entomology* 21, 240–246.

Dickens, J.C. (1979) Electrophysiological investigations of olfaction in bark beetles. *Mitteilungen der Schweizerishce Entomologische Gesellschaft* 52, 203–216.

Doane, C.D. (1999) Controlled-release devices for pheromones. In: Scher, H.B. (ed.) *Controlled-release Delivery Systems for Pesticides*. Marcel Dekker, New York, pp. 295–317.

Duan, J.J. and Prokopy, R.J. (1995) Development of pesticide-treated spheres for controlling apple maggot flies (Diptera: Tephritidae): pesticides and residue extending agents. *Journal of Economic Entomology* 88, 117–126.

Epstein, D. and Gut, L. (2000) *A Pocket Guide for IPM Scouting in Michigan Apples*. Extension Bulletin E-2720, Michigan State University, East Lansing, Michigan, 72 pp.

Epstein, D., Waldstein, D. and Edson, C. (2002) *Michigan Apple Integrated Pest Management Implementation Project. Final Narrative Report*. East Lansing, Michigan, 99 pp.

Evenden, M.L., Borden, J.H. and Van Sickle, G.A. (1995) Predictive capabilities of a pheromone-based monitoring system for western hemlock looper (Lepidoptera: Geometridae). *Environmental Entomology* 24, 933–943.

Evenden, M.L., Judd, G.J.R. and Borden, J.H. (2000) Investigations of mechanisms of pheromone communication disruption of *Choristoneura rosaceana* (Harris) in a wind tunnel. *Journal of Insect Behavior* 13, 499–510.

Faccioli, G., Antropoli, A. and Pasqualini, E. (1993) Relationship between males caught with low pheromone doses and larval infestation of *Argyrotainia pulchellana*. *Entomologia Experimentalis et Applicata* 68, 165–170.

Färbert, P., Koch, U.T., Färbert, A. and Staten, R.T. (1997) Measuring pheromone concentrations in cotton fields with the EAG method. In: Cardé, R.T. and Minks, A.K. (eds) *Insect Pheromone Research, New Directions*. Chapman & Hall, New York, pp. 347–358.

Fein, A. (1986) Excitation and adaptation of *Limulus* photoreceptors by light and inositol 1,4,5-triphosphate. *Trends in Neuroscience* 9, 110–114.

Figueredo, A.J. and Baker, T.C. (1992) Reduction of the response to sex pheromone in the Oriental fruit moth, *Grapholita molesta* (Lepidoptera: Torticidae) following successive pheromonal exposures. *Journal of Insect Behavior* 5, 347–361.

Fonta, C. and Masson, C. (1987) Structural and functional studies of the peripheral olfactory nervous system of male and female bumble bees (*Bombus hypnorum* and *B. terrestris*). *Chemical Senses* 12, 53–69.

Furlong, M.J., Pell, J.K., Choo, O.P. and Rahman, S.A. (1995) Field and laboratory evaluation of a sex pheromone trap for the autodissemination of a fungal entomopathogen *Zoophthora radicans* (Entomophthorales) by the diamondback moth *Plutella xylostella* (Lepidoptera: Yponomeutidae). *Bulletin of Entomological Research* 85, 331–337.

Gaston, L.K., Shorey, H.H. and Sario, S.A. (1967) Insect population control by the use of sex pheromones to inhibit orientation between the sexes. *Nature* 213, 1155.

Gordon, F.C. and Potter, D.A. (1985) Efficiency of Japanese beetle (Coleoptera: Scarabaeidae) traps in reducing defoliation of plants in the urban landscape and effect on larval density in turf. *Journal of Economic Entomology* 78, 774–778.

Gordon, F.C. and Potter, D.A. (1986) Japanese beetle (Coleoptera: Scarabaeidae) traps: evaluation of single and multiple arrangements for reducing defoliation in urban landscapes. *Journal of Economic Entomology* 79, 1381–1384.

Gut, L. and Brunner, J. (1996) *Implementing Codling Moth Mating Disruption in Washington Pome Fruit Orchards*. Tree Fruit Research and Extension Center Information Series 1, Washington State University, Wenatchee, Washington, 8 pp.

Gut, L. and Brunner, J. (1998) Pheromone-based management of codling moth (Lepidoptera: Tortricidae) in Washington apple orchards. *Journal of Agricultural Entomology* 15, 387–405.

Gut, L.J., Wise, J., Isaacs, R. and McGhee, P. (1999) Obliquebanded leafroller control tactics and management strategies: 1998 update. *Proceedings of the Michigan State Horticulture Society* 128, 97–106.

Hall, D.R., Nesbitt, B.F., Marrs, G.J., Green, A.StJ., Campion, D.G. and Critchley, B.R. (1982) Development of microencapsulated pheromone formulations. In: Leonhardt, B.A. and Beroza, M. (eds) *Insect Pheromone Technology: Chemistry and Applications*. ACS Symposium Series Number 190. American Chemical Society, Washington, DC, pp. 131–143.

Hardie, J. and Minks, A.K. (eds) (1999) *Pheromones of Non-lepidopteran Insects Associated with Agricultural Plants*. CAB International, Wallingford, UK, 466 pp.

Haynes, K.F., Miller, T.A., Staten, R.T., Li, W.G. and Baker, T.C. (1987) Pheromone trap for monitoring insecticide resistance in the pink bollworm moth (Lepidoptera: Gelechiidae), new tool for resistance management. *Environmental Entomology* 16, 84–89.

Howell, J.F. (1992) Control of codling moth (*Cydia pomonella*) using pheromone to disrupt mating (Lepidoptera: Tortricidae). In: *Proceedings XIX International Congress of Entomology*. Beijing, p. 202.

Hu, X.P., Shasha, B.S., McGuire, M.R. and Prokopy, R.J. (1998) Controlled release of sugar and toxicant from a novel device for controlling pest insects. *Journal of Controlled Release* 50, 257–265.

Il'ichev, A. and Sexton, S.B. (2002) Reduced application rates of mating disruption for effective control of Oriental fruit moth *Grapholita molesta* (Busck) (Lepidoptera: Tortricidae) on pears. *General and Applied Entomology* 31, 47–51.

Il'ichev, A.L., Gut, L.J., Hossain, M.S., Williams, D.G. and Jerie, P.H. (2002) Area-wide approach for improved control of Oriental fruit moth, *Grapholita molesta* (Busck) (Lepidoptera: Tortricidae) by mating disruption. *General and Applied Entomology* 31, 7–15.

Isaacs, R., Ulczynscki, M., Wright, B., Gut, L.J. and Miller, J.R. (1999) Performance of a microsprayer specifically designed for pheromone-mediated control of insect pests. *Journal of Economic Entomology* 92, 1157–1164.

Jackson, D.M., Brown, G.C., Nordin, G.L. and Johnson, D.W. (1992) Autodissemination of a baculovirus for management of tobacco budworms (Lepidoptera: Noctuidae) on tobacco. *Journal of Economic Entomology* 85, 710–719.

James, D.G., Bartelt, R.J. and Moore, C.J. (1996) Mass-trapping of *Carpophilus* spp. (Coleoptera: Nitidulidae) in stone fruit orchards using synthetic aggregation pheromones and a coattractant: development of a strategy for population suppression. *Journal of Chemical Ecology* 22, 1541–1556.

Jenkins, J.W., Doane, C.C., Schuster, D.J., McLaughlin, J.R. and Jimenez, M.J. (1990) Development and commercial application of sex pheromone for control of the tomato pinworm. In: Ridgway, R.L., Silverstein, R.M. and Inscoe, M.N. (eds) *Behavior-modifying Chemicals for Insect Management*. Marcel Dekker, New York, pp. 269–280.

Johnson, R.L. and Schall, R.A. (1989) Early detection of new pests. In: Kahn, R.P. (ed.) *Plant Protection and Quarantine*, Vol. III. CRC Press, Boca Raton, Florida, pp. 105–116.

Jones, O.T. (1998) Practical applications of pheromones and other semiochemicals. In: House, P.E., Stevens, I.D.R. and Jones, O.T. (eds) *Insect Pheromones and Their Use in Pest Management*. Chapman & Hall, London, pp. 261–355.

Jones, O.T. and Casagrande, E.D. (2000) The use of semichemical-based devices and formulations in area-wide programmes: a commercial perspective. In: Tan, K.-H. (ed.) *Area-wide Control of Fruit Flies and Other Insect Pests*. Peneritg Universiti Sains Malaysia, Pulau Pinang, pp. 285–293.

Judd, G.J.R., Gardiner, M.G.T. and Thomson, D.R. (1996) Commercial treals of pheromone-mediated mating disruption with Isomate-C® to control codling moth in British Columbia apple and pear orchards. *Journal of the Entomological Society of British Columbia* 92, 23–34.

Jutsum, A.R. and Gordon, R.F.S. (eds) (1989) *Insect Pheromones in Plant Protection*. Wiley, Chichester, UK, 369 pp.

Kaissling, K.-E. (1986) Chemo-electrical transduction in insect olfactory receptors. *Annual Review of Neuroscience* 9, 121–145.

Karg, G. and Sauer, A.E. (1995) Spatial distribution of pheromones in fields treated for mating disruption of the European grape vine moth *Lobesia botrana* measured with electroantennograms. *Journal of Chemical Ecology* 21, 1299–1314.

Karg, G. and Sauer, A.E. (1997) Seasonal variation of pheromone concentration in mating disruption trials against European grape vine moth, *Lobesia botrana* (Lepidoptera: Tortricidae) measured by EAG. *Journal of Chemical Ecology* 23, 487–501.

Karg, G., Sauer, A.E. and Koch, U.T. (1990) The influence of plants on the development of pheromone atmospheres measured by EAG method. In: Elsner, N. and Roth G. (eds) *Brain-Perception-Cognition, Proceedings of the 18th Gottingen Neurobiology Conference.* Thieme Verlag, Stuttgart, p. 301.

Karg, G., Suckling, D.M. and Bradley, S.J. (1994) Adsorption and release of pheromone of *Epiphyas postvittana* (Lepidoptera: Tortricidae) by apple leaves. *Journal of Chemical Ecology* 20, 1825–1841.

Karlson, P. and Lüscher, M. (1959) 'Pheromones': a new term for a class of biologically active substances. *Nature* 183, 55–56.

Klein, M.G. and Lacey, L.A. (1999) An attractive trap for autodissemination of entomopathogenic fungi into populations of Japanese beetle, *Popillia japonica* (Coleoptera: Scarabaeidae). *Biocontrol Science and Technology* 9, 151–158.

Klein, M.O. (1981) Mass trapping for suppression of Japanese beetles. In: Mitchell, E.R. (ed.) *Management of Insect Pests with Semiochemicals.* Plenum Press, New York, pp. 183–190.

Klein, U. (1987) Sensillum-lymph proteins from antennal olfactory hairs of the moth *Antheraea polyphemus* (Saturniidae): *Insect Biochemistry* 8, 1193–1204.

Knight, A.L. (1995) The impact of codling moth (Lepidoptera: Tortricidae) mating disruption on apple pest management in Yakima Valley, Washington. *Journal of the Entomological Society of British Columbia* 92, 29–38.

Knight, A.L. (1997) Mating disruption of leafrollers in Washington's apple orchards. *La Difesa delle Piante* 20, 73–77.

Knight, A.L. and Croft, B.A. (1991) Modelling and prediction technology. In: van der Geest, L.P.S. and Evenhuis, H.H. (eds) *Tortricid Pests: Their Biology, Natural Enemies and Control.* World Crop Pests Volume 5. Elsevier, Amsterdam, pp. 301–312.

Knight, A.L. and Hull, L.A. (1988) Area-wide population dynamics of *Platynota idaeusalis* (Lepidoptera: Tortricidae) in Southcentral Pennsylvania pome and stone fruits. *Environmental Entomology* 17, 1000–1008.

Knight, A.L. and Hull, L.A. (1989) Use of sex pheromone traps to monitor azinphosmethyl resistance in tufted apple bud moth (Lepidoptera: Tortricidae). *Journal of Economic Entomology* 82, 1019–1026.

Knight, A.L. and Turner, J.E. (1999) Mating disruption of *Pandemis* spp. (Lepidoptera: Torticidae). *Environmental Entomology* 28, 81–87.

Knight, A.L. and Weissling, T.J. (1999) Behavior-modifying chemicals in management of arthropod pests. In: Ruberson, J.R. (ed.) *Handbook of Pest Management.* Marcel Dekker, New York, pp. 521–545.

Knight, A.L., Thomson, D.R. and Cockfield, S.D. (1998) Developing mating disruption of obliquebanded leafroller (Lepidoptera: Tortricidae) in Washington State. *Environmental Entomology* 27, 1080–1088.

Knipling, E.F. (1979) *The Basic Principles of Insect Populations Suppression and Management.* Agriculture Handbook 512, USDA, Washington, DC, 659 pp.

Koch, U.T., Cardé, A.M. and Cardé, R.T. (2002) Calibration of an EAG system to measure airborne concentration of pheromone formulated for mating disruption of the pink bollworm moth, *Pectinophora gossypiella* (Saunders) (Lep., Gelechiidae). *Journal of Applied Entomology* 126, 431–435.

Koutek, B., Hoskovec, M., Vrkocova, P., Konecny, K. and Feltl, L. (1994) Gas chromatographic determination of vapor pressures of pheromone-like compounds II. Alcohols. *Journal of Chromatography* 679, 307–317.

Koutek, B., Hoskovec, M., Vrkocova, P., Konecny, K., Feltl, L. and Vrkoc, J. (1996) Gas chromatographic determination of vapor pressures of pheromone-like compounds III. Aldehydes. *Journal of Chromatography* 719, 391–400.

Koutek, B., Hoskovec, M., Vrkocova, P. and Feltl, L. (1997) Gas chromatographic determination of vapor pressures of pheromone-like compounds IV. Acetates, a reinvestigation. *Journal of Chromatography* 759, 93–109.

Kuenen, L.P.S. and Baker, T.C. (1981) Habituation versus sensory adaptation as the cause of reduced attraction following pulsed and constant sex pheromone pre-exposure in *Trichoplusia ni*. *Journal of Insect Physiology* 27, 721–726.

Kydonieus, A.F. and Beroza, M. (1982) *Insect Suppression with Controlled Release Pheromone Systems*, Vols 1 and 2. CRC Press, Boca Raton, Florida.

Lanier, G.N. (1990) Principles of attraction–annihilation. In: Ridgway, R.L., Silverstein, R.M. and Inscoe, M.N. (eds) *Behavior-modifying Chemicals for Insect Management.* Marcel Dekker, New York, pp. 25–45.

Latheef, M.A., Witz, J.A. and Lopez, J.D. Jr (1991) Relationships among pheromone trap catches of male corn earworm moths (Lepidoptera: Noctuidae), egg numbers, and phenology in corn. *Canadian Entomologist* 123, 271–281.

Lawson, D.S., Reissig, W.H., Agnello, A.M., Nyrop, J.P. and Roelofs, W.L. (1996) Interference with the mate-finding communication system of the obliquebanded leafroller (Lepidoptera: Tortricidae) using synthetic sex pheromones. *Environmental Entomology* 25, 895–905.

Liburd, O.E., Gut, L.J., Stelinski, L.L., Whalon, M.E., McGuire, M.R., Wise, J.C., Willett, J.L., Hu, X.P. and Prokopy, R.J. (1999) Mortality of *Rhagoletis* species encountering pesticide-treated spheres (Diptera: Tephritidae). *Journal of Economic Entomology* 92, 1151–1156.

Light, D.M., Knight, A.K., Henrick, C.A., Rajapaska, D., Lingren, B., Dickens, J.C., Reynolds, K.M., Buttery, R.G., Merrill, G., Roitman, J. and Campbell, B.C. (2001) A pear-derived kairomone with pheromonal potency that attracts male and female codling moth, *Cydia pomonella* (L.). *Naturwissenschaften* 88, 333–338.

Lindgren, B.S. and Borden, J.H. (1993) Displacement and aggregation of mountain pine beetles, *Dendroctonus ponderosae* (Coleoptera: Scolytidae) in response to their antiaggregation and aggregation pheromones. *Canadian Journal of Forest Research* 23, 286–290.

Linn, C.E. Jr and Roelofs, W.L. (1981) Modification of sex pheromone blend discrimination in male Oriental fruit moths by pre-exposure to (E)-8-dodecenyl acetate. *Physiological Entomology* 6, 421–429.

Linn, C.E. Jr and Roelofs, W.L. (1983) Effect on varying proportions of the alcohol component on sex pheromone blend discrimination in male Oriental fruit moths. *Physiological Entomology* 8, 291–306.

Linn, C.E., Bjostad, L.B., Du, J.W. and Roelofs, W.L. (1984) Redundancy in a chemical signal: behavioral responses of male *Trichoplusia ni* to a 6-component sex pheromone blend. *Journal of Chemical Ecology* 10, 1635–1658.

Mafra-Neto, A. and Baker, T.C. (1996) Timed, metered sprays of pheromone disrupt mating of *Cauda cautella* (Lepidoptera: Pyralidae). *Journal of Agricultural Entomology* 13, 149–168.

Mayer, M.S. and McLaughlin, J.R. (1991) *Handbook of Insect Pheromones and Sex Attractants*. CRC Press, Boca Raton, Florida, 1083 pp.

McBrien, H., Judd, G.J.R. and Borden, J.H. (1994) *Campylomma verbasci* (Heteroptera: Miridae): pheromone-based seasonal flight patterns and prediction of nymphal densities in apple orchards. *Journal of Economic Entomology* 23, 1224–1229.

McDonough, L.M., Brown, D.F. and Aller, W.C. (1989) Insect sex pheromones – effect of temperature on evaporation rates of acetates from rubber septa. *Journal of Chemical Ecology* 15, 779–790.

Metcalf, R.L. and Metcalf, E.R. (1992) *Plant Kairomones in Insect Ecology and Control*. Chapman & Hall, New York, 168 pp.

Millar, J.G. (1995) Degradation and stabilization of E8,E10-dodecadienol, the major component of the sex pheromone of the codling moth (Lepidoptera: Tortricidae). *Journal of Economic Entomology* 88, 1425–1432.

Miller, J.R. (2004) Adsorptive interactions of insect pheromone-like compounds with inorganic solid surfaces: quantifying partitioning coefficients by gas–solid chromatography. *Journal of Chemical Ecology* (in review).

Miller, J.R. and Cowles, R.S. (1990) Stimulo-deterrent diversion: a concept and its possible application to onion maggot control. *Journal of Chemical Ecology* 16, 3197–3212.

Mitchell, E.R. (ed.) (1981) *Management of Insect Pests with Semiochemicals*. Plenum, New York, 514 pp.

Mitchell, E.R., Copeland, W.W., Sparks, A.N. and Sekul, A.A. (1974) Fall armyworm: disruption of pheromone communication with synthetic acetates. *Environmental Entomology* 3, 778–780.

Morewood, P., Gries, G., Liska, Kapitola, P., Häußler, D., Möller, K. and Bogenshutz, H. (2000) Towards pheromone-based monitoring in nun moth, *Lymantria monacha* (L.) (Lep., Lymantriidae) populations. *Journal of Applied Entomology* 124, 77–85.

Morrison, R.J. and Boyd, R.N. (1974) *Organic Chemistry*, 3rd edn. Allin and Bacon, Boston, Massachusetts.

Morse, B.W. and Kulman, H.M. (1985) Monitoring damage by yellowheaded spruce sawflies with sawfly and parasitoid pheromones. *Environmental Entomology* 14, 131–133.

Murlis, J., Elkinton, J.S. and Cardé, R.T. (1992) Odor plumes and how insects use them. *Annual Review of Entomology* 37, 505–532.

Nagata, K. (1989) Revue: pest control by mating disruption in Japan. *Japanese Pesticide Information* 54, 3–6.

Noldus, L.P.J., Potting, R.P.J. and Barendrengt, H.E. (1991) Moth sex pheromone adsorption to leaf surfaces: bridges in time for chemical spies. *Physiological Entomology* 6, 71–86.

Nordlund, D.A., Jones, R.L. and Lewis, W.J. (1981) *Semiochemicals: Their Role in Pest Control.* John Wiley & Sons, New York, 306 pp.

Norris, R.F., Caswell-Chen, E.P. and Kogan, M. (2003) *Concepts in Integrated Pest Management.* Prentice Hall, Upper Saddle River, New Jersey, 586 pp.

Novak, M.A., Reissig, W.H. and Roelofs, W.L. (1978) Orientation disruption of *Argyrotaenia velutinana* and *Choristoneura rosaceana* (Lepidoptera: Tortricidae) male moths. *Journal of the New York Entomological Society* 4, 311–315.

O'Connell, R.J. (1972) Response of olfactory receptors to the sex attractant, its synergist and inhibitor in the redbanded leaf roller, *Argyrotaenia velutinana.* In: Schneider, D. (ed.) *International Symposium Olfaction Taste IV*, Wissenschaftliche Verlagsgesellschaft, Stuttgart, pp. 180–186.

O'Connell, R.J. (1981) The encoding of behaviorally important odorants by insect chemosensory neurons. In: Norris, D.M. (ed.) *Perception of Behavioral Chemicals.* Elsevier/North-Holland Biomedical Press, New York, pp. 133–163

O'Connell, R.J. (1985) Responses to pheromone blends in insect olfactory neurons. *Journal of Comparative Physioliology* 156, 747–761.

Ogawa, K. (1990) Commercial development: mating disruption of tea tortrix moths. In: Ridgway, R.L., Silverstein, R.M. and Inscoe, M.N. (eds) *Behavior-modifying Chemicals for Insect Management.* Marcel Dekker, New York, pp. 547–551.

Park, J.D. and Goh, H.G. (1992) Control of beet armyworm, *Spodoptera exigua* Hubner (Lepidoptera: Noctuidae), using synthetic sex pheromone. I. Control by mass trapping in *Allium fistulosum* field. *Korean Journal of Applied Entomology* 31, 45–49.

Pfeiffer, D.G., Kaakeh, W., Killian, J.C., Lachance, M.W. and Kirsch, P. (1993) Mating disruption to control damage by leafrollers in Virginia apple orchards. *Entomologia Experimentalis et Applicata* 67, 47–56.

Prokopy, R.J., Wright, S.E., Black, J.L., Hu, X.P. and McGuire, M.R. (2000) Attracticidal spheres for controlling apple maggot flies: commercial-orchard trials. *Entomologia Experimentalis et Applicata* 97, 293–299.

Reissig, W.H., Novak, M. and Roelofs, W.L. (1978) Orientation disruption of *Argyrotaenia velutinana* and *Choristoneura rosaceana* male moths. *Environmental Entomolology* 7, 631–635.

Rice, R.E. and Kirsch, P. (1990) Mating disruption of Oriental fruit moth in the United States. In: Ridgway, R.L., Silverstein, R.M. and Inscoe, M.N. (eds) *Behavior-modifying Chemicals for Insect Management.* Marcel Dekker, New York, pp. 193–211.

Ridgway, R.L., Silverstein, R.M. and Inscoe, M.N. (eds) (1990) *Behavior-modifying Chemicals for Insect Management.* Marcel Dekker, New York, 761 pp.

Riedl, H., Hoying, S.A., Barnett, W.W. and Detar, J.E. (1979) Relationship of within-tree placement of the pheromone trap to codling moth catches. *Environmental Entomology* 8, 765–769.

Riedl, H., Seaman, A. and Henrie, F. (1985) Monitoring susceptibility to azinphosmethyl in field populations of the codling moth (Lepidoptera: Tortricidae) with pheromone traps. *Journal of Economic Entomology* 78, 692–699.

Riedl, H., Howell, J.F., McNally, P.S. and Westigard, P.H. (1986) *Codling Moth Management, Use and Standardization of Pheromone Trapping Systems.* Bulletin 1918, University of California, Berkeley, California, 23 pp.

Riherd, C., Nguyen, R. and Brazzel, J.R. (1994) Pest free areas. In: Sharpe, J.L. and Hallman, G.J. (eds) *Quarantine Treatments for Pests of Food Plants.* Westview Press, Boulder, Colorado, pp. 213–223.

Ritter, F.J. (eds) (1979) *Chemical Ecology: Odour Communication in Animals.* Elsevier/North-Holland, Amsterdam, 427 pp.

Roelofs, W.H. and Novak, M. (1981) Small-plot disorientation tests for screening potential mating disruptants. In: Mitchell, E.R. (ed.) *Management of Insect Pests with Semiochemicals: Concepts and Practice.* Plenum, New York, pp. 229–242.

Roelofs, W.L., Glass, E.H., Tette, J. and Comeau, A. (1970) Sex pheromone trapping for red-banded leaf roller control: theoretical and actual. *Journal of Economic Entomology* 63, 1162–1167.

Rothschild, G.H.L. (1975) Control of Oriental fruit moth (*Cydia molesta* [Busck][Lepidoptera, Tortricidae]) with synthetic female pheromone. *Bulletin of Entomological Research* 65, 473–490.

Rothschild, G.H.L. (1981) Mating disruption of lepidopterous pests: current status and future prospects. In: Mitchell, E.R. (ed.) *Management of Insect Pests with Semiochemicals: Concepts and Practice.* Plenum Press, New York, pp. 207–228.

Rumbo, E.R. and Vickers, R.A. (1997) Prolonged adaptation as possible mating disruption mechanism in Oriental fruit moth, *Cydia (=Grapholita) molesta. Journal of Chemical Ecology* 23, 445–457.

Rybczynski, R., Reagan, J. and Lerner, M.R. (1989) A pheromone-degrading aldehyde oxidase in the antennae of the moth *Manduca sexta*. *Journal of Neuroscience* 9, 1341–1353.

Sanders, C.J. (1985) Disruption of spruce budworm, *Choristoneura fumiferana* (Lepidoptera: Tortricidae), mating in a wind tunnel by synthetic pheromone: role of habituation. *Canadian Entomologist* 117, 391–393.

Sanders, C.J. (1988) Monitoring spruce budworm population density with sex pheromone traps. *Canadian Entomologist* 120, 175–183.

Sanders, C.J. (1997) Mechanisms of mating disruption in moths. In: Cardé, R.T. and Minks, A.K. (eds) *Insect Pheromone Research, New Directions*. Chapman & Hall, New York, pp. 333–346.

Sanders, C.J. and Lucuik, G.S. (1996) Disruption of male Oriental fruit moth to calling females in a wind tunnel by different concentrations of synthetic pheromone. *Journal of Chemical Ecology* 22, 1971–1986.

Sauer, A.E. and Karg, G. (1998) Variables affecting the pheromone concentration in vineyards treated for mating disruption of grape vine moth. *Journal of Chemical Ecology* 24, 289–302.

Schlyter, F. and Birgersson, G.A. (1999) Forest beetles. In: Hardie, J. and Minks, A.K. (eds) *Pheromones of Non-lepidopteran Insects Associated with Agricultural Plants*. CAB International, Wallingford, UK, pp. 113–148.

Schmitz, V., Renou, M., Roehrich, R., Stockel, J. and Lecharpentier, P. (1997) Disruption mechanisms in the European grape moth *Lobesia botrana* Den & Schiff. III. Sensory adaptation and habituation. *Journal of Chemical Ecology* 23, 83–95.

Schneider, D. (1956) Electrophysical investigation on the antennal receptors of the silk moth during chemical and mechanical stimulation. *Experientia* 13, 89–91.

Schneider, D. (1962) Electrophysical investigation on the olfactory specificity of sexual attracting substance in different species of moths. *Journal of Insect Physiology* 8, 15–30.

Schwalbe, C.P. and Mastro, V. (1990) Use of pheromones and attractants by government agencies in the United States. In: Ridgway, R.L., Silverstein, R.M. and Inscoe, M.N. (eds) *Behavior-Modifying Chemicals for Insect Management*. Marcel Dekker, New York, pp. 619–630.

Sexton, S.B. and Il'ichev, A. (2000) Pheromone mating disruption with reference to Oriental fruit moth *Grapholita molesta* (Busck) (Lepidoptera: Tortricidae): literature review. *General and Applied Entomology* 29, 63–68.

Shapas, T.J., Burkholder, W.E. and Boush, G.M. (1977) Population suppression of *Trigoderma glabrum* by using pheromone luring for protozoan pathogen dissemination. *Journal of Economic Entomology* 70, 469–474.

Shearer, P.W. and Riedl, H. (1994) Comparison of pheromone trap bioassays for monitoring insecticide resistance in *Phyllonorycter elmaella* (Lepidoptera: Gracillariidae). *Journal of Economic Entomology* 87, 1450–1454.

Shearer, P.W. and Usmani, K.A. (2001) Sex-related response to organophosphorous and carbamate insecticides in adult Oriental fruit moth, *Grapholita molesta*. *Pest Management Science* 57, 822–826.

Shorey, H.H. and Gerber, R.G. (1996a) Use of puffers for disruption of sex pheromone communication among navel orangeworm moths (Lepidoptera:Pyralidae) in almonds, pistachios, and walnuts. *Environmental Entomology* 25, 1154–1157.

Shorey, H.H. and Gerber, R.G. (1996b) Use of puffers for disruption of sex pheromone communication of codling moths (Lepidoptera: Tortricidae) in walnut orchards. *Environmental Entomology* 25, 1398–1400.

Shorey, H.H. and McKelvey, J.J. (1977) *Chemical Control of Insect Behavior*. John Wiley & Sons, New York, 414 pp.

Shorey, H.H., Kaae, R.S. and Gaston, L.K. (1974) Sex pheromones of Lepidoptera: development of a method for pheromonal control of *Pectinophora gossypiella* in cotton. *Journal of Economic Entomology* 67, 347.

Shorey, H.H., Sisk, C.B. and Gerber, R.G. (1996) Widely separated pheromone release sites for disruption of sex pheromone communication in two species of Lepidoptera. *Environmental Entomology* 25, 446–451.

Simandl, J. and Anderbrant, O. (1995) Spatial distribution of flying *Neodiprion sertifer* (Hymenoptera, Diprionidae) males in a mature *Pinus sylvestris* stand as determined by pheromone trap catch. *Scandinavian Journal of Forest Research* 10, 51–55.

Snow, J.W. (1990) Peachtree borer and lesser peachtree borer control in the United States. In: Ridgway, R.L., Silverstein, R.M. and Inscoe, M.N. (eds) *Behavior-modifying Chemicals for Insect Management*. Marcel Dekker, New York, pp. 241–253.

Stanley, B.H., Reissig, W.H., Roelofs, W.L., Schwarz, M.R. and Shoemaker, C.A. (1987) Timing treatments for apple maggot (Diptera: Tephritidae) control using sticky sphere traps baited with synthetic apple volatiles. *Journal of Economic Entomology* 80, 1057–1063.

Staten, R.T., El-Lissy, O. and Antilla, L. (1997) Successful area-wide program to control pink bollworm by mating disruption. In: Cardé, R.T. and Minks, A.K. (eds) *Insect Pheromone Research, New Directions*. Chapman & Hall, New York, pp. 383–396.

Stelinski, L.L. and Liburd, O.E. (2001) Evaluation of various deployment strategies of imidacloprid-treated spheres in highland blueberries for control of *Rhagoletis mendax* (Diptera: Tephritidae). *Journal of Economic Entomology* 94, 905–910.

Stelinski, L.L., Liburd, O.E., Wright, S., Prokopy, R.J., Behle, R. and McGuire, M.R. (2001) Comparison of neonicotinoid insecticides for use with biodegradable and wooden spheres for control of *Rhagoletis* species (Diptera: Tephritidae). *Journal of Economic Entomology* 94, 1142–1150.

Stelinski, L.L., Miller, J.R. and Gut, L.J. (2003a) Presence of long-lasting peripheral adaptation in the obliquebanded leafroller, *Choristoneura rosaceana* and absence of such adaptation in the redbanded leafroller, *Argyrotaenia velutinana*. *Journal of Chemical Ecology* 29, 405–423.

Stelinski, L.L., Gut, L.J. and Miller, J.R. (2003b) Concentration of air-borne pheromone required for long-lasting peripheral adaptation in the obliquebanded leafroller, *Choristoneura rosaceana*. *Physiological Entomology* 29, 97–107.

Stengl, M., Hatt, H. and Breer, H. (1992) Peripheral processes in insect olfaction. *Annual Review of Physiology* 54, 665–681.

Suckling, D.M. and Brockerhoff, E.G. (1999) Control of light brown apple moth (Lepidoptera: Tortricidae) using an attracticide. *Journal of Economic Entomology* 92, 367–372.

Suckling, D.M. and Karg, G. (2000) Pheromones and other semiochemicals. In: Rechcigl, J. and Rechcigl, N. (eds) *Biological and Biotechnological Control of Insect Pests*. CRC Press, Boca Raton, Florida, pp. 63–99.

Suckling, D.M. and Shaw, P.W. (1992) Conditions that favor mating disruption of *Epiphyas postvittana* (Lepidoptera: Tortricidae): *Environmental Entomology* 21, 949–956.

Suckling, D.M., Penman, D.R., Chapman, R.B. and Wearing, C.H. (1985) Pheromone use in insecticide resistance surveys of the light brown apple moth *Epiphyas postvittana*. *Journal of Economic Entomology* 78, 204–206.

Suckling, D.M., Karg, G. and Bradley, S.J. (1996) Apple foliage enhances mating disruption of lightbrown apple moth. *Journal of Chemical Ecology* 22, 325–341.

Swenson, D.W. and Weatherston, I. (1989) Hollow-fiber controlled-release systems. In: Jutsum, A.R. and Gordon, R.F.S. (eds) *Insect Pheromones in Plant Protection*. Wiley, Chichester, UK, pp. 173–197.

Tamaki, Y. (1979) Multi-component sex pheromone of Lepidoptera with special reference to *Adoxophyes* sp. In: Ritter, F.J. (ed.) *Chemical Ecology: Odour Communication in Animals*. Elsevier/North-Holland, Amsterdam, pp. 169–180.

Taschenberg, E.F., Cardé, R.T. and Roelofs, W.L. (1974) Sex pheromone mass trapping and mating disruption for control of redbanded leafroller and grape berry moths in vineyards. *Environmental Entomology* 3, 239–242.

Tatsuki, S. (1990) Application of the sex pheromone of the rice stem borer moth, *Chilo suppressalis*. In: Ridgway, R.L., Silverstein, R.M. and Inscoe, M.N. (eds) *Behavior-modifying Chemicals for Insect Management*. Marcel Dekker, New York, pp. 387–406.

Thayer, G.R. (2002) The sampling range of pheromone traps baited with lures of different strengths for monitoring obliquebanded leafroller, *Choristoneura rosaceana* Harris (Lepidoptera: Tortricidae). MS thesis, Washington State University, Pullman, Washington, 55 pp.

Thomson, D.R., Gut, L.J. and Jenkins, J.W. (1998) Pheromones for insect control. In: Hall, F.R. and Menn, J.J. (eds) *Methods in Biotechnology*, Vol. 5: *Biopesticides: Use and Delivery*. Humana Press, Totowa, New Jersey, pp. 385–412.

Todd, J.L. and Baker, T.C. (1997) The cutting edge of insect olfaction. *American Entomologist* 43, 174–182.

Trematerra, P. and Battaini, F. (1987) Control of *Ephestia kuehniella* Zeller by mass-trapping. *Journal of Applied Entomology* 104, 336–340.

Trimble, R.M. (1995) Mating disruption for control of codling moth *Cydia pomonella* (L.) (Lepidoptera: Tortricidae), in organic apple production in Southwestern Ontario. *Canadian Entomologist* 127, 493–505.

Van Steenwyk, R.A., Oatman, E.R. and Wyman, J.A. (1983) Density treatment level for tomato pinworm (Lepidoptera: Gelechiidae) based on pheromone trap catches. *Journal of Economic Entomology* 76, 440–445.

Varela, L.G., Shearer, P.W., Jones, V.P., Riedl, H. and Welter, S.C. (1997) Monitoring of insecticide resistance in *Phyllonorycter mespilella* (Lepidoptera: Gracillariidae) in four western states. *Journal of Economic Entomology* 90, 252–260.

Vega, F.E., Dowd, P.F. and Bartelt, R.J. (1995) Dissemination of microbial agents using an autoinoculating device and several insect species as vectors. *Biological Control* 5, 545–552.

Vickers, R.A. (1990) Oriental fruit moth in Australia and Canada. In: Ridgway, R.L., Silverstein, R.M. and Inscoe, M.N. (eds) *Behavior-modifying Chemicals for Insect Management*. Marcel Dekker, New York, pp. 183–192.

Vickers, R.A., Thwaite, W.G., Williams, D.G. and Nicholas, A.H. (1998) Control of codling moth in small plots by mating disruption: alone and with limited insecticide. *Entomologia Experimentalis et Applicata* 86, 229–239.

Vogt, R.G. and Riddiford, L.M. (1986) Pheromone reception: a kinetic equilibrium. In: Payne, T.L., Birch, M.C. and Kennedy, C.E.J. (eds) *Mechanisms in Insect Olfaction*. Clarendon Press, Oxford, pp. 201–208.

Waldstein, D.W. and Gut, L.J. (2003a) Comparison of microcapsule density with various apple tissues and formulations of Oriental fruit moth (Lepidoptera: Tortricidae) sprayable pheromone. *Journal of Economic Entomology* 96, 58–63.

Waldstein, D.W. and Gut, L.J. (2003b) Effects of rain and sunlight on Oriental fruit moth (Lepidoptera: Tortricidae) microcapsules applied to apple foliage. *Pest Management Science* (in review).

Wall, C. (1989) Monitoring and spray timing. In: Jutsum, A.R. and Gordon, R.F.S. (eds) *Insect Pheromones in Plant Protection*. Wiley, Chichester, UK, pp. 39–66.

Wall, C. and Perry, J.N. (1983) Further observations on the responses of male pea moth, *Cydia nigricana*, to vegetation previously exposed to sex-attractant. *Entomologia Experimentalis et Applicata* 33, 112–116.

Wall, C. and Perry, J.N. (1987) Range of action of moth sex-attractant sources. *Entomologia Experimentalis et Applicata* 44, 5–14.

Wall, C., Sturgeon, D.M., Greenway, A.R. and Perry, J.N. (1981) Contamination of vegetation with synthetic sex attractant released from traps for the pea moth, *Cydia nigricana*. *Entomologia Experimentalis et Applicata* 30, 111–115.

Wawrzynski, R.P. and Ascerno, M.E. (1998) Mass trapping for Japanese beetle (Coleoptera: Scaraaeidae) suppression in isolated areas. *Journal of Arboriculture* 24, 303–307.

Weatherston, I. (1990) Principles of design of controlled-release formulations. In: Ridgway, R.L., Silverstein, R.M. and Inscoe, M.N. (eds) *Behavior-modifying Chemicals for Insect Management*. Marcel Dekker, New York, pp. 93–112.

Weisling, T.J. and Knight, A.K. (1995) Vertical distribution of codling moth adults in pheromone-treated and untreated plots. *Entomologia Experimentalis et Applicata* 77, 271–275.

Welch, S.M., Croft, B.A. and Michels, M.F. (1981) Validation of pest management models. *Environmental Entomology* 10, 425–432.

Whittaker, R.H. and Feeney, P. (1971) Allelochemicals: chemical interactions between species. *Science* 171, 757–770.

Wood, D.L., Silverstein, R.M. and Nakajima, M. (1970) *Control of Insect Behavior by Natural Products*. Academic Press, New York, 345 pp.

Wyatt, T.D. (1997) Putting pheromones to work: paths forward for direct control. In: Cardé, R.T. and Minks, A.K. (eds) *Insect Pheromone Research, New Directions*. Chapman & Hall, New York, pp. 445–459.

Wyman, J.A. (1979) Effect of trap design and sex attractant release on tomato pinworm catches. *Journal of Economic Entomology* 72, 865–868.

Zhang, G.-F., Meng, S.-Z., Han, Y. and Sheng, C.-F. (2002) Chinese tortrix *Cydia trasisas* (Lepidoptera: Olethreutidae): suppression on street-planting trees by mass trapping with sex pheromone traps. *Environmental Entomology* 31, 602–607.

Zufall, F. and Leinders-Zufall, T. (2000) The cellular and molecular basis of odor adaptation. *Chemical Senses* 25, 473–481.

6 Transgenic Insecticidal Cultivars in Integrated Pest Management: Challenges and Opportunities

Julio S. Bernal, Jarrad Prasifka, M. Sétamou and K.M. Heinz
Department of Entomology, Biological Control Laboratory, Texas A&M University, College Station, TX 77843-2475, USA
E-mail: juliobernal@tamu.edu

Introduction

Transgenic insect-resistant cultivars were first developed in the mid-1980s and were commercially available starting in the mid-1990s (Hilder *et al.*, 1987; Vaeck *et al.*, 1987). Transgenic potato cultivars expressing a toxic protein derived from the bacterium *Bacillus thuringiensis* Berliner (hereafter *Bt*) were available to farmers in 1995, while maize and cotton varieties also expressing *Bt* toxins were available 1 year later. All three commercial introductions were made in the USA, but currently a number of transgenic cultivars are planted in different countries worldwide (e.g. cotton in Australia and Mexico). Thus far, development of insect-resistant cultivars has followed two broad approaches. One approach seeks to develop transgenic cultivars by incorporating genes of plant origin that express proteins that interfere with protein and sugar metabolism, whereas the other seeks to incorporate genes of bacterial origin, largely from *Bt*, which are acutely toxic to insects (Hilder *et al.*, 1987; Vaeck *et al.*, 1987). Crop losses due to insect pests account for a substantial portion of total crop losses worldwide. One estimate indicates that *c.* 13% of total crop production is lost annually to insect pests in the USA (Pimentel *et al.*, 1993). However, perhaps more significant than the extent of losses is the failure, despite ample research investments, to make any substantial progress towards reducing these losses. Losses due to insect pests in the first half of the 20th century were on average ~9%, while losses in the latter half were ~13% (Pimentel *et al.*, 1993). Moreover, while losses did not decrease, insecticide use increased tenfold in the period ~1950–1990, although losses during this period were offset by widespread planting of high-yielding varieties and greater use of fertilizer and other inputs (Pimentel *et al.*, 1993). A suite of problems associated with widespread and heavy pesticide use are now well known and have been discussed at length elsewhere (e.g. Perkins, 1982; Regev, 1984; Repetto, 1985; McConnell and Hruska, 1993; NRC, 1993; Bottrell and Weil, 1995; Lichtenberg and Zimmerman, 1999; Porter *et al.*, 1999). Increasing concern over such problems is one of the principal factors driving the development of integrated pest management (IPM) strategies for many crops worldwide.

A formal concept of IPM was first articulated in the late 1950s (Stern *et al.*, 1959). This first conception was reductionistic in that emphasis was largely restricted to integration of chemical- and biological control tactics. Subsequent definitions expanded the concept to include the use of all available tactics in a harmonious manner so that pest populations are maintained below levels causing economic injury (FAO, 1967; Smith and van den Bosch, 1967), while others emphasized natural mortality factors, such as natural enemies, climate and crop management (Board on Agriculture, 1989). The discussion presented in this chapter is based on a concept of IPM in which biological control and plant resistance are fundamental components.

The goal of IPM remains unchanged: maintain pest populations below levels at which they cause economic injury, while maintaining productive, societal and environmental impacts at acceptable levels. Moreover, the IPM concept adhered to herein includes an agroecosystem-level focus. The different crops within an agroecosystem share pest and natural-enemy populations, so pest-management tactics implemented within one crop are likely to have impacts across crops and across seasons. Thus, regional crop management (i.e. distribution in space and time) is also an important component of IPM. Crop-specific IPM strategies should take into account the potential impacts of production practices, including pest-management practices, on pest and natural-enemy populations of neighbouring and subsequent crops, as well as effects of prior crops. IPM with an agroecosystem-level focus is similar to area-wide pest management (Chandler and Faust, 1998) in that pests are managed across large geographical areas, the boundaries of which are defined by pest-colonization and dispersal capabilities. However, the strategies differ in that area-wide pest management typically focuses on single or a few key pests, such as pink bollworm, *Pectinophora gossypiella* (Saunders), in California, codling moth, *Cydia pomonella* (Linnaeus) in north-western USA and maize rootworms, *Diabrotica* spp., in various US states. It frequently relies on single or a few management tactics (e.g. baculovirus sprays against *Helicoverpa* and *Heliothis* and malathion sprays against boll-weevil, *Anthonomus grandis grandis* Boheman, in southern US cotton), and infrequently takes into account natural-enemy populations.

Insecticidal transgenic cultivars, as a form of plant resistance, are attractive novel tactics and may play central roles in IPM strategies. A number of reasons make transgenic cultivars attractive tactical components of IPM strategies, particularly in relation to the use of insecticides: (i) plants are protected throughout the growing season; (ii) pests are treated at their most susceptible stage; (iii) protection is independent of weather; (iv) pests protected from natural enemies and insecticides by living within plant tissues are exposed to the insecticidal effect of plants; (v) only insects feeding on the crop are directly affected; (vi) the insecticidal factor is confined to plant tissues and therefore does not leach into the environment; and (vii) the insecticidal factor is biodegradable and therefore does not accumulate in the environment (Gatehouse *et al.*, 1991). However, several of these advantages may also be disadvantages. For instance, pest-management decisions are made prior to planting and thus without knowledge of subsequent pest-population levels. Subeconomic pest populations are targeted along with economic populations. Such preemptive management tactics may have a significant impact on the sustainability of transgenic cultivars, a situation aggravated by extensive and continuous planting. Additional significant impacts are likely if the resistance factor of a transgenic cultivar is also an important and valuable insecticide, as in the case of *Bt*. Other important and likely impacts concern non-target, secondary pests of transgenic crops. Typically, populations of secondary-pest species are maintained below economic levels by the action of natural-enemy populations. It is conceivable that transgenic crop plants might impair natural enemies of secondary pests to the extent that biological control is relaxed and these pests become economically important.

Commercially available transgenic crop cultivars, such as *Bt*-expressing cotton, maize and potatoes, are now widely grown in the USA and to a lesser extent worldwide.

Recent reports indicate that areas planted in the USA exceed 7.0 million ha for maize, 1.4 million ha for cotton and 20 thousand ha for potatoes and that they are likely to increase in each case (Gianessi and Carpenter, 1999; NRC, 2000). Worldwide, the total area probably exceeds 35 million ha for these crops (ISAAA, 2002). Moreover, novel transgenic cultivars are being developed for numerous crops worldwide and evaluated for introduction in the near term into many countries (McLaren, 2000). The evident widespread planting of transgenic cultivars and projections for greater areas, particularly in the USA, indicate that these cultivars are quickly becoming mainstream IPM tactics. Thus, it is imperative that the compatibility of transgenic cultivars with biological control and regional crop management, both fundamental tactics of IPM strategies, is evaluated. The goals of this chapter are to identify challenges and opportunities pertinent to widescale adoption of transgenic crop cultivars as an IPM tactic and to suggest lines of research that it is critical to address prior to widespread acceptance. Emphasis is placed upon likely interactions of transgenic crop cultivars with biological control and impacts on pest population levels at the agroecosystem level.

Types of Transgenic Plants Currently Available or Being Developed for Pest Management

Development of insect-resistant transgenic crop cultivars has thus far focused on two distinct approaches: (i) integration of bacterial genes encoding for production of toxic proteins, especially from *Bt*; and (ii) integration of plant genes encoding for production of enzyme inhibitors and sugar-binding lectins. Both approaches were pioneered in the mid-1980s and thus have developed in parallel (Hilder *et al.*, 1987; Vaeck *et al.*, 1987). However, the first approach, based in particular on integration of δ-endotoxin genes derived from various subspecies of *Bt*, has undoubtedly received more attention and thus enjoyed greater progress. To date, all commercially available insect-resistant transgenic cultivars express semi-active *Bt* toxins, whereas cultivars expressing insecticidal plant proteins are not currently available outside of research laboratories.

Cultivars expressing *Bacillus thuringiensis* endotoxin genes

Bt is a ubiquitous soil bacterium that was first isolated and described in the early 1900s (Federici, 1999). Commercial products based on *Bt* toxins and consisting of sporulated, lysed cells of fermented isolates have been used for many years (Dulmage, 1981; Federici, 1999). The insecticidal activity of *Bt* depends on intracellular, insecticidal crystal proteins produced by the bacterium during sporulation, which accumulate as parasporal bodies adjacent to spores. The mode of action depends on a complex process, in which, following ingestion by susceptible insects, parasporal bodies are solubilized in the alkaline environment (pH 8–10) of the midgut, thereby releasing large protoxin molecules (130–140 kDa), which are then reduced through proteolytic cleavage to smaller (~55–70 kDa), active toxins (Gill *et al.*, 1992; Cannon, 1996; Schnepf *et al.*, 1998). Activated toxins paralyse and kill insects by destroying the mid-gut epithelium through lysis of epithelial cells, which allows movement of gut contents into haemolymph, leading to an increase in haemolymph pH (Gill *et al.*, 1992; Cannon, 1996; Schnepf *et al.*, 1998; Federici, 1999). Transgenic crop cultivars express semi-activated *Bt* toxins (~69 kDa); thus, toxic activity does not require solubilization in the insect mid-gut and requires relatively little protoxin-to-toxin conversion (Perlak *et al.*, 1990; Koziel *et al.*, 1993). The abridged mode of action of *Bt* expressed in transgenic crop cultivars has significant implications for the selectivity of these cultivars because both solubilization and protoxin–toxin conversion are important for the specificity of *Bt* activity (Visser *et al.*, 1993; Cannon, 1996). For example, it is feasible that insects unable to solubilize parasporal bodies may have appropriate receptors in the mid-gut epithelium for semi-active *Bt* toxins expressed in transgenic plants (Hilbeck *et al.*, 2000).

Cultivars expressing plant-derived enzyme inhibitors and lectins

Transgenic crop cultivars expressing plant genes are a valuable alternative to cultivars expressing *Bt* toxins. One significant advantage of using plant genes as a source of crop resistance is their broad spectrum of activity across several orders, including sap-sucking insect pests, such as Homoptera (Gatehouse and Gatehouse, 1998). Indeed, plant genes expressing chemical products responsible for plant resistance to insects are probably available against all pest species because they presumably evolved in response to herbivory by insects. Thus far, a number of economically important crop plants, such as tobacco, potato, tomato, rice, sugarcane, oilseed rape, among others, have been genetically transformed to express various genes of plant origin (Gatehouse and Gatehouse, 1998; Legaspi *et al.*, 2004). In many of these cases, significant effects of the plant genes were evident on target-pest mortality rates and/or developmental and reproductive parameters (Gatehouse and Gatehouse, 1998; Legaspi *et al.*, 2004).

Numerous plant chemical defensive compounds have been identified to date, and these compounds are the bases for developing transgenic cultivars expressing plant genes. The number of compounds identified to date potentially provides an unending array of opportunities for developing transgenic crop cultivars effective against virtually any pest species. However, many plant defensive compounds are products involving multiple enzyme pathways and genes, and hence transfer and expression of these products in crop plants is beyond current capabilities (Gatehouse and Gatehouse, 1998). None the less, a number of defensive compounds are protein products of single genes and these have been transferred successfully to various crops. Largely, plant transformation involving plant genes has focused on: (i) protease inhibitors; (ii) α-amylase inhibitors; and (iii) lectins (Gatehouse and Gatehouse, 1998). The varied modes of action and levels of specificity of these gene products increase the potential target-pest range of transgenic cultivars and allow the possibility of combining (pyramiding) genes that are active at various target sites within a pest insect or against various pests. However, levels of protection provided by genes of plant origin are typically lower than those provided by genes expressing *Bt* toxins. Frequently, effects on target insects are sublethal, including reductions in feeding, weight gain, developmental rates and fecundity (Gatehouse and Gatehouse, 1998; Legaspi *et al.*, 2004). Yet transgenic cultivars with sublethal or chronic effects on target pests may be more attractive components of IPM strategies than cultivars with acute toxic effects because they are more likely to be compatible or act synergistically with biological control, as discussed below.

Role of Transgenic Plants in Pest Management

Insect resistance via transgenes is a form of plant resistance against insect pests. However, unlike insect-resistant cultivars developed via conventional breeding methods, commercially available transgenic cultivars, all of which express *Bt* toxins: (i) are acutely toxic against their target pests; and (ii) were developed largely by private enterprises and remain within the commercial realm. The acute toxicity of commercially available transgenic cultivars has two significant implications for IPM. First, these cultivars exert strong selection pressures on target-pest populations, which consequently are expected to rapidly evolve resistance in the absence of effective resistance management strategies. Secondly, the acute toxicity of transgenic cultivars in many cases translates into significant indirect effects on upper-trophic-level consumers, such as natural enemies, and such effects will probably affect naturally occurring biological control. Both implications are further discussed below. Private development of commercial transgenic cultivars has meant that publicly supported IPM researchers are largely excluded from evaluating these cultivars prior to their widespread deployment. At present, participation in research on transgenic cultivars by IPM researchers remains largely restricted to evaluating their effectiveness against target pests and developing resistance management strategies intended to delay the evolution of resistant pest populations. The top-down manner in which commercially available transgenic cultivars were deployed and their rapid adoption by farmers left IPM researchers with little time to adequately integrate these cultivars into existing pest-management strategies. Thus, transgenic cultivars are rapidly becoming stand-alone control tactics against their target pests. Many of the potential ecological and environmental problems that have been identified to date associated with transgenic cultivars, and which remain to be addressed, are a direct result of the top-down and rapid deployment of these cultivars. Moreover, because transgenic cultivars thus far have been entirely appropriated by private industry, external costs such as potentially negative environmental and pest-management impacts, will probably tend to be neglected. A number of these problems are discussed below.

Crop-plant resistance against insects involves either physical (e.g. pubescence) or chemical (e.g. DIMBOA) defences. In general, transgenic cultivars rely on chemically based antibiosis as their plant resistance mechanism. Resistant plants expressing antibiosis produce significant negative physiological effects (e.g. slow development, reduced growth and fertility, death of young individuals, etc.) on susceptible herbivores feeding on their tissues. 'Apparency theory' (Feeny, 1976; Rhoades and Cates, 1976) provides a useful framework for discussing plant resistance to insects and the types of transgenic cultivars that may be most useful in agriculture. Chemical defences involved in antibiosis can be broadly categorized as toxins (qualitative defences) or digestibility reducers (quantitative defences), and apparency theory predicts that plants should rely on either type of chemical defence, depending on the probability of herbivore colonization (Feeny, 1976; Rhoades and Cates, 1976). Apparent plants, such as crop species, will predictably be colonized by large numbers of insect herbivores over evolutionary time and, because insects reproduce at a greater rate than plants, reliance on toxins for chemical defence will rapidly lead to the evolution of resistant herbivore populations. Thus, apparency theory predicts that apparent plants will rely on digestibility reducers for chemical defence. In contrast, unapparent or rare plants will be colonized by fewer insect herbivores, will not support high and continuous insect populations and will thus exert relatively low selection pressure on herbivore populations. These plants are predicted to rely on toxins for chemical defence. However, all commercially available transgenic cultivars express toxins and are acutely toxic to target pest species. The prediction of apparency theory is clear in this case, and current widespread and deep concern over the sustainability of commercial transgenic cultivars due to the evolution of resistance in pest populations points to its high likelihood. In short, the strategy of developing transgenic cultivars that express toxins is not evolutionarily sustainable.

Because crop species are akin to apparent plants, one evolutionarily sustainable approach is to develop crop cultivars that rely on digestibility reducers (e.g. plant-derived enzyme inhibitors and lectins). Moreover, digestibility reducers are more likely than toxins to be compatible with biological control (see below). If a long-term objective of developing IPM strategies for specific crops is to rely more on biological control and plant resistance and less on non-biological alternatives, then emphasis should be placed on developing transgenic cultivars that act synergistically or at least additively with biological control. This requires greater interaction during cultivar development between plant breeders and biological control researchers than currently occurs. One promising approach is to develop antibiosis-expressing transgenic cultivars that reduce yield losses while rendering pests more susceptible to parasitism and predation by natural enemies. For example, plant resistance via digestibility reducers that decrease the amount of feeding and prolong the developmental period of pests may act synergistically with biological control by increasing the length of time during which pests are susceptible to parasitism and predation. Examples are documented in which longer developmental times in herbivores led to increased parasitism or predation rates and more effective biological control (Price *et al.*, 1980; Haggstrom and Larsson, 1995; Luck *et al.*, 1995; Benrey and Denno, 1997; Devine *et al.*, 2000). Moreover, existing cultivars that are known to produce alternative foods (e.g. floral nectar, extrafloral nectar, pollen) at high levels or of high quality or that support (and tolerate) substantial populations of alternative hosts or prey may be targeted for transformation to express digestibility reducers active against target pests. Numerous studies have demonstrated the importance of the availability of alternative food sources or hosts/prey in maintaining natural-enemy populations at levels that effect biological control of target pests (Hagen, 1986; Jervis *et al.*, 1993; Bottrell *et al.*, 1998; Ferro and McNiel, 1998; Thompson and Hagen, 1999). An alternative approach is to develop transgenic cultivars that rely on antixenosis as a resistance mechanism. Resistant plants expressing antixenosis are refractory to colonizing herbivores and hence frequently lead to increased levels of activity and movement of susceptible herbivores on and between plants. Examples are documented in which greater herbivore movement led to greater parasitism rates, and such increments may lead to improved biological control (Pair *et al.*, 1986; Annis and O'Keefe, 1987). Other plant traits not directly targeted against insect pests but that may interact positively with biological control if transferred to crop cultivars are discussed by Hoy *et al.* (1998). Some examples include improved tolerance to diseases, which would raise the threshold level for insect vectors, and altered plant architecture and leaf surfaces, which may affect the exposure of pests to natural enemies.

The development and deployment of transgenic cultivars should be pursued within a context of ecological, evolutionary and economic considerations. Ideally, transgenic cultivars should be developed as tactical components of IPM strategies and their compatibility with biological control tactics should be considered a priority during development. The ecological goal should be to promote a transition to true ecologically based pest management, which maintains populations of potential pests at low levels, and in so doing, obviates the need to apply remedial control measures, such as insecticide applications. The evolutionary goal should be to deploy transgenic cultivars that will not rapidly select for resistant pest populations and consequently will be sustainable in the medium to long term with nil or minimal intervention. The economic goal should be to deploy transgenic cultivars that do not significantly add to production costs and therefore have a cost : benefit ratio that is largely independent of market and yield fluctuations. Transgenic cultivars should be a viable alternative for farmers in marginal as well as high-yielding areas. In sum, a long-term goal of developing transgenic cultivars should be to facilitate transition to ecologically based IPM strategies for major crops worldwide, while maintaining the economic and evolutionary sustainability of the cultivars. However, the context in which commercial transgenic cultivars are developed and deployed may not be compatible with such a goal. It remains to be seen whether future development of transgenic cultivars will be dictated by market opportunities and technological limitations (e.g. single-gene transformations), rather than the needs of farmers and interest in the long-term sustainability of agricultural production. Developing and deploying transgenic cultivars simply because the necessary technology is now available is not justifiable if a long-term goal of IPM research is to develop strategies that contribute to agricultural sustainability. Farmers' needs should be considered prior to transgenic-cultivar development, and the likely ecological, environmental and socio-economic impacts closely examined to the extent that current science allows so that scientifically informed decisions supersede those based on commercial interests.

Challenges and Opportunities in Deploying Transgenic Cultivars as Tactical Components of IPM Systems

Interactions of transgenic crop cultivars with IPM tactics

As discussed in this chapter, IPM has its foundations in biological control and plant resistance, and benefits from regional crop management that has an impact on levels of pest and natural-enemy populations within crops. Because transgenic crop cultivars are a form of plant resistance, it is likely that they will increasingly occupy central roles within IPM strategies, a trend that is rapidly becoming evident in the case of *Bt*-transgenic cultivars in the USA. In consequence, it is important that potential impacts, whether positive or negative, of transgenic crop cultivars on herbivore (including non-target species) and natural-enemy populations at the agroecosystem level are identified and measured.

The levels of pest populations found in crops are influenced by movement of pest- and natural-enemy populations between crops and seasons within agroecosystems. Thus, pest levels and biological control are likely to be affected by deployment of transgenic crop cultivars, particularly if these cultivars affect a wide range of herbivores and are widely planted within agroecosystems. Thus far, a number of studies have documented the occurrence of significant negative effects of transgenic cultivars on non-target insects, while others have shown negligible, non-existent, or positive effects (Bell *et al.*, 1999; Birch *et al.*, 1999; Losey *et al.*, 1999; Schuler *et al.*, 1999; Hilbeck *et al.*, 2000; Jesse and Obrycki, 2000; Wraight *et al.*, 2000; Sétamou *et al.*, 2002a,c). The magnitude and direction of potential effects of transgenic crops on IPM strategies will depend on the degree to which herbivore and natural-enemy populations are affected. Transgenic cultivars that have sublethal effects on herbivores and natural enemies will have a different impact on their populations from that of cultivars that have lethal effects. At the same time, transgenic cultivars that have negligible effects on natural enemies will probably have positive impacts on IPM strategies.

Because transgenic crop cultivars are a form of plant resistance, a tactic that occupies a central role in modern IPM strategies, they will interact with other cornerstone IPM tactics. Frequently, IPM strategies rely on biological control and are affected by regional crop management. Thus, biological control and transient populations of natural enemies and herbivores are likely to be most affected by widespread deployment of transgenic crops. Such impacts may be negative, and thus pose challenges to be redressed, or positive, and thus offer opportunities to be exploited. Redressing any challenges and taking advantage of any opportunities are crucial if transgenic cultivars are to be adequately integrated into IPM strategies.

Biological control

Natural enemies important in biological control include parasitoids, predators, and pathogens. Of these, parasitoids are perhaps the most important as they are responsible for most documented examples of biological control. For example, *c.* three-quarters of 1193 species of predators and parasitoids included in a world review of biological control programmes were parasitoids (Gordh *et al.*, 1999). Predators are widely recognized as important mortality factors, particularly in ephemeral crops (Kogan *et al.*, 1999), even if their impact on pest populations is frequently unappreciated or underestimated. Moreover, the importance of predation for pest management is well established in several crops for which transgenic cultivars are commercially available (e.g. cotton: Hagen *et al.*, 1976; González and Wilson, 1982). Pathogens are not widely relied upon for insect-pest management, although in many cases they are believed to be important short-term regulators of insect populations (Federici, 1999). Indeed, only one pathogen, *Bt*, is widely used for pest management. Thus, the discussion that follows focuses largely on likely and known impacts of transgenic cultivars on parasitoids and biological control by natural enemies other than pathogens.

Parasitoids, mostly in the order Hymenoptera, have larvae that typically consume and kill single hosts, while adults are free-living or only secondarily carnivorous, feeding on host-derived haemolymph. The biology of parasitoid larvae and their relationships with hosts have two important implications for the suite of potential impacts of transgenic cultivars on biological control. First, parasitoid larvae are typically unable to search for a new host if the available host is unsuitable. Secondly, parasitoids are frequently capable of exploiting only a small number of closely related species as hosts. Thus, it is likely that, if a transgenic cultivar affects a parasitoid's main host, all alternative host species will be similarly affected. In consequence, parasitoid females searching for hosts within a transgenic crop are likely to encounter only hosts that are susceptible to the transgenic cultivar, and their offspring are restricted to developing on these hosts. The likely effects of transgenic crop cultivars on parasitoids are summarized in Fig. 6.1. The potential effects of transgenic cultivars on parasitoids are divided into individual- and population-level effects. At the individual level, the likely effects are further divided into those affecting parasitoid larvae and adults. At the population level, the likely effects are those relating to impacts on biological control as a population-level process. Specifically, the effects are divided into negative (e.g. impacts of wide-scale depletion of host populations), synergistic (e.g. greater parasitism levels due to weakened host immune responses) and additive effects (e.g. parasitism of individuals surviving the effects of transgenic plants). Each of these likely effects of transgenic cultivars on parasitoids and biological control by parasitoids is briefly discussed below.

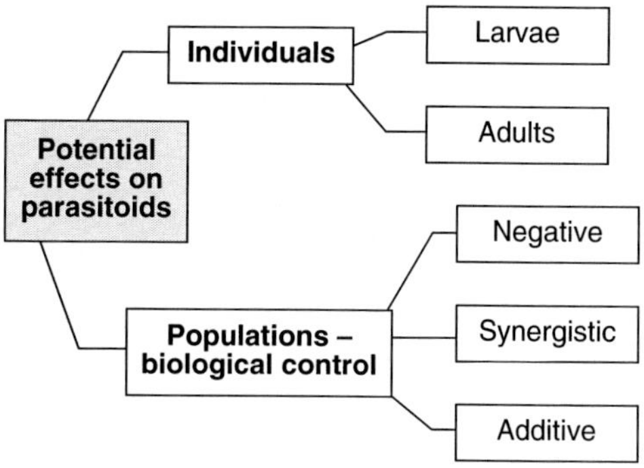

Fig. 6.1. Levels at which transgenic insecticidal crop cultivars may have an impact on parasitoids and some likely directions of impacts.

Individual-level effects

The most direct and obvious effect of transgenic crop cultivars on parasitoid larvae occurs in hosts susceptible to the transgenic cultivar and is the result of the acute toxicity of the cultivar to the host (Fig. 6.2). In this case, parasitoid eggs or larvae are unable to complete their development due to premature death of their host. Such effects on parasitoids are likely to be less important in transgenic cultivars that are not acutely toxic to the targeted pests such as cultivars expressing enzyme inhibitors and lectins or toxins at low doses.

The effects on parasitoid larvae via hosts tolerant of the transgenic cultivar are less obvious and can be either direct or indirect (Fig. 6.2). For example a direct positive effect on parasitoid larvae was evident in a recent study in which it was suggested that the immune-defence reactions of hosts were compromised by feeding on artificial diet containing 2% of the lectin *Galanthus nivalis* agglutinin (GNA) (Bell *et al.*, 1999; Fig. 6.2). Broods of the gregarious parasitoid *Eulophus pennicornis* (Nees) developing on tomato moth larvae, *Lacanobia oleracea* (Linnaeus) fed an artificial diet containing 2% GNA were more than twice as large (21 adult parasitoids per host) as broods developing on hosts fed a diet without GNA (nine adult parasitoids per host) (Bell *et al.*, 1999). However, a similar effect was not evident when hosts were fed transgenic potato leaves expressing GNA at 0.8%. In contrast, direct negative impacts on parasitoid larvae are evident in studies showing decreased survival of parasitoids developing on hosts feeding on transgenic plant tissues due to exposure to toxins (Fig. 6.2). For example, a recent study showed that egg–adult survivorship was *c.* 90% in parasitoids developing on hosts fed non-transgenic maize tissue, while it was *c.* 40% in those whose hosts were intoxicated after feeding on transgenic maize tissue (Bernal *et al.*, 2002). Similar effects on survivorship are evident in predators fed prey intoxicated after feeding on transgenic maize tissue versus those fed healthy prey (Hilbeck *et al.*, 1998).

Indirect effects of transgenic cultivars on parasitoid larvae are probably common as

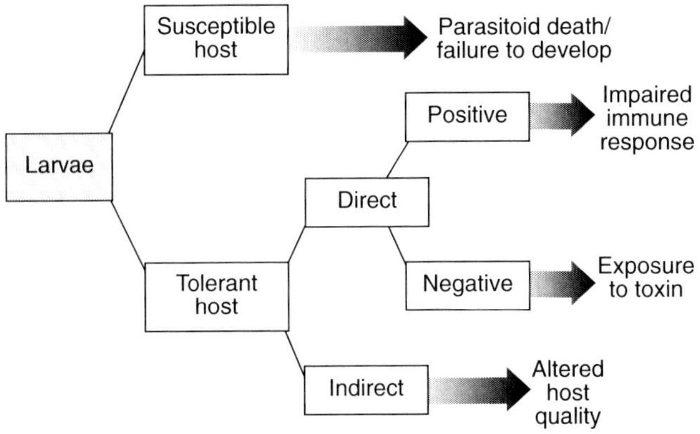

Fig. 6.2. Likely impacts of transgene products on parasitoid larvae developing on/in hosts feeding on transgenic insecticidal crop cultivars.

they derive from effects on a herbivore's quality as a host. For instance, both target and non-target pests of transgenic cultivars are frequently sublethally affected by these cultivars, and negative effects, such as lower weight and nutritional quality, will probably affect developing parasitoid larvae (Fig. 6.2). A number of studies have shown that hosts fed transgenic plant tissue weighed less than hosts fed non-transgenic plant tissue, and that these differences in weight resulted in smaller, less fit parasitoids, smaller parasitoid broods and a lower ratio of female offspring (Adamczyk *et al.*, 1998; Lynch *et al.*, 1999; Couty *et al.*, 2001; Sétamou *et al.*, 2002a,b). In contrast, differences in susceptibility between target and non-target pests of transgenic cultivars may result in indirect positive effects such as improved quality as a host in the latter pests (Fig. 6.2). For example, non-target pests may have greater tolerance than target pests, and may benefit (e.g. increased weight gain) from feeding on transgenic cultivars that express low levels of insecticidal proteins or toxins (De Leo *et al.*, 1998; Sétamou *et al.*, 2002b). Any benefits for parasitoid larvae of feeding on improved-quality hosts, however, will depend on their susceptibility to insecticidal proteins, which may accumulate in their host's gut and haemolymph (Fitches and Gatehouse, 1998; Bell *et al.*, 1999).

Transgenic cultivars may affect parasitoid adults via effects on larvae that develop on tolerant hosts, as discussed above (Fig. 6.2). The effects on parasitoid adults will thus be determined by the susceptibility of the developing larvae to the transgenic cultivar as mediated by the host, or the effects on larvae of developing on hosts of altered quality. In addition, transgenic cultivars may affect parasitoid adults independently of effects on their larvae by interfering with different components of the host-finding process, and these effects may be mediated by the transgenic cultivar itself or by hosts feeding on transgenic plants (Fig. 6.3). For example, plants are known to emit volatiles that attract natural enemies, particularly when they are under attack by herbivores (Turlings and Benrey, 1998; Venzon *et al.*, 1999; Pels and Sabelis, 2000). It is conceivable that volatiles emitted by transgenic cultivars are qualitatively different from those emitted by non-transgenic cultivars. One recent study failed to find differences between herbivore-damaged transgenic and non-transgenic oil-seed rape plants in their attractiveness to the parasitoid *Cotesia plutellae* (Kurdyumov) (Schuler *et al.*, 1999). However, the results of another study showed that when given a choice, sugarcane borer, *Diatraea saccharalis* (Fabricius), and Mexican rice borer, *Eoreuma loftini* (Dyar), adults prefer non-transgenic over transgenic GNA-expressing sugarcane for oviposition, suggesting that differences exist between the sugarcane cultivars in their attractiveness to these herbivores (Bernal and Sétamou, 2003). It is conceivable that these differences may also extend to adult parasitoids searching for

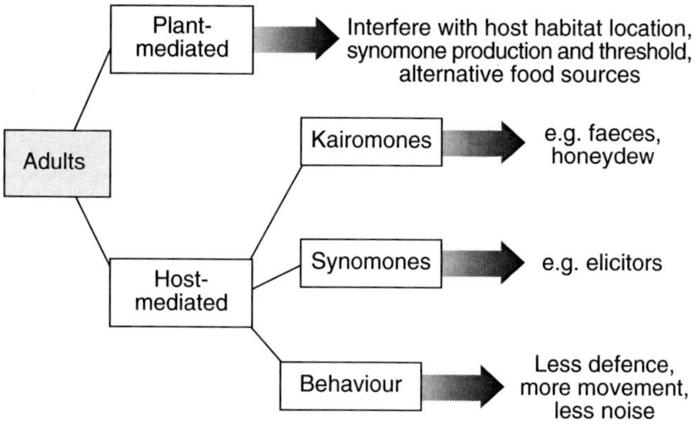

Fig. 6.3. Likely impacts of transgenic insecticidal crop cultivars on parasitoid adults as mediated by the transgenic plant or by hosts feeding on transgenic plants.

sugarcane borer or Mexican rice borer hosts infesting sugarcane plants.

In addition to direct effects, transgenic cultivars may also affect parasitoid adults indirectly via their hosts (Fig. 6.3). Currently, all transgenic cultivars are active on their target pest's gut or digestive processes. Parasitoids in general are known to rely on chemical cues (kairomones) from their hosts (e.g. pheromones) or host products (e.g. frass, honeydew) during the host-searching process. It is conceivable that chemical cues emanating from host products, such as frass or honeydew, may be altered in herbivores feeding on transgenic cultivars. For example, females of the parasitoid *Cotesia marginiventris* (Cresson) were more attracted to frass of fall armyworm, *Spodoptera frugiperda* (J.E. Smith) that had fed on non-transgenic versus transgenic maize tissue (Bernal *et al.*, unpublished data). Also, the lectin GNA was found in similar to higher concentrations in the honeydew relative to the diet of rice brown planthopper, *Nilaparvata lugens* (Stal) between 24 and 72 h after feeding on a GNA diet (Powell *et al.*, 1998). Similarly, production of herbivore-induced chemical cues (synomones) by plants is dependent on elicitors from herbivores, and these elicitors are associated with herbivore feeding (Turlings and Benrey, 1998). It is plausible that the chemical composition of these elicitors is altered in herbivores feeding on transgenic cultivars. However, few data are currently available that address potential differences in herbivore chemical cues, products or elicitors between herbivores feeding on transgenic versus non-transgenic cultivars, or in the attractiveness of these cultivars to parasitoids searching for hosts. In contrast, available data suggest that transgenic cultivars may affect parasitoid adults via alterations in the behaviour of their hosts (Fig. 6.3). The positive effect of increased herbivore movement on parasitism rates (Pair *et al.*, 1986; Annis and O'Keefe, 1987) was discussed above, and it is likely that pests intoxicated or weakened after feeding on transgenic cultivars may be less able to defend themselves against natural-enemy attack. Some studies show that host aggression against parasitoid adults can result in high levels of parasitoid mortality during host handling (Potting *et al.*, 1999). In contrast, natural enemies that rely on airborne or substrate-borne vibratory or visual cues for locating hosts or prey may be less successful if hosts/prey show decreased activity levels when feeding on transgenic cultivars. For example, Mexican rice borer, a stem-boring pest, is less active and produces less noise inside artificial tunnels and consequently is parasitized less frequently when feeding on diet containing GNA versus diet free of GNA (Tomov *et al.*, 2003).

Population-level effects

Finally, transgenic cultivars may affect parasitoids at the population level, and consequently may have an impact on the degrees of biological control they achieve (Fig. 6.4). The effects at the population level may be evident at the agroecosystem level if transgenic cultivars eliminate host populations to the extent that parasitoid populations are unable to persist locally (negative direct) or suffer in terms of their quality (negative indirect), or they will be extensions of effects on parasitoid larvae and adults at the individual level, as discussed above (synergistic or additive) (Fig. 6.4). Agroecosystem-level effects on natural-enemy populations and biological control are further discussed below. Synergistic effects of transgenic cultivars on biological control will arise when pest mortality due to plant resistance and natural enemies within transgenic fields is greater than the expected sum of the mortality from both these factors. However, synergism with biological control is more likely in transgenic cultivars that are sublethal to the targeted pest. For example, greater than expected parasitism of tobacco budworm, *Heliothis virescens* (Fabricius), by the parasitoid *Campoletis sonorensis* (Cameron) was evident in transgenic tobacco expressing a low level of *Bt* endotoxin relative to non-transgenic tobacco (Johnson and Gould, 1992). In this case, the greater parasitism was attributed to longer developmental times for tobacco budworm larvae and the ensuing longer window of opportunity for parasitism, which should be increasingly important where natural enemies are able to exploit only a narrow range of host/prey instars or sizes. Comparable differences in parasitism rates, albeit not significant, were reported in a study involving tobacco budworm, maize earworm, *Helicoverpa zea* (Boddie), and transgenic tobacco (Warren *et al.*, 1992). Similar synergistic interactions are plausible if transgenic cultivars have other sublethal effects, such as weakening a host's immune system or impairing its ability to defend itself from parasitoids, as discussed above.

A significant interaction may be lacking between transgenic cultivars and biological control in the field, with any effects merely being additive (Fig. 6.4). Additivity between transgenic cultivars and biological control through parasitism or predation of pest individuals surviving after feeding on transgenic plants has significant implications for the evolution of resistance in pest populations to transgenic cultivars, because surviving individuals are likely to be genetically resistant to the antibiotic effects of these cultivars. For example, genetic models show that natural enemies may double to quadruple the numbers of generations necessary for the evolution of resistance to transgenic cultivars in pest populations (Johnson and Gould, 1992; Arpaia *et al.*, 1997). However, more comprehensive models indicate that natural enemies may increase, decrease or not affect the rate of resistance evolution in pest populations (Gould *et al.*, 1991).

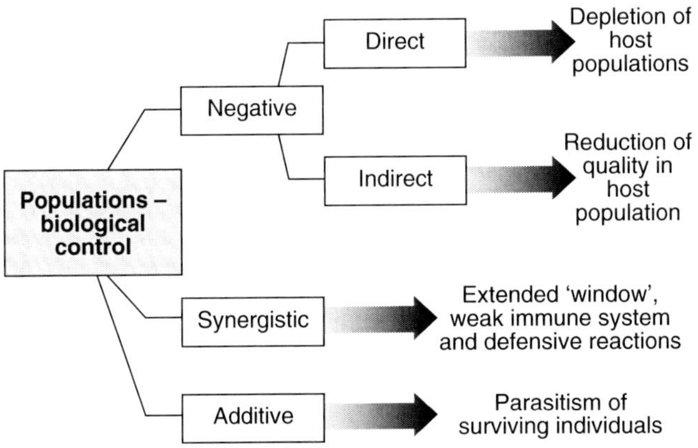

Fig. 6.4. Likely impacts of transgenic insecticidal crop cultivars, and their direction, on parasitoid populations and biological control.

Other important interactions between transgenic cultivars and biological control are likely to occur via secondary pests, which typically are not targeted during transgenic cultivar development. In this case, the most obvious effects will derive from two likely effects of transgenic cultivars. First, secondary pests may be sublethally affected by transgenic cultivars due to greater tolerance of the insecticidal toxins or proteins expressed by these cultivars relative to target (primary) pests. In this case, the expected effects would be similar to those discussed above for sublethally affected primary (i.e. target) pests, as well as the agroecosystem-level effects discussed below. Secondly, a positive interaction between transgenic cultivars and biological control may occur if deployment of these cultivars results in reduced usage of broad-spectrum insecticides against primary pests, which may lead to improved biological control of secondary pests through conservation of natural enemies. A number of examples in which *Bt*-transgenic potato cultivars support greater and more diverse natural-enemy communities are discussed by Hoy *et al.* (1998). However, it will be important that applications of broad-spectrum insecticides are indeed reduced if transgenic cultivars are to enhance conservation biological control of secondary pests. Currently, some studies indicate that lesser amounts of insecticides are applied in transgenic crops, such as *Bt* maize and cotton, whereas others fail to find significant reductions in the use of broad-spectrum insecticides (Fernández-Cornejo and McBride, 2000).

Effects on agroecosystems

Beyond effects on individuals and populations lies the issue of how deployment of transgenic cultivars may alter the agroecosystem outside single transgenic field boundaries. Both contemporaneous and sequential crops affect neighbouring fields, in part because the arthropod faunas of annual crops are often produced by immigration from nearby crops (Kieckhefer and Miller, 1967; Poston and Pedigo, 1975; Ives, 1981; Honêk, 1982; Vorley and Wratten, 1987; Brazzle *et al.*, 1997; Kennedy and Storer, 2000). Indeed, interconnections of both pest and natural-enemy populations in common transgenic crops to other crops in regional production systems are well established.

Common pests of agroecosystems in North Carolina are unquestionably linked. The European corn borer, *Ostrinia nubilalis* (Hübner), colonizes potatoes and wheat (Anderson *et al.*, 1984; Umeozor *et al.*, 1986; Jones, 1994) before infesting maize and cotton (Savinelli *et al.*, 1986, 1988; Umeozor *et al.*, 1986). Maize earworm, *H. zea*, populations initially develop in maize and later colonize cotton and soybean crops in the area (Neunzig, 1963, 1969), while two-spotted spider mites, *Tetranychus urticae* (Koch), from maize become pests in nearby groundnuts (Brandenburg and Kennedy, 1982; Margolies and Kennedy, 1985).

A similar relationship exists for natural enemies in Texas cropping systems. Parasitoids such as *Trichogramma pretiosum* Riley, *C. sonorensis* (Cameron) and *Microplitis croceipes* (Cresson) all parasitize maize earworm or tobacco budworm, *H. virescens* in several crops, including lucerne, maize and potatoes early in the season and lucerne, maize, cotton and grain sorghum later in the season (Puterka *et al.*, 1985). Predators, including spiders, convergent lady bird, *Hippodamia convergens* Guérin-Méneville, and minute pirate bugs, *Orius tristicolor* (White) and *Orius insidiosis* (Say), move between cotton and grain sorghum throughout their coincident cultivation, apparently responding to local prey abundances (Prasifka *et al.*, 1999). Maize and grain sorghum also appear to provide the *H. convergens* colonists that later reproduce and sustain ladybird populations in cotton (J. Prasifka and K.M. Heinz, unpublished data).

Many other examples support the assertion that pest and natural-enemy populations interact across landscapes and seasons, suggesting that interactions between transgenic and non-transgenic crops are likely. However, because current research on the effects of transgenic cultivars is performed on a much smaller scale, agroecosystem effects must be predicted on the bases of limited studies, whose results vary widely (see previous section). In the absence of a clear indication of what can be expected at the individual-field level, possible detrimental and beneficial impacts from the spread of transgenic insecticidal cultivars are considered in turn.

Challenges to IPM in agroecosystems

Transgenic cultivars are typically highly effective against target pests, suggesting that natural-enemy populations using these pests as prey or hosts will also suffer great numerical declines. The results of this across agroecosystems (space and time) could have at least three unwelcome effects, including: (i) target-pest resurgences or secondary-pest outbreaks in non-transgenic cultivars when large areas of transgenic cultivars are concurrently grown; (ii) pest epidemics in other non-transgenic crops; and (iii) chronic secondary-pest outbreaks in transgenic crops through effects on natural-enemy and pest populations.

The potential for pest problems in non-transgenic fields of the same crop is based on the connectivity between fields. Typically, extensive plantings of one annual crop harbour small, non-independent components of mobile pest and natural-enemy metapopulations (Settle *et al.*, 1996; Kennedy and Storer, 2000). Insecticide treatments in one area may decimate natural enemies there, but mobile predators and parasitoids can repopulate treated fields from nearby untreated ones (Wratten and Thomas, 1990). In contrast, transgenic fields in which high pest and natural-enemy mortality occurs represent sinks from which little or no recolonization can be expected. Because of the cost:benefit guessing game that can be required to plant transgenic cultivars (Rice and Pilcher, 1998), some non-transgenic fields will be planted in addition to required refuge plantings. If insecticides are applied to non-transgenic fields, insufficient natural enemies may recolonize these fields, causing pest populations to significantly increase.

Pest epidemics in other crops might also result from reduced abundance or diversity of natural enemies in transgenic fields. If low levels of both pests and natural enemies exist within transgenic fields, the natural process of arthropod colonization of nearby crops will be disrupted and natural biological control compromised. For example, pests may colonize from a weedy host, and a transgenic crop that traditionally supplied natural enemies will probably fail to control the pest, requiring remedial insecticide treatments. Even if the natural-enemy-to-pest ratio remains the same, reduced abundance of colonists favours pests. Herbivorous pests colonize crops in advance of the arrival of natural enemies (Price, 1976), allowing them to exploit abundant resources unchecked. Predators and parasitoids arrive later and ideally halt the exponential growth of pest populations, but they typically operate as density-dependant mortality factors. This argument overlooks the most obvious problem, that some pests will be unaffected by transgenic insecticidal cultivars, while key natural-enemy numbers are reduced.

If the effects of transgenic insecticidal cultivars commonly include those proposed above, then by extension, there will probably be problems throughout the agroecosystem, including chronic secondary pest problems in the transgenic cultivars themselves. Predator and parasitoid populations are not only linked spatially (among crops), but are related temporally (over seasons), with a cycle of natural-enemy colonization repeating annually (Wissinger, 1997). The combination of pest outbreaks and insecticide treatments outside transgenic fields and the recurrent natural-enemy sink caused by transgenic cultivars could reduce the available natural-enemy pool each year until an entirely new pest complex emerges, as has occurred in the past due to overuse of conventional insecticides.

Opportunities for IPM in agroecosystems

While the potential for misuse and unintended side effects of widespread plantings of transgenic crops unquestionably exists, transgenic cultivars potentially represent a tremendous advance in crop protection. In the best case, transgenic cultivars offer not only a novel tool for managing target pests, but also a chance to incorporate biological controls into IPM to an unprecedented level. Among the possible benefits to agroecosystems are: (i) a reduction of pest outbreaks through conservation of natural enemies; (ii) successful area-wide management of mobile, polyphagous crop pests; and (iii) incidental control of various susceptible, non-target pests.

Apart from control of target pests, the effect most often touted by proponents of transgenic insecticidal cultivars is conservation of natural enemies. Insecticidal toxins, such as those produced by *Bt*-transgenic cultivars, are relatively specific and may reduce the overall need for broad-spectrum insecticide applications. With this reduction in insecticide applications, predator and parasitoid (particularly generalists and those not dependent on the target pests of transgenic cultivars) populations should increase and be better able to prevent pest outbreaks, especially of pesticide-induced secondary pests. When the natural-enemy faunas of multiple crops are intimately linked, the presence of one widely grown transgenic cultivar could have benefits across the agroecosystem by providing greater numbers of colonizing natural enemies for other crops.

Transgenic cultivars also provide a unique potential to manage mobile, polyphagous pests on an area-wide basis. Extensive use of transgenic cultivars is advantageous for this application because extensive plantings can exert continuous pressure over both space and time. Area-wide control of European corn borer through increased areas planted to *Bt* cultivars has been proposed (Rice and Pilcher, 1998), but only potential benefits to maize were considered. However, if few non-crop hosts were available, one or more transgenic crops could eliminate this and other polyphagous pests as problems in entire multiple-crop agroecosystems. In cases where non-target pest populations are also reduced by transgenic cultivars, improved control and reduced outbreaks in other susceptible crops (e.g. due to reduced maize earworm movement between maize and cotton) could also be realized.

As alluded to above, the use of transgenic crop cultivars can have unanticipated management benefits against non-target pests. Depending on the levels of toxicity against these pests, other primary or secondary pests could be reduced in status or completely eliminated. For example, *Bt* cotton cultivars developed against bollworm, tobacco budworm and pink bollworm also showed a larval mortality of > 85% against cabbage looper, *Trichoplusia ni* (Hübner), salt-marsh caterpillar, *Estigmene acrea* (Drury), cotton leaf perforator, *Bucculatrix thurberiella* Busck, and European corn borer. Minor insecticidal effects were also shown against beet armyworm and fall armyworm, with larval mortality of 20–25% (Wilson *et al.*, 1992, 1994; Bradley, 1995; Moore *et al.*, 1999).

Potential Role in Developing-country Agriculture

Developing-country agriculture is characterized by a mosaic of production systems, ranging from high-input agriculture that seeks maximum economic returns to subsistence agriculture that seeks maximum yields with minimal variance. The former is typical of export and plantation crops (e.g. cotton, bananas, tomatoes), whereas the latter is typical of staple crops (e.g. maize, potatoes, rice). The focus of this section is on subsistence agriculture, but cotton is included because transgenic cotton cultivars are currently available and because cotton cultivation is frequently viewed as a source of foreign currency in developing countries and figures prominently in agricultural development projects (Murray, 1994).

Several characteristics of subsistence agriculture should be highlighted prior to discussing the potential role of transgenic crop cultivars within pest-management strategies. First, capital available to subsistence farmers for investing in crop production in general and pest management in particular is limited. Thus, the IPM tactics most readily accessible to subsistence farmers are those that do not entail direct economic expenditures. Biological control by natural-enemy conservation and plant resistance (exclusive of commercial transgenic crops) are both free of cost

and fundamental tactics of IPM strategies and should form the bases of IPM systems in subsistence agriculture. Secondly, subsistence agriculture is, by definition, not typically a major source of income, and subsistence farmers must engage in other activities to procure expendable income (e.g. day labour). Thus, pest-management tactics that require minimal time investments will be more appropriate than and favoured over tactics requiring greater time investments. For example, the opportunity costs associated with monitoring pest populations as a basis for insecticide applications will detract from the appeal of insecticide-based IPM strategies even when insecticides are a tenable alternative. Again, plant resistance and biological control are especially appropriate because they require minimal to nil time investments by farmers. Moreover, IPM strategies centred upon insecticidal control tactics may not be appropriate in the context of subsistence farming. Insecticide use within IPM systems is based on the concepts of economic threshold and economic injury levels. These concepts are based on the difference between the costs of intervention (i.e. insecticide application) and non-intervention (i.e. yield loss due to pests), and may not be useful in the context of subsistence agriculture because losses associated with non-intervention are typically not quantified in economic units.

It is clear that plant resistance and biological control are particularly appropriate tactics for pest management in subsistence agriculture and that they should form the bases of IPM strategies in that context. Thus, transgenic cultivars have a great potential to play significant roles in developing-country agriculture. However, it will be crucial that the interactions between transgenic cultivars and biological control are closely examined prior to cultivar deployment. Specifically, breeding efforts should seek to develop cultivars that interact positively (i.e. synergistically) with biological control. This requires greater than current degrees of interaction between plant breeders and biological control researchers. Cultivars that have additive interactions with biological control should be left as second alternatives, while those with negative interactions should be avoided prima facie. IPM systems that are proactive should be developed by seeking positive interactions between transgenic cultivars and biological control. Examples of plant resistance mechanisms likely to result in positive interactions between transgenic cultivars and biological control were discussed above. Positive interactions between transgenic cultivars and biological control will contribute to maintaining pest populations at low levels, which should minimize the need for remedial control tactics such as insecticide applications.

Transgenic cultivars have the potential to generate significant benefits for developing-country agriculture. Among the main benefits lies the potential for reducing pesticide use in cash crops, such as cotton, which are notorious for their reliance on insecticidal control of major pests. For example, reductions in the numbers of insecticide applications to cotton have been documented where *Bt* cotton is planted in the USA, and similar reductions can be expected in developing countries (ERS-USDA, 1999; but see Fernández-Cornejo and McBride, 2000). Moreover, transgenic cultivars may allow recultivation of crops in areas where they were discontinued due to severe pest-management problems. For example, cotton production was largely discontinued in north-eastern Mexico by 1970 and Central America by 1990 following upsets of secondary pests, including species of *Heliothis*, *Spodoptera* and *Trichoplusia*, and the inability to manage these pests economically based on unilateral use of insecticides (DeBach and Rosen, 1991; Murray, 1994). *Bt*-transgenic cotton may significantly contribute to expanding cotton production in those and other areas where lepidopteran pests are major factors limiting cotton production (e.g. Africa; Silvie *et al.*, 2001). However, it is uncertain whether mere transfer of *Bt*-transgenic cultivars from developed to developing countries will lead to substantial benefits in the latter unless locally important pests are targeted and locally adapted crop varieties are transformed. Pest complexes and their susceptibility to *Bt* toxins, as well as desirable agronomic characters, probably differ between most regions.

Transgenic cultivars expressing plant-derived enzyme inhibitors and lectins are particularly promising alternatives to *Bt*-transgenic cultivars in developing countries. Unlike *Bt*-transgenic cultivars, those expressing enzyme inhibitors and lectins are largely being developed by national and international public research institutions and universities and thus are more likely to remain outside the commercial realm. For example, numerous lectin-expressing transgenic crop lines are currently in various stages of development and evaluation (Gatehouse *et al.*, 1991; Legaspi *et al.*, 2004). Moreover, transgenic lines of staple crops of major importance in developing countries, such as potatoes and rice, are in various stages of development and numerous genes expressing activity against pests of these crops have been isolated and are currently being evaluated (Gatehouse *et al.*, 1998; Shu *et al.*, 2000; Machuka, 2001; Legaspi *et al.*, 2004). As discussed above, transgenic cultivars expressing plant-derived genes are more likely than *Bt*-expressing cultivars to act synergistically with biological control, thus potentially decreasing the need for remedial control tactics, such as insecticide applications.

Along with the likely benefits derived from deployment of transgenic cultivars in developing countries are potentially important disadvantages. A few are briefly discussed here in reference to commercially available *Bt*-expressing cultivars. One major concern associated with *Bt*-transgenic cultivars in countries where they are available commercially is the development of resistant pest populations (Gould, 1998), and this concern should be greater in developing countries. In the absence of adequate knowledge and regulation, transgenic cultivars could become a transitional technology, available only as long as resistance development can be postponed, which could ultimately lead to higher levels of pesticide use (Hubbell and Welsh, 1998). For instance, recent development projects in Central America emphasize the production of non-traditional export crops, such as fruits, vegetables and cut flowers (Thrupp, 1995; Nicholls and Altieri, 1997), while cotton remains an attractive and potentially important source of foreign income (Murray, 1994). Production of these crops is heavily dependent on the use of broad-spectrum insecticides, which has led to serious environmental, ecological and human health problems (McConnell and Hruska, 1993; Murray, 1994; Thrupp, 1995; Nicholls and Altieri, 1997). A promising alternative to broad-spectrum insecticides is the use of *Bt*-based insecticides, which have been increasingly employed against cotton pests in Central America, starting in the 1980s (Murray, 1994). Resistance to *Bt* toxins could rapidly evolve if *Bt*-cotton cultivars are widely deployed and planted in Central America, precluding further use of *Bt*-based insecticides in non-traditional export crops and guaranteeing in the short term a continued heavy dependence on broad-spectrum insecticides for pest control.

A number of likely obstacles must be overcome for widespread and sustainable use of transgenic cultivars in developing countries, and these are particularly important in the case of existing *Bt*-transgenic cultivars. One obstacle is the low adoption rate of commercial seed varieties in many developing countries. For example, the areas planted with improved hybrid maize seed in Mexico and Central America are currently below 20% of the total areas planted to maize (Morris and López-Pereira, 1999). Similarly, it remains to be seen whether resistance management programmes, based on non-transgenic refuge crops and designed for developed-country agriculture, will be appropriate for developing-country agriculture. A pervasive feature of subsistence agriculture is the relatively small size of land allocations per farmer. It is uncertain whether land-limited subsistence farmers will be willing to set aside a portion of their crop as a non-transgenic refuge susceptible to high levels of pest damage. Moreover, resistance-management programmes require active industry, government and farmer participation to be effective. Past failures, despite strong national and international interests, in maintaining regional IPM programmes in various crops in developing countries caution against placing too much confidence in the long-term viability of regional resistance-management programmes (Holl *et al.*, 1990; Murray, 1994; Thrupp, 1995; Nicholls and

Altieri, 1997). Finally, other potential problems associated with the deployment of transgenic cultivars in developing countries relate to the unknown consequences of potential gene flow between crops and wild relatives and within crops, and intellectual- and genetic-property issues. These issues have been discussed elsewhere and are not addressed here (e.g. Shand, 1991; De Souza Silva, 1995; Bhat, 1999; Ellstrand *et al.*, 1999; King and Eyzaguirre, 1999). Major staple crops, such as maize, rice, cowpeas and potatoes, have their centres of origin and genetic diversity in developing countries, and therefore it is essential that these issues are addressed in advance of transgenic-cultivar deployment rather than in retrospect.

Conclusions

Transgenic insecticidal cultivars are a novel form of host-plant resistance and, as such, may play major roles in future IPM strategies in many crops worldwide, in both developed and developing countries. Transgenic cultivars are novel because foreign genes confer their resistance, but host-plant resistance is a long-standing and fundamental IPM tactic. They may play major roles in future IPM strategies because, unlike traditionally bred insect-resistant cultivars, they rely on genes and gene products transferred between species, which substantially broadens the opportunities for developing new transgenic cultivars. The development of new transgenic cultivars effective against specific pests appears limited only by our ability to discover and successfully transfer genes between species. As discussed above, genes conferring resistance against all major pests are probably available because insects as a group are major herbivores and plants have evolved diverse chemical defences against herbivory by insects. However, because of their tremendous promise, it is imperative that transgenic cultivars are designed to be sustainable. Transgenic cultivars should be developed within an IPM context, rather than as stand-alone technologies against target pests because of a number of potential problems, discussed above. Specifically, transgenic cultivars should be designed to act synergistically with other IPM tactics, especially biological control.

Some of the major challenges facing transgenic-cultivar development in the future stem from the need for developing cultivars that are compatible with biological control and consequently are more likely to be sustainable bases for IPM strategies. A first challenge will be to develop a sound understanding of the interactions between transgenic plants, herbivores and natural enemies, particularly parasitoids and predators. Moreover, interactions should be studied at the individual, population and community levels. This should facilitate the development of transgenic cultivars that act synergistically with biological control and thus are compatible with IPM. A second challenge will be to better understand the effects of transgenic cultivars on the movement and colonization patterns of herbivore and natural-enemy populations across entire agroecosystems. This will facilitate the development of regional crop-management strategies that include transgenic cultivars and contribute to regional pest-management efforts by managing the movement and colonization patterns of pest and natural-enemy populations. Finally, a third challenge will be to develop and deploy transgenic cultivars against major pests of crops that have little commercial potential, such as pests of staple crops in developing countries and of low-value crops in developed countries. One model which if emulated may prove useful is that followed to develop β-carotene-rich, 'golden' rice through joint public and philanthropic funding, which allowed scientists to develop a (non-insecticidal) transgenic cultivar with little commercial potential in developing countries (Ye *et al.*, 2000).

In conclusion, transgenic cultivars, in general, are a valuable tool in our arsenal of pest-management technologies and therefore it is imperative that they are used according to established IPM principles. However, recent experiences with commercial *Bt* cultivars suggest that basic IPM principles are ignored following deployment of these cultivars: *Bt* toxins are used prophylactically as a first resort, with no regard to established pest threshold levels, and the control of key pests relies on a single tactic. This disregard for IPM principles has serious implications for the sustainability of these cultivars and for IPM strategies that depend on *Bt* toxins. Tacit acceptance that *Bt* cultivars are being used in a manner that is unsustainable is implicit in our deep preoccupation with *Bt*-resistance management. It will be imperative for the compatibility of transgenic cultivars in general with biological control, and therefore IPM, to be addressed during the development process so transgenic cultivars are designed to be sustainable. Increasing our understanding of the interactions between transgenic cultivars and herbivore and natural-enemy populations is necessary and urgent in order to develop transgenic cultivars that play central roles, along with biological control, within IPM strategies.

References

Adamczyk, J.J., Holloway, J.W., Church, G.E., Leonard, B.R. and Graves, J.B. (1998) Larval survival and development of the fall armyworm (Lepidoptera: Noctuidae) on normal and transgenic cotton expressing the *Bacillus thuringiensis* CryIA(c) δ-endotoxin. *Journal of Economic Entomology* 91, 539–545.

Anderson, T.E., Kennedy, G.G. and Stinner, R.E. (1984) Distribution of the European corn borer, *Ostrinia nubilalis* (Hübner) (Lepidoptera: Pyralidae), as related to oviposition preference of the spring-colonizing generation in eastern North Carolina. *Environmental Entomology* 13, 248–251.

Annis, B. and O'Keefe, L.E. (1987) Influence of pea genotype on parasitization of the pea weevil, *Bruchus pisorum* (Coleoptera: Bruchidae) by *Eupteromalus leguminis* (Hymenoptera: Pteromalidae). *Environmental Entomology* 16, 653–655.

Arpaia, S., Gould, F. and Kennedy, G.G. (1997) Potential impact of *Coleomegilla maculata* predation on adaptation of *Leptinotarsa decemlineata* to *Bt*-transgenic potatoes. *Entomologia Experimentalis et Applicata* 82, 91–100.

Bell, H.A., Fitches, E.C., Down, R.E., Marris, G.C., Edwards, J.P., Gatehouse, J.A. and Gatehouse, A.M.R. (1999) The effect of snowdrop lectin (GNA) delivered via artificial diet and transgenic plants on *Eulophus pennicornis* (Hymenoptera: Eulophidae), a parasitoid of the tomato moth *Lacanobia oleracea* (Lepidoptera: Noctuidae). *Journal of Insect Physiology* 45, 983–991.

Benrey, B. and Denno, R.F. (1997) The slow growth-high mortality hypothesis: a test using the cabbage butterfly. *Ecology* 78, 987–999.

Bernal, J.S. and Sétamou, M. (2003) Fortuitous antixenosis in transgenic sugarcane: antibiosis-expressing cultivar deters oviposition by herbivore pests. *Environmental Entomology* 32, 886–894.

Bernal, J.S., Griset, J.G. and Gillogly, P.O. (2002) Impacts of developing on *Bt* maize-intoxicated hosts on fitness parameters of a stemborer parasitoid. *Journal of Entomological Science* 37, 27–40.

Bhat, M.G. (1999) On biodiversity access, intellectual property rights, and conservation. *Ecological Economics* 29, 391–403.

Birch, A.N.E., Geoghegan, I.E., Majerus, M.E.N., McNicol, J.W., Hackett, C., Gatehouse, A.M.R. and Gatehouse, J.A. (1999) Tri-trophic interactions involving pest aphids, predatory 2-spot ladybirds and transgenic potatoes expressing snowdrop lectin for aphid resistance. *Molecular Breeding* 5, 75–83.

Board on Agriculture (1989) *Alternative Agriculture*. National Academy of Sciences, Washington, DC.

Bottrell, D.G. and Weil, R.R. (1995) Protecting crops and the environment: striving for durability. In: Juo, A.S.R. and Freed, R.D. (eds) *Agriculture and the Environment: Bridging Food Production and Environmental Protection in Developing Countries*. American Society of Agronomy, Madison, Wisconsin, pp. 55–73.

Bottrell, D.G., Barbosa, P. and Gould, F. (1998) Manipulating natural enemies by plant variety selection and modification: a realistic strategy? *Annual Review of Entomology* 43, 347–367.

Bradley, J.R. (1995) Expectations for transgenic *Bt* cotton: are they realistic? In: Herber, D.J. and Richter, D.A. (eds) *Proceedings of the Beltwide Cotton Conference*. National Cotton Council of America, Memphis, Tennessee, pp. 763–765.

Brandenburg, R.L. and Kennedy, G.G. (1982) Intercrop relationships and spider mite dispersal in a corn/peanut agro-ecosystem. *Entomologia Experimentalis et Applicata* 32, 269–276.

Brazzle, J.R., Heinz, K.M. and Parrella, M.P. (1997) Multivariate approach to identifying patterns of *Bemisia argentifolii* (Homoptera: Aleyrodidae) infesting cotton. *Environmental Entomology* 26, 995–1003.

Cannon, R.J.C. (1996) *Bacillus thuringiensis* use in agriculture: a molecular perspective. *Biological Reviews* 71, 561–636.

Chandler, L.D. and Faust, R.M. (1998) Overview of area wide management of insects. *Journal of Agricultural Entomology* 15, 319–325.

Couty, A., de la Viña, G., Clark, S.J., Kaiser, L., Pham-Delegue, M.-H. and Poppy, G.M. (2001) Direct and indirect sublethal effects of *Galanthus nivalis* agglutinin (GNA) on the development of a potato-aphid parasitoid, *Aphelinus abdominalis* (Hymenoptera: Aphelinidae). *Journal of Insect Physiology* 47, 553–561.

DeBach, P. and Rosen, D. (1991) *Biological Control by Natural Enemies.* Cambridge University Press, Cambridge.

De Leo, F., Bonadé-Bottino, M.A., Ceci, L.R., Gallerani, R. and Jouanin, L. (1998) Opposite effects on *Spodoptera littoralis* larvae of high expression level of a trypsin proteinase inhibitor in transgenic plants. *Plant Physiolology* 118, 997–1004.

De Souza Silva, J. (1995) Plant intellectual property rights: the rise of nature as commodity. In: Peritore, N.P. and Galve-Peritore, A.K. (eds) *Biotechnology in Latin America: Politics, Impacts, and Risks.* SR Books, Wilmington, North Dakota, pp. 57–67.

Devine, G.J., Wright, D.J. and Denholm, I. (2000) A parasitic wasp (*Eretmocerus mundus* Mercet) can exploit chemically induced delays in the development rates of its whitefly host (*Bemisia tabaci* Genn.). *Biological Control* 19, 64–75.

Dulmage, H.T. (1981) Insecticidal activity of isolates of *Bacillus thuringiensis* and their potential for pest control. In: Burgess, H.D. (ed.) *Microbial Control of Pests and Plant Diseases 1970–1980.* Academic Press, London, pp. 193–223.

Ellstrand, N.C., Prentice, H.C. and Hancock, J.F. (1999) Gene flow and introgression from domesticated plants into their wild relatives. *Annual Review of Ecology and Systematics* 30, 539–563.

ERS-USDA (1999) *Genetically Engineered Crops for Pest Management.* Available at http://www.econ.ag.gov/whatsnew/issues/biotech/

FAO (1967) *Report of the First Session of the FAO Panel of Experts on Integrated Pest Control.* FAO, Rome.

Federici, B.A. (1999) *Bacillus thuringiensis* in biological control. In: Bellows, T.S. and Fisher, T.W. (eds) *Handbook of Biological Control.* Academic Press, San Diego, California, pp. 575–593.

Feeny, P. (1976) Plant apparency and chemical defence. In: Wallace, J.W. and Mansell, R.L. (eds) *Recent Advances in Phytochemistry*, Vol. 10, *Biochemical Interaction Between Plants and Insects.* Plenum Press, New York, pp. 1–40.

Fernández-Cornejo, J. and McBride, W.D. (2000) *Genetically Engineered Crops for Pest Management in US Agriculture: Farm-level Effects.* Agricultural Economic Report No. 786, Economic Research Service, US Department of Agriculture, Washington, DC.

Ferro, D.N. and McNiel, J.N. (1998) Habitat enhancement and conservation of natural enemies of insects. In: Barbosa, P. (ed.) *Conservation Biological Control.* Academic Press, San Diego, California, pp. 123–132.

Fitches, E. and Gatehouse, J.A. (1998) A comparison of the short and long term effects of insecticidal lectins on the activities of soluble and brush border enzymes of tomato moth larvae (*Lacanobia oleracea*). *Journal of Insect Physiology* 44, 1213–1224.

Gatehouse, A.M.R. and Gatehouse, J.A. (1998) Identifying proteins with insecticidal activity: use of encoding genes to produce insect-resistant transgenic crops. *Pesticide Science* 52, 165–175.

Gatehouse, A.M.R., Boulter, D. and Hilder, V.A. (1991) Novel insect resistance using protease inhibitor genes. In: Dennis, E.S. and Llewellyn, D.J. (eds) *Molecular Approaches to Crop Improvement.* Springer-Verlag, Vienna, pp. 63–77.

Gianessi, L.P. and Carpenter, J.E. (1999) *Agricultural Biotechnology: Insect Control Benefits.* National Center for Food and Agricultural Policy. Available at: http://www.bio.org/food&ag/bioins01.html

Gill, S.S., Cowles, E.A. and Pietrantonio, P. (1992) The mode of action of *Bacillus thuringiensis* endotoxins. *Annual Review of Entomology* 37, 615–636.

González, D. and Wilson, L.T. (1982) A food-web approach to economic thresholds: a sequence of pests/predaceous arthropods on California cotton. *Entomophaga* 27, 31–43.

Gordh, G., Legner, E.F. and Caltagirone, L.E. (1999) Biology of parasitic Hymenoptera. In: Bellows, T.S. and Fisher, T.W. (eds) *Handbook of Biological Control.* Academic Press, San Diego, California, pp. 355–381.

Gould, F. (1998) Sustainability of transgenic insecticidal cultivars: integrating pest genetics and ecology. *Annual Review of Entomology* 43, 701–726.

Gould, F., Kennedy, G.G. and Johnson, M.T. (1991) Effects of natural enemies on the rate of herbivore adaptation to resistant plants. *Entomologia Experimentalis et Applicata* 58, 1–14.
Hagen, K.S. (1986) Ecosystem analysis: plant cultivars (HPR), entomophagous species and food supplements. In: Boethel, D. and Eikenbary, R.D. (eds) *Interactions of Plant Resistance and Parasitoids and Predators of Insects*. Ellis Horwood, New York, pp. 151–198.
Hagen, K.S., Bombosch, S. and McMurtry, J.A. (1976) The biology and impact of predators. In: Huffaker, C.B. and Messenger, P.S. (eds) *Theory and Practice of Biological Control*. Academic Press, New York, pp. 93–142.
Haggstrom, H. and Larsson, S. (1995) Slow larval growth on a suboptimal willow results in high predation mortality in the leaf beetle *Galerucella lineola*. *Oecologia* 104, 308–315.
Hilbeck, A., Baumgartner, M., Fried, P.M. and Bigler, F. (1998) Effects of transgenic *Bacillus thuringiensis* corn-fed prey on mortality and development time of immature *Chrysoperla carnea* (Neuroptera: Chrysopidae). *Environmental Entomology* 27, 480–487.
Hilbeck, A., Meier, M.S. and Raps, A. (2000) *Review on Non-target Organisms and Bt-plants*. Report to Greenpeace International, Amsterdam. Ecostrat, Zurich.
Hilder, V.A., Gatehouse, A.M.R., Sheerman, S.E., Barker, F. and Boulter, D. (1987) A novel mechanism of insect resistance engineered into tobacco. *Nature* 330, 160–163.
Holl, K., Daily, G. and Ehrlich, P.R. (1990) Integrated pest management in Latin America. *Environmental Conservation* 17, 341–350.
Honêk, A. (1982) The distribution of overwintered *Coccinella septempunctata* L. (Col., Coccinellidae) adults in agricultural crops. *Zeitschrift für Angewandte Entomologie* 94, 311–319.
Hoy, C.W., Feldman, J., Gould, F., Kennedy, G.G., Reed, G. and Wyman, J.A. (1998) Naturally occurring biological controls in genetically engineered crops. In: Barbosa, P. (ed.) *Conservation Biological Control*. Academic Press, San Diego, California, pp. 185–205.
Hubbell, B.J. and Welsh, R. (1998) Transgenic crops: engineering a more sustainable agriculture? *Agriculture and Human Values* 15, 43–56.
ISAAA (International Service for the Acquisition of Agri-biotech Applications) (2002) *Global GM Crop Area Continues to Grow and Exceeds 50 Million Hectares for First Time in 2001*. Available at: http://www.isaaa.org/press%20release/Global%20Area_Jan2002.htm
Ives, P.M. (1981) Estimation of coccinellid numbers and movement in the field. *Canadian Entomologist* 113, 981–997.
Jervis, M.A., Kidd, N.A.C., Fitton, M.G., Huddleston, T. and Dawah, H.A. (1993) Flower-visiting by hymenopteran parasitoids. *Journal of Natural History* 27, 67–105.
Jesse, L.C.H. and Obrycki, J.J. (2000) Field deposition of *Bt* transgenic corn pollen: lethal effects on the monarch butterfly. *Oecologia* 125, 241–248.
Johnson, M.T. and Gould, F. (1992) Interaction of genetically engineered host plant resistance and natural enemies of *Heliothis virescens* (Lepidoptera: Noctuidae) in tobacco. *Environmental Entomology* 21, 586–597.
Jones, K.D. (1994) Aspects of the biology and biological control of the European corn borer in North Carolina. PhD thesis, North Carolina State University, Raleigh, North Carolina 127 pp.
Kennedy, G.G. and Storer, N.P. (2000) Life systems of polyphagous arthropod pests in temporally unstable cropping systems. *Annual Review of Entomology* 45, 467–493.
Kieckhefer, R.W. and Miller, E.L. (1967) Trends of populations of aphid predators on South Dakota cereal crops – 1963–1965. *Annals of the Entomological Society of America* 60, 516–518.
King, A.B. and Eyzaguirre, P.B. (1999) Intellectual property rights and agricultural biodiversity: literature addressing the suitability of IPR for the protection of indigenous resources. *Agriculture and Human Values* 16, 41–49.
Kogan, M., Gerling, D. and Maddox, J.V. (1999) Enhancement of biological control in annual agricultural environments. In: Bellows, T.S. and Fisher, T.W. (eds) *Handbook of Biological Control*. Academic Press, San Diego, California, pp. 789–818.
Koziel, M.G., Carozzi, N.B., Currier, T.C., Warren, G.W. and Evola, S.V. (1993) The insecticidal crystal proteins of *Bacillus thuringiensis*: past, present and future uses. In: Tombs, M.P. (ed.) *Biotechnology and Genetic Engineering Reviews*, Vol. II. Intercept Press, Andover, Massachusetts, pp. 171–228.
Legaspi, J.C., Legaspi, B.C., Bernal, J.S. and Sétamou, M. (2004) Insect resistant transgenic crops expressing plant lectins. In: Koul, O. and Dhaliwal, G.S. (eds) *Transgenic Crop Protection: Concepts and Strategies*. Science Publishers, New Hampshire.
Lichtenberg, E. and Zimmerman, R. (1999) Adverse health experiences, environmental attitudes, and pesticide usage behavior of farm operators. *Risk Analysis* 19, 283–294.

Losey, J.E., Rayor, L.S. and Carter, M.E. (1999) Transgenic pollen harms monarch larvae. *Nature* 399, 214.

Luck, R.F., Tauber, M.J. and Tauber, C.A. (1995) The contributions of biological control to population and evolutionary ecology. In: Nechols, J.R., Andres, L.A., Beardsley, J.W., Goeden, R.D. and Jackson, C.G. (eds) *Biological Control in the Western United States, Accomplishments and Benefits of Regional Research Project W-84, 1964–1989*. Division of Agriculture and Natural Resources, University of California, Oakland, California, pp. 25–45.

Lynch, R.E., Wiseman, B.R., Plaisted, D. and Warnick, D. (1999) Evaluation of transgenic sweet corn hybrids expressing CryIA(b) toxin for resistance to corn earworm and fall armyworm (Lepidoptera: Noctuidae). *Journal of Economic Entomology* 92, 246–252.

Machuka, J. (2001) Agricultural biotechnology for Africa: African scientists and farmers must feed their own people. *Plant Physiology* 126, 16–19.

Margolies, D.C. and Kennedy, G.G. (1985) Movement of the twospotted spider mite, *Tetranychus urticae*, among hosts in a corn–peanut agroecosystem. *Entomologia Experimentalis et Applicata* 36, 193–196.

McConnell, R. and Hruska, A. (1993) An epidemic of pesticide poisoning in Nicaragua: implications for prevention in developing countries. *American Journal of Public Health* 83, 1559–1562.

McLaren, J.S. (2000) The importance of genomics to the future of crop production. *Pest Management Science* 56, 573–579.

Moore, C.C., Benedict, J.H., Fuchs, T.W., Friesen, R.D. and Payne, C. (1999) *Bt Cotton Technology in Texas: a Practical View*. Bulletin B-6107, Texas Agricultural Extension Service, College Station, Texas.

Morris, M.L. and López-Pereira, M.A. (1999) *Impacts of Maize Breeding Research in Latin America 1966–1997*. CIMMYT, Mexico.

Murray, D.L. (1994) *Cultivating Crisis: the Human Cost of Pesticides in Latin America*. University of Texas Press, Austin, Texas.

Neunzig, H.H. (1963) Wild host plants of the corn earworm and the tobacco budworm in eastern North Carolina. *Journal of Economic Entomology* 56, 135–139.

Neunzig, H.H. (1969) *The Biology of the Tobacco Budworm and the Corn Earworm in North Carolina with Particular Reference to Tobacco as a Host*. Technical Bulletin 196, North Carolina Agricultural Experiment Station, Raleigh, North Carolina.

Nicholls, C.I. and Altieri, M.A. (1997) Conventional agricultural development models and the persistence of the pesticide treadmill in Latin America. *International Journal of Sustainable Development and World Ecology* 4, 93–111.

NRC (Committee on Pesticides in the Diets of Infants and Children) (1993) *Pesticides in the Diets of Infants and Children*. National Academy Press, Washington, DC.

NRC (Committee on Genetically Modified Pest-protected Plants) (2000) *Genetically Modified Pest-protected Plants: Science and Regulation*. National Academy Press, Washington, DC.

Pair, S.D., Wiseman, B.R. and Sparks, A.N. (1986) Influence of four corn cultivars on fall armyworm (Lepidoptera: Noctuidae) establishment and parasitization. *Florida Entomologist* 69, 566–570.

Pels, B. and Sabelis, M.W. (2000) Do herbivore-induced plant volatiles influence predator migration and local dynamics of herbivorous and predatory mites? *Experimental and Applied Acarology* 24, 427–440.

Perkins, J.H. (1982) *Insects, Experts, and the Insecticide Crisis: the Quest for New Pest Management Strategies*. Plenum Press, New York.

Perlak, F.J., Deaton, R.W., Armstrong, T.A., Fuchs, R.L., Sims, S.R., Greenplate, J.T. and Fischoff, D.A. (1990) Insect resistant cotton plants. *Bio/Technology* 8, 939–943.

Pimentel, D., McLaughlin, L., Zepp, A., Lakitan, B., Kraus, T., Kleinman, P., Vancini, F., Roach, W.J., Graap, E., Keeton, W.S. and Selig, G. (1993) Environmental and economic effects of reducing pesticide use in agriculture. *Agriculture, Ecosystems and Environment* 46, 273–288.

Porter, W.P., Jaeger, J.W. and Carlson, I.H. (1999) Endocrine, immune, and behavioral effects of aldicarb (carbamate), atrazine (triazine), and nitrate (fertilizer) mixtures at groundwater concentrations. *Toxicology and Industrial Health* 15, 133–150.

Poston, F.L. and Pedigo, L.P. (1975) Migration of plant bugs and the potato leafhopper in a soybean–alfalfa complex. *Environmental Entomology* 4, 8–10.

Potting, R.P.J., Vermeulen, N.E. and Conlong, D.E. (1999) Active defence of herbivorous hosts against parasitism: adult parasitoid mortality risk involved in attacking a concealed stemboring host. *Entomologia Experimentalis et Applicata* 91, 143–148.

Powell, K.S., Spence, J., Bharathi, M., Gatehouse, J.A. and Gatehouse, A.M.R. (1998) Immunohistochemical and developmental studies to elucidate the mechanism of action of the snowdrop lectin on the rice brown planthopper, *Nilaparvata lugens* (Stal). *Journal of Insect Physiology* 44, 529–539.

Prasifka, J.R., Krauter, P.C., Heinz, K.M., Sansone, C.G. and Minzenmayer, R.R. (1999) Predator conservation in cotton: using grain sorghum as a source for insect predators. *Biological Control* 16, 223–229.
Price, P.W. (1976) Colonization of crops by arthropods: non-equilibrium communities in soybean fields. *Environmental Entomology* 5, 605–611.
Price, P.W., Bouton, C.E., Gross, P., McPheron, B.A., Thompson, J.N. and Weiss, J.E. (1980) Interactions among three trophic levels: influence of plants on interactions between insect herbivores and natural enemies. *Annual Review of Ecology and Systematics* 11, 41–65.
Puterka, G.J., Slosser, J.E. and Price, J.R. (1985) Parasites of *Heliothis* spp. (Lepidoptera: Noctuidae): parasitism and seasonal occurrence for host crops in the Texas Rolling Plains. *Environmental Entomology* 14, 441–446.
Regev, U. (1984) An economic analysis of man's addiction to pesticides. In: Conway, G.R. (ed.) *Pest and Pathogen Control: Strategic, Tactical, and Policy Models*. John Wiley & Sons, New York, pp. 441–453.
Repetto, R. (1985) *Paying the Price: Pesticide Subsidies in Developing Countries*. World Resources Institute, Washington, DC.
Rhoades, D.F. and Cates, R.G. (1976) Towards a general theory of plant antiherbivore chemistry. In: Wallace, J.W. and Mansell, R.L. (eds) *Recent Advances in Phytochemistry*, Vol. 10, *Biochemical Interaction Between Plants and Insects*. Plenum Press, New York, pp. 168–213.
Rice, M.E. and Pilcher, C.D. (1998) Potential benefits and limitations of transgenic *Bt* corn for management of the European corn borer (Lepidoptera: Crambidae). *American Entomologist* 44, 75–78.
Savinelli, C.E., Bacheler, J.S. and Bradley, J.R. Jr (1986) Nature and distribution of European corn borer (Lepidoptera: Pyralidae) larval feeding damage to cotton in North Carolina. *Environmental Entomology* 15, 399–402.
Savinelli, C.E., Bacheler, J.S. and Bradley, J.R. Jr (1988) Ovipositional preferences of the European corn borer (Lepidoptera: Pyralidae) for field corn and cotton under field cage conditions in North Carolina. *Environmental Entomology* 17, 688–690.
Schnepf, E., Crickmore, N., Van Rie, J., Lereclus, D., Baum, J. and Feitelson, J. (1998) *Bacillus thuringiensis* and its pesticidal proteins. *Microbiology and Molecular Biology Reviews* 62, 775–806.
Schuler, T.H., Poppy, G.M., Kerry, B.R. and Denholm, I. (1999) Potential side effects of insect-resistant transgenic plants on arthropod natural enemies. *Trends in Biotechnology* 17, 210–216.
Sétamou, M., Bernal, J.S., Legaspi, J.C. and Mirkov, T.E. (2002a) Effects of snowdrop lectin (GNA) expressed in transgenic sugarcane on fitness of *Cotesia flavipes* (Cameron), a parasitoid of the non-target pest *Diatraea saccharalis* F. *Annals of the Entomological Society of America* 95, 75–83.
Sétamou, M., Bernal, J.S., Legaspi, J.C., Mirkov, T.E. and Legaspi, B.C. Jr (2002b) Evaluation of lectin-expressing transgenic sugarcane against stalkborers (Lepidoptera: Pyralidae): effects on life history parameters and damage. *Journal of Economic Entomology* 95, 469–477.
Sétamou, M., Bernal, J.S., Legaspi, J.C. and Mirkov, T.E. (2002c) Parasitism and location of sugarcane borer hosts by *Cotesia flavipes* (Cameron) (Hymenoptera: Braconidae) on transgenic and conventional sugarcane. *Environmental Entomology* 31, 1219–1225.
Settle, W.H., Ariawan, H., Astuti, E.T., Cahyana, W., Hakim, A.L., Hindayana, D., Lestari, A.S. and Sartanto, P. (1996) Managing tropical rice pests through conservation of generalist natural enemies and alternative prey. *Ecology* 77, 1975–1988.
Shand, H. (1991) There is a conflict between intellectual property rights and the rights of farmers in developing countries. *Journal of Agricultural and Environmental Ethics* 4, 131–142.
Shu, Q., Ye, G., Cui, H., Cheng, X., Xiang, Y., Wu, D., Gao, M., Xia, Y., Hu, C., Sardana, R. and Altosaar, I. (2000) Transgenic rice plants with a synthetic *cry1Ab* gene from *Bacillus thuringiensis* were highly resistant to eight lepidopteran rice pest species. *Entomologia Experimentalis et Applicata* 6, 433–439.
Silvie, P., Deguine, J.P., Nibouche, S., Michel, B. and Vaissayre, M. (2001) Potential of threshold-based interventions for cotton pest control by small farmers in West Africa. *Crop Protection* 20, 297–301.
Smith, R.F. and van den Bosch, R. (1967) Integrated control. In: Kilgore, W.W. and Doutt, R.L (eds) *Pest Control: Biological, Physical, and Selected Chemical Methods*. Academic Press, New York, pp. 295–340.
Stern, V.M., Smith, R.F., van den Bosch, R. and Hagen, K.S. (1959) The integrated control concept. *Hilgardia* 29, 81–101.
Thompson, S.N. and Hagen, K.S. (1999) Nutrition of entomophagous insects and other arthropods. In: Bellows, T.S. and Fisher, T.W. (eds) *Handbook of Biological Control*. Academic Press, San Diego, California, pp. 594–692.
Thrupp, L.A. (1995) *Bittersweet Harvests for Global Supermarkets: Challenges in Latin America's Agricultural Export Boom*. World Resources Institute, Washington, DC.

Tomov, B.W., Bernal, J.S. and Vinson, S.B. (2003) Impacts of transgenic sugarcane expressing GNA lectin on parasitism of Mexican rice borer by *Parallorhogas pyvalophagus* (Marsh) (Hymenoptera: Braconidae). *Environmental Entomology* 32, 866–872.

Turlings, T.C.J. and Benrey, B. (1998) Effects of plant metabolites on the behaviour and development of parasitic wasps. *Ecoscience* 5, 321–333.

Umeozor, O.C., Van Duyn, J.W., Bradley, J.R. Jr and Kennedy, G.G. (1986) Intercrop effects on the distribution of populations of the European corn borer, *Ostrinia nubilalis,* in maize. *Entomologia Experimentalis et Applicata* 40, 293–296.

Vaeck, M., Reynaerts, A., Höfte, H., Jansens, S., De Beuckeleer, M., Dean, C., Zabeau, M., Monatgu, M.V. and Leemans, J. (1987) Transgenic plants protected from insect attack. *Nature* 328, 33–37.

Venzon, M., Janssen, A. and Sabelis, M.W. (1999) Attraction of a generalist predator towards herbivore-infested plants. *Entomologia Experimentalis et Applicata* 93, 305–314.

Visser, B., Bosch, D. and Honée, G. (1993) Domain function studies of *Bacillus thuringiensis* crystal proteins: a genetic approach. In: Entwistle, P.F., Cory, J.S., Bailey, M.J. and Higgs, S. (eds) Bacillus thuringiensis, *an Environmental Biopesticide: Theory and Practice.* John Wiley & Sons, Chichester, UK, pp. 71–88.

Vorley, V.T. and Wratten, S.D. (1987) Migration of parasitoids (Hymenoptera: Braconidae) of cereal aphids (Hemiptera: Aphididae) between grassland, early-sown cereals and late-sown cereals in southern England. *Bulletin of Entomological Research* 77, 555–568.

Warren, G.W., Carozzi, N.B., Desai, N. and Koziel, M.G. (1992) Field evaluation of transgenic tobacco containing a *Bacillus thuringiensis* insecticidal protein gene. *Journal of Economic Entomology* 85, 1651–1659.

Wilson, F.D., Flint, H.M., Deaton, W.R., Fischhoff, D.A., Perlak, F.J., Armstrong, T.A., Fuchs, R.L., Berberich, S.A., Parks, N.J. and Stapp, B.R. (1992) Resistance of cotton lines containing a *Bacillus thuringiensis* toxin to pink bollworm (Lepidoptera: Gelechiidae) and other insects. *Journal of Economic Entomology* 85, 1516–1521.

Wilson, F.D., Flint, H.M., Deaton, W.R. and Buehler, R.E. (1994) Yield, yield components, and fiber properties of insect-resistant cotton lines containing a *Bacillus thuringiensis* toxic gene. *Crop Science* 34, 38–41.

Wissinger, S.A. (1997) Cyclic colonization in predictably ephemeral habitats: a template for biological control in annual crop systems. *Biological Control* 10, 4–15.

Wraight, C.L., Zangerl, A.R., Carroll, M.J. and Berenbaum, M.R. (2000) Absence of toxicity of *Bacillus thuringiensis* pollen to black swallowtails under field conditions. *Proceedings of the National Academy of Science USA* 97, 7700–7703.

Wratten, S.D. and Thomas, C.F.G. (1990) Farm-scale spatial dynamics of predators and parasitoids in agricultural landscapes. In: Bunce, R.G.H. and Howard, C.G. (eds) *Species Dispersal in Agricultural Habitats.* Belhaven Press, London, pp. 219–237.

Ye, X.D., Al-Babili, S., Kloti, A., Zhang, J., Lucca, P., Beyer, P. and Potrykus, I. (2000) Engineering the provitamin A (beta-carotene) biosynthetic pathway into (carotenoid-free) rice endosperm. *Science* 287, 303–305.

7 Plant Resistance against Pests: Issues and Strategies

C. Michael Smith
Department of Entomology, Kansas State University, Manhattan, KS 66506–4004, USA
E-Mail: cmsmith@ksu.edu

Introduction

The production of crop plants with heritable arthropod-resistant traits has been recognized for more than 100 years as a sound approach to crop protection (Painter, 1951; Smith, 1999). Today, hundreds of arthopod-resistant crop cultivars are grown globally, representing the products of many successful cooperative research efforts between entomologists and plant breeders. These efforts have significantly improved world food production, helped to alleviate hunger, improved the nutrition of many populations and transformed many food-importing nations into food exporters. Numerous authors have chronicled the development of plant resistance as a science and as a valuable tool in integrated pest management (IPM) (Snelling, 1941; Painter, 1951; Chesnokov, 1953; Russell, 1978; Lara, 1979; Panda, 1979; Maxwell and Jennings, 1980; Smith, 1989; Dhaliwal and Dilawari, 1993; Smith *et al.*, 1994; Panda and Khush, 1995; Dhaliwal and Singh, 2004).

Benefits of Resistance

Insect-resistant cultivars are highly cost-effective components of IPM systems, and many examples demonstrate how they provide substantial returns on economic investment. In the USA, insect-resistant sorghum cultivars increase producer profits by several hundred million dollars each year (Eddleman *et al.*, 1999). The use of *Bacillus thuringiensis* (*Bt*) transgenic maize hybrids (see below and Fig. 7.1) currently increases US maize producer profits by approximately 7% per year. In Morocco, Hessian-fly-resistant bread wheats have been shown to provide a 9:1 return on investment of research (Azzam *et al.*, 1997). The economic value of genetic resistance in wheat to all major worldwide arthropod pests amounts to more than US$250 million/year (Smith *et al.*, 1999). The multiple insect-resistant rice cultivar IR36 provided approximately US$1 billion of additional annual income to rice producers and processors in Asia for over 20 years (Khush and Brar, 1991). The returns on resistant-cultivar research compared with insecticides range from 100:1 to 10:1, making them a valuable component of crop production and greatly improving the competitiveness of crop producers in many countries. Insect resistant-cultivars also provide substantial environmental benefits and contribute to reduced crop insecticide use and insecticide residues, thus improving the food, health and safety of consumers. The production of insect-resistant cultivars also helps to protect groundwater from pesticide contamination.

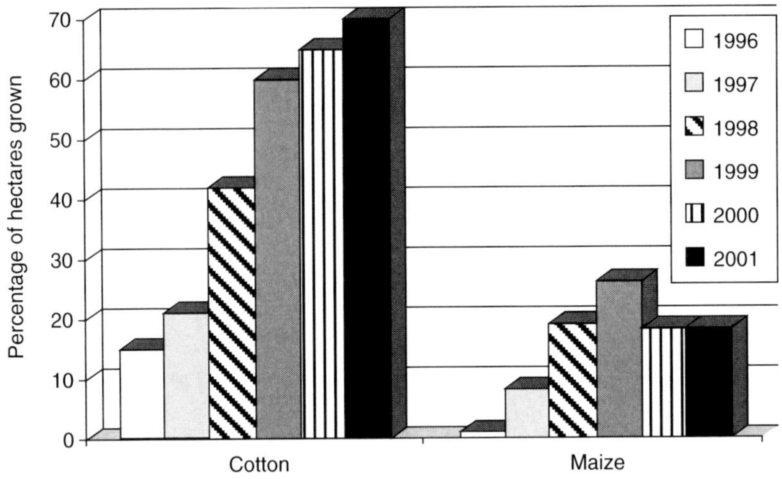

Fig. 7.1. Percentage of cotton and maize cultivars produced in the USA from 1996 to 2001 containing *Bacillus thuringiensis* delta endotoxins (from James, 2000; Anon., 2001).

Issues and Strategies: Recent and Future

There were three predominant scientific issues that affected the thoughts and actions of plant-resistance scientists in the decade of the 1990s. These issues were the development and deployment of the first transgenic insect-resistant cultivars, the discovery of the first molecular markers linked to plant genes for arthropod resistance and the first cloning and sequencing of a plant gene expressing insect resistance.

Transgenes from the bacterium, *Bt*, which encode δ-endotoxin insecticidal proteins, were expressed in the first commercial transgenic cotton, maize and potato cultivars. Although controversial, *Bt* crops are currently marketed and produced in Australia, Canada, China, India, South Africa and the USA. There is a trend towards *Bt* crops becoming more prevalent in global agriculture. Many other proteins toxic to arthropods have been identified and transgenes encoding several of these inhibitors have been used to transform plants expressing insect resistance.

Molecular markers have been used to map conventional insect-resistance genes in several crops (Yencho *et al.*, 2000). Nevertheless, the *Meu-1.2* gene of wild tomato, *Lycopersicon peruvianum*, which expresses resistance to the potato aphid, *Macrosiphum euphorbiae* (Thomas), and three *Meloidogyne* spp. nematodes, is the only insect-resistance gene to be sequenced (Milligan *et al.*, 1998). Bioinformatic computational tools to analyse, interpret and utilize huge amounts of data being generated by genomics research on several major crop plants have already provided genetic maps, physical maps and expressed sequence tag (EST) complementary DNA (cDNA) libraries of several major crop-plant genomes. Future insect-resistance gene cloning and sequence determination will probably proceed by 'data-mining' the genomic information from plant resistance-gene analogues (RGAs), defence response (DR) genes and EST libraries.

Each of these issues has been researched, debated and implemented to varying degrees. In the following sections, information is presented to demonstrate how each issue has become or potentially will become a strategy used by many plant-resistance practitioners in the 21st century.

Transgenic Insect Resistance

A major change occurred in the development of insect-resistant varieties near the end of the 20th century, when *Agrobacterium* transformation systems and biolistic projectile devices were used to transfer genes encoding insecticidal crystal (cry) toxins from the soil bacterium *Bt* into the genome of naïve plant cells. Transformed plants resulting from cell and tissue culture were grown to maturity and produced transgenic seed. When transgenic plant foliage is fed upon by pest insects, ingested crystals are solubilized in the alkaline gut environment, where active toxic fragment(s) are released by insect digestion, and these fragments bind to specific receptors on the midgut cells of susceptible larvae, causing colloid osmotic lysis of those cells, resulting in insect death.

Numerous other proteins toxic to insects have also been identified and expressed in transgenic plants (see reviews by Sharma *et al.*, 2000; Oppert, 2001; Lawrence and Koundal, 2002). These include the carbohydrate-binding proteins lectins; proteinase inhibitors from maize, potato, rice and tomato; proteinase inhibitors from insects; chymotrypsin and trypsin inhibitors from cowpea and sweet potato; and α-amylase inhibitors from common bean. Transgenes encoding several of these inhibitors have been transferred into various crop plants, including bean, cotton, potato and rice. Despite these impressive accomplishments, the general usefulness of transgenic plants containing non-*Bt* toxic proteins in plant protection remains to be implemented.

Many transgenic *Bt* crop plants have been developed (see review by Huang *et al.*, 1999), but only cotton, maize and potato cultivars with transgenes expressing resistance to Coleoptera and Lepidoptera have been produced and marketed. These crops are essentially insecticidal plants, which has complicated their deployment. *Bt* crop-plant production and use have met with strong opposition by environmentalists, primarily in Europe and the USA. The marketing of the first *Bt* cultivars in 1994 launched an intense scientific debate on how much *Bt* is sufficient for effectiveness without selecting for pest population resistance to *Bt*. By 1998, *Bt* maize production had increased to approximately 20% of the US crop and the first *Bt* educational publications were published to provide a scientific basis for the debate (Rissler and Mellon, 1996; Wayland *et al.*, 1998).

Initial experiments by Losey *et al.* (1999) indicated that *Bt* maize pollen applied at very high concentrations to leaves of the milkweed plant was toxic to larvae of the monarch butterfly, *Danaus plexippus* (Linnaeus), feeding on milkweed. Wraight *et al.* (2000) reported no mortality of the larvae of black swallowtail, *Papilio polyxenes* Fabricius, on food plants located at varying distances from field plantings of *Bt* maize, no matter how close the larval food plants were to the pollen-shedding *Bt* maize plants. Subsequent studies by Sears *et al.* (2001) and Stanley-Horn *et al.* (2002) concluded that the risk from *Bt* maize to monarchs is not significant. One *Bt* maize cultivar on only 2% of the annual US crop hectarage was shown to be toxic to monarch larvae and has been eliminated from production. Pimentel and Raven (2000) classified the effects of *Bt* pollen on the food plants of several non-target US butterfly species as relatively insignificant, in comparison with maize pesticide applications and butterfly abiotic mortality factors, such as habitat destruction. Several studies have shown that *Bt* maize has limited effects on beneficial arthropods in maize agroecosystems (Johnson and Gould, 1992; Pilcher *et al.*, 1997; Al-Deeb *et al.*, 2001).

By 1999, the American Phytopathological Society proclaimed that threats to human health were reduced by *Bt* maize production because of reduced incidence of potentially dangerous mycotoxins (APS, 1999). Nevertheless, differences in media coverage of transgenic crops, European public perceptions of transgenic crops and greater European cultural sensitivities about transgenic crops than those of the US public brought out strong European concerns about transgenic crops (Gaskell *et al.*, 1999). The concerns resulted in the US government developing an independent scientific approval process for *Bt* crops and other genetically modified organisms. Public activism peaked between 1999 and 2001, when US government and private agricultural research facilities valued at more than US$4 million were destroyed by bioterrorists in California, Hawaii, Maine, Michigan, Minnesota, New York, Oregon, Washington and Wisconsin. In addition, greenhouse and field experiments involving genetically modified rape, flowers, fruits, maize, oats, onions, trees and wheat were destroyed (Service, 2001).

Consumer support for transgenic food crops in the USA is strong despite these bioterrorism events. A consumer survey conducted in August 2002 indicated that 71% of the US population favoured purchasing produce that had been enhanced through biotechnology, in order for it to be protected from insect damage and require fewer pesticide applications (IFIC, 2002). Nevertheless, the majority of European countries continue to oppose the use of *Bt* (and other genetically modified) crops. Exceptions are occurring in several European countries, such as the Czech Republic, where *Bt* maize was field-tested in 2002.

Improved governmental decision-making processes, better genetically modified food risk/benefit communication, an increasing volume of research data and an endorsement by the United Nations Food and Agriculture Organization (FAO) have led to increased production of *Bt* cotton and maize in countries other than the USA, including Argentina, China, India and South Africa (see below). In spite of these successes, the FAO has voiced concerns that the majority of transgenic crops focus on reducing chemical inputs and labour costs in large corporate farms of developed countries, and not on increasing food supplies for the populations of underdeveloped countries. To date, there has been little corporate or public investment in important food crops of the semi-arid tropics, such as sorghum, millet, pigeon pea, chickpea or groundnut.

Over the past several years, farm trials in India show *Bt* cotton yield increases of approximately 60% more than those of conventional non-*Bt* cultivars (Qaim and Zilberman, 2003). In 2001, more than a quarter of the maize produced in the USA, and over half of the cotton produced was from planting of *Bt* cultivars (Anon., 2001; Fig. 7.1). Data from a 10-year study conducted by Carrière *et al.* (2003) in Arizona indicate that production of *Bt* cotton significantly suppressed populations of the pink bollworm, *Pectinophora gossypiella* Saunders, and that the deployment of *Bt* cotton cultivars contributed to reducing the need for insecticide sprays. Similar results have been reported for bollworm control in South Africa, where *Bt* cotton use shifted from 7% in the 1997/98 growing season to 90% in the 2001/02 growing season on both small and large farms (Kirsten and Gouse, 2002). The primary benefits have been increased yields from improved bollworm control and related decreased production costs from greatly reduced insecticide usage. Sachs *et al.* (1996) transformed a high-terpenoid-content cotton cultivar with the CryIA(b) protein for increased resistance to the tobacco budworm, *Heliothis virescens* (Fabricius). However, no such conventional gene–transgene combinations have been marketed commercially.

Insecticide use against one US insect pest of maize, the European corn borer, *Ostrinia nubilalis* (Hubner), has dropped by approximately 30% since after the commercialization of *Bt* maize in North America. *Bt* maize has proved to be a particularly effective means of borer control, because larvae feed inside maize stalks and are impossible to kill by conventional foliar insecticide-spray applications (Rice and Pilcher, 1998). In contrast, demand for *Bt* potatoes resistant to the Colorado potato beetle, *Leptinotarsa decemlineata* (Say), peaked in 1995, when growers planted them on 22,260 ha in North America. By 2000 that hectarage had declined by 50%, and in 2001 sales of *Bt* potato seed ceased in the USA and Canada, primarily because food-processing companies were concerned about consumer food preferences and reluctantly chose not to market transgenic foods.

The successes of *Bt* maize and cotton globally in both developed and several developing countries, as well as the lack of success of *Bt* potatoes, point to the critical importance of risk communication in developing consumer and producer understanding of new technologies. Abbott *et al.* (2001) have shown that the media coverage of the development and use of *Bt* maize has followed a predictable pattern similar to that of many other risk issues. The pattern is one in which scientific developments are initially communicated by the media to the public in a positive way, albeit at low levels, and research and industry sources are stressed, without perceptions of public advocacy groups. A 'triggering event', in the case of *Bt* maize the initial Losey *et al.* (1999) data, caused a dramatic increase in media coverage. These events allow increased inclusion of information from advocacy groups, who elect (in the case of *Bt*) to exploit negative scientific data, both real and perceived. A trend to report 'for and against' information rather than scientific truth follows, deluging the public with conflicting information in a short period of time. Public attention to these sources then declines as other issues come forward or new information is publicized. The current US–European dichotomy over the acceptance of biotechnology illustrates this point well, with the perceived benefits of *Bt* crops allowing US consumer acceptance and the perceived detriments continuing European rejection of them.

Future transgenic-technology education and risk communication efforts may benefit from integrating the educational capabilities of different agencies involved in food production, processing and distribution. Such initiatives are illustrated by the development of the International Food Information Council- and the US Environmental Protection Agency (EPA)-led development of *Bt* integrated risk-management programmes (see below).

In order for transgenic-crop risk-assessment training to be effective, however, educators may consider using hazard analysis critical control point (HACCP) principles, described by Cuperus *et al.* (1991) for use in food-safety education. HACCP principles have been used extensively to reduce microbiological contamination in the food-processing industry. The HACCP approach assesses hazards and risks associated with growing, harvesting, processing, manufacturing, marketing and distribution of food products and determines critical control points (CCPs) (a point where loss of control may result in an unacceptable health risk). For a given CCP, critical limits are identified, monitoring procedures are established and corrective actions are taken if necessary. Application of HACCP principles may serve to demonstrate how certain perceived risks are in fact not significant.

The history of the scientific development of *Bt*-crop technology is much less complicated, although at some points involving controversy. Before the development of *Bt* transformants, numerous studies determined that insects became resistant to the *cry* toxin gene after prolonged exposure to a high dose of *Bt* (Huang *et al.*, 1999), in the same manner that insect biotypes develop resistance to high doses of conventional pesticides or high levels of conventional gene expression (Llewellyn *et al.*, 1994). Insect biotypes are well documented in the interactions between genes of the gall midge, *Orseolia oryzae* (Wood-Mason), and brown planthopper, *Nilaparvata lugens* (Stal), and rice (Tanaka, 1999; Pani and Sahu, 2000) and the Hessian fly and wheat (Ratcliffe and Hatchett, 1997). Biotype occurrence is influenced by the genetic plasticity of the pest insect, its ecological fitness, the number of resistance genes expressed, the resistance category expressed, the fraction of the crop cultivated in an insect-resistant cultivar and the overall efficacy of the IPM programme used to control the pest.

Some of these same concepts have been used to develop *Bt* crop-plant deployment strategies. In order to obtain their maximum longevity, *Bt* insect-resistant transgenes in maize and cotton are deployed with non-*Bt* plant 'refuges' that enable the survival of pest moths from susceptible larvae to mate with moths produced from larvae resistant to *Bt*. Shifting the mortality of larvae heterozygous for resistance from 50 to 95% provides a tenfold delay in time before the development of resistance (Gould, 1998). A refuge portion of the crop was not commercially or sociologically acceptable at first, but the US EPA coordinated industry and academic research efforts to establish a mechanism to prevent the development of resistance to *Bt* by key maize insect pests.

In 2001, these efforts led to the creation of an insect resistance management (IRM) compliance assurance programme to promote grower compliance and preserve the effectiveness of *Bt* maize. Manufacturers sponsor an annual survey of *Bt* maize growers, con-

Table 7.1. Comparisons of conventional and transgenic resistance in crop plants (from Daly and Wellings, 1996).

Criteria	Conventional	Transgenic
Resistance category	Antibiosis, antixenosis, tolerance	Antibiosis
Mechanism(s)	Chemical and physical	Chemical
Efficacy	Moderate	High
Expression	Constitutive and induced	Constitutive
Management	Optional	Required
Sociology	Simple	Complex
Stability	High	High
Technology transfer	Moderate	Fast

ducted by an independent third party, and growers not in compliance with IRM requirements over 2 consecutive years are denied additional access to *Bt*-maize seed. *Bt*-crop producers must plant at least a 20% non-*Bt* maize refuge, except in certain cotton-growing areas, where at least a 50% non-*Bt* maize refuge is required. Refuge-planting options include blocks within fields, strips across fields or separate fields. *Bt*-maize fields must be planted within 0.8 km of a refuge. The IRM programmes are a first ever type of government–industry-regulated IPM tecnology. As such, US crop-management programmes relying on transgenic technology have entered a new era in crop production. The IRM programmes are an overall success, although some producers have failed to meet the minimum requirements for non-*Bt* refuge plantings. In an initial 2000 survey, 29% of producers were not in compliance, but since that time participation has improved. In 2001 only 13% of the producers surveyed were not in compliance and in 2002 14% were non-compliant (Byrne *et al.*, 2003). The continued (and improved) successes of these producer–regulatory–industry partnerships will depend on a combination of good science, communication and common sense in making decisions about *Bt*-crop cultivar selection and refuge composition.

Daly and Wellings (1996) contrasted aspects of conventional and transgenic plant resistance to insects (Table 7.1). Conventional plant resistance genes are expressed as antibiosis or antixenosis effects on insects and tolerance of the resistant cultivar to an insect pest. Many biochemical and biophysical mechanisms have been identified to explain resistance (Smith, 1989). Transgenic resistance is expressed solely as antibiosis, due to a digestive toxin. While conventional resistance may have both constitutive and induced components, transgenic resistance is fully constitutive. Both types of resistance genes have high stability and, although transgenes have been part of IPM systems for less than 10 years, the transfer of transgenic technology has occurred very quickly.

The major difference in the two types of resistance genes is the management (IRM) plan for transgenes. Although the initial IRM schemes for *Bt* maize are functional, we do not know if they will continue to be effective. *Bt* maize with resistance to the western maize rootworm, *Diabrotica virgifera* LeConte, a far more damaging pest than the European corn borer, has recently been approved by the US EPA (Knight, 2003). A 20% refuge similar to that for European corn borer has been adopted for initial production purposes, but an advisory panel had recommended a much larger 50% refuge to sufficiently dilute *Bt* virulence alleles in surviving maize rootworm larvae. European corn borer-resistant *Bt*-maize cultivars produce high doses of toxin, but the rootworm-resistant maize cultivars cause only 50% rootworm larval mortality. This reduced maize rootworm *Bt* efficiency and decreased refuge size may lead to a different outcome for maize rootworm-resistant *Bt* maize from that currently developing for *Bt* cultivars resistant to lepidopterous larvae. If the Lepidoptera IRM schemes (and others) continue to function successfully, they may serve as workable models for the deployment of conventional plant resistance genes that are expressed as high levels of antibiosis (insect mortality), in order to delay the development of resistant biotypes.

Molecular Marker-assisted Selection of Plant Genes for Insect Resistance

The tagging and mapping of plant genes for insect resistance has accelerated tremendously since the mid-1990s. This progress has been facilitated by the construction of high-density genetic maps of barley, maize, rye, soybean and wheat (Cregan *et al.*, 1999; Hernández *et al.*, 2001; Korzun *et al.*, 2001; Boyko *et al.*, 2002; Sharopova *et al.*, 2002). Yencho *et al.* (2000) reviewed molecular markers in many of these same crops linked to genes expressing resistance to several major insect pests. In the premolecular age of plant resistance to insects, phenotypic evaluations determined the initial identity of a source of resistance or progeny from crosses made between resistant and susceptible parents. The marker-assisted selection (MAS) of plants based on genotype, before the phenotypic trait for resistance is expressed, is now being used in many plant-improvement programmes.

Demonstrating that a molecular marker is linked to a plant resistance gene, however, involves identifying a phenotypic source of resistance, isolating DNA from resistant and susceptible parent plants, hybridizing or amplifying DNA of resistant and susceptible plants with molecular markers from known chromosome locations to identify those that differentially hybridize or amplify DNA in a polymorphic (informative) pattern via gel electrophoresis, and genotypically screening individual plants from segregating populations for linkage to putative molecular markers. Molecular markers have been identified that are linked both to single major genes for resistance and to groups of loci controlling the expression of quantitative resistance, known as quantitative trait loci (QTLs).

Linkages of a resistance gene and a molecular marker may vary greatly. They may be completely linked, where no crossing over occurs between the resistance gene and the marker during meiosis, and the gene and marker are always linked together from one generation to another. They may be incompletely linked and crossing over may occur between the gene and the marker during meiosis. They may have no linkage, because the gene and the marker are located on different chromosomes or are far apart on the same chromosome. Estimates of the recombination between the resistance gene and a linked marker are measured as the recombination frequency (RF). RF values are measured among segregating F_2 plants or $F_{2:3}$ families by matching the phenotype and genotype of each progeny and subjecting the paired data to MAPMAKER (Lander *et al.*, 1987), an interactive computer package for calculating genetic distance and constructing genetic-linkage maps. The linkage between QTLs and marker loci is determined by the way distribution patterns for the resistance character(s) are linked with the segregation of the resistance gene and the molecular marker at each locus.

Why molecular markers?

Many practitioners ask 'Why use molecular markers?', given the additional time and labour required to define a molecular marker linked to an insect resistance gene. The answer lies in the fact that there are several advantages to adopting this technology. Restriction fragment length polymorphism (RFLP) markers and microsatellite or simple-sequence repeat (SSR) markers (see below) behave in a codominant manner to detect heterozygotes in segregating populations of progeny from crosses between resistant and susceptible parents. In contrast, morphological markers behave in a dominant/recessive manner and do not detect heterozygotes (Staub *et al.*, 1996). In general, the allelic variation detected by molecular markers in natural plant populations is considerably greater than that detected by morphological markers. Molecular markers are unaffected by environment and thus phenotype-neutral, while several examples exist to indicate that morphological markers are highly affected by environmental variations (see review by Smith, 1999). Finally, morphological markers may interact epistatically and molecular markers do not, which greatly increases the number screened in a single population.

Types of molecular markers

Several types of molecular markers have been used to determine the locations of insect-resistance genes. They include RFLP markers, mentioned previously, sequence-tagged site (STS) markers, random amplified polymorphic DNA (RAPD) markers, amplified fragment length polymorphism (AFLP) markers and SSRs or microsatellite markers. RFLP markers detect differences between genotype DNA when restriction enzymes cut genomic DNA at specific nucleotide sequences (binding sites) to yield variable-size fragments of DNA base pairs. The digested DNA is electrophoresed and transferred to a nylon membrane via Southern blotting, and the membrane is probed with a ^{32}P–labelled dinucleotide sequence of known chromosome location. The membrane-bound DNA is denatured by heat, allowing probe sequences to bind with complementary sites in the restriction digest and providing information about the putative location of the resistant gene. RFLP probes allow very fine mapping of loci linked to resistance genes, since initially they were more numerous in some crop genomes than other types of molecular markers. Disadvantages of RFLP linkage analysis include the time required to complete (7–10 days) and the use of radioactive isotopes. RFLP analysis has been used to map insect-resistance gene loci in barley, cowpea, mung bean, sorghum, rice and wheat (Moreira *et al.*, 1999; Smith, 1999; Huang *et al.*, 2001; Katsar *et al.*, 2002; Xu *et al.*, 2002).

DNA analysis was revolutionized with the discovery of the polymerase chain reaction (PCR) technique (Mullis, 1990). PCR allows the *in vitro* enzymatic amplification of specific DNA sequences present between two convergent oligonucleotide primers hybridizing to opposite DNA strands. The commercialization of PCR resulted in the thermal cycler, in which PCR primers of known chromosome location are reacted with template DNA, and the amplification products are electrophoresed to identify primers (markers) yielding polymorphisms. Compared with RFLP hybridization, PCR reactions are much faster (2–3 h) and non-radioactive. However, compared with many RFLP genomic maps, there are fewer PCR primers.

Several types of PCR primers identify insect-resistance genes in plants. RAPD PCR primers are short random DNA sequences that alone do not reveal heterozygotes and chromosome-linkage information *per se*. However, RAPD-generated DNA polymorphic bands can be end-sequenced to design location-specific sequence-characterized amplified regions (SCARs). SCARs have been used to identify and map genes for resistance to the rice gall midge, *O. oryzae* (Sardesai *et al.*, 2001), and the brown planthopper, *N. lugens* (Renganayaki *et al.*, 2002).

AFLPs are based on the selective PCR amplification of restriction enzyme-digested DNA fragments, as in RFLP analysis (Vos *et al.*, 1995). However, the DNA bands generated in each amplification contain DNA markers of random origin, which result in many more amplified DNA bands. AFLP markers have been used successfully to identify insect-resistance genes in apple, rice and wheat (Murai *et al.*, 2001; Cevik and King, 2002; Weng and Lazar, 2002).

SSR or microsatellite primers are tandem arrays of 2–5 base repeat units (particularly dinucleotide repeats), which have been found to be widely distributed in eukaryotic DNA. Microsatellite primers have proved to be very useful in crops where they have been developed, such as maize, soybean and wheat. As a result of a rapidly expanding library of microsatellite markers, SSRs are being used with increasing frequency to identify and map genes for insect resistance in wheat (Liu *et al.*, 2001, 2002; Miller *et al.*, 2001). STS markers – PCR markers sequenced from ends of RFLP sequences – have been used to map genes for insect resistance in barley (Nieto-Lopez and Blake, 1994) and rice (Katiyar *et al.*, 2001).

MAS of plants is accelerating the accuracy and rate at which a resistance gene can be tracked in the development of arthropod-resistant cultivars. There are few comparisons of the efficiency of MAS of resistant plants with phenotypic selection. MAS of genes for Russian wheat aphid, *Diuraphis noxia* Mordvilko, resistance in wheat and cereal cyst nematode, *Heterodera avenae* Woll. resistance in barley can be accomplished approximately 30 times faster for approximately 75% more cheaply per evaluation compared to plant phenotypes (Kretshmer *et al.*, 1997; C.M. Smith and X.E. Liu, unpublished). A US MAS genotyping centre for barley and wheat is developing QTLs linked to resistance for fusarium head blight, caused by *Fusarium graminearum* Schwabe (teleomorph *Gibberella zeae* (Schwein.)), in order to genotype plants in breeding populations (Van Sanford *et al.*, 2001).

QTL analysis has successfully identified loci containing insect-resistance genes in maize (Cardinal *et al.*, 2001; Jampatong *et al.*, 2002), rice (Huang *et al.*, 1997, 2001; Xu *et al.*, 2002), soybean (Rector *et al.*, 1998, 2000), tomato (Moreira *et al.*, 1999) and wheat (Castro *et al.*, 2001). Narvel *et al.* (2001) used SSR markers to assess US soybean breeding lines and cultivars developed over a 30-year period using conventional phenotypic selection for resistance to foliar feeding by several Lepidoptera. Although some resistance has been transferred, very few minor resistance QTLs have been transferred. MAS has been used to develop near-isogenic soybean lines with multiple insect-resistant QTLs, suggesting that the use of MAS in soybean is justified. Extensive research comparing the use of conventional selection of phenotypic resistance in maize to feeding damage by the southwestern corn borer, *Diatraea grandiosella* Dyar, to QTL mapping of resistance-gene loci suggests that QTL-MAS and conventional selection methods are equivalent in their ability to improve the level of resistance (Willcox *et al.*, 2002). Although the cost of MAS alone is approximately one-tenth the cost of conventional selection, the accurate identification of QTL position and the cost of generating these initial data as the first step in the MAS process of QTL analysis makes conventional selection more cost effective.

Cloning and Sequencing Plant Resistance Genes

The examples described above demonstrate how plant resistance to insects is mediated by constitutive gene effects. As mentioned previously however, the only insect-resistance gene identified to date is the *Meu1.2* gene from wild tomato, *L. peruvianum*, which confers resistance to the potato aphid, *M. euphorbiae* (Kaloshian *et al.*, 1995, 1997; Rossi *et al.*, 1998; Vos *et al.*, 1998) and to three species of the root-knot nematode, *Meloidogyne* spp. (Roberts and Thomason, 1986). *Meu1.2* is a member of the nucleotide-binding site–leucine-rich region (NBS–LRR) family of disease- and nematode-resistance genes (Milligan *et al.*, 1998). The LRR region of *Meu1.2* functions to signal localized cell death and programmed cell death (Hwang *et al.*, 2000; Wang *et al.*, 2001). Similarities in the sequence and function of other pest-resistance genes are beginning to show patterns. RGA sequences (see below) from barley map to loci in regions involved in resistance to the maize leaf aphid, *Rhopalosiphum maidis* (Fitch). These same sequences are similar to the wheat NBS-LRR *Cre3* gene for resistance to the cereal cyst nematode, *H. avenae* (Lagudah *et al.*, 1997). Other NBS–LRR-related sequences have also been mapped to regions controlling resistance to the melon aphid, *Aphis gossypii* Glover (Brotman *et al.*, 2002).

Functional genomics

During the past decade, the genomes of *Arabidopsis* and rice were sequenced, opening huge opportunities for in-depth studies of the molecular bases of plant resistance. The sequencing of extremely large genomes such as wheat, however, remains well in the future. In the interim, plant resistance researchers are data-mining information about *Arabidopsis* and rice gene sequence, function, and expression, in order to provide new information about the biochemical and physiological pathways involved in the resistance of other plants to insects.

RGAs are conserved amino acid motifs (such as NBS and LRR motifs) derived from sequence comparisons of predominant classes of insect-, disease-, and nematode-resistance genes. RGAs have been isolated in *Arabidopsis*, barley, lettuce, maize, rice, soybean and wheat (Seah *et al.*, 1998; Shen *et al.*, 1998; Speulman *et al.*, 1998; Leister *et al.*, 1999; Mago *et al.*, 1999; Tada, 1999; Graham *et al.*, 2000). Many cereal-crop RGAs map to orthologous positions in different cereal species. The fact that *Meu-1.2* and *Mi* are active against two organisms as distantly related as aphids and nematodes supports the hypothesis that RGAs can also be used to clone or design genes for insect resistance in crops. Map positions of RGAs in the *Triticeae* indicate that these genes occur in clusters and are more closely linked physically than those in other regions with similar genetic distances (Feuillet and Keller, 1999; Li *et al.*, 1999; Boyko *et al.*, 2002). For this reason, knowledge of the chromosome locations and genome organizations of RGAs in wheat and other crops will be of great value in candidate resistance-gene analyses.

The genomes of cereal crops, such as barley, maize, rice, rye, sorghum and wheat, are highly conserved, i.e. the arrangement of many of the genes in a region of a chromosome of one species is similar to that of a chromosome region of another plant species in DNA sequence (Ahn *et al.*, 1993; Paterson *et al.*, 1995; Boyko *et al.*, 1999). Resistance-gene maps in barley, sorghum, rice and wheat demonstrate the synteny among loci of these crops linked to genes expressing resistance to several species of pest aphids and planthoppers (Fig. 7.2). The exploitation of such conserved gene order to identify pest-resistance loci of interest will greatly stimulate efforts to clone insect-resistance genes in cereals and other crops as functional genomics becomes more of a reality in agricultural research. For example, using syntenous areas of barley, rice and wheat chromosomes where resistance genes have

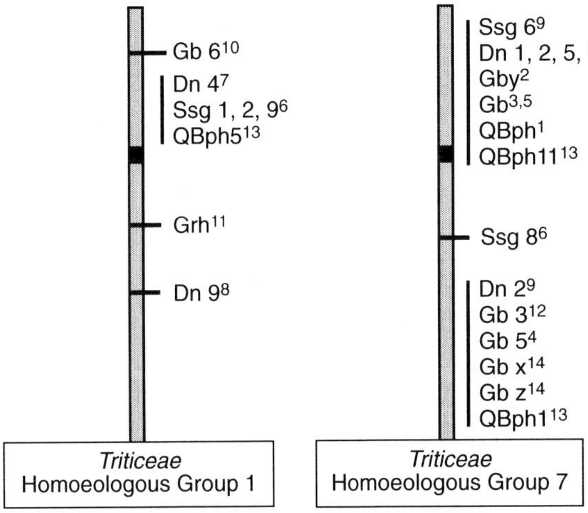

Fig. 7.2. Map positions of Heteroptera-resistance gene loci in barley, rice, sorghum and wheat on *Triticeae* homoeologous chromosome groups 1 and 7 (relative loci positions for illustration only, not ordered). *Gb,* greenbug, *Schizaphis graminum* (Rondani), resistance gene in wheat; *Dn*, Russian wheat aphid, *Diuraphis noxia* (Mordvilko), resistance gene in wheat; *Ssg,* greenbug-resistance gene in sorghum; *QBph,* main effect brown planthopper, *Nilapartava lugens* (Stal), resistance QTL in rice; *Grh,* green rice leafhopper, *Nephotettix cincticeps* (Uhler), resistance gene in rice. [1]Alam and Cohen, 1998; [2] Boyko *et al.*, 2002; [3]Castro *et al.*, 2001; [4] Dubcovsky *et al.*, 1998; [5] Moharramipour *et al.*, 1998; [6] Katsar *et al.*, 2002; [7] Liu *et al.*, 2002; [8] Liu *et al.*, 2001; [9] Miller *et al.*, 2001; [10] C.M. Smith and S. Starkey, unpublished; [11] Tamura *et al.*, 1999; [12] Weng and Lazar, 2002; [13] Xu *et al.*, 2002; [14] L. Zhu, C.M. Smith, E. Boyko and S. Starkey, unpublished.

been mapped, specific candidate rice bacterial artificial chromosome (BAC) contiguous segments can be subjected to *in silico* analyses to identify sequences similar to those of known resistance genes.

Expressed plant resistance genes

Plants use both constitutive and induced defences to protect themselves from insect attack. From a plant-breeding standpoint, plant resistance genes are viewed as constitutively active (always transcribed) in order to identify them in breeding programmes. With the advent of cDNA technologies, however, information about plant genes expressed in reaction to disease and insect attack has exploded, with more than 3000 articles published since 1995 (see reviews of Walling, 2000; Kessler and Baldwin, 2001; Heil and Bostock, 2002). When plant tissues are damaged, messenger RNA (mRNA) signals are translated to proteins. Unique mRNA gene transcripts expressed in resistant plants can now be identified by reverse transcription, where an RNA molecule is copied back into its cDNA by reverse transcriptases. cDNA populations from infested and uninfested plants can be subjected to subtractive suppressive hybridization to remove the hybridized sequences common to both populations. The unhybridized sequences unique to the resistant plant then become a 'subtracted' library of resistant cDNAs, which is then sequenced to determine the function of the putative resistance genes. Several studies of expressed insect-resistance genes are currently in progress.

Messenger RNA differential display, a related technique, was used by Hermsmeier *et al.* (2001) to study *Nicotiana attenuata* responses to feeding by tobacco hornworm, *Manduca sexta* (Johannsen), in the first experiment to identify mRNA transcripts produced after insect attack. Over 500 genes were involved in plant response, and 27 were verified as differentially expressed and sequenced. Transcripts encoding a Thr deaminase gene and a pathogen-inducible α-dioxygenase gene (see below), both involved in the plant defence response (Fig. 7.3), were expressed at significantly increased levels.

Complementary DNA libraries can also be probed with molecular markers of known genome function and location, in order to determine the degree of involvement of the expressed gene(s) in resistance. Finally, unique cDNAs can be used to probe oligonucleotide microarrays (gene chips) as a means of determining resistance-gene function based on mRNA expression levels. Commercial oligonucleotide microarrays now allow rapid screening of plant cDNAs expressing potential resistance, with over 14,000 expressed *Arabidopsis thaliana* sequences. Reymond *et al.* (2000) constructed a small-scale microarray of 150 ESTs implicated in *Arabidopsis* defence to demonstrate differences in genes activated by *Pieris rapae* (Linneaus) feeding, mechanical wounding, and water stress. Several hundred thousand cereal cDNA ESTs are currently being produced publicly for macro- or microarrays that will soon allow searches for expressed genes related to or involved in insect resistance in cereal crops (http://www.ncbi.nlm.nih. gov/dbEST/).

Insect-resistance elicitors

Plant reactions to both insect and disease attack may include hypersensitive cell death, as in the case of *Meu1.2*, activation of DR genes and the redirection of normal cell-maintenance genes to plant defence. In DR-gene activation, plants produce elicitors that activate plant gene expression and the synthesis of volatile and non-volatile allelochemicals. The similarities of plant elicitors in response to attacks by different insect species may be the result of common insect salivary enzymes, although some elicitors regulate very insect species-specific responses (van de Ven *et al.*, 2000; Walling, 2000). In addition, chewing insects cause extensive plant-tissue damage, which elicits different plant responses from those induced in response to feeding by piercing/sucking insects, which cause comparatively less tissue damage (Fidantsef *et al.*, 1999; Stout *et al.*, 1999; Walling, 2000). Plant responses to

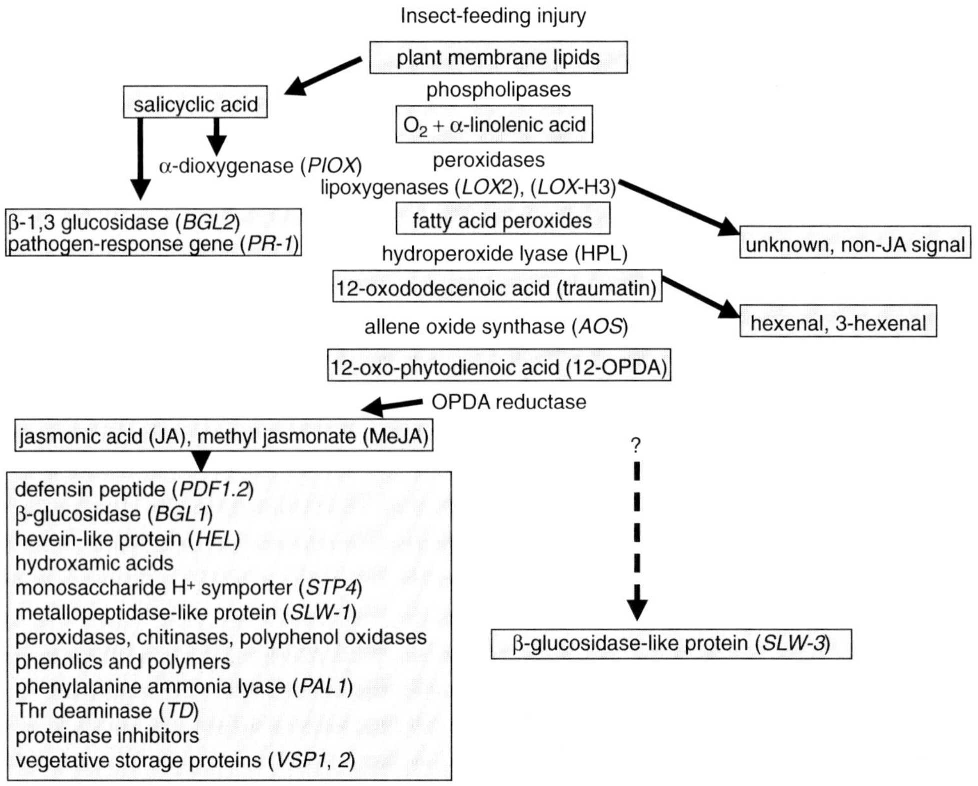

Fig. 7.3. A generalized diagram of known induced plant-resistance elicitors, genes and gene products produced by insect damage to plant tissues. Substrates shown in shaded grey boxes, enzymes in between. Italics indicate gene(s) expression of compounds involved in insect resistance. For detailed explanations, see individual references (Bergvinson *et al.*, 1994; Bell *et al.*, 1995; Lee *et al.*, 1997; Botha *et al.*, 1998; Fidantsef *et al.*, 1999; Royo *et al.*, 1999; Stout *et al.*, 1999; Forslund *et al.*, 2000; Reymond *et al.*, 2000; Stotz *et al.*, 2000, 2002; van de Ven *et al.*, 2000; Walling, 2000; Halitschke *et al.*, 2001; Hermsmeier *et al.*, 2001; Kessler and Baldwin, 2001; Moran and Thompson, 2001; Slesak *et al.*, 2001; Vancanneyt *et al.*, 2001.)

mechanical damage also differ from those involved in response to feeding damage (Botha *et al.*, 1998; Forslund *et al.*, 2000; Halitschke *et al.*, 2001; Winz and Baldwin, 2001).

When insects damage plant tissues during feeding, the death of tissues results in the production of an oxidative 'burst'. This event triggers the degradation of linolenic acid, which involves elicitors that signal the need of the production of allelochemical plant defences, such as proteinase inhibitors, phenolics and enzymes involved in the eventual production of plant structural defences (Fig. 7.3).

Lipoxygenases involved in cell membrane lipid degradation contribute to the production of jasmonic acid (JA) signals. Transcripts encoding *LOX* and *LOX2* genes have been shown to be strongly induced by feeding of the potato aphid, *M. euphorbiae*, on tomato (Fidanstef *et al.*, 1999) and the green peach aphid, *Myzus persicae* (Sulzer), feeding on *Arabidopsis* (Moran and Thompson, 2001). *LOX H3* genes in potato regulate resistance to the beet armyworm, *Spodoptera exigua* (Hubner), and Colorado potato beetle, but the resistance is not regulated by JA production (Royo *et al.*, 1999). The related enzyme hydroperoxide lyase plays a role in the resistance of potato to the green peach aphid (Vancanneyt *et al.*, 2001). In some plants, wounding also induces increased ethylene production, which blocks the JA signal but induces allene oxide synthase (AOS) production. Increased AOS levels induced by tobacco hornworm feeding are thought to help sustain JA production (Ziegler *et al.*, 2001).

Pathways involving jasmonates (JA and methyl jasmonate (MeJA)), ethylene (ET) and salicylic acid (SA) induce plant defences after insect attack, and some types of induced resistance are elicited by unknown and as yet unexplained types of elicitors. For example, both JA and ET induce the squash gene *SLW-1* for resistance to feeding by the silverleaf whitefly, *Bemisia argentifolii* Bellows and Perring, but an additional gene (*SLW-3*) is regulated by an unknown elicitor (van de Ven *et al.*, 2000).

The majority of induced responses identified to date in plants resulting from insect attack involve the jasmonate pathway. Methyl JA-treated wheat plants produce increased amounts of the defensive compound hydroxamic acid and sustain reduced phloem ingestion by bird cherry oat aphid, *Rhopalosiphum padi* (L.), compared with control plants (Slesak *et al.*, 2001). Methyl JA-induced accumulation of ferulic acid and phenolic polymers leads to cell-wall strengthening and increased insect resistance in barley and maize (Bergvinson *et al.*, 1994; Lee *et al.*, 1997). As cDNA library techniques have become employed, genes encoding specific defence compounds have been identified in the transcriptomes of insect-challenged plants of several families of plants. These include genes for a hevein-like protein (*HEL*) (Reymond *et al.*, 2000), a metallopeptidase-like protein (*SLW-1*) (van de Ven *et al.*, 2000), vegetative storage proteins (*VSP1, 2*) (Bell *et al.*, 1995), a novel β-glucosidase gene (*BGL1*) (Stotz *et al.*, 2002) and genes encoding the defensin peptide (*PDF1.2*), phenylalanine ammonia lyase (*PAL1*) and a monosaccharide symporter (*STP4*) (Moran and Thompson, 2001).

SA promotes the development of systemic acquired resistance, a broad-range resistance against pathogens and some insects. Hermsmeier *et al.* (2001) found that *N. attenuata* plants fed on by tobacco hornworm were strongly induced to produce high amounts of transcripts encoding the gene for α-dioxygenase (*PIOX*), a pathogen-inducible plant enzyme. The same plants expressed high levels of Thr deaminase (*TD*), a metabolic catalyst in the production of structural and chemical defences (Fig. 7.3). Interestingly, the results of Moran and Thompson (2001) also demonstrated that green peach aphid feeding on *Arabidopsis* induces major increases in the expression of the *PR-1* and *BGL2* genes, both of which are associated with the SA defence signalling pathway. In related defence signal 'cross-talk' studies, several authors have noted that defence responses induced by JA and ET may be antagonized by those induced by SA (Dong, 1998; Reymond and Farmer, 1998; Bostock, 1999; Pieterse and van Loon, 1999; Stotz *et al.*, 2002).

Herbivore-specific elicitors that induce plant defence responses have also been isolated from oral secretions of some lepidopterous larvae. These include the lytic enzyme β-glucosidase, isolated from salivary secretions of larvae of the imported cabbageworm, *Pieris brassicae* (Linnaeus) (Mattiacci *et al.*, 1995) and fatty acid conjugates isolated from the larval regurgitant of the tobacco hornworm (Halitschke *et al.*, 2001) and the beet armyworm (Alborn *et al.*, 1997). In the tobacco hornworm–*Arabidopsis* interaction, the application of the caterpillar fatty acid conjugates to wounded leaf tissue elicits a JA burst and the production of volatile plant defence compounds (Turlings and Benrey, 1998; Halitschke *et al.*, 2000, 2001). For additional discussion of the interactions between different plant-defence elicitors, readers are referred to the excellent reviews of Kessler and Baldwin (2001) and Heil and Bostock (2002).

Elicitor-induced responses play a role in induced plant resistance to insects. However, JA-induced responses also lower plant fitness and reduce seed yields, suggesting that plants bred to respond with heightened levels of insect resistance when attacked may be counter-productive in relation to efficient crop production (Baldwin *et al.*, 1997;

Baldwin, 1998). There are many gaps in the level and extent of knowledge about this exciting area of resistance-expression pathways. Additional research at both the biochemical and molecular levels will be critical to a better understanding of how different species of plants integrate separate and multiple elicitor signals generated as part of the defences against both insects and diseases.

Recommendations and Future Strategies

Resistant cultivars have proved to be ecologically and socially acceptable to consumers and economically feasible for producers for over 100 years. Future insect-resistant crops will play a very important role in world sustainable agricultural systems, and the benefits of their use will become more prominent as world food needs increase, especially in the developing countries of the semi-tropics.

In spite of their successful use in developed countries, the high level of *Bt* expression, similar to high doses of conventional pesticide or high levels of conventional gene expression, may promote the development of insects resistant to *Bt* crop plants. The longevity of *Bt* transgenes should be extended however, due to the advent of the use of IRM programmes centred on non-transgene refuges that allow the survival of homozygous susceptible pest individuals. In addition, current *Bt* cultivars are based on the differences in only four active protein domains of toxin expression. This should be a concern of future plant-resistance research efforts, and strategies need to be developed that will allow the development and commercialization of transgenic plants containing non-*Bt* toxic proteins in IPM systems. Future insect-resistant (primarily transgenic) crop-development strategies and efforts need to address the growing need to increase food supplies for the populations of underdeveloped countries. Conventional resistance-breeding efforts have made strides to improve crops such as sorghum, millet, pigeon pea and chickpea during the past several decades. However, major corporate and public investments are urgently needed now to use conventional and molecular methods to develop increased levels of insect resistance in these important food crops of the semi-arid tropics.

There are hundreds of insect-resistance genes deployed in improved cultivars globally, but the continual evolution of virulent biotypes dictates the need for the identification of new sources of resistance and for MAS systems to identify and track these genes. The refinement and increased use of MAS techniques and MAS centres should be encouraged in order to accelerate the rate and accuracy of breeding crop plants for insect resistance.

Our knowledge of how plants recognize insect-feeding attacks and the elicitors they produce in response to insect feeding is increasing rapidly. The evolving model of the differences in plant defence-response elicitors must be researched, challenged and modified to better understand induced resistance function and how plant metabolism can possibly be modified to use induced crop-plant resistance in insect pest-management programmes.

There is also a great need for additional information about the sequence and function of expressed cDNAs unique to both resistant and susceptible plants under attack. The use of mRNA differential display and subtractive suppressive hybridization studies should be encouraged to accomplish this goal. In addition to the sequence information provided by unique cDNAs, they can be used to probe oligonucleotide macro- and microarrays, in order to determine expressed resistance-gene function based on mRNA expression levels.

The existence of RGAs in many crop plants suggests that current and future plant-resistance researchers should increasingly utilize these genetic resources to provide *in silico* information about the location and function of candidate resistance genes. As a more complete knowledge (and eventual sequencing) of the genomes of additional crop plants develops, the use of genomic macro- and microarrays will also become valuable tools to answer questions about where resistance genes are located and what biochemical and biophysical gene products mediate their function.

The ultimate goal of resistance-gene expression studies, genomic studies and MAS systems should be to identify plant genes that can be cloned and used to transform crop plants for insect resistance in the same way that they have been transformed with *Bt* and other related transgenes toxic to insects.

References

Abbott, E.A., Lucht, T., Jensen, J. and Jordan-Conde, Z. (2001) Handling of GM crop issues by the mass media. In: *Proceedings Illinois Crop Protection Technology Conference*. University of Illinois, Urbana, Illinois, pp. 11–12.

Ahn, S.N., Anderson, J.A., Sorrells, M.E. and Tanksley, S.D. (1993) Homoeologous relationships of rice, wheat and maize chromosomes. *Molecular and General Genetics* 241, 483–490.

Alam, S.N. and Cohen, M.B. (1998) Detection and analysis of QTLs for resistance to the brown planthopper, *Nilapavata lugens*, in a doubled-haploid rice population. *Theoretical and Applied Genetics* 97, 1370–1379.

Alborn, T., Turlings, T.C.J., Jones, T.H., Stenhagen, G., Loughrin, J.H. and Tumlinson, J.H. (1997) An elicitor of plant volatiles from beet armyworm oral secretion. *Science* 276, 945–949.

Al-Deeb, M.A., Wilde, G.E. and Higgins, R.A. (2001) No effect of *Bacillus thuringiensis* corn and *Bacillus thuringiensis* on the predator *Orius insidiosus* (Say) (Hemiptera: Anthocoridae). *Environmental Entomology* 30, 625–629.

American Phytopathological Society (APS) (1999) Threats to health reduced with *Bt* corn hybrids. Available at: http://www.scisoc. org/opae/media/btcorn.htm

Anon. (2001) *Corn and Biotechnology Special Analysis*. National Agricultural Statistic Service, USDA, Washington, DC.

Azzam, A., Azzam, S., Lhaloui, S., Amri, A., El Bouhssini, M. and Moussaoui, M. (1997) Economic returns to research in Hessian fly (Diptera: Cecidomyidae) resistant bread-wheat varieties in Morocco. *Journal of Economic Entomology* 90, 1–5.

Baldwin, I.T. (1998) Jasmonate-induced responses are costly but benefit plants under attack in native populations. *Proceedings of the National Academy of Sciences of the USA* 95, 8113–8118.

Baldwin, I.T., Zhang, Z.-P., Diab, N., Ohnmeiss, T.E., McCloud, E.S., Lynds, G.Y. and Schmelz, E.A. (1997) Quantification, correlations and manipulation of wound-induced changes in jasmonic acid and nicotine in *Nicotiana sylvestris*. *Planta* 210, 397–404.

Bell, E., Creelman, R.A. and Mullet, J.E. (1995) A chloroplast lipoxygenase is required for wound-induced jasmonic acid accumulation in *Arabidopsis*. *Proceedings of the National Academy of Sciences of the USA* 92, 8675–8679.

Bergvinson, D.J., Arnason, J.T. and Pietrzak, L.N. (1994) Localization and quantification of cell wall phenolics in European corn borer resistant and susceptible maize inbreds. *Canadian Journal of Botany* 72, 1243–1249.

Bostock, R.M. (1999) Signal conflicts and synergies in induced resistance to multiple attackers. *Physiological and Molecular Plant Pathology* 55, 99–109.

Botha, A.M., Nagel, M.A.C., Van der Westhuizen, A.J. and Botha, F.C. (1998) Chitinase isoenzymes in near-isogenic wheat lines challenged with Russian wheat aphid, exogenois ethylene and mechanical wounding. *Botanical Bulletin of the Academica Sinica* 39, 99–106.

Boyko, E.V., Gill, K.S., Mickelson-Young, L., Nasuda, S., Raupp, W.J., Ziegler, J.N., Singh, S., Hassawi, D.S., Fritz, A.K., Namuth, D., Lapitan, N.L.V. and Gill, B.S. (1999) A high-density genetic linkage map of *Aegilops tauschii*, the DS-genome progenitor of bread wheat. *Theoretical and Applied Genetics* 99, 16–26.

Boyko, E.V., Kalendar, R., Korzun, V., Korol, A., Schulman, A. and Gill, B.S. (2002) A high density genetic map of *Aegilops tauschii* includes genes, retro-transposons, and microsatellites which provide unique insight into cereal chromosome structure and function. *Plant Molecular Biology* 48, 767–790.

Brotman, Y., Silberstein, L., Kovalski, I., Perin, C., Dogimont, C., Pitrat, M., Klingler J., Thompson, G.A. and Perl-Treves, R. (2002) Resistance gene homologues in melon are linked to genetic loci conferring disease and pest resistance. *Theoretical and Applied Genetics* 104, 1055–1063.

Byrne, P., Ward, S., Harrington, J. and Fuller, L. (2003) Transgenic plants: an introduction and resource guide. Available at: http://www.colostate.edu/programs/lifesciences/TransgenicCrops/news.html#stillbreaking

Cardinal, A.J., Lee, M., Sharopova, N., Woodman-Clikeman, W.L. and Long, M.J. (2001) Genetic mapping and analysis of quantitative tratit loci for resistance to stalk tunneling by the European corn borer in maize. *Crop Science* 41, 835–845.

Carrière, Y., Ellers-Kirk, C., Sisterson, M., Antilla, L., Whitlow, M., Dennehy, T.J. and Tabashnik, B.E. (2003) Long-term regional suppression of pink bollworm by *Bacillus thuringiensis* cotton. *Proceedings of the National Academy of Sciences of the USA* 100, 1519–1523.

Castro, A.M., Worland, A., Vasicek, A., Clua, A.A., Gimenez, D., Ellerbrook, C. and Tacaliti, M.S. (2001) Mapping a QTL involved with greenbug and Russian wheat aphid resistance. In: *Proceedings 4th International Triticeae Symposium*. Consejenia de Agricultura y Pesca, Cordoba, Spain, pp. 345–349.

Cevik, V. and King, G.L. (2002) High-resolution genetic analysis of the *Sd*-1 aphid resistance locus in *Malus* spp. *Theoretical and Applied Genetics* 105, 346–354.

Chesnokov, P.G. (1953) *Methods of Investigating Plant Resistance to Pests*. National Science Foundation, Washington, DC (Israel Program for Scientific Translation).

Cregan, P., Jarvik, T., Bush, A.L., Shoemaker, R.C., Lark, K.G., Kahler, A.L., Kaya, N., VanToai, T.T., Lohnes, D.G., Chung, J. and Specht, J.E. (1999) An integrated genetic linkage map of the soybean genome. *Crop Science* 39, 1464–1490.

Cuperus, G.W., Kendall, P., Rehe, S., Frisbee, R., Hall, K., Bruhn, C., Deer, D., Woods, F., Branthaver, B., Weber, G., Poli, B., Buege, D., Linker, M., Andress, E., Wintersteen, W., Dost, F., Hegley, F., Aselhage, J., Stelltflug, L., Doe, R. and Murray, G. (1991) *Integration of Food Safety and Water Quality Concepts Throughout the Food Production, Processing and Distribution Educational Programs using HACCP Philosophies*. Circular E-903, Cooperative Extension Service, Oklahoma State University, Stillwater.

Daly, J.C. and Wellings, P.W. (1996) Ecological constraints to the deployment of arthropod resistant crop plants: a cautionary tale. In: Floyd, R.B., Shepard, A.W. and De Barro, P.J. (eds) *Frontiers of Population Ecology*. CSIRO Publishing, Melbourne, pp. 311–323.

Dhaliwal, G.S. and Dilawari, V.K. (eds) (1993) *Advances in Host Plant Resistance to Insects*. Kalyani Publishers, New Delhi.

Dhaliwal, G.S. and Singh, R. (eds) (2004) *Host Plant Resistance to Insects: Concepts and Applications*. Panima Publishing Corporation, New Delhi.

Dong, X. (1998) SA, JA, ethylene, and disease resistance in plants. *Current Opinions in Plant Biology* 1, 316–323.

Dubcovsky, J., Lukaszewski, A.J., Echaide, M., Antonelli, E.F. and Porter, D.R. (1998) Molecular characterization of two *Triticum speltoides* interstitial translocations carrying leaf rust and greenbug resistance genes. *Crop Science* 38, 1655–1660.

Eddleman, B.R., Chang, C.C. and McCarl, B.A. (1999) Economic benefits from grain sorghum variety improvement in the United States. In: Wiseman, B.R. and Webster, J.A. (eds) *Economic, Environmental, and Social Benefits of Resistance in Field Crops*. Thomas Say Publications, Entomological Society of America, Lanham, Maryland, pp. 17–44.

Feuillet, C. and Keller, B. (1999) High genome density is conserved at syntenic loci of small and large grass genomes. *Proceedings of the National Academy of Sciences of the USA* 96, 8265–8270.

Fidantsef, A.L., Stout, M.J., Thaler, J.S., Duffey, S.S. and Bostock, R.M. (1999) Signal interactions in pathogen and insect attack: expression of lipoxygenase, proteinase inhibitor II, and pathogenesis-related protein P4 in the tomato, *Lycopersicon esculentum*. *Physiological and Molecular Plant Pathology* 54, 97–114.

Forslund, K., Perrersson, J., Bryngelsson, T. and Jonnson, L. (2000) Aphid infestation induces PR-proteins differentially in barley susceptible or resistant to the birdcherry-oat aphid. *Physiologia Plantarum* 110, 496–502.

Gaskell, G., Bauer, M.W., Durant, J. and Allum, N.C. (1999) Worlds apart? The reception of genetically modified foods in Europe and the U.S. *Science* 285, 384–387.

Gould, F. (1998) Sustainability of transgenic insecticidal cultivars: integrating pest genetics and ecology. *Annual Review of Entomology* 43, 701–726.

Graham, M.A., Marek, L.F., Lohnes, D., Cregan, P. and Shoemaker, R. (2000) Expression and genome organization of resistance gene analogs in soybean. *Genome* 43, 86–90.

Halitschke, R., Kessler, A., Kahl, J., Lorenz, A. and Baldwin, I.T. (2000) Ecophysiological comparison of direct and indirect defences in *Nicotiana attenuata*. *Oecologia* 124, 408–417.

Halitschke, R., Schittko, U., Pohnert, G., Boland, W. and Baldwin, I.T. (2001) Molecular interactions between the specialist herbivore *Manduca sexta* (Lepidoptera, Sphingidae) and its natural host *Nicotiana attenuata*. III. Fatty amino-acid conjugates in herbivore oral secretions are necessary and sufficient for herbivore-specific plant responses. *Plant Physiology* 125, 711–717.

Heil, M. and Bostock, R.M. (2002) Induced systemic resistance (ISR) against pathogens in the context of induced plant defences. *Annals of Botany* 89, 503–512.

Hermsmeier, D., Schittko, U. and Baldwin, I.T. (2001) Molecular interactions between the specialist herbivore *Manduca sexta* (Lepidoptera, Sphingidae) and its natural host *Nicotiana attenuata*. I. Large-scale changes in the accumulation of growth-and defence-related plant mRNAs. *Plant Physiology* 125, 683–700.

Hernández, P., Dorado, G., Prieto, P., Giménez, M.J., Ramírez, M.C., Laurie, D.A., Snape, J.W., and Martín, A. (2001) A core genetic map of *Hordeum chilense* and comparisons with maps of barley (*Hordeum vulgare*) and wheat (*Triticum aestivum*). *Theoretical and Applied Genetics* 102, 1259–1264.

Huang, F., Higgins, R.A. and Buschman, L.L. (1999) Transgenic *Bt*-plants: successes, challenges, and strategies. *Pestology* 23, 2–29.

Huang, N., Parco, A., Mew, T., Magpantay, G., McCouch, S., Guiderdoni, E., Xu, J.C., Subudhi, P., Angeles, E.R. and Khush, G.S. (1997) RFLP mapping of isozymes, RAPD and QTLs for grain shape, brown planthopper resistance in a doubled haploid rice population. *Molecular Breeding* 3, 105–113

Huang, Z., He, G., Shu, L., Li, X. and Zhang, Q. (2001) Identification and mapping of two brown planthopper resistance genes in rice. *Theoretical and Applied Genetics* 102, 929–934.

Hwang, C.F., Bhakta, A.V., Truesdell, G.M., Pudlo, W.M. and Williamson, V.M. (2000) Evidence for a role of the N terminus and leucine-rich repeat region of the *Mi* gene product in regulation of localized cell death. *Plant Cell* 12, 1319–1329.

International Food Information Council (IFIC) (2002) U.S. consumer attitudes toward food biotechnology survey: 23 September, 2002. Available at: http://ific.org/relatives/17860. pdf

James, C. (2000) *Global Status of Commercialized Transgenic* Crops. ISAAA, Briefs 21, Preview. ISAAA Ithaca, NY. Available at: http://www.isaaa.org/publications/briefs/Brief_21.htm

Jampatong, C., McMullen, M.D., Barry, D.B., Darrah, L.L., Byrne, P.F. and Kross, H. (2002) Quantitative trait loci for first- and second-generation European corn borer resistance derived from the maize inbred Mo47. *Crop Science* 41, 584–593.

Johnson, M.T. and Gould, F. (1992) Interaction of genetically engineered host plant resistance and natural enemies of *Heliothis virescens* (Lepidoptera: Noctuidae) in tobacco. *Environmental Entomology* 21, 586–597.

Kaloshian, I., Lange, W.H. and Williamson, V.M. (1995) An aphid-resistance locus is tightly linked to the nematode-resistance gene, *Mi*, in tomato. *Proceedings of the National Academy of Sciences of the USA* 92, 622–625.

Kaloshian, I., Kinser, M.G., Ullman, D.E. and Williamson, V.M. (1997) The impact of *Meu1*-mediated resistance in tomato on longevity, fecundity, and behaviour of the potato aphid, *Macrosiphum euphorbiae*. *Entomologia Experimentalis et Applicata* 83, 181–187.

Katiyar, S.K., Tan, Y., Huang, B., Chandel, G., Xu, Y., Zhang, Y., Xie, Z. and Bennett, J. (2001) Molecular mapping of gene *Gm-6*(*t*) which confers resistance against four biotypes of Asian rice gall midge in China. *Theoretical and Applied Genetics* 103, 953–961.

Katsar, C.S., Paterson, A.H., Teetes, G.L. and Peterson, G.C. (2002) Molecular analysis of sorghum resistance to the greenbug (Homoptera: Aphididae). *Journal of Economic Entomology* 95, 448–457.

Kessler, A. and Baldwin, I.T. (2001) Defensive function of herbivore-induced plant volatile emissions in nature. *Science* 291, 2141–2144.

Khush, G.S. and Brar, D.S. (1991) Genetics of resistance to insects in crop plants. *Advances in Agronomy* 45, 223–274.

Kirsten, J. and Gouse, M. (2002) *Bt* cotton in South Africa: adoption and impact on farm incomes amongst small- and large-scale farmers. ISB News Report. Available at: http://www.isb.vt.edu/articles/oct0204.htm

Knight, J. (2003) Agency ignoring its advisers over *Bt* maize. *Nature* 422, 5.

Korzun, V., Malyshev, S., Voylokov, A.V. and Börner, A. (2001) A genetic map of rye (*Secale cereale* L.) combining RFLP, isozyme, protein, microsatellite and gene loci. *Theoretical and Applied Genetics* 102, 709–717.

Kretshmer, J.M., Chalmers, K.J., Manning, S., Karakousis, A., Barr, A.R., Islam, M.R., Logue, S.J., Choe, Y.W., Barker, S.J., Lance, R.C.M. and Langridge, P. (1997) RFLP mapping of the *Ha2* cereal cyst nematode resistance gene in barley. *Theoretical and Applied Genetics* 94, 1060–1064.

Lagudah, E.S., Moullet, O. and Appels, R. (1997) Map-based cloning of a gene sequence encoding a nucleotide binding domain and a leucine-rich region at the *Cre3* nematode resistance locus of wheat. *Genome* 40, 659–665.

Lander, E.S., Green, P., Abrahamson, J., Barlow, A., Daly, M.J., Lincoln, S.E. and Newburg, L. (1987) MAPMAKER: an interactive computer package for constructing primary genetic maps of experimental and natural populations. *Genomics* 1, 174–181.
Lara, F.M. (1979) *Principios de Resistancia de Plantas a Insectos* – (in Portugese). Piracicaba, Livroceres, Brazil.
Lawrence, P.K. and Koundal, K.R. (2002) Plant protease inhibitors in control of phytophagous insects. *Electronic Journal of Biotechnology* 5, 15 April. Available at: http://www.ejbiotechnology.info/content/vol5/issue1/full/3/index.html
Lee, J.E., Vogt, T., Hause, B. and Lëbler, M. (1997) Methyl jasmonate induces an *O*-methyltransferase in barley. *Plant Cell Physiology* 38, 851–862.
Leister, D., Kurth, J., Laurie, D.A., Yano, M., Sasaki, T., Graner, A. and Schulze-Lefert, P. (1999) RFLP- and physical mapping of resistance gene homologues in rice (*O. sativa*) and barley (*H. vulgare*). *Theoretical and Applied Genetics* 98, 509–520.
Li, W.L., Faris, J.D., Chitmoor, J.M., Leach, J.F., Hulbert, S.H., Liu, D.J., Chen, P.D. and Gill, B.S. (1999) Genomic mapping of defence response genes in wheat. *Theoretical and Applied Genetics* 98, 226–233.
Liu, X.M., Smith, C.M., Gill, B.S. and Tolmay, V. (2001) Microsatellite markers linked to six Russian wheat aphid resistance genes in wheat. *Theoretical and Applied Genetics* 102, 504–510.
Liu, X.M., Smith, C.M. and Gill, B.S. (2002) Mapping of microsatellite markers linked to the *Dn4* and *Dn6* genes expressing Russian wheat aphid resistance in wheat. *Theoretical and Applied Genetics* 104, 1042–1048.
Llewellyn, D., Cousins, Y., Mathews, A., Hartweck, L. and Lyon, B. (1994) Expression of *Bacillus thuringensis* insecticidal protein genes in transgenic crop plants. *Agriculture, Ecosystems & Environment* 49, 85–93.
Losey, J.E., Rayor, L.S. and Carter, M.E. (1999) Transgenic pollen harms monarch larvae. *Nature* 399, 214.
Mago, R., Nair, S. and Mohan, M. (1999) Resistance gene analogues from rice: cloning, sequencing and mapping. *Theoretical and Applied Genetics* 99, 50–57.
Mattiacci, L., Dicke, M. and Posthumas, M.A. (1995) Beta-glucosidase – an elicitor of herbivore-induced plant odor that attracts host-searching parasitic wasps. *Proceedings of the National Academy of Sciences of the USA* 92, 2036–2040.
Maxwell, F.G. and Jennings, P.R. (1980) *Breeding Plants Resistant to Insects.* John Wiley & Sons, New York.
Miller, C.A., Altinkut, A. and Lapitan, N.L.V. (2001) A microsatellite marker for tagging *Dn2*, a wheat gene conferring resistance to the Russian wheat aphid. *Crop Science* 41, 1584–1589.
Milligan, S., Bodeau, J., Yaghoobi, J., Kaloshian, I., Zabel, P. and Williamson, V. (1998) The root-knot nematode resistance gene *Mi* from tomato is a member of the leucine zipper, nucleotide binding, leucine-rich repeat family of plant genes. *Plant Cell* 10, 1307–1319.
Moharramipour, S., Tsumki, H., Sato, K. and Yoshida, H. (1997) Mapping resistance to cereal aphids in barley. *Theoretical and Applied Genetics* 94, 592–596.
Moran, P.J. and Thompson, G.A. (2001) Molecular responses to aphid feeding in *Arabidopsis* in relation to plant defence pathways. *Plant Physiology* 125, 1074–1085.
Moreira, L.A., Mollema, C. and van Heusden, S. (1999) Search for molecular markers linked to *Liriomyza trifolii* resistance in tomato. *Euphytica* 109, 149–156.
Mullis, K. (1990) The unusual origin of the polymerase chain reaction. *Scientific American* April, 262, 56–65.
Murai, H., Hashimoto, Z., Sharma, P.N., Shimizu, T., Murata, K., Takumi, S., Mori, N., Kawasaki, S. and Nakamura, C. (2001) Construction of a high resolution linkage map of a rice brown planthopper (*Nilaparvata lugens* Stal) resistance gene *bph2*. *Theoretical and Applied Genetics* 103, 526–532.
Narvel, J.A., Walker, D.R., Rector, B.G., All, J.N., Parrott, W.A. and Boerma, R. (2001) A retrospective DNA marker assessment of the development of insect resistant soybean. *Crop Science* 41, 1931–1939.
Nieto-Lopez, R.M. and Blake, T.K. (1994) Russian wheat aphid resistance in barley: inheritance and linked molecular markers. *Crop Science* 34, 655–659.
Oppert, B.S. (2001) Transgenic plants expressing enzyme inhibitors and the prospects for biopesticide development. In: Koul, O. and Dhaliwal, G.S. (eds) *Phytochemical Biopesticides.* Harwood Academic Publishers, Amsterdam, pp. 83–95.
Painter, R.H. (1951) *Insect Resistance in Crop Plants.* University of Kansas Press, Lawrence, Kansas.
Panda, N. (1979) *Principles of Host-Plant Resistance to Insect Pests.* Allanheld, Osmun & Co. and Universe Books, New York.
Panda, N. and Khush, G.S. (1995) *Host Plant Resistance to Insects.* CAB International, Wallingford, UK.
Pani, J. and Sahu, S.C. (2000) Inheritance of resistance against biotype 2 of the Asian rice gall midge, *Orseolia oryzae. Entomologia Experimentalis et Applicata* 95, 15–19.

Paterson, A., Lin, Y.-R., Li, Z., Scherta, K.F., Doebley, J.F., Pinson, S.R.M., Liu, S.-C., Stansel, J.W. and Irvine, J.E. (1995) Convergent domestication of cereal crops by independent mutations at corresponding genetic loci. *Science* 269, 1714–1718.

Pieterse, C.M and van Loon, L.C. (1999) Salicylic acid-independent plant defence pathways. *Trends in Plant Science* 4, 52–58.

Pilcher, C.D., Obrycki, J.J., Rice, M.E. and Lewis, L.C. (1997) Preimaginal development, survival, field abundance of insect predators on transgenic *Bacillus thuringiensis* corn. *Environmental Entomology* 26, 446–454.

Pimentel, D.S. and Raven, P.H. (2000) *Bt* corn pollen impacts on nontarget Lepidoptera: assessment of effects in nature. *Proceedings of the National Academy of Sciences of the USA* 97, 8198–8199.

Qaim, M. and Zilberman, D. (2003) Yield effects of genetically modified crops in developing countries. *Science* 299, 900–902.

Ratcliffe, R.H. and Hatchett, J.H. (1997) Biology and genetics of the Hessian fly and resistance in wheat. In: Bobdari, K. (ed.) *New Developments in Entomology*. Research Signpost, Scientific Information Guild, Trivandrum, pp. 47–67.

Rector, B.G., All, J.N., Parrott, W.A. and Boerma, H.R. (1998) Identification of molecular markers linked to quantitative trait loci for soybean resistance to corn earworm. *Theoretical and Applied Genetics* 96, 786–790.

Rector, B.G., All, J.N., Parrott, W.A. and Boerma, H.R. (2000) Quantitative trait loci for antibiosis resistance to corn earworm in soybean. *Crop Science* 40, 233–238.

Renganayaki, K., Fritz, A.K., Sadasivam, S., Pammi, S., Harrington, S.E., McCouch, S.R., Kumar, S.M. and Reddy, A.S. (2002) Mapping and progress toward map-based cloning of brown planthopper biotype-4 resistance gene introgressed *Oryza officinalis* into cultivated rice, *O. sativa*. *Crop Science* 42, 2112–2117.

Reymond, P. and Farmer, E.E. (1998) Jasmonate and salicylate as global signals for defence gene expression. *Current Opinions in Plant Biology* 1, 404–411.

Reymond, P., Weber, H., Damond, M. and Farmer, E.E. (2000) Differential gene expression in response to mechanical wounding and insect feeding in *Arabidopsis*. *Plant Cell* 12, 707–719.

Rice, M.E. and Pilcher, C.D. (1998) Potential benefits and limitations of transgenic *Bt* corn for management of the European corn borer (Lepidoptera: Crambidae). *American Entomologist* 44, 75–78.

Rissler, J. and Mellon, M. (1996) *The Ecological Risks of Engineered Crops*. MIT Press, Cambridge, Massachusetts.

Roberts, P.A. and Thomason, I.J. (1986) Variability in reproduction of isolates of *Meloidogyne incognita* and *M. javanica* on resistant tomato genotypes. *Plant Disease* 70, 547–551.

Royo, E., Leon, J., Vancanneyt, G., Albar, J.P., Rosahl, S., Ortego, F., Castanera, P. and Sanchez-Serrano, J.J. (1999) Antisense-mediated depletion of a potato lipoxygenase reduces wound induction of proteinase inhibitors and increases weight gain of insect pests. *Proceedings of the National Academy of Sciences of the USA* 96, 1146–1151.

Rossi, M., Goggin, F.L., Milligan, S.B., Klaoshian, I., Ullman, D.E. and Williamson, V.M. (1998) The nematode resistance gene *Mi* of tomato confers resistance against the potato aphid. *Proceedings of the National Academy of Sciences of the USA* 95, 9750–9754.

Russell, G.E. (1978) *Plant Breeding for Pest and Disease Resistance*. Butterworth Publishers, Boston, Massachusetts.

Sachs, E.S., Benedict, J.H., Taylor, J.F., Stelly, D.M., Davis, S.K. and Altman, D.W. (1996) Pyramiding *CryIA(b)* insecticidal protein and terpenoids in cotton to resist tobacco budworm (Lepidoptera: Noctuidae). *Environmental Entomology* 25, 1257–1266.

Sardesai, N., Kumar, A., Rajyashri, K.R., Nair, S. and Mohan, M. (2001) Identification and mapping of an AFLP marker linked to *Gm7*, a gall midge resistance gene and its conversion to a SCAR marker for its utility in marker aided selection in rice. *Theoretical and Applied Genetics* 105, 691–698.

Seah, S., Sivasithamparam, K., Karalousis, A. and Lagudah, E.S. (1998) Cloning and characterisation of a family of disease resistance gene analogs from wheat and barley. *Theoretical and Applied Genetics* 97, 937–945.

Sears, M.K., Hellmich, R.L., Stanley-Horn, D.E., Oberhauser, K.S., Pleasants, J.M., Mattila, H.R., Siegfried, B.D. and Dively, G.P. (2001) Impact of *Bt* corn pollen on monarch butterfly populations: a risk assessment. *Proceedings of the National Academy of Sciences of the USA* 98, 11937–11942.

Service, R.F. (2001) Arson strikes research labs and tree farm in Pacific Northwest. *Science* 292, 1622–1623.

Sharma, H.C., Sharma, K.K., Seetharama, N. and Ortiz, R. (2000) Prospects for using transgenic resistance to insects in crop improvement. *Electronic Journal of Biotechnology* 3, 15 August. Available at: http://www.ejbiotechnology.info/content/vol3/issue2/full/3/index.html

Sharopova, N., McMullen, M.D., Schultz, L., Schroeder, S., Sanchez-Villeda, H., Gardiner, J., Bergstrom, D., Houchins, K., Melia-Hancock, S., Musket, T., Duru, N., Polacco, M., Edwards, K., Ruff, T., Register, J.C., Brouwer, C., Thompson, R., Velasco, R., Chin, E., Lee, M., Woodman-Clikeman, W., Long, M.J., Liscum, E., Cone, K., Davis, G. and Coe, E.H. Jr (2002) Development and mapping of SSR markers for maize. *Plant Molecular Biology* 48, 463–481.

Shen, K.A., Meyers, B.C., Islam-Faridi, M.N., Chin, D., Stelly, D.M. and Michelmore, R.W. (1998) Resistance gene candidates identified by PCR with degenerate oligonucleotide primers map to clusters of resistance genes in lettuce. *Molecular Plant–Microbe Interaction* 11, 815–823.

Slesak, E., Slesak, M. and Gabrys, B. (2001) Effect of methyl jasmonate on hydroxamic acid, protease activity, and bird cherry-oat aphid *Rhoplaosiphum padi* L. probing behavior. *Journal of Chemical Ecology* 12, 2529–2543.

Smith, C.M. (1989) *Plant Resistance to Insects – A Fundamental Approach.* John Wiley & Sons, New York.

Smith, C.M. (1999) Plant resistance to insects. In: Rechcigl, J. and Rechcigl, N. (eds) *Biological and Biotechnological Control of Insects.* Lewis Publishers, Boca Raton, Florida, pp. 171–205.

Smith, C.M., Khan, Z.R. and Pathak, M.D. (1994) *Techniques for Evaluating Insect Resistance in Crop Plants.* Lewis Publishers, Boca Raton, Florida.

Smith, C.M., Quisenberry, S.S. and du Toit, F. (1999) The value of conserved wheat germplasm possessing arthropod resistance. In: Clement, S.L. and Quisenberry, S.S. (eds) *Global Plant Genetic Resources for Insect Resistant Crops.* CRC Press, Boca Raton, Florida, pp. 25–49.

Snelling, R.O. (1941) Resistance of plants to insect attack. *Botany Reviews* 7, 543–586.

Speulman, E., Bouchez, D., Holub, E. and Beynon, J.L. (1998) Disease resistance gene homologs correlate with disease resistance loci of *Arabidopsis thaliana. Plant Journal* 14, 467–474.

Stanley-Horn, D.E., Dively, G.P., Hellmich, R.L., Mattila, H.R., Sears, M.K., Rose, R., Jesse, L.C.H., Losey, J.E., Obrycki, J.J. and Lewis, L. (2002) Assessing the impact of Cry1Ab-expressing corn pollen on monarch butterfly larvae in field studies. *Proceedings of the National Academy of Sciences of the USA* 98, 11931–11936.

Staub, J.E., Serquen, F.C. and Gupta, M. (1996) Genetic markers, map construction, and their application in plant breeding. *HortScience* 31, 729–741.

Stotz, H.U., Pittendrigh, B.R., Kroyman, J., Weniger, K., Fritsche, J., Bauke, A. and Mitchell-Olds, T. (2000) Induced plant defense responses against chewing insects. Ethylene signaling reduces resistance of *Arabidopsis* against Egyptain cotton worm but not diamond back moth. *Plant Physiology* 124, 1007–1017.

Stotz, H.U., Koch, T., Biedermann, A., Weniger, K., Boland, W. and Mitchell-Olds, T. (2002) Evidence for regulation of resistance in *Arabidopsis* to Egyptian cotton worm by salicylic and jasmonic acid signaling pathways. *Planta* 214, 648–652.

Stout, M.J., Fidantsef, A.L., Duffey, S.S. and Bostock, R.M. (1999) Signal interactions in pathogen and insect attack: systemic plant-mediated interactions between pathogens and herbivores of the tomato, *Lycopsericon esculentum. Physiological and Molecular Plant Pathology* 54, 115–130.

Tada, T. (1999) PCR-amplified resistance gene analogs link to resistance loci in rice. *Breeding Science* 49, 267–273.

Tamura, K., Fukuta, Y., Hirae, M., Oya, S., Ashikawa, I. and Yagi, T. (1999) Mapping of the *Grh1* locus for green rice leafhopper resistance in rice using RFLP markers. *Breeding Science* 49, 11–14.

Tanaka, K. (1999) Quantitative genetic analysis of biotypes of the brown planthopper *Nilaparvata lugens*: heritability of virulence to resistant rice varieties. *Entomologia Experimentalis et Applicata* 90, 279–287.

Turlings, T.C.J. and Benrey, B. (1998) Effects of plant metabolites on the behavior and development of parasitic wasps. *EcoScience* 5, 321–333.

Vancanneyt, G., Sanz, C., Farmaki, T., Paneque, M., Ortego, F., Castañera, P. and Sánchez-Serrano, J. (2001) Hydroperoxide lyase depletion in transgenic potato plants leads to an increase in aphid performance. *Proceedings of the National Academy of Sciences of the USA* 98, 8139–8144.

van de Ven, W.T.G., LeVesque, C.S., Perring, T.M. and Walling, L.L. (2000) Local and systemic changes in squash gene expression in response to silverleaf whitefly feeding. *Plant Cell* 12, 1409–1424.

Van Sanford, D.V., Anderson, J., Campbell, K., Costa, J., Cregan, P., Griffey, C., Hayes, P. and Ward, R. (2001) Discovery and deployment of molecular markers linked to fusarium head blight resistance: an integrated system for wheat and barley. *Crop Science* 41, 638–644.

Vos, P., Hogers, R., Bleeker, M., Rijans, M., Van de Lee, T., Hornes, M., Frijters, A., Pot, J., Kuiper, M. and Zabeau, M. (1995) AFLP: a new technique for DNA fingerprinting. *Nucleic Acids Research* 23, 4407–4414.

Vos, P., Simons, G., Jesse, T., Wijbrandi, J., Heinen, L., Hogers, R., Frijters, A., Groenendijk, J., Diergaarde, P., Reijans, M., Fierens-Onstenk, J., de Both, M., Peleman, J., Liharska, T., Hontelez, J. and Zabeau, M. (1998) The tomato *Mi-1* gene confers resistance to both root-knot nematodes and potato aphids. *Nature Biotechnology* 16, 1315–1316.

Walling, L.L. (2000) The myriad plant responses to herbivores. *Journal of Plant Growth Regulators* 19, 195–216.

Wang, Y.H., Garvin, D.F. and Kochian, L.V. (2001) Nitrate-induced genes in tomato roots: array analysis reveals novel genes that may play a role in nitrogen nutrition. *Plant Physiology* 127, 345–359.

Wayland, S., Mulkey, M., Johnson, S., Green, J., Kearnes, D., Barnes, D., Ellis, V., Anderson, J., Hutton, P. and Dorsey, L. (1998) *The Environmental Protection Agency's White Paper on Bt Plant-pesticide Resistance Management*. US Environmental Protection Agency, Washington, DC, 84 pp.

Weng, Y. and Lazar, M.D. (2002) Amplified fragment length polymorphism- and simple sequence repeat-based molecular tagging and mapping of greenbug resistance gene *Gb3* in wheat. *Plant Breeding* 121, 218–223.

Willcox, M.C., Khairallah, M.M., Bergvinson, D., Crossa, J., Deutsch, J.A., Edmeades, G.O., Gonzalez-de-Leon, D., Jiang, C., Jewell, D.C., Mihm, J.A., Williams, W.P. and Hoisington, D. (2002) Selection for resistance to southwestern corn borer using marker-assisted selection and conventional backcrossing. *Crop Science* 42, 1516–1528.

Winz, R.A. and Baldwin, I.T. (2001) Molecular interactions between the specialist herbivore *Manduca sexta* (Lepidoptera, Sphingidae) and its natural host *Nicotiana attenuata*. IV. Insect-induced ethylene reduces jasmonate-induced nicotine accumulation by regulating *N*-methyltransferase transcripts. *Plant Physiology* 125, 2189–2202.

Wraight, C.L., Zangerl, A.R., Carroll, M.J. and Berenbaum, M.R. (2000) Absence of toxicity of *Bacillus thuringiensis* pollen to black swallowtails under field conditions. *Proceedings of the National Academy of Sciences of the USA* 97, 7700–7703.

Xu, X.F., Mei, H.W., Luo, L.J., Cheng, X.M. and Li, Z.K. (2002) RFLP-facilitated investigation of the quantitative resistance of rice to brown planthopper (*Nilaparvata lugens*). *Theoretical and Applied Genetics* 104, 248–253.

Yencho, G.C., Cohen, M.B. and Byrne, P.F. (2000) Applications of tagging and mapping insect resistance loci in plants. *Annual Review of Entomology* 45, 393–422.

Ziegler, J., Keinanen, M. and Baldwin, I.T. (2001) Herbivore-induced allene oxide synthase transcripts and jasmonic acid in *Nicotiana attenuata*. *Phytochemistry* 58, 729–738.

8 The Pesticide Paradox in IPM: Risk–Benefit Analysis

Paul Guillebeau
Department of Entomology, University of Georgia Cooperative Extension Service, Athens, GA 30602, USA
E-mail: pguillebeau@bugs.ent.uga.edu

Introduction

By definition, integrated pest management (IPM) combines a variety of tactics into a comprehensive system to manage pest populations. As much as possible, it is important to use components that are compatible with one another. Otherwise, the pest-management programme may become overly reliant on a single tactic. Such a programme is not IPM, and the strategy carries greater risks. The failure of a single tactic may allow a rapid increase in the pest population. In a truly integrated programme, the failure of a single component is less likely to be catastrophic because pest populations are controlled through a variety of techniques.

Pesticides are one of the tools available to IPM practitioners, and many pest-management programmes depend on the efficient use of pesticides. Unfortunately, some pesticides impair or eliminate other components of an IPM system. This occurs most commonly when broad-spectrum insecticides destroy populations of beneficial arthropods. The primary pest population may rebound or secondary pests may become a problem.

In many situations, a marketable product cannot be produced economically without pesticides, but the focus of modern pest control is IPM. This chapter discusses the role of pesticides in IPM programmes; situations that result in incompatibility; and ways to better incorporate pesticides into an IPM system.

Integrated pest management and pesticides, defining the terms

A number of people consider pesticides and IPM to be incompatible. In many ways, however, this apparent paradox is a function of the broad and dynamic definitions of both 'pesticide' and 'integrated pest management'. The Environmental Protection Agency (EPA) regulates pesticides in the USA; the Agency defines a pesticide as 'any substance or mixture of substances intended for preventing, destroying, repelling, or mitigating any pest' (US EPA, 2002b). This definition comprises many highly toxic, broad-spectrum chemicals, but it also includes materials with a non-toxic mode of action, such as pheromones. Likewise, IPM is broadly defined, and the definition has changed substantially over the last few decades.

The idea of integrated control is generally credited to Hoskins *et al.* (1939). They stated:

> [B]iological and chemical control are considered … as the two edges of the same sword … nature's own balance provides the major part

 Integrated Pest Management: Potential, Constraints and Challenges (eds O. Koul, G.S. Dhaliwal and G.W. Cuperus)

> of protection … insecticides should be used so as to interfere with natural control of pests as little as possible.

This definition sounds much like some current definitions discussed below, but one must remember that Hoskins *et al.* practised pest management before the era of modern pesticides. Because the pesticides available at that time had limited value in many situations, pest management relied more heavily on non-chemical options.

By the 1960s, definitions of IPM reflected the use of more effective pesticides:

> Integrated control is a pest population management system that utilizes all suitable techniques either to reduce pest populations and maintain them at levels below those causing economic injury or to so manipulate the populations that they are prevented from causing such injury.
>
> (Smith and Van den Bosch, 1967)

The focus of IPM was agricultural efficiency and avoiding economic losses, both strong incentives for grower acceptance of IPM.

Two decades later, Flint and van den Bosch (1981) published a definition that acknowledged the role of pesticides while making it clear that chemicals should be the secondary line of defence:

> IPM is an ecologically based pest control strategy that relies heavily on natural mortality factors … and seeks out control tactics that disrupt these factors as little as possible. IPM uses pesticides, but only after systematic monitoring of pest populations and natural control factors indicates a need.

By the 1990s, the reduction of pesticide use had become a basic tenet of IPM definitions:

> Integrated pest management, or IPM, is an approach to pest control that utilizes regular monitoring to determine if and when treatments are needed and employs physical, mechanical, cultural, biological and educational tactics to keep pest numbers low enough to prevent intolerable damage or annoyance. Least-toxic chemical controls are used as a last resort.
>
> (Olkowski and Daar, 1991)

In 1993, the EPA, the US Department of Agriculture (USDA), and the US Food and Drug Administration (FDA) pledged to have 75% of the US agricultural acreage under IPM by the year 2000 and to reduce the use of pesticides (US EPA, 2002a).

A government report criticized IPM because there has been no significant reduction in the use of pesticides in US agriculture (US Congress Office of Technology Assessment, 1995). Part of the criticism may be an artefact of the imprecise measurement of IPM adoption. Monitoring and application thresholds are key elements of most IPM definitions, and these parameters have been commonly used to measure IPM adoption (Vandeman *et al.*, 1994). This method of measuring IPM provides little information about the intensity of pesticide application.

Because of the myriad definitions of IPM and the apparent lack of pesticide reduction, some authors have suggested new terms. Frisbie and Smith (1991) proposed 'biointensive IPM' with its focus on biological controls, host-plant resistance and cultural controls. Likewise, the National Research Council (1996) published a report on *Ecologically Based Pest Management* that would mitigate environmental, economic and safety risks. New terms have limited value, however, because these same concepts are expressed in many of the earlier definitions for IPM.

Although groups differ in their interpretations, IPM remains a useful compromise between environmental/human-health advocacy groups and industries that may need chemicals to manage out-of-control pest populations. Additionally, modern practitioners agree that pesticides must be used judiciously if biological control is to play a significant role in IPM. In this context, the working definition of IPM includes the goal of reducing pesticide risks to non-target organisms, rather than focusing strictly on reducing pesticide use. Beyond that point, however, generalizations about pesticides and IPM are of limited value. The impact of pesticides in a particular situation depends on the pesticide, the pest and the management situation. Ironically, widespread application of an organophosphate insecticide made possible the greatest reduction of pesticide use and concomitant increase in biological control for an IPM programme.

Pesticides and Cotton IPM

The boll-weevil invasion was the beginning of a pesticide treadmill for cotton that lasted for more than 50 years. By 1917, every Georgia county that produced cotton reported boll-weevils (Hunter, 1917), and cotton yields had already decreased by 32% compared with pre-weevil production (Anon., 1917; Floyd and Treanor, 1944). In 1920, cotton growers across the South applied 10 million lb. of calcium arsenate to control boll-weevil (Coad, 1920). Growers in 1940 were advised to apply calcium arsenate based on a threshold of weevil infestation. Even with a threshold, cotton growers in Georgia alone applied 1.5 million lb. of calcium arsenate (Anon., 1940).

Organophosphate insecticides were introduced for boll-weevil control in the 1950s (Rainwater and Gaines, 1951). By 1968, cotton growers in the South were applying from ten to 18 treatments of insecticides per season (Martin *et al.*, 1968). Boll-weevil was a major target; however, growers were also using multiple pesticide applications to control other pests, primarily the bollworm complex, namely, *Helicoverpa zea* (Boddie) and *Heliothis virescens* (Fabricius) (Haney *et al.*, 2001). Numerous arthropod predators and parasitoids attack bollworms (Whitcomb and Bell, 1964); cotton was clearly on a pesticide treadmill as bollworm populations were released from biological controls by early-season sprays. Cotton production in the southern USA remained heavily dependent on insecticides until the boll-weevil eradication programme.

Intense pesticide application was a key element of the boll-weevil eradication programme (Haney *et al.*, 2001). In 1987, the Georgia programme added an average of 8.4 treatments per acre of azinphos-methyl to the normal in-season spray schedule. In total, more than 287,000 lb. of azinphos-methyl was applied. In 1988 and 1989, each cotton acre received an average of 9.1 and 12.4 additional applications, respectively, of malathion; nearly 350,000 lb. of malathion was applied each year.

After the first 3 years, pesticide applications to control boll-weevil dropped substantially. In 1990, the average number of pesticide applications to control boll-weevil had fallen to 2.7; in 1991, the average number of applications was 0.5. By 1996, pesticide applications were no longer made to control boll-weevil except for spot treatments made when boll-weevils were detected in traps (Haney *et al.*, 2001).

The overall pesticide reductions associated with boll-weevil eradication are enormous. In 1971, growers applied 73 million lb. of insecticide active ingredient (Ridgeway *et al.*, 1983) to 11.5 million acres of cotton in the USA (Anon., 1993), with an average of 6.8 lb. of active ingredient per acre. In 1992, 20 million lb. of insecticide active ingredient was applied to 11.1 million acres of cotton (Anon., 1993; Gianessi and Anderson, 1995), or 1.8 lb. of active ingredient per acre. The average number of insecticide applications per acre in Georgia dropped from 16.5 from 1987 to 1992 to less than 3.5 from 1993 to 1999 (Haney *et al.*, 2001).

The eradication of boll-weevil allowed cotton farmers to place greater emphasis on biological control agents. Prior to boll-weevil eradication, the economic impact (yield losses + cost of control) of *Spodoptera exigua* (Hubner) in Georgia averaged nearly 2% of the cotton crop's value (Haney *et al.*, 2001). During the pesticide-intensive eradication period, the economic impact of *S. exigua* increased to nearly 7%. Since the boll-weevil programme ended, the economic impact has remained at less than 1% (Haney *et al.*, 2001). Ruberson *et al.* (1994) concluded that the pesticidal disruption of biological controls was a key factor in the outbreaks of this typically minor and sporadic pest.

The IPM advantages of boll-weevil eradication are also clear for major cotton pests like bollworms, *H. zea* and *H. virescens*. Before boll-weevil eradication, annual economic losses from these two pests averaged more than 26% of the cotton value in Georgia. After boll-weevil eradication, the economic impact of bollworms fell to approximately 8.5% annually (Haney *et al.*, 2001). A key reason for this decline is the abundance of natural control agents that attack bollworms; until the eradication programme, insecticides applied for boll-weevil also eliminated most biocontrol agents.

Pesticides and Biological Controls

Every pesticide is compatible with some other IPM techniques. Pesticide applications do not preclude cultural controls, such as ploughing, host-plant resistance or physical barriers, such as screens or mesh. As the definition and techniques of IPM have evolved, however, there is greater emphasis on biological control organisms. Currently, biological control relates chiefly to the management of insects although there are examples of biological control agents for weeds and plant diseases.

Pesticides and biological control of insects

Many pesticides have broad-spectrum activity against a variety of insects. A 1956 review of the literature found a large number of studies that discuss the effects of pesticides on populations of beneficial arthropods (Ripper, 1956). Similar reports throughout the years have demonstrated that insecticides can devastate populations of beneficial insects (Turnipseed *et al.*, 1975; Wilkenson *et al.*, 1979; Roach and Hopkins, 1981). However, even the impact of broad-spectrum insecticides is unpredictable.

Predator resistance to pesticides has been documented and even selected in some situations. Guillebeau and All (1989) noted similar-sized populations of striped lynx spider, *Oxyopes salticus* Hentz, in control plots and plots treated with the organophosphate insecticide, methyl parathion. Redmond and Brazzel (1968) reported methyl-parathion resistance in populations of *O. salticus* in an area with a long history of methyl-parathion use. Predatory mites have been selected in laboratory and field trials for resistance to a variety of pesticides, including carbamates and organophosphates (Croft and Stickler, 1983; Hoy, 1985). Trials with other predators have not produced stable resistance at levels that are valuable for field application (Adams and Cross, 1967; Grafton-Cardwell and Hoy, 1986).

Other studies reported predator/parasitoid tolerance of some non-selective insecticides even when resistance was not suspected. In a pecan study, pyrethroids were not toxic to larvae and adult stages of the predatory lacewing, *Chrysoperla rufilabris* (Bermeister), but organophosphates and carbamates were (Mizzell and Schiffhauer, 1990). Conversely, pyrethroids were toxic to the predatory lady beetle, *Olla v-nigrum* Say, but the organophosphates phosalone, methidathion, ethion and malathion were relatively non-toxic. All of the pyrethroids and the organophosphates tested were toxic to another lady beetle, *Hippodamia convergens* (Guerin-Meneville), but lindane was not.

The impacts upon beneficial populations vary even within a single class of pesticides. Among the pyrethroids, Wright and Verkerk (1995) report that cypermethrin is generally less toxic than permethrin to parasitoids; however, the reverse is true against predators. Similarly, the organophosphates comprise many of the compounds that have the greatest impact on beneficial populations, but some of the compounds are much less damaging to natural enemies (Theiling and Croft, 1988). A new class of chemicals, the neonicotinoids, also vary in effects on predatory arthropods (Mizell and Sconyers, 1992).

Introduced in the 1990s, the neonicotinoid imidacloprid is widely used, with activity against sucking insects (e.g. aphids) and some species of beetles, flies and moths (Elbert *et al.*, 1990, 1991). Some spiders, some predatory Coleoptera and some predatory Heteroptera are tolerant of imidacloprid (Kunkel *et al.*, 1999; Elzen, 2001), but other closely related Coleoptera and Heteroptera are highly susceptible to imidacloprid (Delbeke *et al.*, 1997; Sclar *et al.*, 1998).

Bacillus thuringiensis is recommended for organic production and IPM because of its low risk to non-target organisms, including biological control organisms (Abbott Laboratories, 1982). However, even a biorational like *B. thuringiensis* can interfere with IPM. The introduced cinnibar moth, *Tyria jacobaeae* (Linnaeus), has been used successfully as a biocontrol agent for the noxious weed tansy ragwort, *Senecio jacobaea* L. (McEvoy *et al.*, 1991). *B. thuringiensis* may also be used in the same areas to control gypsy moth, *Lymantria dispar* (Linnaeus), and western spruce budworm, *Choristoneura occidentalis* Freeman

(Morris, 1982). Some instars of cinnabar moth are killed by *B. thuringiensis*, and field experiments suggest that *B. thuringiensis* applications could interfere with IPM programmes to manage tansy ragwort (James *et al.*, 1993).

As a group, insecticides have the greatest impact on beneficial arthropod populations, but the potential effects of other types of pesticides should also be considered in IPM programmes. The fungicide zineb is toxic to the parasitoid *Trichogramma cacoeciae* Marchal (Franz and Fabrietius, 1971; Van Driesche *et al.*, 1998). Theiling and Croft (1988) and Mizell and Schiffhauer (1990) report, however, that commonly used fungicides and acaricides were compatible with arthropod predators observed in a pecan study. Some herbicides have also been shown to affect beneficial populations (Theiling and Croft, 1988).

Although the greatest impacts of pesticides have been reported on arthropod populations, pesticides may also affect populations of vertebrate predators that feed on insects. The US EPA restricted the insecticides carbofuran and diazinon because of adverse effects on bird populations (EXTOXNET, 2002a,b). Triazine herbicides have recently been shown to emasculate frogs and toads (Hayes *et al.*, 2002). Other broad-spectrum insecticides, such as aldicarb and methyl parathion, are toxic to a wide range of vertebrate insect predators occurring in agricultural production systems (EXTOXNET, 2002c).

Sublethal effects of pesticides further complicate the impact of pesticides on beneficial arthropods. *Bracon hebetor* Say laid fewer viable eggs after sublethal exposure to the carbamate insecticide carbaryl (Grosch, 1975). Parker *et al.* (1976) reported decreased fecundity for the coccinellid *Menochilus sexmaculatus* (Fabricius) after sublethal doses of malathion, an organophosphate. Repellent effects of pyrethroid insecticides have been reported since the 1980s (Riedl and Hoying, 1983; Jacobs *et al.*, 1984). Sublethal exposures to imidacloprid, azinphos-methyl (organophosphate), carbaryl or malathion have increased egg production in some pest arthropods and some predatory arthropods (Lowery and Sears, 1986; Morse and Zareh, 1991; James, 1997).

The discussion of sublethal effects continues indefinitely with many unanswered questions if multitrophic effects, ecological interactions, behavioural effects, etc. are included. An exhaustive treatise is beyond the scope of this chapter, but Croft (1990) edited a comprehensive review of the relationship between arthropod biological controls and pesticides. Wright and Verkerk (1995) discuss multitrophic evaluation of pesticide applications.

Fungicides and biological control of insects

Insect pests are attacked by naturally occurring fungal pathogens. Early in the 20th century, researchers noted that fungicide applications could result in the rebound of arthropod pest populations (Rolfs and Fawcett, 1908). Since that time, several studies have linked increases in pest populations to the fungicide effects on fungal entomopathogens; Olmert and Kenneth (1974) provide a review of early experiments.

Modern pesticides can have similar effects. *Nomuraea rileyi* Farlow fungus is a naturally occurring control agent for velvet bean caterpillar, *Anticarsia gemmatalis* (Hubner). Johnson *et al.* (1976) showed that applications of benomyl alone and in combination with insecticides reduced the infection rate of velvet bean caterpillar up to tenfold, with a concomitant reduction in yield.

Aphids are a serious pest of pecans and cotton; naturally occurring epizootics caused by fungi help control populations. This phenomenon is of particular interest because pecan aphid pests are resistant to many insecticides (Dutcher, 1983; Dutcher and Htay, 1985). Pickering *et al.* (1990) reported that aphid mortality due to fungal infection was significantly reduced in pecans by the application of triphenyltin hydroxide fungicide. Smith and Hardee (1996) discovered that the application of a granular fungicide at planting could reduce the prevalence of the entomopathogenic fungus *Neozygites fresenii* Batkow, which helps control populations of the cotton aphid *Aphis gossypii* Glover.

In many areas of potato (*Solanum tuberosum* L.) production, growers must protect the crop from a major insect pest, Colorado potato beetle, *Leptinotarsa decemlineata* (Say), and fungal diseases (primarily early blight, *Alternaria solani* (Ell. and Mart.) Jones and Grout, and late blight, *Phytophthora infestans* (Mont. de Bary). The Colorado potato beetle is of particular concern because this pest has become resistant to nearly every available insecticide (Insecticide Resistance Action Committee, 2003). Additionally, environmental risks escalate as growers apply higher rates of pesticides to control resistant populations of Colorado potato beetle.

Research suggests that the fungal pathogen *Beauvaria bassiana* (Balsamo) Vullemin may be a useful biological control agent for Colorado potato beetle (Boiteau, 1988). If *B. bassiana* can be used effectively, growers would have an additional tool to manage resistance. Furthermore, *B. bassiana* would not have many of the undesirable environmental consequences associated with broad-spectrum conventional insecticides.

Using *B. bassiana* to control Colorado potato beetle in potatoes is challenging because disease management in potatoes relies on regular applications of fungicides (University of Idaho, 1999). Todorova *et al.* (1998) reported that field applications of six common fungicides (chlorothalonil, maneb, thiophanate-methyl, mancozeb, metalaxyl + mancozeb, and zineb) inhibited the growth and sporulation of *B. bassiana*; a herbicide, glufosinate-ammonium, also inhibited *B. bassiana*. Surprisingly, the herbicide diquat synergized the insecticidal activity of *B. bassiana*. Other researchers found similar results using chlorothalonil and mancozeb (Jaros-Su *et al.*, 1999). However, they reported that applications of the fungicide copper hydroxide had relatively little effect on *B. bassiana*, suggesting that it may be possible to integrate *B. bassiana* with some types of fungicides.

Pesticides and biological control of weeds

The US federal government and numerous state agencies have programmes to actively control noxious weeds (USDA-APHIS, 2002). An IPM strategy is the focus of most noxious-weeds programmes, including pesticides and biological controls. It is important to use pesticides selectively to avoid disruption of the biological components of the IPM programme.

The California Department of Food and Agriculture has several projects to control noxious weeds in non-agricultural settings. As part of their programme to control purple loosestrife (*Lythrum salicaria* L.), the Department releases two species of leaf beetle (*Galerucella* spp.) that attack this weed. However, these beetles are considered to be very susceptible to pesticides, and newly released colonies should be protected from broad-spectrum insecticides (Pitcairn, California, 2003, personal communication).

Several releases of *Galerucella* spp. were made into riparian zones, but the beetles could not be found shortly after release (Pitcairn, California, 2003, personal communication). Repeated, similar incidents caused the project team to investigate and speculate about possible causes. Although the disappearance of the beetles was not resolved with certainty, the team conjectured that malathion applications made to control mosquitoes in the riparian zone also killed the beetles. The North Dakota State University Extension Service (1997) and bulletins from other states caution that insecticides used for adult mosquito control will also kill *Galerucella* beetles.

Water hyacinth, *Eichhornia crassipes* Mart., is the target of another California weed control project. *Neochetina* spp. weevils attack water hyacinth, and they are a component of the California strategy. However, herbicides are also widely used to control hyacinth. If the hyacinth is killed before the weevils complete development, the weevils also die. During the 1980s and 1990s, the programme to release *Neochetina* spp. weevils was nearly scuttled because other programmes were applying herbicides to water hyacinth in the weevil-release areas (Pitcairn, California, 2003, personal communication). Eventually, better cooperation and communication between the various programmes fostered the use of both herbicides and weevils as part of an IPM programme.

Many states have IPM programmes to control the invasive musk thistle, *Carduus nutans* L. The head weevil, *Rhinocyllus conicus* Froelich, and the rosette weevil, *Trichosirocalus horridus* (Panzer), are biological control agents released to help control musk thistle. Herbicide applications are also commonly used to manage musk thistle. If the herbicides are applied while the weevils are still in the larval stage, the weevils are killed along with the plant. To provide maximum protection for the weevils, herbicide should not be applied to musk thistle in Oklahoma from mid-May to mid-July. However, late May is a good time for growers to apply a single application of herbicide to control a spectrum of weeds (Medlin *et al.*, 2003). Because musk thistle is a ubiquitous problem, conflict may arise between adjacent properties if all parties do not control for thistle. Unlike herbicide application, weevil release does not produce an immediate, obvious effect on musk thistle. The Oklahoma Extension Service has developed a sign to indicate a weevil-release site, so that nearby property owners will not think that no action was taken against musk thistle (Bolin, Oklahoma, 2003, personal communication).

Making Pesticides More Compatible with IPM

Clearly, pesticides can have a variety of negative impacts on populations of biocontrol agents. However, the effects are largely unpredictable, even within closely related groups of organisms or chemical classes. Even with risk and uncertainty, however, pesticides remain a critical component of IPM. Van Emden (2002) calls chemical control, biological control, cultural control and host-plant resistance the four main building-blocks of IPM. Pesticides will continue to be an integral and necessary part of IPM for the foreseeable future, especially in cropping systems with multiple pests (Graves *et al.*, 1999). The challenge for pest managers is to use pesticides in ways that maintain the value of the other IPM components.

To many people, organic farming is the antithesis of conventional farming in terms of pesticide use. However, organic growers still need pesticides, although they are limited to a list of pesticides approved for organic production. The particular chemicals vary somewhat with the government or organization certifying organic production. Insecticidal soaps and oils are permitted in most programmes. Natural products, such as pyrethrum and rotenone, are also widely accepted (Frick, 2002). In the USA, there is an official *National List of Allowed and Prohibited Substances* for growers that wish to be certified as organic by USDA (USDA, 2002). Like conventional operations, organic production tries to use pesticides in ways that are compatible with other pest-management techniques.

In his review, Ripper (1956) points out that biological controls and pesticides are important components of pest-management systems. He suggests that research should aim to reduce the negative impacts of pesticides in two ways: (i) manipulate application techniques for non-selective pesticides; and (ii) develop selective pesticides with fewer adverse effects on beneficial populations.

The quarantine use of pesticides is an important tool for IPM because new pest species do not become established. Quarantines often mandate the use of pesticides. For example, nursery stock, grass sod and other regulated articles cannot be shipped out of the US quarantine area for imported fire ants until the materials have been treated with an approved insecticide (USDA-APHIS, 1997). Japan requires fumigation of plant materials with methyl bromide or approved substitutes if foreign pests are detected (Japanese Market Information, 2002). Although quarantine use of pesticide is a critical tool for IPM, there is little or no impact on beneficial populations of arthropods.

Even though pesticides are the primary method of control recommended for many turf pests, producers are implementing new ideas to attenuate the unwanted effects of pesticides. Sampling schemes and degree-day models have been developed for the chinch bug complex, *Blissus leucopterus* Montandon, *Blissus inularis* Barber and *Blissus leucopterus* (Say), and identification of the species combination determines the need for insecticide applications (Reinert *et al.*, 1995). The *Handbook of Turfgrass Insect Pests*, edited by Brandenburg and Villani (1995), provides an overview of turf IPM and the role of pesticides for management of other turf insect pests.

Pesticide selectivity can be improved by manipulating spray parameters, such as placement or timing. Watson (1975) reported that applying azinphos-methyl to the lower two-thirds of the plant could control cotton pests as predatory anthocorids (*Orius* spp.) increased in the upper third of the plant. Applying herbicides around the base of cotton plants conserved a complex of beneficial arthropods (Stam *et al.*, 1978).

Careful timing of pesticide applications minimizes contact between the beneficial organism and the pesticide. It was already recognized in the 1950s that the proper timing of dichlorodiphenyltrichloroethane (DDT) applications minimized negative impacts on beneficials in walnut and melon pest-management systems (Michelbacher and Middlekauff, 1950; Bartlett and Ortega, 1952). Applications of paraquat or glyphosate to control orchard weeds during the spring harms populations of the predatory mite *Neoseiulus fallacies* Garman occupying the ground cover (Pfeiffer, 1986); herbicide applications made in the autumn have much less impact on the predatory mites because most of them have moved into the tree canopy.

Selective techniques enhance the role of pesticides in IPM for apples, peaches and other stone fruits. Alternate-row middle spraying is a recommended practice in some circumstances (Horton *et al.*, 2002; Washington State University, 2002). Growers apply pesticide only to the middle of every other row; this method reduces pesticide impacts on beneficial organisms. Ironically, increasing herbicide use improves insect-pest management. Stinkbugs and plant bugs are key pests of stone fruits in the southern USA because these insects scar the fruit. This complex of pests is less common in orchards with effective broadcast control of broad-leaved weeds (Horton *et al.*, 2002).

Pesticides are often applied at planting to control early-season pests. Typically, the pesticide is applied in a continuous stream in the row, while the seeds are placed at discrete intervals within the row. A new technique limits the pesticide application to the row space adjacent to the seeds. Lohmeyer *et al.* (2003) report that precision placement of the pesticide may reduce the pesticide rate by more than 50% without sacrificing efficacy. Growers still gain the benefits of the pesticides in their IPM programmes, but the impacts on beneficial populations are greatly reduced.

Spot applications, alternate-row middle applications, precision applications and selective spray timing all create spatial or temporal refugia for beneficials where they do not come in contact with the pesticide. Effective use of refugia requires thorough knowledge of the interactions between the plant, the pest and the biocontrol agent, and information about how the interactions change over time (Verkerk, 2002). Few studies of tritrophic interactions have been published; more data are needed to use refugia more effectively (Verkerk *et al.*, 1998).

Scientists recognized early that pesticide selection was an important factor in preserving beneficial populations (Lord, 1949; Ripper *et al.*, 1951; Bartlett, 1952; van den Bosch and Stern, 1962). Growers now have greater flexibility as pesticide companies introduce products with fewer impacts on beneficial arthropods. Rapid-CP (2002) uses data from the USDA Interregional Research Project Number 4 and USDA Office of Pest Management policy to create a comprehensive database to track potential, pending and recently registered new products. Pymetrozine, indoxacarb, spinosad, bifenzate, methoxyfenozide, tebufenozide and pyriproxyfen are new insecticides that are reported to have low toxicity for beneficial insects. Because of EPA incentives to register reduced-risk pesticides, growers will probably have an increased number of options with fewer effects on biological controls.

Formulation can also improve the selectivity of pesticides. Ripper *et al.* (1948) demonstrated that a coating of cellulose made insecticide significantly less toxic to beneficials, although there was only a slight reduction in efficacy against the target pests. Franz and Fabrietius (1971) reported that rates of parasitism varied from 2 to 89% in the presence of various formulations of zineb. Parasitism may be affected by either the inert ingredients or some interaction between the pesticide and the inert ingredients.

In general, dust formulations are more toxic to beneficials than wettable powders or emulsifiable concentrates. Bartlett (1951) demonstrated that dusts can kill hymenopteran parasitoids even in the absence of a pesticide. Carbamate and organophosphate insecticides were more toxic to spiders when applied as dusts instead of emulsion (Yoo *et al.*, 1984). Encapsulation of pesticide can increase selectivity, but results vary. Dahl and Lowell (1984) reported 70–100% mortality for nabids and coccinellids in lucerne sprayed with encapsulated methyl parathion but little or no mortality for spiders. They further recounted low mortality for coccinellids and high mortality for hymenopterans in cotton treated with encapsulated methyl parathion. Asquith *et al.* (1976) and Hull (1979) found that encapsulated methyl parathion had little or no effect on mite predators found on apple trees.

Greater pesticide selectivity is sometimes gained with granular formulations. Foliar pesticide applications to control lesser cornstalk borer on groundnuts disrupted natural controls and promoted secondary outbreaks of mites and lepidopteran pests, but granular formulations controlled the pest without secondary problems (Newsom *et al.*, 1976). Granular applications of carbofuran and thiofanox did not significantly reduce numbers of coccinellids, chrysopids or parasitized aphids in barley, but foliar applications of the same materials caused substantial reductions (Ba-Angood and Stewart, 1980).

Using pesticides systemically will protect populations of some kinds of beneficial arthropods. Ripper (1957) showed that systemic use of schradan and oxydemeton-methyl had little or no effect on predatory coccinellids or syrphids. Populations of the spider *Lycosa pseudoannulata* Boesenberg were not greatly reduced by systemic use of carbofuran on rice, but populations of the predatory bug, *Cyrtorhinus lividipennis* Reuter were nearly eliminated (Dyck and Orlido, 1977). Systemic use of aldicarb on cotton caused high mortality for the heteropteran *Nabis* and *Geocoris* spp., but populations of spiders, *Chrysoperla* larvae and hymenopteran parasitoids were nearly unaffected (Ridgeway *et al.*, 1967). In rice and cotton, the affected predator species were facultative plant feeders.

The EPA classifies pheromones as pesticides even though they often have a nontoxic mode of action. Additionally, pheromones typically affect a single species or a small group of closely related species (Nordland *et al.*, 1981). Consequently, pheromones have little or no effect on biological control agents in most cases. Pheromones are used directly to disrupt mating in a number of lepidopteran pest species, including codling moth, *Cydia pomonella* (Linnaeus), and pink bollworm, *Pectinophora gossypiella* Saunders (Carde and Minks, 1995). Pheromones are commonly used to monitor pest populations so that pesticide applications can be made at the optimum time and place. Pheromones and/or food baits are also used in combination with toxic agents. Programmes to contain or eradicate Mediterranean fruit fly populations typically include a combination of an attractant and a poison (Roessler, 1989). As more pheromones become available, pest managers will have greater flexibility to integrate pesticides into IPM.

Baits do not always provide as much selectivity as pheromones because they can attract a broader range of arthropods. Nasca *et al.* (1983) reported that malathion bait applications to control fruit flies were less toxic to predacious lacewings. Surprisingly, greater selectivity was gained by applying the bait to the entire tree instead of offering the bait on plastic strips. Conversely, Ehler *et al.* (1984) found that a malathion-bait spray applied for fruit flies greatly reduced populations of gall-midge parasitoids, allowing gall-midge populations to increase 90-fold.

Policies to Improve Compatibility of Pesticides and IPM

Pest Management and the Environment in 2000 was an international symposium to discuss the future of pest management (Aziz *et al.*, 1992). One conclusion was that pesticides would continue to be an essential component of IPM (Perfect, 1992). He also points out a common paradox of pesticides and IPM. Misapplication of pesticides is the greatest threat to many IPM programmes; however, the overuse of pesticides may also be the impetus for IPM with greater emphasis on non-chemical controls. Resistance to pyrethroid insecticides caused *Helicoverpa armigera* (Hubner) to become one of the most economically damaging pests in Indian agriculture (Sharma, 1991). In response, the Indian Council of Agricultural Research instigated a major programme for natural controls of *H. armigera*, including nuclear polyhedrosis virus and parasitoid wasps (Perfect, 1992).

Although the future of IPM includes pesticides, new chemicals will be quite different from the broad-spectrum pesticides first developed for agricultural-pest management. The EPA Office of Pesticide Programs (OPP) has several programmes to accelerate the introduction of pesticides with fewer risks to the environment, including biological control organisms (Andersen *et al.*, 1996). The EPA 25b rule exempts very low-risk products, including some biological agents and common food substances (e.g. eugenol and soybean oil), from regulation; a complete list of 25b-exempted materials can be found on the Internet (http://www.epa.gov/oppbppd1/biopesticides/regtools). Regulatory relief will encourage research and marketing of these products because a large investment will not be necessary to fulfil EPA requirements for registration of conventional pesticides.

Andersen *et al.* (1996) report that the Biopesticides and Pollution Prevention Division (BPPD) of OPP streamlines the registration process for biological pesticides, including microbials, plant-incorporated protectants and biochemicals active through a non-toxic mode of action (e.g. pheromone traps). Biopesticides generally affect only the target pest and closely related organisms (US EPA BPPD, 2002b). As part of an IPM programme, biopesticides can greatly decrease the use of conventional pesticides and maintain high crop yields. At the end of 2001, the EPA had listed approximately 195 registered biopesticide active ingredients and 780 products (US EPA BPPD, 2002a).

The OPP reduced-risk programme encourages the registration of conventional pesticides that fit low-risk criteria, including compatibility with IPM (OPP, 2002). The EPA offers greatly reduced registration time as an economic incentive. Reduced-risk pesticides are registered in an average of 16 months versus 38 months for a conventional pesticide not classified as reduced risk. The pesticide can be marketed sooner, and the registrant has several more growing seasons with their pesticide under patent.

The USDA developed a strategic plan for IPM that was released in 2003 (IPM, 2002). Among other goals, the plan defines specific targets to reduce levels of hazardous pesticides detected in surface drinking-water supplies and pesticide residues in the major foods consumed by infants and children. The strategy calls for alternatives to pesticides that result in unacceptable residue levels in food-crop commodities. The USDA plan clearly recognizes the role of pesticides in IPM. One target calls for improvement of pesticide application methods, timing and placement that results in improved efficacy with reduced pesticide residues in raw agricultural commodities.

In some European countries, laws were not specifically crafted to promote IPM, but growers are likely to use more non-chemical pest-control methods as pesticides become less available. Sweden, Denmark and The Netherlands mandated a 50% or greater reduction in total pesticide use by 2000 (Matteson, 1995). Sweden's plan included regional plant-protection centres promoting IPM.

A number of developing countries promote IPM or pesticide reduction through official policies or regulation (Elkstrom, 2002). China introduced the Green Certificate programme and banned highly toxic pesticides from vegetable crops. Biological control is a national priority for Cuba; the new policy is intended to make IPM biointensive, with 80% of pests managed through biological control. Iran formed the High Council of

Policy and Planning for Reduction of Agricultural Pesticides. A National IPM Committee in Malaysia is reducing pesticide use and increasing farmer knowledge of other pest-management techniques. Nepal, the Philippines, Sri Lanka and Indonesia have adopted national IPM policies. The Food and Agriculture Organization and the World Health Organization work with the international community and developing nations to develop pesticide policy, but many developing countries have only rudimentary regulation of pesticide use (Schaefers, 1996).

Although most growers incorporate some elements of IPM (e.g. monitoring for pests), many modern production systems are overly dependent on pesticides that disrupt natural controls. Reliance on pesticides has undoubtedly slowed the progress of IPM. Pest management should not be chemically dependent, but it is imprudent to ignore the benefits associated with the judicious use of pesticides. The most efficient and sustainable IPM programmes will recognize a critical role for pesticides.

As IPM and pesticides evolve, the use of pesticides is likely to increase as new chemicals and new application techniques make pesticides more compatible with biological controls and other IPM components. The pesticide industry is emphasizing new products with greater selectivity for natural enemies and minimal environmental impact (Sengonca, 2002). Governmental programmes and regulations encourage the introduction of pesticides with fewer non-target effects. Improved application techniques allow us to use pesticides more precisely, with less off-target deposition. Research is providing greater insight into interactions between biological controls, pests, hosts and pesticides. This knowledge will help IPM practitioners use pesticides in ways that conserve natural controls.

References

Abbott Laboratories (1982) *Toxicology Profile: Dipel,* Bacillus thuringiensis *Insecticide.* Chemical and Agricultural Products Division, Chicago, Illinois.

Adams, C.H. and Cross, W.H. (1967) Insecticide resistance in *Bracon mellitor,* a parasite of the boll weevil. *Journal of Economic Entomology* 60, 1016–1020.

Andersen, J., Leslie, A., Matten, S. and Kumar, R. (1996) The Environmental Protection Agency's programs to encourage the use of safer pesticides. *Weed Technology* 10, 966–968.

Anon. (1917) Annual Report of the Georgia Department of Agriculture. *Georgia Department of Agriculture Monthly Bulletin* 5, 130–131.

Anon. (1940) *Annual Report of the Georgia Agriculture Extension Service.* Georgia Agriculture Extension Service, pp. 5–7

Anon. (1993) *Agricultural Statistics 1993.* National Agricultural Statistics Service, USDA, Washington, DC, 57 pp.

Asquith, D., Hull, L.A. and Mowry, P.D. (1976) Apple, tests of insecticides (1975). *Insecticide and Acaracide Tests* 1, 17–19.

Aziz, A., Kadir, S.A. and Barlow, H.S. (eds) (1992) *Pest Management and the Environment in 2000.* CAB International, Wallingford, UK, 401 pp.

Ba-Angood, S.A. and Stewart, R.K. (1980) Effect of granular and foliar insecticides on cereal aphids and their natural enemies on field barley in southwestern Quebec. *Canadian Entomologist* 112, 1309–1313.

Bartlett, B.R. (1951) The action of certain inert dust materials on parasitic hymenoptera. *Journal of Economic Entomology* 44, 891–896.

Bartlett, B.R. (1952) A study of insecticide resistance in strains of *Drosophila melanogaster. Canadian Entomologist* 84, 189.

Bartlett, B.R. and Ortega, J.C. (1952) Relation between natural enemies and DDT-induced increases in frosted scale and other pests of walnuts. *Journal of Economic Entomology* 45, 783–785.

Boiteau, G. (1988) Control of the Colorado potato beetle *Leptinotarsa decemlineata*: learning from the Soviet experience. *Bulletin of the Entomological Society of Canada* 20, 9–15.

van den Bosch, R. and Stern, V. (1962) The integration of chemical and biological control of arthropod pests. *Annual Review of Entomology* 7, 367–368.

Brandenburg, R.L. and Villani, M.G. (eds) (1995) *Handbook of Turfgrass Insect Pests*. Entomological Society of America Press, Lanham, Maryland, 140 pp.
Carde, R.T. and Minks, A.K. (1995) Control of moth pests by mating disruption: successes and constraints. *Annual Review of Entomology* 40, 559–586.
Coad, B.R. (1920) Killing boll weevils with poison dust. *United States Department of Agriculture Yearbook* 1920, 241–252.
Croft, B.A. (1990) *Arthropod Biological Control Agents and Pesticides*. John Wiley & Sons, New York, 723 pp.
Croft, B.A. and Stickler, K. (1983) Natural enemy resistance to pesticides: documentation, characterization, theory and application. In: Georghiou, G.P. and Saito, T. (eds) *Pest Resistance to Pesticides*. Plenum, New York, pp. 669–702.
Dahl, G.H. and Lowell, J.R. (1984) Microencapsulated pesticides and their effects on nontarget insects. In: Scher, H.B. (ed.) *Advances in Pesticide Formulation Technology*. American Chemical Society, Washington, DC, pp. 141–150.
Delbee, F., Vercruysse, P., Tirry, L., DeClerq, P. and Degheele, D. (1997) Toxicity of diflubenzuron, pyriproxyfen, imidacloprid, and diafenthruion to the predatory bug, *Orius laevigatus* (Heteroptera: Anthocoridae). *Entomophaga* 42, 349–358.
Dutcher, J.D. (1983) Pecan pest management – where are we? In: Payne, J.A. (ed.) *Pecan Pest Management – Are We There?* Miscellaneous Publication 13, Entomological Society of America, College Park, Maryland, pp. 133–140.
Dutcher, J.D. and Htay, U.T. (1985) Impact assessment of carbaryl, dimethoate, and dialifor on foliar pecan and nut pests of pecan orchards. *Journal of the Georgia Entomological Society* 18, 495–507.
Dyck, V.A. and Orlido, G.C. (1977) Control of the brown planthopper, *Nilaparvata lugens*, by natural enemies and timely application of narrow-spectrum insecticides. *International Rice Research Institute Annual Report* 1976, 28–72.
Ehler, L.E., Endicott, P.C., Herlein, M.B. and Alvarado-Rodriguez, B. (1984) Medfly eradication in California: impact of malathion bait sprays on an endemic gall midge and its parasitoids. *Entomologia Experimentalis et Applicata* 36, 201–208.
Elbert, A., Overbeck, H., Iwaya, K. and Tsuboi, S. (1990) Imidacloprid, a novel systemic nitromethylene analogue insecticide for crop protection. *Proceedings Brighton Crop Protection Conference on Pests and Diseases* 1, 21–28.
Elbert, A., Becker, B., Hartwig, J. and Erdelen, C. (1991) Imidacloprid: a new systemic insecticide. *Pfanzenshutz Nachrichten Bayer* 44, 113–116.
Elkstrom, G. (2002) Pesticide reduction in developing countries. In: Pimentel, D. (ed.) *Encyclopedia of Pest Management*. Marcel Dekker, New York, pp. 598–605.
Elzen, G.W. (2001) Lethal and sublethal effects of insecticide residues on *Orius insidiosus* (Hemiptera: Anthocoridae) and *Geocoris punctipes* (Hemiptera: Lygaeidae). *Journal of Economic Entomology* 94, 55–59.
EXTOXNET (2002a) Pesticide information profiles: diazinon. Available at: http://ace.orst.edu/cgi-bin/mfs/01/pips/diazinon.htm
EXTOXNET (2002b) Pesticide information profiles: carbofuran. Available at: http://ace.orst.edu/cgi-bin/mfs/01/pips/carbofur.htm
EXTOXNET (2002c) Pesticide information profiles: aldicarb. Available at: http://ace.orst.edu/cgi-bin/mfs/01/pips/aldicarb.htm
Flint, M.L. and van den Bosch, R. (1981) *Introduction to Integrated Pest Management*. Plenum Press, New York, 240 pp.
Floyd, D.L. and Treanor, K. (1944) Georgia agricultural facts. *University of Georgia Agriculture Extension Service Bulletin* 511, 28–29.
Franz, J.M. and Fabrietius, K. (1971) Testing the sensitivity to pesticides of entomophagous arthropods – trials using *Trichogramma*. *Zeitschrift für Angewandte Entomologie* 68, 278–288.
Frick, B. (2002) Organic farming. In: Pimentel, D. (ed.) *Encyclopedia of Pest Management*. Marcel Dekker, New York, pp. 554–557.
Frisbie, R.E. and Smith, J.W. Jr (1991) Biologically intensive integrated pest management: the future. In: Menn, J.J. and Steinhauer, A.L. (eds) *Progress and Perspectives for the 21st Century*. Entomological Society of America Press, Lanham, Maryland, pp. 151–64.
Gianessi, L.P. and Anderson, J.E. (1995) *Pesticide Use in the U.S. Crop Production: National Summary Report*. The National Center for Food and Agricultural Policy, Washington, DC, 280 pp.

Grafton-Cardwell, E.E. and Hoy, M.A. (1986) Genetic improvement of common green lacewing, *Chrysoperla carnea* (Neuroptera: Chrysopidae): selection for carbaryl resistance. *Environmental Entomology* 15, 1130–1136.

Graves, J.B., Leonard, B.R. and Ottea, J.A. (1999) Chemical approaches to managing arthropod pests. In: Ruberson, J. (ed.) *Handbook of Pest Management*. Marcel Dekker, New York, pp. 449–486.

Grosch, D.S. (1975) Reproductive performance of *Bracon hebetor* after sublethal doses of carbaryl. *Journal of Economic Entomology* 68, 659–662.

Guillebeau, L.P. and All, J.N. (1989) Impact of weekly applications of selected new pyrethroids or improved *Bacillus thuringiensis* Berliner on the arthropod predator complex in cotton. PhD dissertation, University of Georgia, Athens, Gorgia.

Haney, P.B., Herzog, G. and Roberts, P.M. (2001) Boll weevil eradication in Georgia. In: *Boll Weevil Eradication in the U.S. through 1999*. Cotton Foundation Reference Book Series No. 6, Memphis, Tennessee, pp. 258–88.

Hayes, T., Haston, K., Tsui, M., Hoang, A., Haeffele, C. and Vonk, A. (2002) Feminization of male frogs in the wild. *Nature* 419, 895–896.

Horton, D., Bellinger, B., Gorsuch, C. and Ritchie, D. (2002) *Southeastern Peach, Nectarine, and Plum Pest Management and Culture Guide*. Bulletin 1171, Cooperative Extension Service, University of Georgia, Athens, Georgia, 42 pp.

Hoskins, W.M., Borden, A.D and Michelbacher, A.E. (1939) Recommendations for a more discriminating use of insecticides. In: *Proceedings of the 6th Pacific Scientific Congress*, Vol. 5, Davis, California, pp. 119–123.

Hoy, M.A. (1985) Recent advances in genetics and genetic improvement of the Phytoseiidae. *Annual Review of Entomology* 30, 345–370.

Hull, L.A. (1979) Apple, tests of insecticides 1978. *Insecticide and Acaricide Tests* 4, 20–22.

Hunter, W.D. (1917) The boll weevil program with special reference to means of reducing damage. *USDA Farmers' Bulletin* 848, 1–40.

Insecticide Resistance Action Committee (2003). Resistance depleting potato growers' arsenal. Available at: http://plantprotection.org/irac/Growers/Potato.htm

IPM (2002) IPM roadmap will be unveiled at meeting. Issue no. 105. Available at: http://www.ipmnet.org/IPMnet_NEWS/news105.htm

Jacobs, R.J., Kouskolekas, C. and Gross, H. (1984) Responses of *Trichogramma pretiosum* (Hymenoptera: Trichogrammatidae) to residues of permethrin and endosulfan. *Environmental Entomology* 13, 355–358.

James, D.G. (1997) Imidacloprid increases egg production in *Amblyseius victoriensis* (Acari: Phytoseiidae). *Experimental and Applied Acarology* 21, 75–82.

James, R.R., Miller, J.C and Lighthart, B. (1993) *Bacillus thuringiensis* var. *kurstaki* affects a beneficial insect, the cinnabar moth (Lepidoptera: arctiidae). *Journal of Economic Entomology* 86, 334–339.

Japanese Market Information (2002) Fresh fruits. Available at: http://www.pic.or.jp/jp/jmi/007.htm

Jaros-Su, J., Groden, E. and Zhang, J. (1999) Effects of selected fungicides and the timing of fungicide application on *Beauvaria bassiana* induced mortality of the Colorado potato beetle. *Biological Control* 15, 259–269.

Johnson, D.W., Kish, L.P. and Allen, G.E. (1976) Field evaluation of selected pesticides on the natural development of the entomopathogen, *Nomuraea rileyi*, on the velvet bean caterpillar in soybean. *Environmental Entomology* 5, 964–966.

Kunkel, B.A., Held, D.W. and Potter, D.A. (1999) Impact of halofenozide, imidacloprid, and bendiocarb on beneficial invertebrates and predatory activity in turfgrass. *Journal of Economic Entomology* 92, 922–930.

Lohmeyer, K.H., All, J.N., Roberts, P.M. and Bush, P. (2003) Precision application of aldicarb to enhance efficiency of thrips (Thysanoptera: Thripidae) management in cotton. *Journal of Economic Entomology* 96, 748–754.

Lord, F.T. (1949) The influence of spray programs on the fauna of apple orchards in Nova Scotia II: oystershell scale. *Canadian Entomologist* 79, 196–209.

Lowery, D.T. and Sears, M.K. (1986) Stimulation of reproduction of the green peach aphid (Homoptera: Aphididae) by azinphosmethyl applied to potatoes. *Journal of Economic Entomology* 79, 1530–1533.

Martin, N.R., Nix, J.E., McArthur, W.C. and Brannen, S.J. (1968) Effects of alternative production practices on costs and returns in producing cotton in selected areas of Georgia. *University of Georgia College of Agriculture Experimental Station–USDA Research Bulletin* 34, 1–38.

Matteson, P.C. (1995) The 50% pesticide cuts in Europe: a glimpse of our future. *American Entomologist* 41, 210–220.
McEvoy, P., Cox, C. and Coombs, E. (1991) Successful biological control of ragwort, *Senecio jacobaea*, by introduced insects in Oregon. *Ecological Applications* 4, 430–442.
Medlin, C., Bolin, P., Roduner, M. and Stritske, J. (2003) *Integrated Control of Invasive Thistles in Oklahoma.* Bulletin F-7318 Stillwater, Oklahoma, Oklahoma Cooperative Extension Service, Oklahoma State University.
Michelbacher, A.E. and Middlekauff, W.W. (1950) Control of the melon aphid in northern California. *Journal of Economic Entomology* 43, 444–447.
Mizell, R.F. and Schiffhauer, D.E. (1990) Effects of pesticide on pecan aphid predators *Chrysoperla rufliabris* (Neuroptera: Chrysopidae), *Hippodamia convergens, Cylconeda sanguinea, Olla v-nigrum* (Coleoptera: Coccinellidae), and *Aphelinus perpallidus* (Hymenoptera: Encyrtidae). *Journal of Economic Entomology* 83, 1806–1812.
Mizell, R.F. and Sconyers, M.C. (1992) Toxicity of imidacloprid to selected arthropod predators. *Florida Entomologist* 75, 277–280.
Morris, O.N. (1982) Bacteria as pesticides: forest applications. In: Kurstak, E. (ed.) *Microbial and Viral Pesticides.* Marcel Dekker, New York, pp. 239–287.
Morse, J.G. and Zareh, N. (1991) Pesticide induced hormoligosis of citrus thrips (Thysanoptera: Thripidae) fecundity. *Journal of Economic Entomology* 84, 1169–1174.
Nasca, A.J., Fernandez, R.V., De Herrero, A.J. and Manzur, B.E. (1983) Incidence of chemical treatment for control of fruit flies on chrysopids and hemerobiids on citrus trees. *CITRON* 1, 47–73.
National Research Council (1996) *Ecologically Based Pest Management: New Solutions for a New Century.* National Academy, Washington, DC, 35 pp.
Newsom, L.D., Smith, R.F. and Whitcomb, W.H. (1976) Selective pesticides and selective use of pesticides. In: Huffaker, C.B. and Messenger, P.S. (eds) *Theory and Practice of Biological Control.* Academic Press, New York, pp. 565–591.
Nordland, D.A., Jones, R.L. and Lewis, W.J. (1981) *Semiochemicals: Their Role in Pest Control.* Wiley, New York, 306 pp.
North Dakota State University Extension Service (1997) Biological control agents released for purple loosestrife control. *North Dakota Pesticide Quarterly* 15, 9.
Olkowski, W. and Daar, S. (1991) *Common Sense Pest Control.* Taunton Press, Newtown, Connecticut, 715 pp.
Olmert, I. and Kenneth, R.G. (1974) Sensitivity of the entomopathogenic fungi to fungicides and insecticides. *Journal of Economic Entomology* 31, 371–372.
OPP (2002) General overview: reduced-risk pesticide program. Available at: http://www.epa.gov/oppfod01/trac/safero.htm
Parker, F.D., Ming, N., Pend, T. and Singh, G. (1976) The effect of malathion on fecundity, longevity, and geotropism of *Menochilus sexmaculatus. Environmental Entomology* 5, 495–501.
Perfect, T.J. (1992) IPM in 2000. In: Aziz, A., Kadir, S.A. and Barlow, H.S. (eds) *Pest Management and the Environment in 2000.* CAB International, Wallingford, UK, pp. 47–53.
Pfeiffer, D.G. (1986) Effects of field applications of paraquat on densities of *Panonychus ulmi* Koch and *Neoseiulus fallacies* Garman. *Journal of Agricultural Entomology* 3, 322–325.
Pickering, J., Dutcher, J.D. and Ekbom, B.S. (1990) The effect of a fungicide on fungal-induced mortality of pecan aphids in the field. *Journal of Economic Entomology* 83, 1801–1805.
Rainwater, C.F. and Gaines, J.C. (1951) Seasonal decline in the effectiveness of the certain insecticides against boll weevil. *Journal of Economic Entomology* 44, 971–974.
Rapid-CP (2002) Recent Ag Products in Defence of Crop Plants. Available at: http://cipmtest.ent.ncsu.edu/rapidcp/
Redmond, K.R. and Brazzel, B.R. (1968) Response of the striped lynx spider, *Oxyopes salticus*, to commonly used pesticides. *Journal of Economic Entomology* 61, 327–328.
Reinert, J.A., Heller, P.R. and Crocker, R.L. (1995) Pest information: chinch bugs. In: Brandenburg, R.L. and Villani, M.G. (eds) *Handbook of Turfgrass Insect Pests.* Entomological Society of America Press, Lanham, Maryland, pp. 38–42.
Ridgeway, R.L., Lingren, P.D., Cowan, C.B. and Davis, J.W. (1967) Populations of arthropod predators and *Heliothis* spp. after applications of systemic insecticides to cotton. *Journal of Economic Entomology* 60, 1012–1016.

Ridgeway, R.L., Lloyd, E.P. and Cross, W.H. (1983) *Cotton Insect Management with Special Reference to the Boll Weevil*. Agriculture Handbook 589, Agricultural Research Service, United States Department of Agriculture, Washington, DC.

Riedl, H. and Hoying, S. (1983) Toxicity and residual activity of fenvalerate to *Tryphlodromus occidentalis* (Acari: Tetranychidae) on pear. *Canadian Entomologist* 115, 807–813.

Ripper, W.E. (1956) Effect of pesticides on balance of arthropod populations. *Annual Review of Entomology* 1, 403–439.

Ripper, W.E. (1957) Selective insecticides and the balance of arthropod populations. *Agricultural Chemistry* 12, 36–37, 103, 105.

Ripper, W.E., Greenslade, R.M., Health, J. and Barker, K. (1948) New formulation of DDT with selective properties. *Nature* 161, 484.

Ripper, W.E., Greenslade, R.M. and Hartley, G.S. (1951) Selective insecticides and biological control. *Journal of Economic Entomology* 44, 448–459.

Roach, S.H. and Hopkins, A.R. (1981) Reduction in arthropod predator populations in cotton fields treated with insecticides for *Heliothis* spp. control. *Journal of Economic Entomology* 74, 454–457.

Roessler, Y. (1989) Insecticidal bait and cover sprays. In: Robinson, A.S. and Hooper, G. (eds) *Fruit Flies, Their Biology, Natural Enemies, and Control*. Elsevier, Amsterdam, pp. 329–336.

Rolfs, P.H. and Fawcett, H.S. (1908) Fungus diseases of scale insect and whitefly. Bulletin, Florida Agriculture Experiment Station, Gainesville, Florida, 17 pp.

Ruberson, J.R., Herzog, G.A., Lambert, W.R. and Lewis, W.J. (1994) Management of the beet armyworm in cotton: role of natural enemies. *Florida Entomologist* 77, 440–453.

Schaefers, G.A. (1996) Status of pesticide policy and regulations in developing countries. *Journal of Agricultural Entomology* 13, 213–222.

Sclar, D., Gerace, D. and Crenshaw, W.S. (1998) Observations of population increases and injury by spider mites (Acari: Tetranychidae) on ornamental plants treated with imidacloprid. *Journal of Economic Entomology* 91, 250–255.

Sengonca, L. (2002) Conservation of natural enemies. In: Pimentel, D. (ed.) *Encyclopedia of Pest Management*. Marcel Dekker, New York, pp. 141–143.

Sharma, D. (1991) India battles to eradicate major crop pest. *New Scientist* 150, 10 August, 15.

Smith, M.T. and Hardee, D.D. (1996) Influence of fungicide on development of an entomopathogenic fungus in the cotton aphid. *Environmental Entomology* 25, 677–687.

Smith, R.F. and van den Bosch, R. (1967) Integrated control. In: Wendell, W., Kilglore, W.W. and Doutt, R.L. (eds) *Pest Control: Biological, Physical and Selected Chemical Methods*. Academic Press, New York, pp. 295–340.

Stam, P.A., Clower, D.F., Graves, J.B. and Schilling, P.E. (1978) Effects of certain herbicides on some insects and spiders found in Louisiana cotton fields. *Journal of Economic Entomology* 71, 477–480.

Theiling, K.M. and Croft, B.A. (1988) Pesticide side-effects on arthropod natural enemies: a database summary. *Agriculture, Ecosystems, and the Environment* 21, 191–218.

Todorova, S.I., Coderre, D., Duchesne, R. and Cote, J. (1998) Compatibility of *Beauvaria bassiana* with selected fungicides and herbicides. *Biological Control* 27, 427–433.

Turnipseed, S.G., Todd, J.W. and Campbell, W.V. (1975) Field activity of selected foliar insecticides against geocorids, nabids, and spiders in soybeans. *Journal of Entomological Science* 10, 272–277.

United States (US) Congress Office of Technology Assessment (1995) *Biologically Based Technologies for Pest Control*. US Government Printing Office, New York, 204 pp.

United States Department of Agriculture (USDA) (2002) *National List of Allowed and Prohibited Substances*. The National Organic Program. Available at: http://www.ams.usda.gov/nop/

United States Department of Agriculture Animal Plant Health Inspection Service (USDA-APHIS) (1997) *Federal Domestic Quarantines*. Available at: http://www.aphis.usda.gov/npb/F&SQS/usdaqua.html

United States Department of Agriculture Animal Plant Health Inspection Service (USDA-APHIS) (2002) *Federal Noxious Weed Program*. Available at: http://www.aphis.usda.gov/ppq/weeds/

United States Environmental Protection Agency (US EPA) (2002a) *Inception of Pesticide Environmental Stewardship Program – IPM and Reduced Risk*. Available at: http://www.epa.gov/oppbppd1/PESP/inception.htm

United States Environmental Protection Agency (US EPA) (2002b) What is a pesticide? Available at: http://www.epa./pesticides.

United States Environmental Protection Agency Biopesticides and Pollution Prevention Division (US EPA BPPD) (2002a) Pesticides: biopesticides. Available at: http://www.epa.gov/oppbppd1/biopesticides/

United States Environmental Protection Agency Biopesticides and Pollution Prevention Division (US EPA BPPD) (2002b) What are biopesticides? Available at: http://www.epa.gov/oppbppd1/biopesticides

University of Idaho (1999) Potato late blight. Available at: http://www.uidaho.edu/ag/plantdisease/plbchc.htm

Vandeman, A.J., Fernandez-Cornejo, S., Jans, S. and Lin, B.H (1994) *Adoption of Integrated Pest Management in United States Agriculture.* Bulletin No. 707, Economic Research Service Agriculture Information, United States Department of Agriculture, 26 pp.

Van Driesche, R.G., Mason, J.L., Wright, S.E. and Prokopy, R.J. (1998) Effect of reduced insecticide and fungicide on parasitism of leafminers (*Phyllonorycter* spp.) (Lepidoptera: Gracillariidae) in commercial apple orchards. *Environmental Entomology* 27, 578–582.

Van Emden, H.F. (2002) Integrated pest management. In: Pimentel, D. (ed.) *Encyclopedia of Pest Management.* Marcel Dekker, New York, pp. 413–415.

Verkerk, R.H.J. (2002) Refugia for pests and natural enemies. In: Pimentel, D. (ed.) *Encyclopedia of Pest Management.* Marcel Dekker, New York, pp. 685–688.

Verkerk, R.H.J., Leather, S.R. and Wright, D.J. (1998) The potential for manipulating crop–pest–natural enemy interaction for improved insect pest management. *Bulletin of Entomological Research* 88, 493–501.

Washington State University (2002) *Crop Protection Guide for Tree Fruits in Washington.* Washington State University Cooperative Extension Service Bulletin EB0914, 84 pp. Available at: http://cru.cahe.wsu.edu/CEPublications/eb0419/eb0419.pdf

Watson, T.F. (1975) Practical consideration in the use of selective insecticides against major crop pests. In: Street, J.C. (ed.) *Pesticide Selectivity.* Marcel Dekker, New York, pp. 47–65.

Whitcomb, W.H. and Bell, K. (1964) Predaceous insects, spiders, and mites of Arkansas cotton fields. *Arkansas Agriculture Experiment Station Bulletin* 690, 1–84.

Wilkenson, J.D., Biever, K.D. and Ignoffo, C.M. (1979) Synthetic pyrethroids and organophosphates against the parasitoid, *Apanteles marginventris* and the predators *Georcoris punctipes, Hippodamia convergens*, and *Podisus maculiventris. Journal of Economic Entomology* 72, 473–475.

Wright, D.J. and Verkerk, R.H.J. (1995) Integration of chemical and biological control systems for arthropods: evaluation in a multitrophic context. *Pesticide Science* 44, 207–218.

Yoo, J.K., Kwon, U.W., Park, H.M. and Lee, H.R. (1984) Studies on the selective toxicity of insecticides for rice insect pests and a predacious spider, *Pirata subpiraticus. Korean Journal of Plant Protection* 23, 166–171.

9 Manipulation of Host-finding and Acceptance Behaviours in Insects: Importance to IPM

Richard S. Cowles
Connecticut Agricultural Experiment Station, Valley Laboratory, PO Box 248, Windsor, CT 06095, USA
E-mail: Richard.Cowles@po.state.ct.us

Introduction

Manipulating host-finding and acceptance behaviours can be used to shift much of a highly mobile and discriminating insect population to plants or traps outside our valued crops. Using a trap crop, insect-refuge crop, diversionary crop or mass trapping may directly reduce the population density in the valued crop below an economic threshold. Behavioural manipulation alone may often be inadequate for plant protection, but, when combined with appropriately chosen antibiosis factors (insecticides or resistant varieties), a pest-management system incorporating a behavioural dimension can yield evolutionarily stable and favourable results. The ability not only to prevent the evolution of insecticide resistance but to select against physiological resistance (both to insecticides and to varietal resistance) is an important outcome that could make behavioural manipulation an essential tool in modern agriculture. Behavioural manipulation can also lead to greater sustainability in agricultural systems by improving the efficiency of biologically based population suppression by concentrating the insect-pest population in a pesticide-free diversionary crop.

Host Finding and Acceptance

The adult female of most insects is responsible for finding appropriate hosts and laying eggs where the resources can sustain her young. With a few notable exceptions (such as ballooning larvae), the developing larvae of holometabolous insects have limited dispersal capabilities; therefore the host-seeking and host-acceptance 'decisions' made by the mother determine to a large degree the spatial distribution of larvae within patchy and unpredictable environments and the success or failure in larval development (Courtney and Kibota, 1989; Mayhew, 1997). For other mobile pests, such as adult hemipterans, alate aphids and thrips, arrestment on hosts and subsequent feeding may lead to direct crop injury and virus transmission (Kennedy *et al.*, 1961). Insect behaviour adaptive for selecting favourable hosts should involve the relatively simple algorithm: move if the conditions are poor, and stop (and either feed or lay eggs) if the conditions are good.

The study of insect behaviour, physiology and chemical ecology contributes to the understanding of the principles of host-finding and acceptance by insects. With herbivorous insects, this has been a rich area for study in the field of insect–plant

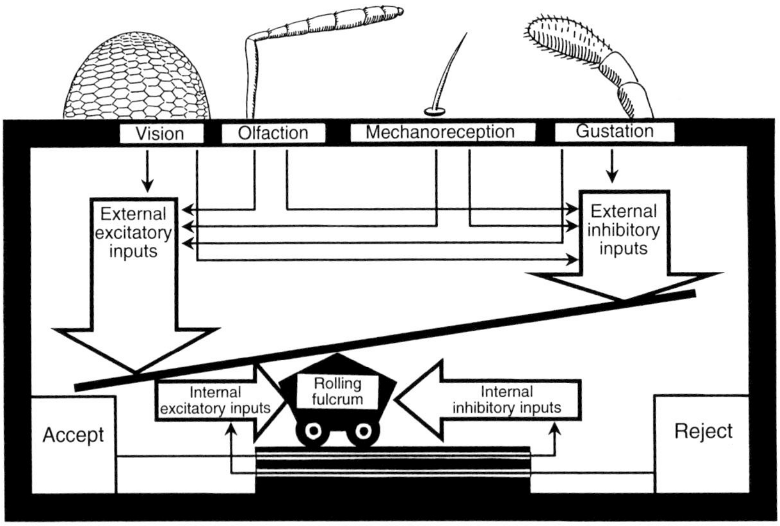

Fig. 9.1. The rolling-fulcrum model for host-plant acceptance or rejection by insects. Sensory stimuli from each sense may contain a mixture of positive and negative factors, which, when taken in balance, determine the overall quality of the host. Acceptance or rejection events (for feeding or oviposition) are also mediated by the physiological state of the insect, which is determined by internal hormonal factors and by the negative feedback from host-acceptance or -rejection events. (Redrawn with permission from Miller and Strickler, 1984.)

interactions. Insects use intensive examining behaviours, during which cues from multiple sensory modalities combine either synchronously or sequentially, while sampling the environment to measure the potential host's qualities. All sensory modalities may participate: olfaction, taste (comprised of any contact chemoreception), vision and tactile stimulation (Miller and Strickler, 1984; Harris and Miller, 1986; Papaj, 1986; Prokopy, 1986; Harris and Rose, 1990). At the same time that host examining is taking place, the internal physiology of the insect determines an acceptance threshold, which governs whether those external stimuli are sufficient to trigger consummatory behaviours, such as sustained feeding or egg laying.

The interaction between external stimuli and internal physiological conditions was investigated in pioneering work by Dethier (1982), and later described by the physical analogy, the 'rolling-fulcrum model' (Miller and Strickler, 1984). This physical model depicts a seesaw (Fig. 9.1), where the sum of behavioural inputs registering positive host stimuli (activants) contributes a downward 'force' on the left side. If there is sufficient force, the left side of the balance is depressed so that it triggers a host-acceptance event. However, there are also sensory stimuli from deterrents or repellents that contribute a downward force on the opposing or inhibitory side. If these inhibitory stimuli have sufficient 'force', the balance will tip sufficiently to trigger a host-rejection event. The physiological state influences the outcome by determining the position of the fulcrum, which rolls to the right under a state of deprivation (hunger or absence of an ovipositional outlet), thus making it easier for fewer or less intensely positive stimuli to trigger a host-acceptance event. With satiety or depleted ovaries, the fulcrum would be expected to roll far to the left, making host-acceptance events unlikely. The arrows beneath the rolling fulcrum indicate a negative-feedback control mechanism: each acceptance event contributes a signal that increases the internal inhibitory inputs, while rejection events increase internal excitatory inputs.

Insect behaviour cannot be expected to work exactly like the physical model of the rolling fulcrum, because individuals may change their response to the same set of cues over time due to learning (Papaj, 1986; Papaj and Prokopy, 1988; Bernays, 2001). The rolling-fulcrum model may apply to two or more levels of sensory integration, which introduces ambiguity into our understanding of how sensory inputs affect behavioural outcomes. Even at the level of the sensory neurone, a well-defined stimulus can result in different output depending on its complex interaction with other positive and negative external stimuli and the insect's physiological condition. For example, in the generalist-feeding gypsy moth, *Lymantria dispar* (Linneaus), and tent caterpillars, *Malacosoma americanum* (Fabricius), the firing rate of sugar receptors found on the maxillae was modulated by the presence of tannins. Thus, the caterpillar might sense the substrate containing the mixture of phagostimulants and deterrents as simply being less sweet than the same substrate without the presence of tannins (Dethier, 1982). Therefore, the 'weight' of the sugar stimulus apparently changes, depending on the combination of other stimuli present, making a peripheral sense organ respond virtually instantaneously, like the rolling-fulcrum model. The ability within a single sensory modality to integrate stimuli in a complex manner can make behavioural dissection of the response to host chemical stimuli very difficult. For example, green-leaf volatiles are important in the response of Colorado potato beetle to hosts, and yet each component individually has little effect on beetle behaviour (Visser and Ave, 1978). In another example, a combination of chemical compounds is required to evoke maximal upwind anemotactic response to host volatiles (Evans and Allen-Williams, 1998).

Stimulus interaction can occur across several modalities over time and space with the whole animal, too. This is most easily observed with behavioural assays measuring the response of individuals or groups of insects to host models. The simplest quantitative method is to count the behavioural end result, such as numbers of eggs laid or feeding bouts. With examples ranging from Thysanoptera to Orthoptera, Lepidoptera, Diptera, Coleoptera and others, insects appear to usually respond synergistically to combinations of visual, chemical and tactile host cues (Harris and Miller, 1982, 1986; Papaj, 1986; Gray and Borden, 1989; Harris and Rose, 1990; Judd and Borden, 1991; Szentesi *et al.*, 1996). It is unlikely that all host cues are simultaneously processed to generate an instantaneous '*Gestalt*' (total image) of host-plant quality; rather, the insect is selectively 'attentive' to particular subsets of stimuli during sequential habitat-seeking, host-finding and examining behaviours (Bernays, 1996; Spencer *et al.*, 1999). It follows, then, that several examining behaviours may be required to fully assess host qualities. For larger, strongly flying insects, detection by receptive insects of host odours can lead to odour-conditioned anemotaxis, which may take place as sustained flight or short, hopping flights (Visser and de Jong, 1987; Haynes and Baker, 1989; Evans and Allen-Williams, 1993, 1998; Byers, 1996; Szentesi *et al.*, 1996; McDonald and Borden, 1997; Petterson *et al.*, 1998; Barata and Araujo, 2001). Response to these odours may in turn lead to heightened responsiveness to appropriate visual cues (Bernays, 1996). Weaker fliers, such as aphids and thrips, may fly toward visual targets and simply settle on to plants as a chemokinetic response (flight arrestment) upon intercepting appropriate odour cues (Teulon *et al.*, 1999).

Sequential processing of different aspects of host stimuli could explain how information integrated over time leads to synergistic responses, and gives some insight into the evolution of efficient information processing in neurally constrained systems (Bernays, 2001). Synergism in these instances at least partly results from the effect a specific stimulus has on the insect's probability to transition to the next finding or examining behaviour. For example, in my laboratory comparison of onion-fly finding and examining behaviours on host models, landing on red foliage was reduced by approximately 45% compared with the green foliar model, equivalent to the per cent reduction in oviposition. The presence of cinnamaldehyde reduced oviposition by 86%, irrespective of foliar colour, and did not measurably influence behavioural transition probabilities prior to substrate probing. The placement of this ovipositional deterrent was at the base of the foliage model, allowing interruption of the transition from foliar-examining behaviours to ovipositor probing. When considered together, these two effects, foliar colour and the presence of deterrents, are indeed synergistic, but the mechanism is through independent influence on the probability that one examining behaviour will proceed to another (Cowles, 1990). Harris and Miller (1983) also investigated the influence of foliar colour on host-acceptance behaviours, but found that, in addition to influencing foliar landing, yellow colour also influenced the post-alighting behaviours of stem-runs and substrate probing with the ovipositor. Whether an insect is attentive to one or a few host characteristics at any given time, when several steps in a sequence of behaviours lead to the insect accepting the resource, the end result is a function of the joint probability of the transition from each behaviour to another, a multiplicative function. Using Bernays's (2001) concept of 'selective attention', onion flies do seem attentive first to host-associated odours, which stimulate upwind movement, followed temporally by responsiveness to visual cues (Judd and Borden, 1989, 1991). Thus, it seems parsimonious that each examining behaviour is likely to be associated with assessment of a particular subset of sensory cues from the potential host.

Another example of temporal synergism is given by the study of stimulus interaction within the chemical senses of *Plutella xylostella* (Linnaeus) (Spencer *et al.*, 1999). During short-term evaluation of moth behaviours, neither alkanes nor sinigrin alone stimulates much oviposition. Individual moths on substrates with stimulatory alkanes (and no sinigrin) spent a longer time on potential host substrates: affected behaviours in the presence of alkanes included arrestment components, including longer residence time, reduced movement rates and increased turning on the substrate. Sinigrin alone did not evoke arrestment or additional host-examining. There was increased oviposition when evaluating sinigrin in overnight studies with caged moths, which can probably be attributed to their increased ovipositional deprivation. Alkanes and sinigrin in combination led to enhanced arrestment of moths on the substrate and rapidly proceeded to a critical examining behaviour and antennal swabbing (seen only in the presence of joint presentation of alkane and sinigrin), which usually immediately preceded oviposition. Perhaps additional high-resolution behavioural observation studies will lead to similar understanding of the particular cues to which insects are 'attentive' during specific host-examining behaviours (Spencer *et al.*, 1999).

Investigators need to be careful when conducting studies of stimulus interactions to avoid confounding host stimuli with additional, semiochemical-based cues. For example, in the onion fly example, visual and chemical resources (presented alone) each received 2.8% of the total eggs deposited in this laboratory choice test, but the combination visual + chemical resources received 78% of the eggs (Miller and Harris, 1985). Part of this extreme degree of synergism may have been due to the additional responsiveness of examining flies to previously laid eggs and aggregation pheromones (Judd and Borden, 1992), causing a positive-feedback dynamic. Thus, the presence of egg-associated pheromones may have interfered with isolating the response to host cues, and each subsequent oviposition event was probably influenced by prior egg-laying events. Since the presence of eggs acts as an additional activant for gravid onion flies, the only way to tease out the effect of host stimuli would be to present a fly with a new host model following each egg-laying event (Judd and Borden, 1992). This example also demonstrates the importance of non-host cues (in this case, the presence of conspecifics) in the calculus of host-quality assessment.

Manipulation of Host Acceptance Behaviour: the Push–Pull Approach

Insect behaviours may follow generalizable patterns that can allow us to better understand how either to 'jam' their signals by making otherwise acceptable hosts deterrent or to key in on specific cues that lead to increased resource acceptance. The focus of these efforts should be to maximize the differential between those resources attracting insects and those that are deterrent. As I clarify below, neither approach alone can be expected to be entirely satisfactory, but the combined use might be successful if deterrents were applied to valued crops and attractants applied to diversionary (or trap) crops. This strategy has variously been called the 'push–pull' or 'stimulo-deterrent' approach, and was independently conceptualized in the USA by Miller (1986) and in Australia by Rice (1986). The efforts of these groups were focused on manipulating the behaviour of the onion fly (Miller and Cowles, 1990; Cowles and Miller, 1992) and of *Heliothis* spp. (Pyke *et al.*, 1987).

A good example for the failure of unilaterally presented attractants exists with efforts to mass-trap Japanese beetles to protect adjacent host plants. Japanese beetles appear to follow the commonly observed host-seeking *modus operandi* of strongly flying insects with the following features: (i) host-finding in a patchy environment probably uses upwind anemotaxis, triggered by detection of host odours by a physiologically receptive beetle; (ii) upwind flight probably increases the responsiveness of the beetle to visual cues; and (iii) the beetle lands on visually detected potential hosts within the odour plume (Bernays, 2001). Placed in this context, we can expect potent chemical attractants to lead not only to excellent trap catches, but also to increased damage by the pest in host vegetation downwind from the odour source. This was observed for Japanese beetle adult feeding close to floral-lure-baited traps, in which defoliation of hosts increased in relation to intensity of attractant used (the number of traps), proximity of the traps to the hosts and the location of the hosts downwind from the attractant odour source (Gordon and Potter, 1985). The same principle in host-finding behaviours is seen with the use of aggregation pheromones on trees to concentrate the attack of bark beetles (Borden *et al.*, 1983; Gray and Borden, 1989). The release of the mountain pine beetle's three-component semiochemical blend resulted in attraction of beetles to the vicinity of the pheromone source and achieved the objective: to concentrate beetle attack in a small area suitable for removal of the population through logging. However, the beetles attacked not only the tree upon which the pheromone was released but also adjacent trees, presumably because these trees were downwind within the pheromone plume. This result was expected, because beetles responding to aggregation pheromones orient visually to trees while flying within the plume, with greater response towards larger-diameter trees.

Arrestment prior to arriving at the source of supernormal odour stimuli (stimuli more active than those found in nature) poses a potential for damage anywhere downwind from the odour source where a responding insect may become arrested at a host. Supernormal visual stimuli, on the other hand, should not have this characteristic, because insects can directly locate the source of the visual signal (Miller and Strickler, 1984). However, even if insects take flight in response to supernormal visual stimuli, they may become arrested by odour cues prior to reaching that visual target, such as was seen with thrips (Teulon *et al.*, 1999). These observations suggest that an especially valuable strategy would be to apply deterrents to hosts close to attractive traps or crops. Such a placement could elicit pest movement until the pest encounters the trap or attractive crop.

Unilateral deployment of deterrents, repellents or crops resistant via antixenosis can also be expected to fail. The most important prediction from the rolling-fulcrum model is that insects will become accepting of deterrent, repellent or resistant hosts when deprived of highly acceptable hosts. Acceptance by insects confined to deterrent plants has been observed by many workers, and fits a theoretical framework for individual females to maximize their lifetime fitness (Roitberg and Prokopy, 1983; Fitt, 1986; Mangel, 1989; Mangel and Roitberg, 1989; Aluja and Boller, 1992).

Behavioural models also predict that, if readily accepted hosts are made available, insects will release their consummatory behaviour on these hosts, the insect may never enter a deprived physiological state and deterrent hosts will continue to be rejected. This phenomenon can be measured in the field and the laboratory as the difference between no-choice and choice environments in the acceptance of deterrent hosts. As early as 1871, Colorado potato beetle was found to accept several resistant varieties of potato grown as monocultures (a no-choice environment), whereas in small-plot trials (with several varieties in proximity) these varieties were rejected (Casagrande, 1987). Therefore, superior ovipositional or feeding resources may have to be provided as part of the crop system if repellents or deterrents are to be useful.

The practical aspects of deploying the push–pull strategy will involve: (i) finding and using stimuli that cause host rejection; (ii) finding and using stimuli that attract the pest, especially if they can be presented as supernormal stimuli; and (iii) understanding the effect of these factors on pest movement so that the two alternatives can be spatially presented in an optimal manner.

Potential Tools for Implementing the 'Push–Pull' Strategy

What tools have potential for manipulating insect behaviour in an agricultural setting? The answer is that there are almost limitless possibilities waiting for creative investigators to exploit. The interaction of multiple sensory modalities in host recognition by insects may allow the manipulation of a very wide menu of plant or environmental characteristics to affect host finding and recognition. In a crop system designed to avoid pest deprivation, the full potential for manipulating pest behaviour has to take into account attracting and arresting pests at traps, trap crops, refuge crops or diversionary crops, while at the same time deterring or repelling the same pests from valued crops. An entire suite of cues could be used to maximize the differential movement of pests away from the valued crop and into diversionary crops, an approach not yet fully tested.

Deterrents and activants can be considered opposite sides of the same coin – investigation of host-acceptance behaviours often discloses features of hosts that are suboptimal or even deterrent. For example, fractionation of host-plant chemicals can yield both activants and deterrents (Lundgren, 1975; Doss, 1983; Renwick and Radke, 1987; Scarpati *et al.*, 1993; Huang *et al.*, 1994; Grant and Langevin, 1995; Honda, 1995). These examples suggest that manipulation of host biochemistry could lead to changes in the balance of attractants, phagostimulants and ovipositional deterrents in plants. Plant breeding is an obvious option for manipulation of host-plant chemistry. However, even a subtle shift in host chemistry can yield unexpected changes in acceptance by insects. For example, cucumber plants inoculated with plant growth-promoting rhizosphere bacteria emit fewer cucurbitacins, elicit reduced feeding by cucumber beetle vectors and thereby decrease the transmission of *Erwinia tracheiphila* (Zehnder *et al.*, 2001). In another example, varying the nitrogen fertilization of crops can also influence the acceptance of strawberry foliage by adult black vine weevils, which probably also influences their fecundity and the spatial pattern of oviposition (Cram, 1965; Hesjedal, 1984; R.S. Cowles, unpublished data).

Other chemical cues may be indirectly derived from the host plant through its interaction with insect herbivores. Olive fruit flies, *Bactrocera oleae* (Gmelin), cause sap to exude from olive fruits into which they have laid eggs. This sap contains several compounds, both hydrophilic and lipophilic, which act as a marking pheromone that reduces the likelihood of subsequent oviposition into that fruit (Girolami *et al.*, 1981).

Bacteria colonizing plant tissues can also modify insect response. Chemical factors maximally stimulating onion-fly oviposition are elicited not by healthy onion plants, but by onion tissue colonized by soft-rot bacteria, particularly *Erwinia carotovora* (Dindonis and Miller, 1980, 1981; Hausmann and Miller, 1989). If bacterial growth could be prevented in this tissue, then emission of these potent volatile cues could be reduced. While some of the chemical stimuli have been characterized, the full complement of chemical stimuli that release onion-fly examining and oviposition is not yet completely understood (Hausmann and Miller, 1989).

Perhaps the most behaviourally potent chemical cues can be expected from the pheromones and allomones of conspecifics and competitors, respectively. As recently reviewed by Borden (1997), engraver beetles are a group rich in pheromonal communication systems that can be manipulated. Complexes of species have co-evolved and compete with each other for the limited resource of susceptible phloem tissue in trees. Colonizing beetles produce volatile aggregation pheromones, which at some concentrations, along with host kairomones, attract additional conspecifics to the same tree. Mass attack of the tree, up to a certain point, is adaptive because it allows the beetles to overwhelm the tree's defensive mechanisms (sap or resin flow). At higher concentrations, the same aggregation pheromone (with certain species) may act as an anti-aggregation pheromone, and there are several examples where the aggregation pheromone of one species can act as a deterrent for another species. These pheromones have been successfully deployed in push–pull strategies to cause outbreaks to implode rather than continuing to expand.

Two examples suggest that synthetically derived attractants may perform as well as or better than compounds found naturally occurring in hosts. Males of several tephritid species are attracted to the components of trimedlure, compounds for which the natural significance remains unknown (Foster and Harris, 1997). Structure–activity relationship work with cinnamyl compounds discovered potent attractants for *Diabrotica* spp. that probably do not occur naturally in cucurbit flowers (Metcalf and Lampman, 1989). Thus, synthetic compounds may fit insects' receptors and could take the place of natural compounds for use in behavioural manipulation.

Lepidopteran eggs are usually deposited on the leaf surface, often accompanied by pheromones deterring oviposition by conspecifics (Behan and Schoonhoven, 1978; Klijnstra, 1986; Schoonhoven, 1990; Schoonhoven *et al.*, 1990; Poirier and Borden, 1991; Dempster, 1992; Gabel and Thiery, 1992; Thiery *et al.*, 1992). Ovipositional deterrents associated with eggs have also been found with weevils (Stansly and Cate, 1984; Messina *et al.*, 1987; Kozlowski, 1989; Credland and Wright, 1990; Ferguson and Williams, 1991; Mbata and Ramaswamy, 1995). Tephritid eggs deposited internally in fruit perhaps may not be detectable by conspecifics; however, adult females of this group commonly lay down marking pheromones on the fruit surface, which usually deters subsequent oviposition (Prokopy, 1981; Hurter *et al.*, 1989; Straw, 1989; Aluja and Boller, 1992; Papaj *et al.*, 1992). Oviposition-deterring pheromone has also been detected with cecidomyiids (Quiring *et al.*, 1998), and agromyzids (Quiring and McNeil, 1984). Hessian flies may not use oviposition-deterring pheromones, but do appear to assess previous colonization of hosts and avoid infested plants when laying eggs (Kanno and Harris, 2002).

The presence of other cues associated with the presence of conspecifics, such as frass (Renwick and Radke, 1980; Dittrick *et al.*, 1983; Anderson and Lofqvist, 1996), pulverized larvae (Gross, 1984) and oral secretions (Poirier and Borden, 2000), can act as deterrents for oviposition or feeding. Presence of other herbivores on the host can also deter oviposition or feeding (Jones *et al.*, 1988; Finch and Jones, 1989; Schoonhoven *et al.*, 1990). Each of these examples suggests that application of synthetic pheromone could be used to prevent feeding or oviposition on plants we wish to protect.

Wilson and Bossert (1963) provide a framework for understanding the behavioural consequences and longevity of pheromone cues related to their physical chemical properties. We should expect that, unlike social-insect alarm pheromones, which are small molecules that dissipate rapidly (Wilson and Bossert, 1963), oviposition-deterring pheromones will have relatively long persistence and low volatility. In one instance, the oviposition-deterring pheromone of *Pieris brassicae* (Linnaeus) was found to persist for 14 days on cabbage leaves under laboratory conditions

(Schoonhoven *et al.*, 1981). In another example, the ovipositional deterrent associated with *Callosobruchus subinnotatus* eggs dissipated over 4 weeks following the removal of those eggs (Mbata and Ramaswamy, 1995). Long-residual properties of synthetic ovipositional deterrents could make their application useful in agricultural systems (Averill and Prokopy, 1987; Klijnstra and Schoonhoven, 1987).

Experience from field tests of specific semiochemical-based ovipositional deterrents has varied from suggesting potential for economically relevant suppression of oviposition with the host-marking pheromone of cherry fruit fly (Boller *et al.*, 1987) to being rather disappointing with *P. brassicae* (Klijnstra and Schoonhoven, 1987). *P. brassicae* responded to its own oviposition-deterring pheromone by decreasing contact with treated plants and flying to the edge of the field cage, where they then proceeded to lay their eggs. Unfortunately, this did not observably reduce the number of eggs laid.

Together, these experiments highlight one important feature of deterrents: movement can be expected to exhibit fewer turns and longer bouts of straight travel following encounter with deterrents; this can be expected to increase the insect's displacement from the deterrent source (Roitberg *et al.*, 1984). Furthermore, the effectiveness of the deterrent can be expected to decrease over time if exposure is continuous and the insect becomes deprived (Roitberg and Prokopy, 1983; Aluja and Boller, 1992).

As reviewed by Gould (1984) and Lockwood *et al.* (1984), one consequence of insecticide application is the evolution not only of physiological resistance but also of behaviours that allow the insect to avoid toxic residues. Therefore, insecticides applied to crops may already in some cases be acting as repellents or deterrents. The unfortunate side to using insecticides for deterrent or repellent properties is that these products would still have the same negative environmental consequences as if they had been applied to directly kill pests. Perhaps structure–activity relationship studies could be undertaken, using bioassays and pest populations that have evolved behavioural avoidance responses, to find non-toxic alternatives to mimic the pesticides' deterrent properties.

Many workers have investigated plant-protective chemistry, such as essential oils, to deter or repel insects (e.g. Cowles *et al.*, 1990). This idea is appealing, for, if we can make a plant appear to the insect sufficiently like a non-host, oviposition or feeding should be averted. One of the challenges in working with repellents is that, because they are volatile, controlled-release formulations might be necessary to allow more than a few days of effect. Larger compounds with low volatility should have better residual properties. Therefore, deterrents may have more practical value. Two compounds are particularly interesting and may play useful roles: azadirachtin has behavioural as well as insect growth-regulator properties, and so can be used as a feeding deterrent (Simmonds and Blaney, 1996). Capsaicin has nearly unique pungent flavour qualities – organisms ranging from barnacles to bears are deterred by this compound. Insects are also sensitive to capsaicin (Cowles *et al.*, 1989). However, application of such a powerful irritant can be very unpleasant unless the operator is wearing full protective clothing.

Prokopy and Owens (1983) reviewed disruption of visual stimuli used in host finding. Foliage colour may be a useful trait for plant breeders to protect some crops. For example, red varieties of cabbage were less accepted by ovipositing *Delia radicum* (Linnaeus) than they were of green varieties (Prokopy *et al.*, 1983), just as red-leaf cotton varieties were less preferable to boll-weevil compared with those with green leaves (Maxwell, 1977). Reflective mulches have been investigated extensively, and have been successful in reducing transmission of viruses by aphids and thrips (Prokopy and Owens, 1983). The recent registration of kaolin-based plant protectants should give a tool for changing not only the colour characteristics but also the surface texture of plants, and perhaps even the ability of an insect to sense the plant's surface chemistry. Kaolin is highly reflective of ultraviolet light; as such it can be used to deter aphid landing in the same way that reflective mulches have been used in the past (Bar-Joseph and Frenkel, 1983; McBride, 2000). Other aspects of 'jamming' the visual system could include camouflaging plants with interplanted vegetation to obscure hosts, or even the use of lights to attract night-flying moths (Prokopy and Owen, 1983).

Physical characteristics of the leaf, particularly the presence of trichomes, can dramatically affect insect feeding or oviposition. For example, black vine-weevil adult feeding preference for strawberry foliage of the commercial variety 'Totem' vs. the species *Fragaria chiloensis*, was experimentally demonstrated to be mediated by leaf hairs (Doss *et al.*, 1987). In a similar way, obscure root-weevil adult feeding preference for different species of rhododendron is largely mediated by trichome characteristics (Doss, 1983). Pubescence of leaves can also influence ovipositional preference, as demonstrated with *Chilo partellus* (Swinhoe) on maize (Kumar, 1992). Leaf texture also influences pre-ovipositional behaviour in diamondback moths and Hessian flies, but these characteristics may be difficult to manipulate (Harris and Rose, 1990; Spencer *et al.*, 1999). Perhaps application of temporarily sticky or filamentous barriers on the surfaces of plants will offer a direct method for interfering with host-examining barriers by insects on the leaf surface, and thereby prevent egg laying or feeding. Ongoing work with ethylene vinyl acetate fibres, in appearance like candy floss, has shown promise as a deterrent for onion and cabbage maggots (Friedlander, 2002).

The fact that several chemicals can act together synergistically to increase host acceptance should give us hope that the opposite may also be true: that chemical deterrents in concert may synergistically decrease oviposition or feeding. Studies of onion-fly deterrents (Cowles, 1990) were inconclusive and could not reject the hypothesis that different compounds were acting in an additive fashion. Chemical deterrents may affect herbivore specialist and generalist species differently (Bernays *et al.*, 2000; Bernays, 2001); so controlled studies investigating this hypothesis should ideally include close relatives representing both.

How might combinations of visual and chemical deterrents perform? Gravid onion flies were tested in the laboratory with various negative (or simply less positive) ovipositional stimuli, with the general resulting observation that stimuli tended to have a multiplicative effect on egg-laying behaviour (Cowles, 1990). For example, there were twice as many eggs deposited around the green as around the red foliar models and ten times as many eggs laid around models without cinnamaldehyde (a deterrent) as the one with cinnamaldehyde. The green model without deterrent elicited 33 times as much oviposition as the red model combined with cinnamaldehyde, not significantly different from the 20-fold difference predicted by a behavioural model using independent reductions in behavioural transitions. A multiplicative model for interaction between deterrents of different sensory modalities leads to an interesting result: it may be difficult to completely deter consummatory behaviours on host plants. If, for example, we could decrease oviposition with a visual deterrent by 60%, and with a chemical deterrent by 25%, the combination might be expected to reduce oviposition not by 85%, but by 70% ($1 - [1 - 0.6] \times [1 - 0.25]$).

Certain materials could dramatically change the tactile, flavour or visual appearance of plants for many species of insects, such as ethylene vinyl acetate fibres, capsaicin and kaolin, respectively. These tools may have commercial advantages for disrupting host finding and acceptance by insects in agriculture over specific deterrents (e.g. oviposition-deterring pheromones), because, like conventional pesticides, they could provide 'broad-spectrum' aspects for managing pest insects.

Demonstrated Application of Behavioural Manipulation in Trap Crops

Manipulation of host-finding and acceptance behaviours has been practised with trap cropping for many years. Trap crops, reviewed by Hokkanen (1991), provide a resource that is more attractive to mobile pests so that they are partitioned away from a valued crop, thereby providing some degree of protection to the valued crop. The trap crop may be the same variety as the main crop, but planted earlier so that the older, larger plants can produce more attractants, be more visible and thus be more apparent to the mobile pest. Alternatively, the crop may be of a different species or variety from that of the main crop, and be selected for its use because it is highly attractive and receives more eggs or feeding in choice tests with the main crop. Usually, reproduction of the pest in the trap crop is prevented, either through insecticide use or crop destruction. Occasionally the trap crop is justified as a means for aggregating the pest in a crop for which control options are available, whereas these options cannot be used on the valued crop (Hunt and Whitfield, 1996; Pair, 1997). A variant of the trap crop concept is the attract–annihilate or attracticide approach, which uses behavioural manipulation to bring insects to traps, baits or insecticide deposits, with or without the presence of hosts. Behavioural manipulation for the attracticide approach has recently been reviewed (Foster and Harris, 1997).

Trap crops have been used to concentrate the pest population on fewer plants, allowing a smaller area to be treated with insecticides to reduce the pest population (e.g. Pair, 1997). In rare examples, growers do not need to apply insecticides, either in the trap crop or in the valued crop. A recent example used every fifth row planted to *Nicotiana kawakamii* to divert oviposition by *Heliothis virescens* (Fabricius) from *Nicotiana tabacum*. The *N. kawakamii* appeared to not only divert oviposition from the interplanted tobacco, but may also have diverted oviposition from nearby monoculture tobacco plots, making comparison of the trap-crop effect difficult. The cultural practice of topping plants could make insecticide applications unnecessary for control of *H. virescens* in a valued crop of tobacco or a diversionary crop of *N. kawakamii*, perhaps permitting biological control agents to become better established (Jackson and Sisson, 1998). Radin and Drummond (1994) used squash as a diversionary crop to protect cucumbers from striped cucumber beetles. A diversionary crop area of 15 and 50% of the area grown in the valued crop was equivalent for reducing pest populations in the valued crop, and prevented the cucumbers from having to be sprayed. Potatoes have been used to divert Colorado potato beetles from feeding and reproducing on tomatoes (Hunt and Whitfield, 1996). In this instance, one row of potatoes for every eight of tomatoes provided protection, as long as the potato plants were able to grow and provide sufficient resources to accommodate the beetles. Plantings of wheat have been effective as a diversionary crop for concentrating wireworms, *Agriotes obscurus* (Linnaeus), in strips displaced laterally from where strawberries were being planted (Vernon *et al.*, 2000). The interplanted wheat reduced the mortality of strawberry plants from 43 to 5.3%. In the strip cropping system developed to prevent lygus damage to cotton, lucerne was kept growing in a lush manner. Preventing senescence of the lucerne required mowing two alternate strips on a 14-day schedule, which coincidentally also caused significant mortality to lygus-bug immature stages. The lygus population remained within the lucerne crop to the degree that cotton bolls were protected from feeding. An additional benefit was the reproduction of several species of generalist predators within the lucerne crop, which then aided management of other insect pests in cotton (Godfrey and Leigh, 1994).

In some respects, insect pests diverted away from the valued crop system could be thought of in the same way as agronomists think of weeds. A weed can be defined as a plant growing where it is not wanted, whereas the same plant growing outside a crop might be considered a wild flower. In a parallel reasoning, populations of insects normally considered to be 'pests' that develop in a place or on hosts that are not intended for harvest may not be considered an economic threat, and so are not truly pests. These insect populations developing outside the valued crop may serve important ecological functions, especially as a nursery for the propagation of natural enemies and as a population reservoir that is not exposed to insecticides. Under most circumstances, some intervention to prevent unconstrained reproduction of a pest in a refuge crop is necessary, otherwise a burgeoning population could have devastating consequences due to their movement from the diversionary crop to other crops more sensitive to pest feeding (Kennedy *et al.*, 1987). However, if a pest population is highly concentrated in a diversionary crop, many density-dependent population-regulation mechanisms, including intraspecific competition, more efficient predation, parasitization and disease epizootics, may be able to suppress the population without additional intervention. Physical destruction of part of the diversionary crop to manipulate pest survival may have greater pest-management value than treating it with insecticides, as the former strategy will minimize the selection for insecticide resistance and the environmental impact on natural enemies, which may disperse to the diversionary crop remnant.

Implications for Insecticide-resistance Management

Manipulation of insect behaviour in the push–pull strategy is immediately relevant to issues of insecticide resistance. Partitioning the insect population into two habitats permits survival of a non-selected population, which may, through interbreeding with the selected population, maintain the resistance allele(s), principally in a heterozygous condition (Comins, 1977). As a practical issue, these factors have been the basis for the attempt to prevent the development of resistance to insecticidal transgenic crops (valued crops) through mandated planting of non-transgenic crops as refuges (Gould, 1998). Including behavioural manipulation as part of the valued crop/refuge crop design could optimize the use of refugia by minimizing the amount of space planted to the refuge crop, while at the same time reducing the proportion of the population that ends up being selected within the valued crop.

Genetic models needed for predicting the evolutionary outcome when there are interactions between behavioural traits and physiological resistance can be quite complex. Gould (1984) discusses a two-gene, two-allele model with selection on behavioural traits associated with deterrency vs. physiological resistance (such as to insecticides or a crop with antibiosis-based insect resistance). Factors that need to be taken into consideration include: the quality of the treated and untreated habitats for pest survival (calculated as though the antibiosis-related mortality factor were not present), mortality due to pesticide exposure, the manner in which insect behaviour partitions the population into the two habitats and possible pleiotropic effects of carrying physiological-resistance or behavioural-avoidance alleles (i.e. fitness costs for these alleles).

I included each of the components described by Gould's (1984) model, with some modifications, into my own simulation model. For example, I incorporated a user-defined dosage-dependent mortality function for each genotype, so that manipulation of pesticide concentration on the resulting selection could be compared (Taylor and Georghiou, 1979). Gould showed the results as adaptive landscapes, in which the mean fitness for every gene-frequency combination is calculated. Theoretically, selection forces populations up the nearest 'slope' to increase their mean fitness. Unfortunately, the fitness for each gene-frequency combination is not mathematically unique, because each gene-frequency combination can actually be represented by a large number of combinations of genotype frequencies, each with its own mean fitness value (Lewin, 1988). To display the results in my model, each combination of physiological-resistance allele frequency and behavioural-avoidance allele frequency is characterized by its evolutionary outcome. Thus, each figure is divided into four regions (at most), showing those combinations of initial gene frequencies for which subsequent generations are 'attracted' by selection to a 'corner' where the alleles become fixed. To make these figures unambiguous, a modified program tracked the gene frequencies for sample populations, starting at specific gene frequency combinations. Selection in these sample populations is shown with arrows that track the population through successive generations. This deterministic model operates according to the following principles:

1. The behavioural trait and physiological resistance alleles are not genetically linked.
2. There is random mating for the entire population.
3. The first generation in the model assigns Hardy–Weinberg equilibrium frequencies for the nine genotypes.
4. Each genotype's frequency is governed by a random combination of the haploid gametes according to their proportion following selection.
5. The insects partition themselves between habitats based on the strength of habitat discrimination as determined by the avoidance trait.
6. The survival to the next generation is determined by the action on immature stages of the antibiosis factor (pesticide concentration), the quality of the habitat and the action of any pleiotropic fitness cost associated with carrying the physiological-resistance or behavioural-avoidance allele.

The user of this model defines the following parameters: (i) the slope and median lethal dose (LD_{50}) for the three physiological-resistance genotypes; (ii) the treated and untreated habitat suitability; (iii) the probability that the *AA* and *aa* genotypes are found in the treated habitat; and (iv) the fitness cost for carrying *R* or *a* alleles. The fitness costs were not used in the following simulation results.

Results from this model are shown in the following figures, with the accompanying parameter values for running the models given in Table 9.1. Variations tested with this simulation model included the following pest-management strategies: a conventional insecticide-only crop with minimal pest development in a non-treated refuge with both very low (Fig. 9.2A) and high (Fig. 9.2B) dose uses of insecticides, an attracticide or lethal trap-crop approach (Fig. 9.3), a refuge-crop system (Fig. 9.4) and push–pull systems investigating variations in diversionary-crop findability (Fig. 9.5) and suitability for insect development (Fig. 9.6).

In the insecticide-only simulation (Fig. 9.2), the crop is assumed to be somewhat more suitable for insect development than the difficult-to-find alternative hosts, which are envisioned perhaps to consist of weeds outside the field. A consequence of the alternative crop being difficult to find, even for potential 'avoiding' genotypes, is that selection rapidly favours the evolution of physiological resistance. The extreme selection for physiological resistance in the low-dose situation (Fig. 9.2A) results from the disproportionate removal of homozygous susceptible individuals from the population. In many respects the high-dose strategy results in a variation of the refuge-crop principle: as

Table 9.1. Parameters used for running the population-genetics simulation model presented in Figs 9.2–9.6. The use of the parameters is as described by Gould (1984). No pleiotropic fitness costs were used.

	Habitat quality		Mortality from insecticide in treated habitat			Proportion in treated habitat		
Fig. no.	Untreated	Treated	*rr*	*Rr*	*RR*	*aa*	*Aa*	*AA*
9.2A	0.4	0.5	0.91	0.49	0.09	1.00	0.95	0.90
9.2B	0.4	0.5	1.00	0.99	0.79	1.00	0.95	0.90
9.3	0.5	0.5	1.00	0.99	0.79	0.90	0.50	0.10
9.4	0.4	0.5	1.00	0.99	0.79	0.90	0.85	0.80
9.5A	0.5	0.4	1.00	0.99	0.79	0.90	0.70	0.50
9.5B	0.5	0.4	1.00	0.99	0.79	0.95	0.50	0.05
9.6A	0.3	0.5	1.00	0.99	0.79	0.90	0.60	0.30
9.6B	0.7	0.5	1.00	0.99	0.79	0.90	0.60	0.30
9.6C	0.2	0.1	1.00	0.99	0.79	0.90	0.50	0.10

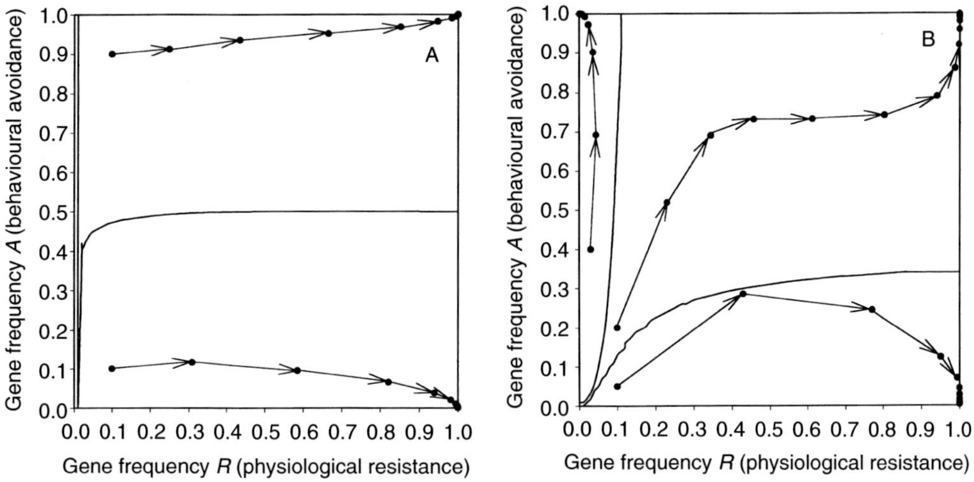

Fig. 9.2. Population-genetics simulation for an insecticide-based strategy for controlling the population in the valued crop and with no refuge designed in the crop system. (A) Low-dose strategy, (B) high-dose strategy. See Table 9.1 for parameter values used in the simulation.

long as there is some alternative untreated habitat available, selection to a non-physiologically resistant, behaviourally avoiding population is possible. If the gene frequency for resistance to an antibiosis factor is initially too high (Fig 9.2B), the population evolves to an avoiding and physiologically resistant genotype. In these conventional

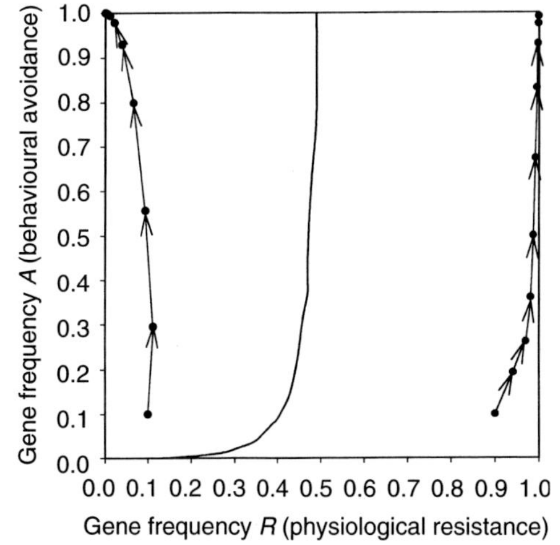

Fig. 9.3. Population-genetics simulation results for an attracticide strategy, which could entail either a kairomone plus insecticide bait system, a pheromone plus insecticide or an insecticide-treated trap crop. The population would initially respond to the attractant (an *aa* population) and evolves to either not respond to or to avoid the attractant. See Table 9.1 for parameter values used in the simulation.

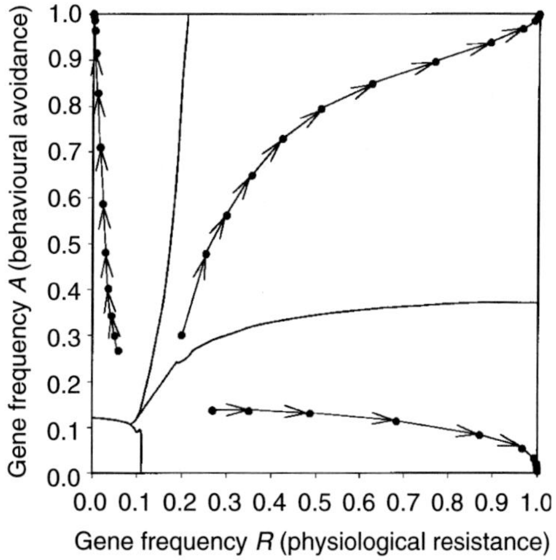

Fig. 9.4. Population-genetics simulation results for the high-dose refuge-crop strategy for controlling the population in the valued crop and with no behavioural modifiers used to direct the pest population to the refuge designed in the crop system. See Table 9.1 for parameter values used in the simulation.

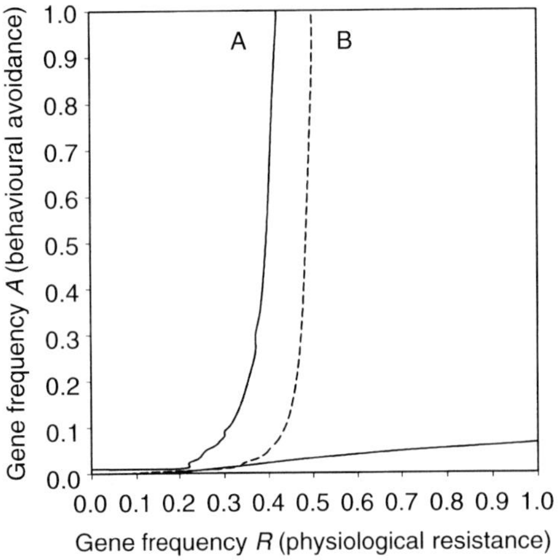

Fig. 9.5. Population-genetics simulation results for the push–pull strategy. The high-dose approach controls the population in the valued crop, a refuge (diversionary crop) is designed into the crop system and behavioural modifiers direct the pest population to the refuge. (A) Lower findability for diversionary crop, (B) higher findability for diversionary crop. See Table 9.1 for parameter values used in the simulation.

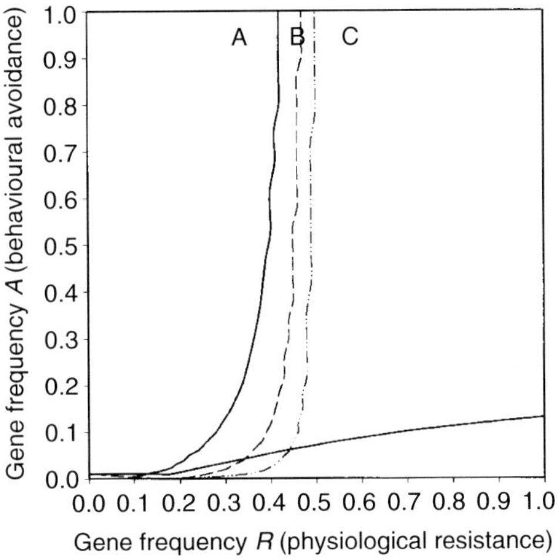

Fig. 9.6. Population-genetics simulation results for the push–pull strategy. The high-dose approach controls the population in the valued crop, a refuge (diversionary crop) is designed into the crop system and behavioural modifiers direct the pest population to the refuge. (A) Low suitability for diversionary crop, (B) higher suitability for diversionary crop, (C) highly findable diversionary crop, with low suitability for both the valued crop and the diversionary crop. See Table 9.1 for parameter values used in the simulation.

insecticide systems, the avoidance factor is considered to be any behavioural trait that can permit the insect to detect differences between the insecticidal and untreated habitats. As an unmanipulated component to the management system, the insect's population could easily start out either as principally avoiding (high frequency of *A* allele) or non-avoiding (high frequency of *a* allele). As Gould (1984) points out, deterrents could be added to insecticide formulations to be certain that the insect population can detect insecticides, and so will automatically start out in the upper left corner, which can be selected to an *AArr* population, rather than the unfavourable *AARR* or even less favourable *aaRR* genotypes.

Attracticidal approaches are represented in Fig. 9.3. Representative crop systems would include trap crops, in which the pest is killed with insecticides in the trap crop, and also kairomonal insecticide (e.g. Pair, 1997) or pheromone fibre-insecticide combinations (e.g. Butler and Las, 1983). The assumption in running this model is that initial conditions for the insect population are a high degree of response to the attractant (starting predominantly as the *aa* genotype), and that evolution of avoidance behaviours would imply loss of the attraction. While genetic variation might be presumed to be minimal for altered response to sex pheromones or host kairomones, this model does raise the concern that the evolutionary outcome when using the attracticide system is unfavourable, and that resistance to such a strategy is likely if genetic variation for response to the attractant exists. The complex blends of pheromone components in some natural communication systems suggest that insects may already have adapted to a crowded chemical communication channel. Species using complex blends may thus be more of a concern than those using single compounds. Insect attraction to kairomones is also complex, so there ought to be many ways in which insect populations can adapt to kairomone-based attracticides.

In the high-dose refuge crop system (Fig. 9.4), populations may be selected to any of the four 'corners' depending on the initial gene frequencies. This model was run with the assumption that the refuge crop is not easily distinguished by the insect from the insecticidal crop, and that, if the insecticidal trait were not present, insect survival in the valued crop would be slightly better than in the refuge crop (perhaps due to a greater presence of natural enemies in the non-insecticidal habitat). By including an alternative habitat in which the pest population can develop, the gene frequency would have to be quite high to lead to selection for the physiological-resistance trait.

The push–pull approach (Figs 9.5 and 9.6) is different from the refuge crop system principally in the insects' ability to detect differences and discriminate between habitats. The pest population would initially have a high frequency for the *A* or avoidance allele because this would be the character(s) we would manipulate to effect discrimination between habitats (deterrents, repellents and attractants). Parameters affecting the evolutionary outcome of the push–pull approach are given, with simulations that assess combinations of the diversionary crop findability (low vs. high) and suitability (low vs. high), relative to the valued crop, which is always presumed to be protected with insecticides or genetically based antibiosis traits.

In all of these simulations, the advantage to using a diversionary crop becomes apparent: behavioural avoidance is maintained and gene frequencies need to be exceedingly high to allow selection of the population to fix the physiologically resistant *R* allele. One of the surprises in performing these simulations was that fairly major differences in diversionary-crop findability (Fig. 9.5) and suitability (Fig. 9.6) had little effect on the resulting selection outcomes. Although the transition line shows the expected shift to the right with improved findability and with high suitability of the diversionary crop, the overall outcome is virtually the same. Nearly any initial gene frequency will eventually lead to fixation of the *A* or avoidance trait, which is favourable for pest management because it maintains the pest population in the diversionary crop.

I have also included an example of the kinds of parameters that I feel would be most favourable to try to design into a crop system (Fig. 9.6C). In this simulation, there is poor survival of the pest in the valued crop, slightly greater survival in the diversionary crop and high discrimination between the two habitats. With these conditions, we would have low overall survival, maintenance of behavioural preference for the diversionary crop and continued susceptibility of the population to the physiological antibiosis factor.

With all simulation efforts, the user should beware of assumptions underlying the results. The assumptions of large populations mating randomly, used in these simulations, may not reflect reality. However, I hope that the overall favourable evolutionary results suggested here will spur others to investigate behavioural manipulation as a necessary component for developing sustainable agricultural systems.

References

Aluja, M. and Boller, E.F. (1992) Host marking pheromone of *Rhagoletis cerasi*: foraging behavior in response to synthetic pheromonal isomers. *Journal of Chemical Ecology* 18, 1299–1311.

Anderson, P. and Lofqvist, J. (1996) Oviposition deterrents from potato, wheat germ, larval frass, and artificial diet for *Agrotis segetum* (Lepidoptera: Noctuidae). *Environmental Entomology* 25, 653–658.

Averill, A.L. and Prokopy, R.J. (1987) Residual activity of oviposition-deterring pheromone in *Rhagoletis pomonella* (Diptera: Tephritidae) and female response to infested fruit. *Journal of Chemical Ecology* 13, 167–177.

Barata, E.N. and Araujo, J. (2001) Olfactory orientation responses of the eucalyptus woodborer, *Phoracantha semipunctata*, to host plant in a wind tunnel. *Physiological Entomology* 26, 26–37.

Bar-Joseph, M. and Frenkel, H. (1983) Spraying citrus plants with kaolin suspensions reduces colonization by the spirea aphid (*Aphis citricola* van der Goot). *Crop Protection* 2, 371–374.

Behan, M. and Schoonhoven, L.M. (1978) Chemoreception of an oviposition deterrent associated with eggs in *Pieris brassicae*. *Entomologia Experimentalis et Applicata* 24, 163–179.

Bernays, E.A. (1996) Selective attention and host plant specialization. *Entomologia Experimentalis et Applicata* 80, 125–131.

Bernays, E.A. (2001) Neural limitations in phytophagous insects: implications for diet breadth and evolution of host affiliation. *Annual Review of Entomology* 46, 703–727.

Bernays, E.A., Oppenheim, S., Chapman, R.F., Kwon, H. and Gould, F. (2000) Taste sensitivity of insect herbivores to deterrents is greater in specialists than in generalists: a behavioral test of the hypothesis with two closely related caterpillars. *Journal of Chemical Ecology* 26, 547–563.

Boller, E.F., Schöni, R. and Bush, G.L. (1987) Oviposition deterring pheromone in *Rhagoletis cerasi:* biological activity of a pure single compound verified in semi-field test. *Entomologia Experimentalis et Applicata* 45, 17–22.

Borden, J.H. (1997) Disruption of semiochemical-mediated aggregation in bark beetles. In: Cardé, R.T. and Minks, A.K. (eds) *Insect Pheromone Research: New Directions*. Chapman & Hall, New York, pp. 421–438.

Borden, J.H., Chong, L.J. and Fuchs, M.C. (1983) Application of semiochemicals in post-logging manipulation of the mountain pine beetle, *Dendroctonus ponderosae* (Coleoptera: Scolytidae). *Journal of Economic Entomology* 76, 1428–1432.

Butler, G.D. and Las, A.S. (1983) Predaceous insects: effect of adding permethrin to the sticker used in gossyplure applications. *Journal of Economic Entomology* 76, 1448–1451.

Byers, J.A. (1996) Temporal clumping of bark beetle arrival at pheromone traps: modelling anemotaxis in chaotic plumes. *Journal of Chemical Ecology* 22, 2133–2155.

Casagrande, R.A. (1987) The Colorado potato beetle: 125 years of mismanagement. *Bulletin of the Entomological Society of America* 33, 142–150.

Comins, H.N. (1977) The development of resistance in the presence of migration. *Journal of Theoretical Biology* 64, 177–197.

Courtney, S.P. and Kibota, T.T. (1989) Mother doesn't always know best: selection of hosts by ovipositing insects. In: Bernays, E.A. (ed.) *Insect–Plant Interactions*, Vol. 2. CRC Press, Boca Raton, Florida, pp. 161–188.

Cowles, R.S. (1990) Manipulating oviposition of the onion fly, *Delia antiqua* (Meigen): a stimulo-deterrent diversionary approach. Doctoral Dissertation, Michigan State University, East Lansing, Michigan.

Cowles, R.S. and Miller, J.R. (1992) Diverting *Delia antiqua* (Diptera: Anthomyiidae) oviposition with cull onions: field studies on planting depth and a greenhouse test of the stimulo-deterrent concept. *Environmental Entomology* 21, 453–460.

Cowles, R.S., Keller, J.E. and Miller, J.R. (1989) Pungent spices, ground red pepper, and synthetic capsaicin as onion fly ovipositional deterrents. *Journal of Chemical Ecology* 15, 719–730.

Cowles, R.S., Miller, J.R., Hollingworth, R.M., Abdel-aal, M.T., Szurdoki, F., Baur, K. and Matolcsy, G. (1990) Cinnamyl derivatives and monoterpenoids as nonspecific ovipositional deterrents of the onion fly. *Journal of Chemical Ecology* 16, 2401–2428.

Cram, W.T. (1965) Fecundity of the root weevils, *Brachyrhinus sulcatus* and *Sciopithes obscurus* on strawberry at different conditions of host plant nutrition. *Canadian Journal of Plant Science* 45, 219–225.

Credland, P.F. and Wright, A.W. (1990) Oviposition deterrents of *Callosobruchus maculatus* (Coleoptera: Bruchidae). *Physiological Entomology* 15, 285–298.

Dempster, J.P. (1992) Evidence of an oviposition-deterring pheromone in the orange-tip butterfly, *Anthocharis cardamines* (L). *Ecological Entomology* 17, 83–85.

Dethier, V.G. (1982) Mechanism of host-plant recognition. *Entomologia Experimentalis et Applicata* 40, 49–56.

Dindonis, L.L. and Miller, J.R. (1980) Host finding responses of onion and seedcorn flies to healthy and decomposing onions and several synthetic constituents of onion. *Environmental Entomology* 9, 467–472.

Dindonis, L.L. and Miller, J.R. (1981) Onion fly and little house fly host finding selectively mediated by decomposing onion and microbial volatiles. *Journal of Chemical Ecology* 7, 419–426.

Dittrick, L.E., Jones, R.L. and Chiang, H.C. (1983) An oviposition deterrent for the European corn borer, *Ostrinia nubilalis* (Lepidoptera: Pyralidae), extracted from larval frass. *Journal of Insect Physiology* 29, 119–121.

Doss, R.P. (1983) Root weevil feeding on rhododendron: a review. *Journal of Environmental Horticulture* 1, 67–71.

Doss, R.P., Shanks, C.H. Jr, Chamberlain, J.D. and Garth, J.K.L. (1987) Role of leaf hairs in resistance of a clone of beach strawberry, *Fragaria chiloensis*, to feeding by adult black vine weevil, *Otiorhynchus sulcatus* (Coleoptera: Curculionidae). *Environmental Entomology* 16, 764–768.

Evans, K.A. and Allen-Williams, L.J. (1993) Distant olfactory responses of the cabbage seed weevil, *Ceutorhynchus assimilis*. *Physiological Entomology* 18, 251–256.

Evans, K.A. and Allen-Williams, L.J. (1998) Response of cabbage seed weevil (*Ceutorhynchus assimilis*) to baits of extracted and synthetic host-plant odor. *Journal of Chemical Ecology* 24, 2101–2114.

Ferguson, A.W. and Williams, I.H. (1991) Deposition and longevity of oviposition-deterring pheromone in the cabbage seed weevil. *Physiological Entomology* 16, 27–33.

Finch, S. and Jones, T.H. (1989) An analysis of the deterrent effect of aphids on cabbage root fly (*Delia radicum*) egg-laying. *Ecological Entomology* 14, 387–391.

Fitt, G.P. (1986) The influence of a shortage of hosts on the specificity of oviposition behaviour in species of *Dacus* (Diptera: Tephritidae). *Physiological Entomology* 11, 133–143.

Foster, S.P. and Harris, M.O. (1997) Behavioral manipulation methods for insect pest-management. *Annual Review of Entomology* 42, 123–146.

Friedlander, B.P. (2002) Cornell entomologist uses 'cotton candy' to protect crops as maggots and worms develop resistance to insecticides. *Cornell News*, 13 Feb., 2002. Available at: http://www.news.cornell.edu/releases/Feb02/Hoffman-IPM.bpf.html

Gabel, B. and Thiery, D. (1992) Biological evidence of an oviposition-deterring pheromone in *Lobesia botrana* Den. et Schiff. (Lepidoptera, Tortricidae). *Journal of Chemical Ecology* 18, 353–358.

Girolami, V., Vianello, A., Strapazzon, A., Ragazzi, E. and Veronese, G. (1981) Ovipositional deterrents in *Dacus oleae*. *Entomologia Experimentalis et Applicata* 29, 177–188.

Godfrey, L.D. and Leigh, T.F. (1994) Alfalfa harvest strategy effect on lygus bug (Hemiptera: Miridae) and insect predator population density: implications for use as a trap crop in cotton. *Environmental Entomology* 23, 1106–1118.

Gordon, F.C. and Potter, D.A. (1985) Efficiency of Japanese beetle (Coleoptera: Scarabaeidae) traps in reducing defoliation of plants in the urban landscape and effect on larval density in turf. *Journal of Economic Entomology* 78, 774–778.

Gould, F. (1984) Role of behavior in the evolution of insect adaptation to insecticides and resistant host plants. *Bulletin of the Entomological Society of America* 30, 34–51.

Gould, F. (1998) Sustainability of transgenic insecticidal cultivars: integrating pest genetics and ecology. *Annual Review of Entomology* 43, 701–726.

Grant, G.G. and Langevin, D. (1995) Oviposition deterrence, stimulation, and effect on clutch size of *Choristoneura* (Lepidoptera: Tortricidae) species by extract fractions in host and nonhost foliage. *Environmental Entomology* 24, 1656–1663.

Gray, D.R. and Borden, J.H. (1989) Containment and concentration of mountain pine beetle (Coleoptera: Scolytidae) infestations with semiochemicals, validation by sampling of baited and surrounding zones. *Journal of Economic Entomology* 82, 1399–1405.

Gross, H.R. Jr (1984) *Spodoptera frugiperda* (Lepidoptera: Noctuidae): deterrence of oviposition by aqueous homogenates of fall armyworm and corn earworm larvae applied on whorl-stage corn. *Environmental Entomology* 13, 1498–1501.

Harris, M.O. and Miller, J.R. (1982) Synergism of visual and chemical stimuli in the oviposition behavior of *Delia antiqua*. In: Visser, J.H. and Minks, A.K. (eds) *Proceedings 5th International Symposium on Insect–Plant Relationships*. Pudoc, Wageningen, The Netherlands, pp. 117–122.

Harris, M.O. and Miller, J.R. (1983) Color stimuli and oviposition behaviour of the onion fly, *Delia antiqua* (Meigen) (Diptera: Anthomyiidae). *Annals of the Entomological Society of America* 76, 766–771.

Harris, M.O. and Miller, J.R. (1986) Host-acceptance behavior in an herbivorous fly, *Delia antiqua*. *Journal of Insect Physiology* 34, 179–190.

Harris, M.O. and Rose, S. (1990) Chemical, color, and tactile cues influencing the oviposition behavior of the Hessian fly (Diptera: Cecidomyiidae). *Environmental Entomology* 19, 303–308.

Hausmann, S.M. and Miller, J.R. (1989) Production of onion fly attractants and ovipositional stimulants by bacterial isolates cultured on onion. *Journal of Chemical Ecology* 15, 905–916.

Haynes, K.F. and Baker, T.C. (1989) An analysis of anemotactic flight in female moths stimulated by host odour and comparison with the males' response to sex pheromone. *Physiological Entomology* 14, 279–289.

Hesjedal, K. (1984) Influence of the nitrogen content in strawberry leaves on the fecundity of the vine weevil, *Otiorhynchus sulcatus* F. (Coleoptera: Curculionidae). *Acta Agricultura Scandinavica* 34, 188–192.

Hokkanen, H.M.T. (1991) Trap cropping in pest management. *Annual Review of Entomology* 36, 119–138.

Honda, K. (1995) Chemical basis of differential oviposition by lepidopterous insects. *Archives of Insect Biochemistry and Physiology* 30, 1–23.

Huang, X.P., Renwick, J.A.A. and Sachdev-Gupta, K. (1994) Oviposition stimulants in *Barbarea vulgaris* for *Pieris rapae* and *P. napi oleracea*: isolation, identification and differential activity. *Journal of Chemical Ecology* 20, 423–438.

Hunt, D.W.A. and Whitfield, G. (1996) Potato trap crops for control of Colorado potato beetle (Coleoptera: Chrysomelidae) in tomatoes. *Canadian Entomologist* 128, 407–412.

Hurter, J., Boller, E.F., Stadler, E., Raschdorf, F. and Schreiber, J. (1989) Oviposition-deterring pheromone in *Rhagoletis cerasi* L.: purification and determination of the chemical constitution. In: Cavalloro, R. (ed.) *Fruit Flies of Economic Importance. Proceedings of the CEC/IOBC International Symposium, 7–10 April 1987, Rome*, pp. 147–148.

Jackson, D.M. and Sisson, V.A. (1998) Potential of *Nicotiana kawakamii* (Solanaceae) as a trap crop for protecting flue-cured tobacco from damage by *Heliothis virescens* (Lepidoptera: Noctuidae) larvae. *Journal of Economic Entomology* 91, 759–766.

Jones, T.H., Cole, R.A. and Finch, S. (1988) A cabbage root fly oviposition deterrent in the frass of garden pebble moth caterpillars. *Entomologia Experimentalis et Applicata* 49, 277–282.

Judd, G.J.R. and Borden, J.H. (1989) Distant olfactory response of the onion fly, *Delia antiqua*, to host-plant odor in the field. *Physiological Entomology* 14, 429–441.

Judd, G.J.R. and Borden, J.H. (1991) Sensory interaction during trap-finding by female onion flies: implications for ovipositional host-plant finding. *Entomologia Experimentalis et Applicata* 58, 239–249.

Judd, G.J.R. and Borden, J.H. (1992) Aggregated oviposition in *Delia antiqua* (Meigen): a case for mediation by semiochemicals. *Journal of Chemical Ecology* 18, 621–635.

Kanno, H. and Harris, M.O. (2002) Avoidance of occupied hosts by the Hessian fly: oviposition behaviour and consequences for larval survival. *Ecological Entomology* 27, 177–188.

Kennedy, G.G., Gould, F., DePonti, O.M.B. and Stinner, R.E. (1987) Ecological, agricultural, genetic, and commercial considerations in the deployment of insect-resistant germplasm. *Environmental Entomology* 16, 327–338.

Kennedy, J.S., Booth, C.O. and Kershaw, W.J.S. (1961) Host finding by aphids in the field. III. Visual attraction. *Annals of Applied Biology* 49, 1–21.

Klijnstra, J.W. (1986) The effect of an oviposition deterring pheromone on egglaying in *Pieris brassicae*. *Entomologia Experimentalis et Applicata* 41, 139–146.

Klijnstra, J.W. and Schoonhoven, L.M. (1987) Effectiveness and persistence of the oviposition deterring pheromone of *Pieris brassicae* in the field. *Entomologia Experimentalis et Applicata* 45, 227–235.

Kozlowski, M.W. (1989) Oviposition and host object marking by the females of *Ceutorhynchus floralis* (Coleoptera: Curculionidae). *Entomologia Generalis* 14, 197–201.

Kumar, H. (1992) Inhibition of ovipositional responses of *Chilo partellus* (Lepidoptera: Pyralidae) by the trichomes on the lower leaf surface of a maize cultivar. *Journal of Economic Entomology* 85, 1736–1739.

Lewin, R. (1988) The uncertain perils of an invisible landscape. *Science* 240, 1405–1406.

Lockwood, J.A., Sparks, T.C. and Story, R.N. (1984) Evolution of insecticide resistance to insecticides: a reevaluation of the roles of physiology and behavior. *Bulletin of the Entomological Society of America* 30, 41–50.

Lundgren, L. (1975) Natural plant chemicals acting as oviposition deterrents on cabbage butterflies, *Pieris brassicae* (L.), *P. rapae* (L.), and *P. napi* (L.). *Zoologica Scripta* 4, 253–258.

Mangel, M. (1989) An evolutionary interpretation of the 'motivation to oviposit.' *Journal of Evolutionary Biology* 2, 157–172.

Mangel, M. and Roitberg, B.D. (1989) Dynamic information and host acceptance by a tephritid fruit fly. *Ecological Entomology* 14, 181–189.

Maxwell, F.G. (1977) Plant resistance to cotton insects. *Bulletin of the Entomological Society of America* 23, 199–203.

Mayhew, P.J. (1997) Adaptive patterns of host-plant selection by phytophagous insects. *Oikos* 79, 417–428.

Mbata, G.N. and Ramaswamy, S.B. (1995) Factors affecting the stability and recognition of the oviposition marker pheromone of *Callosobruchus subinnotatus* (Pic). *Journal of Stored Products Research* 31, 157–163.

McBride, J. (2000) Whitewashing agriculture. *Agriculture Research* 48, 14–17.

McDonald, R.S. and Borden, J.H. (1997) Host-finding and upwind anemotaxis by *Delia antiqua* (Diptera: Anthomyiidae) in relation to age, ovarian development, and mating status. *Environmental Entomology* 26, 624–631.

Messina, F.J., Barmore, J.L. and Renwick, J.A.A. (1987) Oviposition deterrent from eggs of *Callosobruchus maculatus*: spacing mechanism or artifact? *Journal of Chemical Ecology* 13, 219–226.

Metcalf, R.L. and Lampman, R.L. (1989) Estragole analogues as attractants for corn rootworms (Coleoptera: Chrysomelidae). *Journal of Economic Entomology* 82, 123–129.

Miller, J.R. (1986) *Cull Onions as a Trap Crop for Onion Maggot.* Funded proposal of USDA Competitive Research Grants Office, Washington, DC.

Miller, J.R. and Cowles, R.S. (1990) Stimulo-deterrent diversion: a concept and its possible application to onion maggot control. *Journal of Chemical Ecology* 16, 3197–3212.

Miller, J.R. and Harris, M.O. (1985) Viewing behavior-modifying chemicals in the context of behavior: lessons from the onion fly. In: Acree, T.E. and Soderlund, D.M. (eds) *Semiochemistry: Flavors and Pheromones.* Walter de Gruyter, Berlin, pp. 3–31.

Miller, J.R. and Strickler, K.L. (1984) Finding and accepting host plants. In: Bell, W.J. and Cardé, R.T. (eds) *Chemical Ecology of Insects.* Chapman & Hall, New York, pp. 127–155.

Pair, S.D. (1997) Evaluation of systemically treated squash trap plants and attracticidal baits for early-season control of striped and spotted cucumber beetles (Coleoptera: Chrysomelidae) and squash bug (Hemiptera: Coreidae) in cucurbit crops. *Journal of Economic Entomology* 90, 1307–1314.

Papaj, D.R. (1986) Conditioning of leaf-shape discrimination by chemical cues in the butterfly *Battus philenor*. *Animal Behavior* 34, 1281–1288.

Papaj, D.R. and Prokopy, R.J. (1988) The effect of prior adult experience on components of habitat preference in the apple maggot fly (*Rhagoletis pomonella*). *Oecologia* 76, 538–543.

Papaj, D.R., Averill, A.L., Prokopy, R.J. and Wong, T.T.Y. (1992) Host-marking pheromone and use of previously established oviposition sites by the Mediterranean fruit fly (Diptera: Tephritidae). *Journal of Insect Behavior* 5, 583–598.

Petterson, J., Karunaratne, S., Ahmed, E. and Kumar, V. (1998) The cowpea aphid, *Aphis craccivora*, host plant odours and pheromones. *Entomologia Experimentalis et Applicata* 88, 177–184.

Poirier, L.M. and Borden, J.H. (1991) Recognition and avoidance of previously laid egg masses by the oblique-banded leafroller (Lepidoptera: Tortricidae). *Journal of Insect Behavior* 4, 501–508.

Poirier, L.M. and Borden, J.H. (2000) Influence of diet on repellent and feeding-deterrent activity of larval oral exudates in spruce budworms (Lepidoptera: Tortricidae). *Canadian Entomologist* 132, 81–89.

Prokopy, R.J. (1981) Oviposition-deterring pheromone system of apple maggot flies. In: Mitchell, E.R. (ed.) *Management of Insect Pests with Semiochemicals.* Plenum Press, New York, pp. 477–494.

Prokopy, R.J. (1986) Visual and olfactory stimulus interaction in resource finding by insects. In: Payne, T.L., Birch, M.C. and Kennedy, C.E.J. (eds) *Mechanisms in Insect Olfaction.* Clarendon Press, Oxford, UK, pp. 81–89.

Prokopy, R.J. and Owens, E.D. (1983) Visual detection of plants by herbivorous insects. *Annual Review of Entomology* 28, 337–364.

Prokopy, R.J., Collier, R.H. and Finch, S. (1983) Leaf color used by cabbage root flies to distinguish among host plants. *Science* 221, 190–192.

Pyke, B., Rice, M., Sabine, G. and Zalucki, M. (1987) The push–pull strategy–behavioral control of *Heliothis*. *The Australian Cotton Grower* 8, 7–9.

Quiring, D.T. and McNeil, J.N. (1984) Intraspecific competition between different aged larvae of *Agromyza frontella* (Rondani) (Diptera: Agromyzidae): advantages of an oviposition-deterring pheromone. *Canadian Journal of Zoology* 62, 2192–2196.

Quiring, D.T., Sweeney, J.W. and Bennett, R.G. (1998) Evidence for a host-marking pheromone in white spruce cone fly, *Strobilomyia neanthracina*. *Journal of Chemical Ecology* 24, 709–721.

Radin, A.M. and Drummond, F.A. (1994) An evaluation of the potential use of trap cropping for control of the striped cucumber beetle, *Acalymma vittatta* (F.) (Coleoptera: Chrysomelidae). *Journal of Agricultural Entomology* 11, 95–113.

Renwick, J.A.A. and Radke, C.D. (1980) An oviposition deterrent associated with frass from feeding larvae of the cabbage looper, *Trichoplusia ni* (Lepidoptera: Noctuidae). *Environmental Entomology* 9, 318–320.

Renwick, J.A.A. and Radke, C.D. (1987) Chemical stimulants and deterrents regulating acceptance or rejection of crucifers by cabbage butterflies. *Journal of Chemical Ecology* 13, 1771–1776.

Rice, M. (1986) Semiochemicals and sensory manipulation strategies for behavioral management of *Heliothis* spp. Oshsenheimer (Lepidoptera: Noctuidae). In: Zalucki, M.P. and Twine, P.H. (eds) *Heliothis Ecology Workshop 1985 Proceedings*. Queensland Department of Primary Industries, Brisbane, Australia, pp. 27–45.

Roitberg, B.D. and Prokopy, R.J. (1983) Host deprivation influence on response of *Rhagoletis pomonella* to its oviposition deterring pheromone. *Physiological Entomology* 8, 69–72.

Roitberg, B.D., Cairl, R.S. and Prokopy, R.J. (1984) Oviposition deterring pheromone influences dispersal distance in tephritid fruit flies. *Entomologia Experimentalis et Applicata* 35, 217–220.

Scarpati, M.L., Lo-Scalzo, R. and Vita, G. (1993) *Olea europaea* volatiles attractive and repellent to the olive fruit fly (*Dacus oleae*, Gmelin). *Journal of Chemical Ecology* 19, 881–891.

Schoonhoven, L.M. (1990) Host-marking pheromones in Lepidoptera, with special reference to two *Pieris* spp. *Journal of Chemical Ecology* 16, 3043–3052.

Schoonhoven, L.M., Sparnaay, T., van Wissen, W. and Meerman, J. (1981) Seven-week persistence of an oviposition-deterrent pheromone of *Pieris brassicae* on cabbage plants. *Journal of Chemical Ecology* 7, 583–588.

Schoonhoven, L.M., Beerling, E.A.M., Klijnstra, J.W. and van Vugt, Y. (1990) Two related butterfly species avoid oviposition near each other's eggs. *Experientia* 46, 526–528.

Simmonds, M.S.J. and Blaney, W.M. (1996) Azadirachtin: advances in understanding its activity as an antifeedant. *Entomologia Experimentalis et Applicata* 80, 23–26.

Spencer, J., Pillai, S. and Bernays, E.A. (1999) Synergism in the ovipositional behavior of *Plutella xylostella*: sinigrin and wax compounds. *Journal of Insect Behavior* 12, 483–500.

Stansly, P.A. and Cate, J.R. (1984) Discrimination by ovipositing boll weevils (Coleoptera: Curculionidae) against previously infested *Hampea* (Malvaceae) flower buds. *Environmental Entomology* 13, 1361–1365.

Straw, N.A. (1989) Evidence for an oviposition-deterring pheromone in *Tephritis bardanae* (Schrank) (Diptera: Tephritidae). *Oecologia* 78, 121–130.

Szentesi, A., Hopkins, T.L. and Collins, R.D. (1996) Orientation responses of the grasshopper, *Melanoplus sanguinipes*, to visual, olfactory and wind stimuli and their combinations. *Entomologia Experimentalis et Applicata* 80, 539–549.

Taylor, C.E. and Georghiou, G.P. (1979) Suppression of insecticide resistance by alteration of gene dominance and migration. *Journal of Economic Entomology* 72, 105–109.

Teulon, D.A., Hollister, B., Butler, R.C. and Cameron, E.A. (1999) Color and odor responses of flying western flower thrips: wind tunnel and greenhouse experiments. *Entomologia Experimentalis et Applicata* 93, 9–19.

Thiery, D., Gabel, B. and Pouvreau, A. (1992) Semiochemicals isolated from the eggs of *Ostrinia nubilalis* as oviposition deterrent in three other moth species of different families. *Series Entomologica* 49, 149–150.

Vernon, R.S., Kabaluk, T. and Behringer, A. (2000) Movement of *Agriotes obscurus* (Coleoptera: Elateridae) in strawberry (Rosaceae) plantings with wheat (Graminae) as a trap crop. *Canadian Entomologist* 132, 231–241.

Visser, J.H. and Ave, D.A. (1978) General green leaf volatiles in the olfactory orientation of the Colorado beetle, *Leptinotarsus decemlineata*. *Entomologia Experimentalis et Applicata* 24, 738–749.

Visser, J.H. and de Jong, R. (1987) Plant odour perception in the Colorado potato beetle: chemoattraction towards host plants. *Series Entomologica* 41, 129–134.

Wilson, E.O. and Bossert, W.H. (1963) Chemical communication among animals. *Recent Progress Hormone Research* 19, 673–716.

Zehnder, G.W., Murphy, J.F., Sikora, E.J. and Kloepper, J.W. (2001) Application to rhizobacteria for induced resistance. *European Journal of Plant Pathology* 107, 39–50.

10 Integrated Pest Management in Forestry: Potential and Challenges

Imre S. Otvos
Natural Resources Canada, Canadian Forest Service, Pacific Forestry Centre, 506 West Burnside Road, Victoria, B.C., V8Z 1M5, Canada
E-mail: iotvos@pfc.cfs.nrcan.gc.ca

Introduction

The forest ecosystem is much more complex and resilient than that of agriculture and the threshold level of damage caused by insects or pathogens is much higher in forestry than what most consumers are willing to accept on or in their fruits or vegetables. When a certain segment of the population is protesting, sometimes unlawfully, even against the proper and judicious use of pesticides in the forest environment, they should stop and think about the number of chemical sprays it takes to keep their apples and tomatoes free of blemish. They should also think of the home gardener and the number of sprays it takes to keep their roses free of aphids or other 'creepy-crawlies'.

This chapter will briefly summarize events leading up to the development of integrated pest management (IPM). Due to the vastness of the candidate subject, only the stages and progression towards IPM in forestry will be illustrated by giving Canadian examples from my perspective. The main objectives that will be dealt with and covered include a review of the biological control of forest insects in Canada (mainly parasitoid introductions and work with insect viruses). The evolution of IPM will be illustrated with three examples, two involving native species and one an introduced species, i.e. spruce budworms (illustrating the transition from the use of chemical to biological pesticides), Douglas fir tussock moth (DFTM) (development of the first truly IPM for a defoliator) and gypsy moth (an introduced species that became established in eastern North America, but is still treated as a quarantine pest in western North America). Management of bark beetles, contributions in forest weed and plant-pathogen control and, finally, a perspective on the future potentials and challenges of IPM in forestry, exotic insects, decreasing pesticide use, genetic engineering of entomopathogens and transgenetic trees will be discussed.

Importance of Forestry in Canada

Forestry is important to the Canadian economy, but this importance is not reflected in the amount of money dedicated to research of forest pest-related problems, which in many cases is lower than that provided in other countries where forestry is also an important part of the economy. Even so, it is postulated that the development of insect control in forests of other countries where forestry is also important to the economy has followed a similar path to that of Canada. The forests of Canada occupy 45.3% (418 million ha) of the country's total land area. Of this, 245 million ha, about 58.6%, is productive forestland (Lowe *et al.*, 1996; FAO, 2002). Canada has 15.6% of the world's softwood timber resource (19.3 billion m^3), and is only exceeded by that of the former Soviet Union (60.4%). Canada is a major supplier of forest products in the world: in 1999, it accounted for 13.5% of the world's total coniferous tree harvest, ranking third after the USA at 26.7% and Europe at 17.9% (Council of Forest Industries, 2000). In 1999, Canada produced 21.2% of the world's softwood lumber, and this represented about 47.8% of the world's softwood lumber exports (Council of Forest Industries, 2000). The value of forest-product exports in 2001 was about CAN$44.1 billion. In the same year, the net foreign-exchange earnings from forestry were about $34.4 billion. This represents the second greatest contribution to the economy of Canada after energy production (Statistics Canada, 2002).

Most of the forests in Canada are publicly owned. As the land base managed for timber and other forestry use is shrinking and the size of the forest set aside for parks and ecological reserves is increasing, the appetite and demand of the increasing population has to be met from a smaller forest land base. Therefore, the losses caused by insects, disease and fire activity must be reduced. The average annual volume loss due to forest insect pests for 1988–1992 is estimated at 5.9 million m^3. This is less than the estimated 20.9 million m^3 volume loss caused by diseases, but is more than double the 2.7 million m^3 burned from 1988–1992 and constitutes 6% of the 102.3 million m^3 annual harvest (Wood and Van Sickle, 1994).

Biological Control of Forest Insects in Canada

Both inoculative and inundative methods of insect control have been used in forestry, with varying degrees of success. Classical biological control has great appeal to forest entomologists because, in forestry, pest control does not always have to be immediate and the threshold level of economic damage is higher than in agriculture or horticulture and some damage can often be tolerated. Classical biological control was first used in Canada's forests against exotic insects, which were almost always introduced without their natural enemies. The use of biological control in Canada has a long history, starting in 1910 with the introduction of over 1000 specimens of the ichneumonid wasp, *Mesoleius tenthredinis* Morley, from England to control the larch sawfly, *Pristiphora erichsonii* Hartig (McGugan and Coppel, 1962). Since that time, considerable use has been made of the natural enemies of insects, using first insect parasitoids and predators and more recently entomopathogens.

The majority of attempts to control insect pests in Canada using biological agents used only one group of organisms – parasitoids. In some cases a combination of two groups – generally parasitoids and nucleopolyhedrovirus (NPV) – were used, and occasionally three groups – parasitoids, NPV and *Bacillus thuringiensis* subsp. *karstäki* (*Btk*). The best known example of the latter is gypsy moth. Biological control attempts against the spruce budworm have even tried to use a fourth agent, microsporidia.

Results of biological control attempts in Canada have been documented in detail in the four volumes of *Biological Control Programmes Against Insects and Weeds in Canada* (McGugan and Coppel, 1962; Canadian Department of Agriculture and Canadian Department of Fisheries and Forestry, 1971; Kelleher and Hulme, 1984; Mason and Huber, 2002). Hulme (1988) has also published a brief summary of some of the highlights. Figure 10.1 shows that the biological control attempts were entomocentric, most of the targets were insects (both agricultural and forest pests) and most of the control agents introduced were also insects. Considerably fewer control programmes were conducted on weeds and there was no work done on plant pathogens until the last two decades (Fig. 10.1). Only control attempts using parasitoids and insect viruses will be discussed in this section.

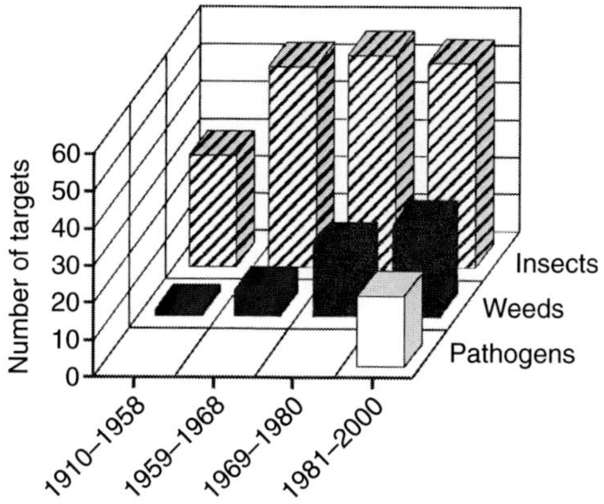

Fig. 10.1. Comparison of the number of biological control projects conducted against insects, weeds and plant pathogens in Canada between 1910 and 2000 (after Mason and Huber, 2002).

Canada was one of the leaders of classical biological control, in both agriculture and forestry. There were three main reasons for this:

- In the past, most invasive species in North America came from Europe, reflecting past trading patterns and routes (Mattson *et al.*, 1994; Niemela and Mattson, 1996).
- The value of biological control in Canada was recognized at an early date. 'Throughout the history of classical biological control in Canada a close link has existed between CAB International (formerly IIBC or CIBC) in Delemont, Switzerland, and Agriculture and Agri-Food Canada (AFC, formerly Agriculture Canada) and Canadian Forest Service laboratories' (Mason *et al.*, 2002: xiii).
- 'The first biological control laboratory was established in Canada … in Fredericton in 1912 … [then] The Dominion Parasite Laboratory was established in 1929 at Belleville [Ontario] …' (Wallace, 1995) and was operating until 1972, when it was closed due to cutbacks and government reorganization. Regarding the Belleville laboratory, Mason *et al.* (2002: xiii) state: 'This laboratory had one of the largest concentrations of biological control specialists in the world.'

Since the early 20th century, biological control attempts have been conducted against 22 exotic forest insects (Table 10.1) and 27 native forest insects (Table 10.2) in Canada. In most cases the biocontrol agents used were parasitoids, pathogens (mainly viruses and *Btk*) and in a few cases predators. Some of the early introductions were only explorative and tentatively assessed (McGugan and Coppel, 1962), and a re-evaluation of the outcomes of all the introductions may be in order.

Of the 22 exotic forest pests, 17 had a single control-agent group (parasitoids) introduced to control them. Of the 17 species, the introduction was considered successful against eight (47%), resulting in long-term control that suppressed the pest, virtually eliminating the damage without need for further direct application of the control. The introduction of parasitoids was partially successful against two species, promising against three species, unknown in two cases and two were classified as failures. In the case of the remaining five exotic species, the introduction of two (usually parasitoid and NPV) or three natural control-agent groups occurred. The outcome of these introductions was successful with at least one of the control-agent groups (Table 10.1).

When one is evaluating the control outcome of pathogen use, it is worth considering the difference between the use and success of NPV and *Btk* in biological control. NPV is to some degree self-propagating after it has been applied, and is usually applied only once during the course of an outbreak to achieve control, while *Btk* almost always has to be applied annually until the outbreak collapses or, in the case of quarantine pests, until the target insect is eradicated. In some cases, *Btk* has to be applied several times in a season and maybe for a number of years to achieve the desired level of control. An example of the latter is gypsy-moth eradication in western North America, where *Btk* is applied three or four times to the susceptible stage in the course of a year for 2–3 years until eradication is achieved. Therefore, it should be noted that although the use of bacterium (*Btk*) application is classified as a success, it should be classified as a 'success' only

Table 10.1. Evaluation of the biological control attempts against exotic forest insect pests in Canada between 1910 and 2000.[a]

Target pest	Common name	Control agent	Outcome[b]
Coleophora laricella (Hübner)	Larch case-bearer	Parasitoids	Success
Eulecanium tiliae (Linnaeus)	Lecanium scale	Parasitoids	Success
Fenusa pusilla (Lepeletier)	Birch leaf-miner	Parasitoids	Success
Leucoma salicis (Linnaeus)	Satin moth	Parasitoids	Success
Diprion similis (Hartig)	Introduced pine sawfly	Parasitoids	Success
Gilipnia hercyniae (Hartig)	European spruce sawfly	Parasitoids	Success
Pristiphora erichsonii (Hartig)	Larch sawfly	Parasitoids	Success
Pristiphora geniculata (Hartig)	Mountain ash sawfly	Parasitoids	Success
Adelges piceae (Ratzeburg)	Balsam woolly adelgid	Predators	Partial success
Rhyacionia buoliana (Denis and Schiffmuller)	European pine-shoot moth	Parasitoids	Partial success
Acantholyda erythrocephala (Linnaeus)	Pine false web-worm	Parasitoids	Promising
Profenusa thomsoni (Konow)	Amber-marked birch leaf-miner	Parasitoids	Promising
Xanthogaleruca luteola (Müller)	Elm-leaf beetle	Bacterium	Promising
Carulaspis juniperi (Bouche)	Juniper scale	Parasitoids	Failure
Coleophora serratella (Linnaeus)	Birch case-bearer	Parasitoids	Failure
Euproctis chrysorrhoea (Linnaeus)	Brown-tail moth	Predators	Unknown
Gilpinia frutetorum (Fabricius)	Nursery pine sawfly	Parasitoids	Unknown
Lymantria dispar (Linnaeus)	Gypsy moth	Parasitoids	[Partial] success
		Bacterium	[Partial] success
		NPV	Partial success
Operophtera brumata (Linnaeus)	Winter moth	Parasitoids	Success
		Bacterium	[Partial] success
		NPV	Partial success
Neodiprion lecontei (Fitch)	Red-headed sawfly	Parasitoids	Success
		Bacterium	[Partial] success
		NPV	Success
Neodiprion sertifer (Geoffroy)	European pine sawfly	Parasitoids	Success
		NPV	Success
Neodiprion swainei (Middleton)	Swaine jack pine sawfly	Predators	Partial success
		Parasitoids	Partial success
		NPV	Success

[a] The list was compiled from the four volumes of the review of biocontrol programmes in Canada (McGugan and Coppel, 1962; Canadian Department of Agriculture and Canadian Department of Fisheries and Forestry, 1971; Kelleher and Hulme, 1984; Mason and Huber, 2002).

[b] It should be noted that the classification categories under 'outcome' are somewhat confusing. Success can mean: (i) the introduction was successful and the organism became established and has an impact on the original target host but it does not necessarily control the pest (e.g. parasitoid introduction or the use of *Btk* against the gypsy moth); or (ii) the target organism is under control now after the introduction (e.g. larch sawfly or the winter moth). These will be clarified in a later re-analysis of the control attempts.

Table 10.2. Evaluation of the biological control attempts against native forest insect pests in Canada between 1910 and 2000.[a]

Target pest	Common name	Control agent	Outcome[b]
Acleris gloverana (Walsingham)	Western black-headed budworm	Bacterium	Partial success
Acleris variana (Fernald)	Eastern black-headed budworm	Bacterium	Partial success
Hemichroa crocea (Geoffroy)	Striped alder sawfly	Parasitoids	Success
Coleotechnites starki (Freeman)	Lodgepole needle miner	Parasitoids	Failure
Cydia strobilella (Linnaeus)	Spruce seed moth	Parasitoids	Failure
Lambdina fiscellaria somniaria (Hulst)	Western oak looper	Predators	Failure
Pissodes strobi (Peck)	White pine weevil	Parasitoids	Failure
Exoteleia pinifoliella (Chambers)	Pine needle miner	Parasitoids	Unknown
Mindarus abietinus Koch	Balsam twig aphid	Predators	Unknown
Neodiprion pratti Dyar	Jack pine sawfly	Parasitoids	Unknown
Neodiprion nanulus Schedl	Red pine sawfly	Parasitoids	Unknown
Neodiprion tsugae Middleton	Hemlock sawfly	Parasitoids	Unknown
Neodiprion virginianus complex	Pine sawfly	Parasitoids	Unknown
Operophtera bruceata (Hulst)	Bruce spanworm	NPV	Unknown
Pineus strobi (Hartig)	Pine bark adelgid	Predators	Unknown
Dendroctonus rufipennis (Kirby)	Eastern spruce beetle	Predators	Failure
		Parasitoids	Failure
Choristoneura fumiferana (Clemens)	Spruce budworm	Parasitoids	Partial success
		Bacterium	Partial success
		NPV	Partial success
		Other viruses (CPV, GV, EV)	Partial success
		Microsporidia	Partial success
Choristoneura occidentalis Freeman	Westerm spruce budworm	Bacterium	Partial success
		NPV	Partial success
		Other viruses (CPV, GV, EV)	Partial success
Choristoneura pinus pinus Freeman	Jack pine budworm	Bacterium	Success
		NPV	Success
Lambdina fiscellaria fiscellaria Guenée	Hemlock looper	Parasitoids	Failure
		Bacterium	Partial success
Lambdina fiscellaria lugubrosa Hulst	Western hemlock looper	Predator	Unknown

Continued on next page

Table 10.2. *Continued from previous page.*

Target pest	Common name	Control agent	Outcome[b]
Malacosoma disstria Hübner	Forest tent caterpillar	Parasitoids	Success
		Bacterium	Partial success
		NPV	Failure
		Microsporidia	Failure
Orgyia leucostigma (J.E. Smith)	White-marked tussock moth	Bacterium	Partial success
		NPV	Success
Orgyia pseudotsugata (McDunnough)	Douglas fir tussock moth	Bacterium	Partial success
		NPV	Success
Zeiraphera canadensis Mutuura and Freeman	Spruce bud moth	Parasitoids	Failure
		Bacterium	Failure
		Nematodes	Partial success
Neodiprion abietis (Harris)	Balsam fir sawfly	Parasitoids	Unknown
		Bacterium	Partial success
		NPV	Unknown
Pikonema alaskensis (Rohwer)	Yellow-headed spruce sawfly	Parasitoids	Unknown
		Bacterium	Partial success
		Nematodes	Success

[a] The list was compiled from the four volumes of the review of biocontrol programmes in Canada (McGugan and Coppel, 1962; Canadian Department of Agriculture and Canadian Department of Fisheries and Forestry, 1971; Kelleher and Hulme, 1984; Mason and Huber, 2002).

[b] It should be noted that the classification categories under 'outcome' are somewhat confusing. Success can mean: (i) the introduction was successful and the organism became established and has an impact on the original target host but it does not necessarily control the pest (e.g. parasitoid introduction or the use of *Btk* against the gypsy moth); or (ii) the target organism is under control now after the introduction (e.g. larch sawfly or the winter moth). These will be clarified in a later re-analysis of the control attempts.

in cases of eradication (gypsy moth from western North America). In other cases, when *Btk* is applied to reduce the pest population, its use should perhaps be classified as partial success, since it generally has to be applied for a number of consecutive years.

Control attempts were made against 27 native forest insects using parasitoids, predators and pathogens: 15 by a single group and 12 by multiple groups of organisms (Table 10.2). Of the single-group introductions against native pests, three (12%) of the 15 were classified as successful, four (26.7%) as failures and the remaining eight (53.3%) as unknown. Of the 12 other native pests targeted with multiple-group introductions, one (eastern spruce beetle) was a failure, and for the other 11 species at least one of the biological control agents tried was successful.

It should be noted that the outcome 'success' means that the biological control agent tried/introduced became established (in the case of parasitoids and predators), maintained its population and attacked and killed a portion of the target pest. In the case of pathogens (either virus or *Btk*) success meant that it infected and killed a proportion of the target insect and the introduced organisms did not necessarily produce an acceptable level of control over the pest. The case of *Choristoneura fumiferana* (Clemens) illustrates this point well. All the control agents listed in Tables 10.1 and 10.2 had an effect and therefore could be classified as a 'success', but in terms of control all five groups had only partial success because the spruce budworm still has to be controlled, i.e. by applying one of the control-agent groups listed.

Data are scarce on the benefits and monetary return of the cost of developing and applying classical biological control measures. No thorough cost–benefit analysis has been done in Canada on the use of biological control. However, one such summary was compiled for biological control activities in Australia. Of the 12 biological control attempts conducted, all were analysed, two-thirds were considered successful and the overall cost:benefit ratio, based on all 12 attempts, was 1:10 (Marsden *et al.*, 1980). Hulme (1988) cited two cost : benefit estimates from Canada. Biological control of the European spruce sawfly cost about CAN$300,000 and saved 8.5 million cords of wood valued at CAN$6 million, giving a cost : benefit ratio of 1:20 (Reeks and Cameron, 1971). In the case of another defoliating insect, the winter moth, the cost of introducing parasitoids into eastern Canada was estimated at $160,000 and the value of the oak trees killed before the introduction was estimated at CAN$2 million. The introduction of parasitoids prevented the loss of another CAN$12 million worth of oak trees. Based on these, the cost:benefit ratio was estimated at 1:12.5 (Embree, 1971).

Viral insecticides

Interest in the use of viruses for forest insect control in Canada began in the late 1930s with the discovery of a polyhedrovirus that was credited with causing the collapse of a large outbreak of the European spruce sawfly, *Gilpinia hercyniae* Hârtig, the most important forest defoliator at that time (Balch and Bird, 1944; Cameron, 1975a). The European spruce sawfly virus had been accidentally introduced from Europe, probably along with one of the imported parasitoids of the sawfly.

The impressive beneficial effect of this accidental virus introduction resulted in accelerated work with insect viruses in the hope that outbreaks of various other pests could also be terminated with viruses, similar to that of the European spruce sawfly. Most of the work to develop viral insecticides for several insect pests has been done on NPVs and granuloviruses (GV), subgroups A and B, respectively, of the family Baculoviridae. Limited work has also been conducted on cytoplasmic polyhedroviruses (CPVs – Reoviridae) and entomopox virus (EPV – Poxviridae) as potential control agents for the spruce budworm (Table 10.3; Cunningham and Kaupp, 1995).

NPVs and GVs are highly host-specific, although some viruses may infect several related species in the same genus (Cunningham and Kaupp, 1995), e.g. *Orgyia* spp. Their high host specificity makes viral insecticides ideal from an ecological point of

Table 10.3. Research and operational applications of insect viruses against the more important forest defoliators in Canada, 1971–2000 (Kelleher and Hulme, 1984; Mason and Huber, 2002).

Target insect				Total area sprayed
Common name	Scientific name	Virus	Year(s)	ha/tp
Spruce budworm	*Choristoneura fumiferana*	NPV	1971–2000	2180[a]
Spruce budworm	*Choristoneura fumiferana*	GV	1979–1980	16
Spruce budworm	*Choristoneura fumiferana*	EPV	1971–1972	519
Spruce budworm	*Choristoneura fumiferana*	EPV + NPV	1972	32[b]
Western spruce budworm	*Choristoneura occidentalis*	NPV	1976–1982	252
Western spruce budworm	*Choristoneura occidentalis*	GV	1982	172
Jack pine budworm	*Choristoneura pinus pinus*	NPV	1985	50
Gypsy moth	*Lymantria dispar*	NPV	1982–1994	1280[c]
Forest tent caterpillar	*Malacosoma disstria*	NPV	1976–1980	45
Balsam fir sawfly	*Neodiprion abietis*	NPV	1999	4
European pine sawfly	*Neodiprion sertifer*	NPV	1975–1993	152
Red-headed pine sawfly	*Neodiprion lecontei*	NPV[d]	1976–1995	5881
White-marked tussock moth	*Orgyia leucostigma*	NPV	1975–1987	43
Douglas fir tussock moth	*Orgyia pseudotsugata*	NPV[e]	1974–1993	2620

[a] Includes 16 ha sprayed with NPV and fenitrothion.
[b] Includes 16 ha sprayed with NPV + EPV and fenitrothion.
[c] Cunningham (1998).
[d] Red-headed pine sawfly NPV was registered in 1987 as Lecontvirus.
[e] Douglas fir tussock moth NPV was registered under the names Virtuss® (*Op*NPV produced in *Orgyia leucostigmata*) and TM-Biocontrol-1 (*Op*MNPV produced in *Orgyia pseudotsugata*).
[f] tp, time period.

view, but it also makes their development and commercialization less attractive for profit-making companies. Consequently, all development work and registration of viruses for forestry use in Canada was done in the laboratories of the Canadian Forest Service, in cooperation with some of the provincial governments.

Field trials using viruses have been conducted on 19 species of forest insect pests in Canada; 11 Lepidoptera and eight Hymenoptera. Both aerial and ground trials were conducted on eight species, while the remainder were tested with ground applications only (Cunningham and Kaupp, 1995). The more important virus tests for ten insects are summarized in Table 10.3.

The control of the European spruce sawfly, *Gilpinia hercyniae* (Hartig), may be the best example of a biological control programme in Canada. It was an important pest of spruce trees in eastern Canada and the USA. Twenty-seven species of parasitoids from Europe and Japan were released between 1933 and 1951, of which nine became established (McGugan and Coppel, 1962). In the late 1930s an NPV was noticed in the sawfly populations and was the key factor in controlling this sawfly (Balch and Bird, 1944). The virus was fortuitously introduced with parasitoids from Europe. After the virus epizootic in central Canada, the virus was purposefully transferred to a number of new locations in eastern Canada (McGugan and Coppel, 1962). The outbreak collapsed in the early 1940s and the European spruce sawfly has been controlled since then by the introduced parasitoids and the virus, as no further outbreaks recurred.

Similarly, the red-headed pine sawfly, *Neodiprion lecontei* Fitch, is one of the most important insects attacking young red pine, *Pinus resinosa* Aiton, plantations in eastern Canada. An NPV was found in red-headed pine sawfly in 1950 (Bird, 1961), and extensive laboratory and ground-spray trials conducted on the virus showed promise (Cunningham and de Groot, 1984;

Cunningham *et al.*, 1986). The virus was tested experimentally between 1978 and 1980 in Quebec in 92 plantations with a combined area of 1051 ha. The virus was also tested in Ontario between 1980 and 1990 in 478 plantations with a combined area of 3546 ha. Based on these successful applications, the virus received temporary registration in 1983 and full registration in 1987 under the trade name Lecontvirus®. The production of this and other similar viruses in colony-feeding sawflies is relatively easy and cost effective. Heavily infested plantations are treated with the virus by mist-blowers and the dead colonies are collected daily (preferably), starting 1 week after treatment (Cunningham and McPhee, 1986).

Orgyia pseudotsugata (McDunnough) is a native defoliator in British Columbia, Canada and the north-western USA. Outbreaks of this defoliator recur periodically and are terminated by an epizootic caused by a native NPV. This naturally occurring virus was considered to have an excellent potential for biological control both in Canada and in the USA. The work on the virus and how the virus was incorporated into an IPM system is described later in this chapter. *C. fumiferana* is the most important forest insect pest in Canada and, as such, has had the greatest number of biological control agents tested against it (Tables 10.2 and 10.3). Because attempts to control it by parasitoid introduction (from Europe and Japan) or relocation (from western Canada to eastern Canada) did not provide the desired control, insect viruses were also investigated. Extensive field trials with NPV, GV and EPV showed NPV to be the most efficacious virus tested (Cameron, 1975b). Consequently, research concentrated on NPV (Cunningham and Howse, 1984; Cunningham, 1985a). NPV and GV were also tested against the western spruce budworm, *Choristoneura occidentalis* Freeman (Otvos *et al.*, 1989) and the jack pine budworm, *Choristoneura pinus* Freeman (Table 10.3). While virus infection occurred in the treated plots, mortality was generally low (40–60%) and not sufficient to control the treated populations. Although virus does occur in budworm populations in the field, natural virus epizootics have never been observed. Because of the economic importance of the spruce budworm, genetic manipulation of its virus to enhance its effectiveness was a high priority of the Canadian government (Cunningham and Kaupp, 1995). However, no significant advancements were made in improving virus efficacy over the last 20 years and *Btk* is still the only pathogen registered for budworm control.

Viruses registered for forest insect control in Canada

Following the more important virus trials described above, three viral insecticides, all NPVs, received temporary registration in 1983 and full registration in Canada in 1987: one for the red-headed pine sawfly, *N. lecontei*, and two for the control of DFTM, *O. pseudotsugata*. The multicapsid isolate of the DFTM virus (*Op*MNPV), produced in DFTM larvae, was registered in the USA in 1976 under the trade name TM Biocontrol-1®. The same product, under the same trade name, TM Biocontrol-1, was also registered for use in Canada in 1987. In addition, the same virus, produced in Canada in the white-marked tussock moth, *Orgyia leucostigma* (J.E. Smith), also received full registration in 1987 under the trade name Virtuss® (Cunningham and Kaupp, 1995). The recommended dosage on the label for both viruses is 2.5×10^{11} polyhedral inclusion bodies (PIB) per hectare (Cunningham and Kaupp, 1995). Negotiations are currently underway in Canada to transfer the production and sale of these three registered viruses to a private company for commercialization. The use of these three registered viruses is insignificant compared with the use of *Btk*.

Management of Spruce Budworms (*Choristoneura* spp.)

Development of the management system currently used for the spruce budworms (*Choristoneura* spp.) illustrates well how the control for these important defoliators evolved, from the use of a single hard chemical to the use of the ecologically less disrup-

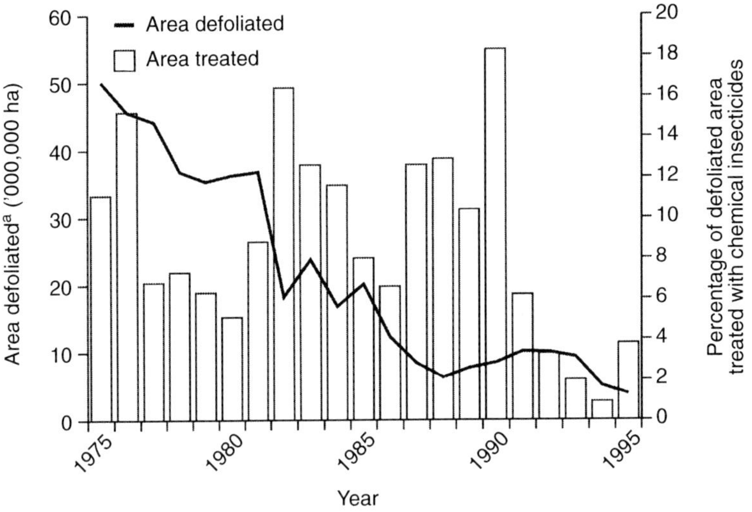

Fig. 10.2. Area defoliated by spruce budworm, *Choristoneura fumiferana,* from 1975 to 1995 and the percentage of defoliated area treated with insecticides for the same years (from Canadian Council of Forest Ministers, 2003).

[a] Only areas that were moderately or severely defoliated were included in the source data.

tive bacterial insecticide *Btk*. Before the management system is discussed, it is worthwhile to sketch the economic importance of the insect over time. The change in economic importance has influenced our response to damage. It is also useful to mention key research activities (including attempts to use biological control agents) that were undertaken in the hope of increasing our control options for these budworm pests. The evolution of budworm control measures in Canada is representative of how control measures evolved for most forest defoliators, not only in Canada, but in other parts of the world as well.

Economic importance of spruce budworms

Among the forest defoliators in Canada, three budworms are the most important. These are the:

- spruce budworm, *C. fumiferana* (Clemens);
- western spruce budworm, *C. occidentalis* Freeman;
- jack pine budworm, *C. pinus pinus* Freeman.

All three are native to North America and outbreaks have occurred repeatedly over the last three centuries (Blais, 1985; Harris *et al.*, 1985a). They pose a threat to approximately 60 million ha of susceptible forests in eastern Canada and the USA (Nigam, 1980; Kettela, 1983; Talerico, 1984). Budworm larvae feed on a number of conifer species, and consecutive years of defoliation result in growth reduction, top kill and, ultimately, tree mortality (Alfaro and Maclauchlan, 1992). Normally, 4–5 years of severe defoliation are required to kill susceptible trees and approximately 7–8 years to kill less vulnerable, immature stands (MacLean, 1980). Prior to the 1940s, spruce budworm damage was of little concern because its principal host, balsam fir, *Abies balsamea* (L.) Miller, was considered a weed species and was not utilized. As the demand for wood increased, balsam fir became an important commodity. Due to this change in forest utilization, the damage caused during the budworm outbreak in the 1940s was considered unacceptable. Extensive spray operations were initiated and had to be continued, unfortunately, almost annually somewhere in Canada

(Prebble, 1975; van Frankenhuyzen, 1990) up to the present (Fig. 10.2).

In fact, several biological control agents have been investigated over the years to evaluate their potential for control of *C. fumiferana*. These control agents included exotic and native parasitoids, viruses microsporidia, fungi and even irradiation. Considerable work has been conducted on mass rearing and release of the native egg parasitoid, *Trichogramma minutum* Riley, against *C. fumiferana*. These studies have shown that parasitism of budworm eggs could be high (about 80%) in the small release plots, but the cost was also high (about CAN$400/ha) (Smith *et al.*, 2002). This approach may be justified in high-value seed orchards but is not practical for large-scale use in forestry.

Western spruce budworm parasitoids, not present in eastern Canada, were introduced from western Canada but did not become established. Exotic parasitoids from two closely related species from Europe (*Choristoneura murinana* (Hubner)) and Japan (*Choristoneura diversana* (Hubner)) were imported and released against *C. fumiferana* in eastern Canada but these did not become established either.

Extensive field trials with four virus types (CPV, NPV, EPV and GV) have been conducted against both *C. fumiferana* (Cunningham and Howse, 1984; Cunningham, 1985a,b; Cunningham and Kaupp, 1995) and *C. occidentalis* (Otvos *et al.*, 1989; Shepherd *et al.*, 1995). The goal of initiating an epizootic to control the budworm populations was not achieved in any of these trials (Cunningham and Kaupp, 1995) and the vertical transmission of the virus decreased in subsequent years (Otvos *et al.*, 1989).

None of these biological control attempts offered a practical alternative to replace the use of insecticide. Consequently, most of the control work has concentrated on aerial treatment of larvae with insecticide, although attempts have also been made to spray spruce budworm adults with chemical insecticides during dispersal, but without the desired results (Kettela, 1995).

History of Aerial Applications of Insecticides in Canada's Forests

Use of chemical insecticides

The history of the early use of insecticides was summarized in various chapters for all of Canada in Prebble (1975). Spruce budworm management, up to 1965, illustrates the first generation of control measures. Defoliator management was exclusively based on the application of chemical insecticides, using the most efficacious products (the more insects killed, the better). The first aerial applications of insecticides in forestry, using calcium arsenate dust, occurred between 1927 and 1930 on about 3200 ha against the spruce budworm, eastern hemlock looper, and the western hemlock looper (Prebble, 1975). Starting in the mid-1940s, the availability of dichlorodiphenyltrichloroethane (DDT), a wide-spectrum insecticide, and surplus aircraft from the Second World War led to the widespread use of aerial spraying to combat insect problems in Canada on a large operational scale (Figs 10.2 and 10.3). DDT and related compounds were adopted for forest insect control because they appeared very promising, based on their use in agriculture and in public health. After the early successes with DDT (i.e. contained a typhus epidemic and controlled malaria), overly optimistic statements, such as 'mosquito-transmitted disease will disappear', 'some pests will become extinct' and 'all major pest problems appeared to be solved or solvable', became common (Casida and Quistad, 1998: 1). DDT became the most commonly used insecticide in forestry because of its high efficacy against all defoliators and in the control of many agricultural insects. The use of DDT in the forests after the Second World War increased, but adversely affected fish populations (Logie, 1975) and DDT accumulated in birds and other components of the environment (Pearce, 1975).

From 1965 to 1980 the second generation of insect control, including the spruce budworm, was based on a reactive response – waiting until there was an insect outbreak and then trying to control it. Control was largely

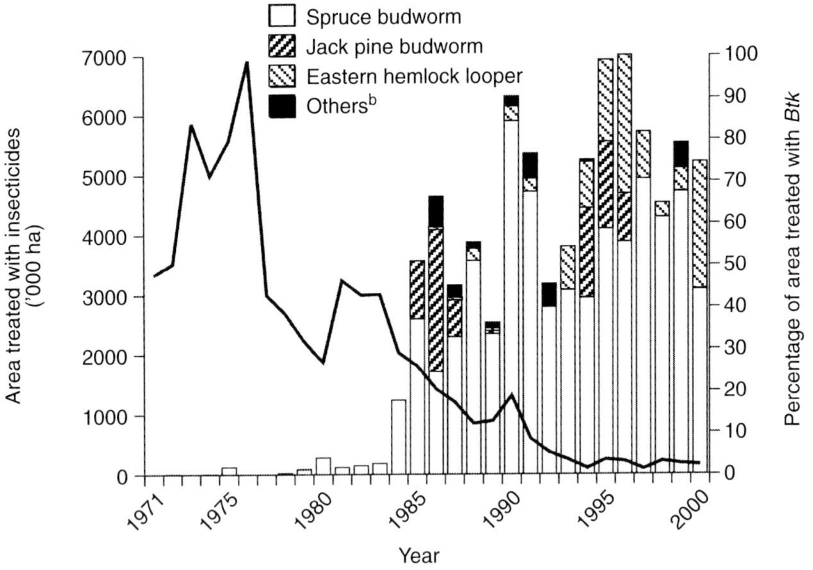

Fig. 10.3. Forest area in Canada treated with all insecticides[a] from 1971 to 2000 (from Canadian Forest Service, 1975, 1976, 1977, 1978; van Frankenhuyzen, 1990; Armstrong and Cook, 1993; Moody, 1993a,b; Hall, 1994, 1995, 1996; Cadogan, 1995; Canadian Council of Forest Ministers, 2002).
[a]For details including what insecticides were used over the years, see the source references. [b]Other insects include gypsy moth, *Lymantria dispar*, and forest tent caterpillar, *Malacosoma disstria*.

achieved by the application of synthetic chemical insecticides, first using wide-spectrum and then, as side effects became known, more selective chemical insecticides. The goal was to suppress the target insect populations with an efficacious product, with minimal side effects, and to protect the infested stands until harvesting could occur or the outbreak collapsed. However, public pressure was mounting to find biological alternatives to chemical insecticides.

In an effort to minimize damage caused by defoliators, considerable research was conducted on their population dynamics, especially that of *C. fumiferana*, to try to understand how and why outbreaks developed and the role of natural control factors. The Green River Project was the first large programme conducted on any insect pest in Canada, and expectations were high. The salient points of the programme were summarized in a volume edited by R.F. Morris (1963). While the programme contributed much new knowledge, it did not solve the management issues posed by the spruce budworm. However, it did provide new, much needed, tools to use. Coarse hazard-rating systems were developed, based on the location and frequency of past outbreaks and on stand susceptibility (stands with a high proportion of balsam fir and > 60 years old) (Regniere and Lysyk, 1995). Sex pheromones were identified and used to monitor population fluctuations of the insect and to warn of impending outbreaks (Sanders, 1988). Methods were developed to sample egg masses in the autumn and the overwintering larval stage in order to forecast expected defoliation (Morris, 1954; Miller, 1957; Miller *et al.*, 1971). The forecasted level of defoliation was, in turn, used to decide which infested areas to protect. There was a shift from treating as much infested area as possible with the available chemicals, aircrafts and funds to a more selective treatment of infested areas.

As the undesirable side effects of hard chemicals (broad-spectrum insecticides) became known, there was a shift towards more benign pesticides with fewer side effects. There was also a trend away from the reactive mode towards a proactive mode of insect management. With the proactive approach, attempts were made to predict when and where outbreaks would occur and to forecast the expected level of damage (defoliation) (Morris, 1954; Miller, 1957; Miller *et al.*, 1971) rather than waiting until the outbreaks developed and severe damage had occurred. This was aided by the Green River Project research.

Despite progressively reducing the dosages of DDT used (from 1–2 lb./acre to 0.25 lb./acre (Nigam, 1975)), effects on non-targets were still noted and deemed unacceptable. The search for more acceptable and ecologically less disruptive substitutes began as early as the 1950s and eventually led to the replacement of DDT by non-persistent organophosphates and carbamates in the late 1960s. Four organophosphates (phosphamidon, fenitrothion, trichlorfon and acephate) and two carbamates (aminocarb and mexacarbate) were registered and used extensively. Phosphamidon was used initially as a buffer spray in watershed areas to prevent DDT contamination of prime salmon fisheries waters. However, the use of DDT was extended to large-scale operations during the mid-1970s, due to a shortage of the other, more selective, insecticides. Further use of phosphamidon was discontinued in 1977 because of its negative impact on birds, even at low dosages. Fenitrothion replaced DDT in 1969 and as of 2003 it is still registered for use in New Brunswick, although there is considerable public pressure to ban its use entirely (J.C. Cunningham, 1996, personal communication). It was considered relatively safe for fish and birds, but had caused some mortality in aquatic insects and pollinators immediately after spraying (Varty, 1978). Trichlorfon was used in place of fenitrothion from 1973 to 1977 near cultivated blueberries to minimize adverse effects on pollinators and was used when non-spray buffer zones were placed around blueberry fields, precluding the use of trichlorfon. Acephate was the 'safest' chemical insecticide developed for aerial application on budworm-infested forests, as it was relatively less toxic to fish, birds and mammals than phosphamidon, fenitrothion and trichlorfon (Nigam, 1975), but it could not compete in cost and efficacy with fenitrothion and aminocarb. Mexacarbate was used in small quantities from 1969 to 1974, until the supply was exhausted (Nigam, 1975, 1980). Fenitrothion and aminocarb were the main insecticides used during the latter part of the 1970s, when approximately 9 million kg of fenitrothion and 0.9 million kg of aminocarb were used (Nigam, 1980).

The area of budworm-infested stands treated with insecticides rose dramatically during the mid-1970s in response to the increased size of budworm outbreaks (Fig. 10.3). However, the proportion of treated to infested forest declined (Figs 10.2 and 10.3) during this period due to a shortage of chemicals, economic constraints, and mounting public opposition to the wide-scale use of chemical insecticides in the forest. During the latter part of the 1970s, political reaction to public pressure over the Reye's-syndrome controversy (Wood and Bogdan, 1986) and suspected carcinogenicity of fuel oil carrier led to a gradual decline in the treatment of infested forests in eastern Canada (Nigam, 1980). This resulted in an increase in the size of the dead or dying forests in the budworm-infested stands in eastern Canada.

Aerial application remained the main method of control; insecticides were applied to reduce budworm populations, protect foliage and prevent tree mortality. Of the 29.9 million ha of forest sprayed from 1944 to 1973, 43.3% was treated with DDT, 55.5% with phosphamidon and fenitrothion, 1.2% with other chemicals and only about 0.04% with biological insecticides (Prebble, 1975; Armstrong and Cook, 1993). Realizing the undesirable side effects of the chemical insecticides, potentially less disruptive alternatives were identified (especially entomopathogens), and research was initiated.

During the major spruce budworm outbreak of the 1970s–1980s in eastern Canada (Kettela, 1983), significant tree mortality was first observed in 1973, when approximately 1 million ha of dead or dying timber was recorded. Tree mortality increased over the years and, by 1987, heavy mortality had occurred over approximately 18 million ha. The outbreak peaked in 1975, when nearly 50 million ha were infested in the eastern Canadian provinces, and close to 9 million ha were treated with phosphamidon and fenitrothion (Fig. 10.2; Nigam, 1980; Kettela, 1983). A total of 13 million kg of chemical insecticides were used to protect trees on 50 million ha, from 1965 to 1980, in eastern Canada (Nigam, 1980; Kettela, 1983). Between 1985 and 1990, approximately 2.8 million ha were treated in eastern Canada to control spruce budworm (Cunningham and van Frankenhuyzen, 1991).

There were large areas defoliated between 1981 and 1990. The areas defoliated annually ranged between a low of about 20 million ha in 1986 to a high of about 55 million ha in 1990 (Fig. 10.2).

A bilateral multi-year project on the spruce budworm was initiated between Canadian and American researchers in 1977 (CANUSA). The results of this project were published in an even larger volume (Sanders *et al.*, 1985) than that of R.F. Morris (1963). However, even these additional research findings failed to provide the knowledge and recommendations needed to manage budworm outbreaks without aerial application of insecticide to prevent budworm damage. Insecticide treatment remained the only effective tool to reduce larval populations, but the use of insecticide shifted away from broad-spectrum chemicals through the more selective chemicals towards bioinsecticides.

Attempts were made to develop viral insecticides. Entomopathogenic viruses had been isolated from field-collected budworm larvae as early as 1958 (Bird, 1959). Extensive laboratory and field trials were conducted with NPVs and GVs to control both the (eastern) spruce budworm and the western spruce budworm between 1974 and 1988. The field trials with these viruses were disappointing (Cunningham, 1985a; Otvos *et al.*, 1989; Cunningham and Kaupp, 1995). Consequently, no further work was done with the naturally occurring virus strains and forest managers were left with the only other microbial agent under investigation, *Btk*.

Use of *Bacillus thuringiensis*

Btk is effective against Lepidoptera (Dulmage, 1982), the order to which most forest defoliators belong. Although *Btk* is commonly found in soil microbiota (Martin and Travers, 1989), it has never been observed to control forest insect-pest populations in nature and must be applied as an insecticidal spray, sometimes more than once, over the infested area. It is now widely used to control several defoliating insects in Canada. Progress in the development of *Btk* as an operational alternative to synthetic chemical pesticides has been reviewed by Cunningham (1985b), van Frankenhuyzen (1990) and Cunningham and van Frankenhuyzen (1991).

The first aerial-spray trials using *Btk* in Canada were against the western black-headed budworm, *Acleris gloverana* (Walsingham), in British Columbia (Kinghorn *et al.*, 1961) and spruce budworm in New Brunswick (Mott *et al.*, 1961). Most of the early research on *Btk* was conducted on spruce budworm in the 1970s, in an effort, in part, to replace chemical insecticides. These early tests gave variable efficacy and population control. It took over 25 years from its first experimental use in the early 1960s to reach large-scale operational use in the mid-1980s (Fig. 10.3). The following were some of the contributing factors that led to the acceptance of *Btk* as a replacement for chemical insecticides in defoliator control.

Btk potency increased from 4 billion international units (BIU) per litre in the early 1970s to 16.9 BIU/l in the mid-1980s, by which time the dosage rate of 30 BIU/ha was routinely applied in 2.4 l of undiluted product for the control of spruce budworm. The cost of *Btk* decreased between 1981 and 1988 from 4.5 times to only 1.2 times more expensive than chemical insecticides. The cost of *Btk* was fairly constant at between 35 and 40¢/BIU from 1986 to 1990 (Cunningham and van Frankenhuyzen, 1991). The higher-potency *Btk* products (containing more BIU per litre) permitted using lower volumes and had the added benefits of reducing transportation costs and increasing spray-plane productivity. At the same time, *Btk* products were providing increased efficacy and reliability of control operations (van Frankenhuyzen, 1990) and foliage protection of the treated trees. *Btk* was finally considered a viable alternative to chemical insecticides for use against spruce budworm by 1981 (Smirnoff and Morris, 1984) and several other major forest defoliators a few years later (van Frankenhuyzen, 1990). These improvements, together with the political decision not to use chemical insecticides, as a result of public pressure and environmental concerns, led to favouring the use of biologicals, resulting in the widespread acceptance of *Btk* as a fully operational control option for forest defoliators.

Btk is considered environmentally benign (Otvos and Vanderveen, 1993) and is the only registered and commercially available microbial agent in Canada for forest insect control. By 1993 there were 18 *Btk* products registered for use in insect control in Canada (Otvos and Vanderveen, 1993). Although three viruses are also registered for forest insect control, they are not available commercially. The operational use of *Btk* for control of spruce budworms increased from about 2% of the total area treated in 1980[1] to 20 in 1984, 63 in 1990 and nearly 100% by 1996 (Cunningham and van Frankenhuyzen, 1991; van Frankenhuyzen, 1993; Fig. 10.3). In most of Canada, *Btk* is now the only insecticide used for budworm control, apart from a recently registered biorational (Mimic). Only New Brunswick continues to use both chemical (fenitrothion) and *Btk* products (Cunningham and van Frankenhuyzen, 1991). By far the greatest use of *Btk* in Canada is for spruce budworm control, with 2.8 million ha treated between 1985 and 1990 (Cunningham and van Frankenhuyzen, 1991). Similar increases in the use of *Btk* have occurred in the USA and Europe (van Frankenhuyzen, 2000).

Potential and constraints

The use of *Btk* is not without problems. Depending on weather conditions in the field, *Btk* in spray droplets is considered to be effective for only 3–5 days following application, after which the larvae may not acquire a lethal dose (van Frankenhuyzen, 1995). The effectiveness of *Btk* is questionable against high densities of spruce budworm (more than 25 larvae per branch sample).

Although *Btk* affects only larval stages of Lepidoptera, some criticism has been raised because of its potential impact on non-target beneficial or desirable Lepidoptera. Non-target Lepidoptera may be important in the food-chain of some insectivorous birds, or they may be rare or endangered species.

The effect of *Btk* on non-target Lepidoptera has been investigated by several authors, including Miller (1990). Studies have shown that both numbers of non-target insects and species richness were depressed for 2–3 years following treatments. However, all but the rare species recolonized the treated areas within 2–4 years after treatment (Miller, 1990). This was confirmed by Boulton *et al.* (2002), who found a significantly lower abundance of non-target Lepidoptera on plants that received an operational *Btk* spray (30 BIU/ha) in a plot treated against western spruce budworm, than on plants that were covered and excluded *Btk*. The two most common insect species on the shrubs made a full recovery within 2 years of the *Btk* spray. However, sparsely distributed species declined on both the treated and covered plants, suggesting a general decline of Lepidoptera species independent of the spray. A different experimental study, where double the regular registered dose was used (60 BIU/ha), yielded similar results. Some species made a full recovery by the end of the second year, but this could not be demonstrated for the sparsely distributed species (Boulton, 1999; T.J. Boulton and I.S. Otvos, unpublished data).

The effects of *Btk* treatment on non-target Lepidoptera may only be a concern when endangered insect species are in the spray area. However, the relative 'value' of the endangered species should be compared with the potential damage the target species will cause if no treatment takes place.

The timing of *Btk* treatment not only is important in terms of efficacy, but may also be a factor in preventing deleterious effects on natural enemies. Conflicting data on the effects of *Btk* treatments on the natural enemies of eastern and western spruce budworm have been published. Most reports indicated no deleterious effects (Buckner *et al.*, 1974; Reardon *et al.*, 1982; Morris, 1983; Niwa *et al.*, 1987). Otvos and Raske (1980a,b) reported an apparent increase in per cent parasitism by the two most common and important budworm parasitoids, i.e. *Apanteles fumiferanae* Viereck and *Glypta fumiferanae* (Viereck), while Hamel (1977) reported a negative impact. Nealis and van Frankenhuyzen (1990) indirectly confirmed the findings reported by Otvos and Raske (1980a,b) and recommended applying *Btk* at the peak of the fourth-instar stage to increase efficacy against the spruce budworm and to minimize or prevent negative effects on these two important budworm parasitoids.

The evolution of the currently used management method to minimize spruce budworm damage is typical of most defoliator control, and is probably representative of the general approach worldwide. At first, damage caused by the insect is not considered economically important until the affected resource is desired by society. Then the most efficacious control method is used – usually chemical insecticide with broad-spectrum effects. As non-target effects become known, alternative control methods are developed. Sometimes these have minimal side effects on the environment. The use of introduced and natural enemies, such as parasitoids and pathogens, can also have some effects, even if only on the biodiversity of the region. When our food and shelter are secured, segments of the population become socially conscientious and concerned about biodiversity. It is ironic, however, that, in the same population segments, some people still insist on buying unblemished fruits and vegetables produced by repeated use of chemical insecticides.

Integrated Pest Management of Douglas Fir Tussock Moth: a Case Study

The management system developed for the DFTM, integrating a pheromone-detection system with early application of the naturally occurring, laboratory-produced virus, illustrates a true IPM system.

Two morphotypes of the virus have been isolated from DFTM larvae and identified as the cause of these epizootics (Hughes and Addison, 1970). One morphotype exhibits singly occluded virus particles (*Op*SNPV) in the PIB. The second has multiple viral particles embedded in bundles within the PIB (*Op*MNPV) (Hughes and Addison, 1970). The majority of the research has been conducted on the multiple-embedded virus. The use of virus to control DFTM was considered in British Columbia as early as 1962, when the first field trial was conducted on individual trees (Morris, O.N., 1963). The first aerial spray trials using *Op*NPV against DFTM in British Columbia were conducted jointly by personnel of the Canadian Forest Service, the US Department of Agriculture (USDA) Forest Service and the British Columbia Ministry of Forests from 1974 to 1976 (Ilnytzky *et al.*, 1977; Stelzer *et al.*, 1977; Cunningham and Shepherd, 1984). Field treatments in 1975 using the laboratory-produced, naturally occurring virus caused high infection and high larval mortality, but the treated stands still sustained considerable damage because the virus was applied late in the declining phase of the outbreak. Therefore, the effect of the virus application on the course of the outbreak could not be evaluated. However, the experiment has shown that the laboratory-produced virus behaves like the naturally occurring one and can be used to cause an epizootic in the field. Based on these results and safety testing of *Op*MNPV conducted in the USA, the USDA Forest Service registered the viral insecticide as TM-Biocontrol-1® in 1976.

The experiment conducted in 1975 raised the following questions:

- Can the virus be introduced into the population at the beginning of the outbreak, and will it cause an epizootic?
- Will such an epizootic reduce damage normally associated with a DFTM outbreak?
- Can the virus dose be reduced to lower the cost of application?
- Can DFTM populations be monitored to predict future outbreaks?

To introduce the virus at the beginning of an outbreak requires a reliable monitoring system. It was necessary to determine where and when outbreaks were likely to occur. In a separate study, concurrent with the virus work, a dependable and sensitive pheromone monitoring system was developed for early warning of impending outbreaks (Shepherd *et al.*, 1985).

Monitoring with pheromone traps

Pheromone-baited traps were placed in susceptible stands, which were defined by overlaying maps of previous outbreaks, forest types and biogeoclimatic zones. The most susceptible stands, revealed by the overlays, tended to be located in the driest part of the range of Douglas fir, where it mixes with ponderosa pine, *Pinus ponderosa* P. Laws. ex C. Laws. Within this forest habitat, permanent monitoring stations were established and pheromone-baited traps were operated annually to monitor changes in male moth density over the course of an outbreak cycle to reveal patterns (Shepherd *et al.*, 1985; Shepherd and Otvos, 1986).

Population trends in pheromone-baited traps were followed from endemic to epidemic levels during the course of one outbreak. The number of successive years of upward trends of male moths caught was used to predict outbreak development. Three consecutive years in which the number of male moths caught increased and exceeded 25 moths per trap indicated that an outbreak was expected within the next 1 or 2 years (Shepherd *et al.*, 1985). The pheromone-trap system only gives advance warning that an outbreak is imminent and signals that another, more precise sampling system should be deployed in the area. Thus, after 2 years of upward trends, additional networks of traps are placed around the indicator stations to locate the foci of the developing outbreak and refine prediction. An egg-mass survey is then initiated during the autumn or winter to determine the insect density at the centre of the developing infestation and to predict the level of potential damage the following year (Shepherd *et al.*, 1984a). If egg masses are present, all available options for managing the insect problem should be examined to select the most appropriate for the situation. These options include protecting infested stands by virus or insecticide application or doing nothing and letting the outbreak run its course. A stepwise selection process for control actions is provided as a guideline (Shepherd and Otvos, 1986). The use of a registered chemical insecticide with a fast knock-down power may be considered appropriate in high-use recreational areas, such as parks, because the hairs of the caterpillars and those on the egg masses may cause a severe allergic reaction called 'tussockosis' in some sensitive people (Perlman *et al.*, 1976).

First experiment – 1981

A developing DFTM outbreak in south-central British Columbia was discovered in 1980 before any defoliation occurred. This allowed an experiment in 1981 to determine if a viral epizootic could be initiated at an early phase of the outbreak, before it would occur naturally, by ground and aerial application of the virus, and whether the application would reduce damage (Shepherd *et al.*, 1984b). Although a natural epizootic also occurred in the control plots containing high and moderate DFTM populations, the incidence of viral infection in the treated plots was considerably earlier and much higher, indicating the beneficial effects of the viral spray. Even at low population density, treatment effects were excellent, and a natural epizootic in the control plot occurred much later. The results showed that the virus can be introduced into DFTM populations at an early phase of the outbreak and that a viral epizootic can be initiated in first- and second-instar larvae by both aerial and ground treatment (Shepherd *et al.*, 1984b). The aerially applied virus caused an epizootic among DFTM larvae at low (41 larvae/m^2), medium (97 larvae/m^2) and high (206 larvae/m^2) population densities. Laboratory rearing of larvae collected from the field and weekly microscopic examination of the dead larvae indicated that a second wave of epizootic occurred among the survivors.

Table 10.4. Douglas fir tussock moth larval and tree mortality in virus[a]-treated and untreated plots, Veasy Lake, British Columbia, 1982 (from Otvos *et al.*, 1995).

Plot	Treatment[b] (PIB/ha in 9.4 l/ha)	1982 pre-spray larvae/m²	% Population reduction 6 weeks post-spray[c]	% Sample trees killed by DFTM 1983	% Sample trees killed by DFTM 1984
T1	1.6×10^{10} Oil	182.8	64.7	0	0
T2	8.3×10^{10} Oil	145.8	90.6	2	7
T3	2.5×10^{11} Oil	302.0	95.1	0	4
T4	2.5×10^{11} Molasses	41.8	86.6	0	0
				$\bar{\chi} = 0.6$	$\bar{\chi} = 2.8$
C1	Control	197.5		53	60
C2	Control	136.9		–[d]	–[d]
C3	Control	360.6		60	62
C4	Control	81.2		0	0
				$\bar{\chi} = 37.8$	$\bar{\chi} = 40.7$

[a] The virus used was *Op*MNPV produced in white-marked tussock moth (Virtuss®).
[b] PIB, polyhedral inclusion bodies; Oil, oil-based formulation containing 25% blank oil carrier and 75% water; Molasses, molasses formulation containing 25% commercial-grade molasses and 75% water.
[c] Per cent reduction was calculated by the modified Abbott's formula (Fleming and Retnakaran, 1985).
[d] Trees in part of the plot were cut down during power-line construction.

Second experiment – 1982

In a separate experiment the following year, water and oil based formulations, and reduced dosages[2] of the virus were compared to determine the most effective dose and formulation and to investigate if the recommended application rate of 2.5×10^{11} PIB/ha could be reduced and still prevent damage (Table 10.4). The initial impact of the 1982 applications was that by 2 weeks post-spray 10–30% of larvae had become infected. These infected larvae died, liberated polyhedra on to the foliage and increased the amount of inoculum. A secondary wave of virus infection then developed among the surviving larvae (DFTM larvae have a long feeding period) and by 6 weeks after spraying the population had collapsed. Per cent infection, development of the epizootic among the larvae and larval mortality, corrected for natural mortality, in the treated plots were related to dosage and the second lowest dose of 8.3×10^{10} PIB/ha gave similar results (about 91% mortality) to the full dose of 2.5×10^{11} PIB/ha (95% mortality). Virus transmission was higher in plots with higher larval densities. Development of the viral disease was temperature-dependent and the spread of the disease was influenced by host density. NPV infection was not detected in the untreated control plots until 5 weeks after spraying and at a much lower level than in the treated plots.

Virus application has additional benefits besides direct larval mortality. Among the surviving larvae, sublethal effects of virus application included higher pupal mortality, lower per cent adult emergence and a shift in sex ratio. More males (about twice as many) than females emerged in the treated plots, while the sex ratio of the adults was close to 1:1 in the control plots (Otvos *et al.*, 1987). This alteration of the sex ratio is probably due to the fact that female DFTM have one more larval instar than males, are exposed to the virus longer and consequently suffer higher mortality. This change in sex ratio among the survivors of virus treatment has been frequently reported in other insects, including the spruce budworm (Duan and Otvos, 2001).

These results indicated that it is possible to control DFTM populations with *Op*MNPV at about one-third of the registered label dose (at least at 8.3×10^{10} PIB/ha) and potentially even lower dosages at an early stage of an outbreak. Treatment at this lower dosage also prevented significant tree mortality in the treated stands (Table 10.4).

Summary and benefits of the system

The results of the experiments with pheromone traps and *Op*MNPV application over the past 20 years have proved that DFTM outbreaks can be prevented by a single application of virus at the beginning phase of an outbreak. Foliage protection may be negligible in the year of application, but acceptable in the following years. Tree mortality can be prevented when the treatment is applied early in the outbreak cycle and to early-instar larvae. The management system developed for DFTM was successfully tested during the 1990–1993 outbreak, and tree mortality was minimal in the stands treated with *Op*MNPV because early infection prevented the development of full-blown outbreaks in the treated stands. The operationally proved management system was accepted by the British Columbia Ministry of Forests, became part of the Forest Practices Code in the province and is used to manage DFTM populations.

The pest-management system for the DFTM (Shepherd and Otvos, 1986) enables forest managers to predict when and where outbreaks are likely to occur so that control measures can be planned and implemented to minimize or prevent damage. It is hoped that the DFTM pest-management system described above will serve as a prototype for the development of pest-management systems for other defoliating forest pests elsewhere. Work is underway to develop a similar management system for other defoliators in British Columbia and other parts of Canada. When using such a system, one may not always be able to use naturally occurring biological control agents to reduce damage; however, it should be possible to identify stands susceptible to attack by various defoliators and this in itself will be useful to forest managers. The steps of the DFTM management system are provided below for possible adaptation and use for other insects.

Steps of the DFTM management system

- Identify stands susceptible to DFTM attack based on overlaying defoliation maps of past outbreaks, forest types and biogeoclimatic zones.
- Within susceptible stands, establish permanent monitoring stations representing the range of past DFTM outbreaks.
- Set up pheromone traps at these locations and monitor male moth catches annually for one outbreak cycle to determine the threshold of male moth catches that indicates impending outbreaks.
- Deploy additional auxiliary pheromone traps to help to locate infested stands when male moth catches reach threshold numbers.
- Search for egg masses near permanent or auxiliary pheromone traps when moth catches reach threshold numbers.
- When egg masses are found, consider all available options to manage DFTM populations and select the most appropriate one for the area.
- Implement the action chosen.

Development of IPM for the Introduced Gypsy Moth in North America

The gypsy moth, *Lymantria dispar* (Linnaeus), presents a unique example of the progression of forest insect-pest control in North America. Probably more research has been done on the gypsy moth in North America, since its introduction *c.* 130 years ago, than on any other forest insect pest, either native or introduced. In spite of all of this research, the gypsy moth continues to spread. Hopefully, though, some important lessons have been learnt.

The gypsy moth is a polyphagous Eurasian forest defoliator that is known to feed on over 300 species of trees and shrubs, with oaks as the most favoured hosts (Leonard, 1981). Two strains of *L. dispar* are commonly recognized, one from Europe and the other from Asia. In Europe, the gypsy moth feeds mainly on deciduous trees, but in Asia it also feeds on conifers, primarily larch. The females of the European strain, although

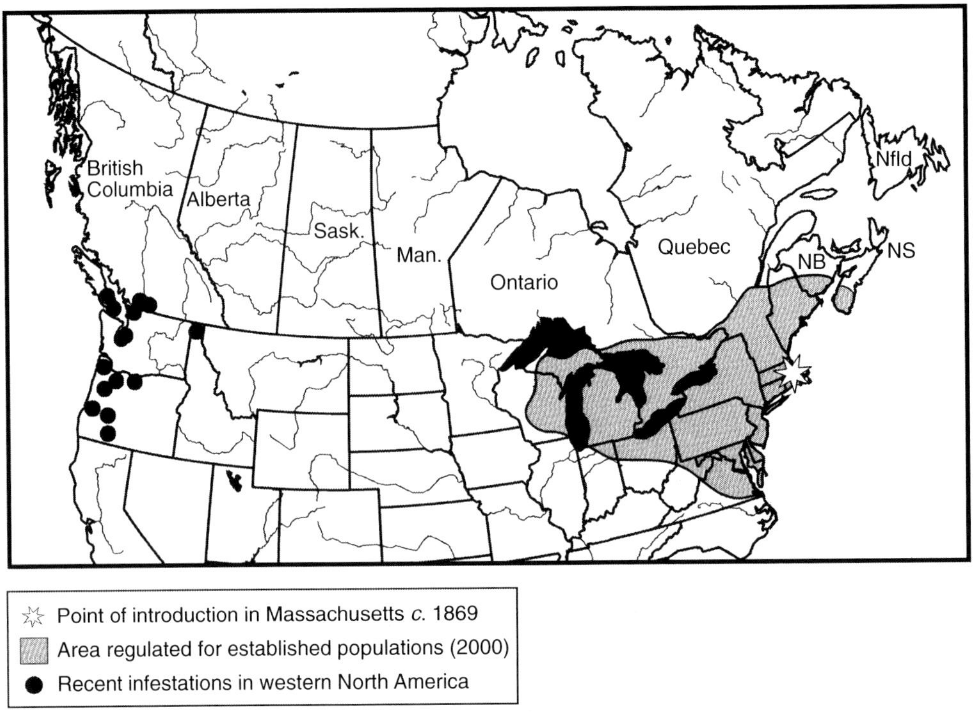

Fig. 10.4. Current extent of infestation and recent introductions of gypsy moth into North America (after Nealis, 2002). Sask., Saskatchewan; Man., Manitoba; Nfld, Newfoundland; NB, New Brunswick; NS, Nova Scotia.

winged, cannot fly, while females of the Asian variety can fly (Baranchikov, 1989; Wallner, 1989). This makes the latter a potentially more destructive strain than the European variety if introduced into Canada (where most of the trees are conifers), in terms of both its potential faster rate of spread and greater potential damage to the forests of Canada. These two strains of gypsy moth hybridize in the laboratory (Keena *et al.*, 1995) and in the field (Prasher and Mastro, 1995), and both sexes of the hybridized offspring (F1 and F2) can fly (Keena *et al.*, 1995).

The European strain of gypsy moth in eastern North America

The European strain was introduced under unusual circumstances at Bedford, near Boston, Massachusetts, USA, in 1869 (McManus and McIntyre, 1981; Liebhold *et al.*, 1989). Leopold Trouvelot, a naturalist, intentionally brought the insect from France to North America and the insect escaped from his laboratory during a storm. Although he reported the incident immediately, nothing was done at the time. The gypsy-moth infestation increased in extent and defoliation began to appear about 10 years after the initial escape. The spread of the gypsy moth since that time is well documented (McManus and McIntyre, 1981; Montgomery and Wallner, 1988). An eradication programme was started in 1889 and appeared to be working, reducing the infestation to such a degree that in 1900 the eradication programme was stopped by the state of Massachusetts (McFadden and McManus, 1991). This was a fatal mistake, with serious unforeseen consequences. Within 5 years of stopping the eradication programme, gypsy-moth infestations increased drastically and new infestations were discovered in three adjacent states (McManus and McIntyre,

1981), by which time the European strain of gypsy moth was well established in eastern North America (Fig. 10.4). A prompter response and eradication programme against this quarantine threat might have prevented the vast harm resulting from this incidence of pest introduction (Dunlap, 1980).

The gypsy moth spread west between 1906 and 1920, at an estimated rate of 9.6 km/year. In an effort to halt the spread of the gypsy moth and the subsequent damage to plants, the Domestic Plants Quarantine Act was created in 1912 in the USA (and in 1924 in Canada) to regulate the movement of plant material from gypsy-moth-infested areas. A barrier zone was established in 1923, from the Canadian border (Quebec) south along the Hudson River Valley to Long Island (New York), to prevent the westward spread of the gypsy moth. Infestations inside and west of this barrier zone were to be eradicated, while infestations to the east were to be controlled or managed by suppression, using various means (McFadden and McManus, 1991). Eradication programmes were designed to eliminate isolated populations of the gypsy moth, while suppression programmes were designed to protect foliage and/or reduce larval populations and slow the spread. The barrier zone became infested in 1939, and eradication attempts were terminated in 1941. However, the barrier zone was reinstituted in 1953 after the gypsy-moth populations exploded in 1951/52.

The European gypsy moth can spread in two ways: natural dispersal or accidental transport of pupae or egg masses by humans. Natural dispersal occurs over only short ranges when first- or second-instar larvae are transported by wind as female moths cannot fly (Mason and McManus, 1981; Elkinton and Liebhold, 1990; McFadden and McManus, 1991). This spread of first and second instar larvae by natural dispersal was estimated to be about 2 km/year (Liebhold *et al.*, 1992). Inadvertent transport of cocoons and/or egg masses from infested areas to uninfested areas by humans (through vacationing, moving and transporting goods) can be over much larger distances and is much more important in the spread of the gypsy moth than natural spread. For example, in 1983, ten gypsy-moth infestations were found in California about 3200 km from the infestation in the east, despite aggressive quarantine inspection at the California State border. In 1983, these infestations were treated with chemical insecticides to eradicate them at an estimated cost of US$1.5 million (McFadden and McManus, 1991).

Chemical control

At first, the broad-spectrum insecticides Paris green, lead arsenate and DDT were used in both eradication and suppression projects. Of the chemical insecticides used, DDT was considered so effective that, between 1949 and 1959, 3.7 million ha of gypsy-moth-infested stands were treated (Liebhold and McManus, 1999). However, the environmental damage caused by the extensive use of DDT against the gypsy moth was cited as a specific example of unacceptable chemical pollution in Rachel Carson's book *Silent Spring* (1962). Even though only about 50 ha of defoliation was noted, the eradication was discontinued in 1958 because of concerns about the environmental persistence of DDT. 'Hopes of eradicating the gypsy moth were abandoned [in 1958] and long overdue emphasis was placed on research' (McFadden and McManus, 1991). In 1959, carbaryl (Sevin®) replaced DDT for use in suppression programmes. Over time, the control products selected and used in suppression projects evolved from broad-spectrum insecticides, such as carbaryl (Sevin®), trichlorfon (Dylox®), acephate (Orthene®) and diflubenzuron (Dimilin®) to the much more selective and environmentally more acceptable biological insecticides, such as *Btk* and virus, as well as the use of pheromones in the 1980s (Liebhold and McManus, 1999). This gradual change in use pattern of controlling/suppressing the gypsy-moth populations in the eastern USA is illustrated in Fig. 10.5A. The use of *Btk*, virus and sex pheromones will be discussed separately under biological control.

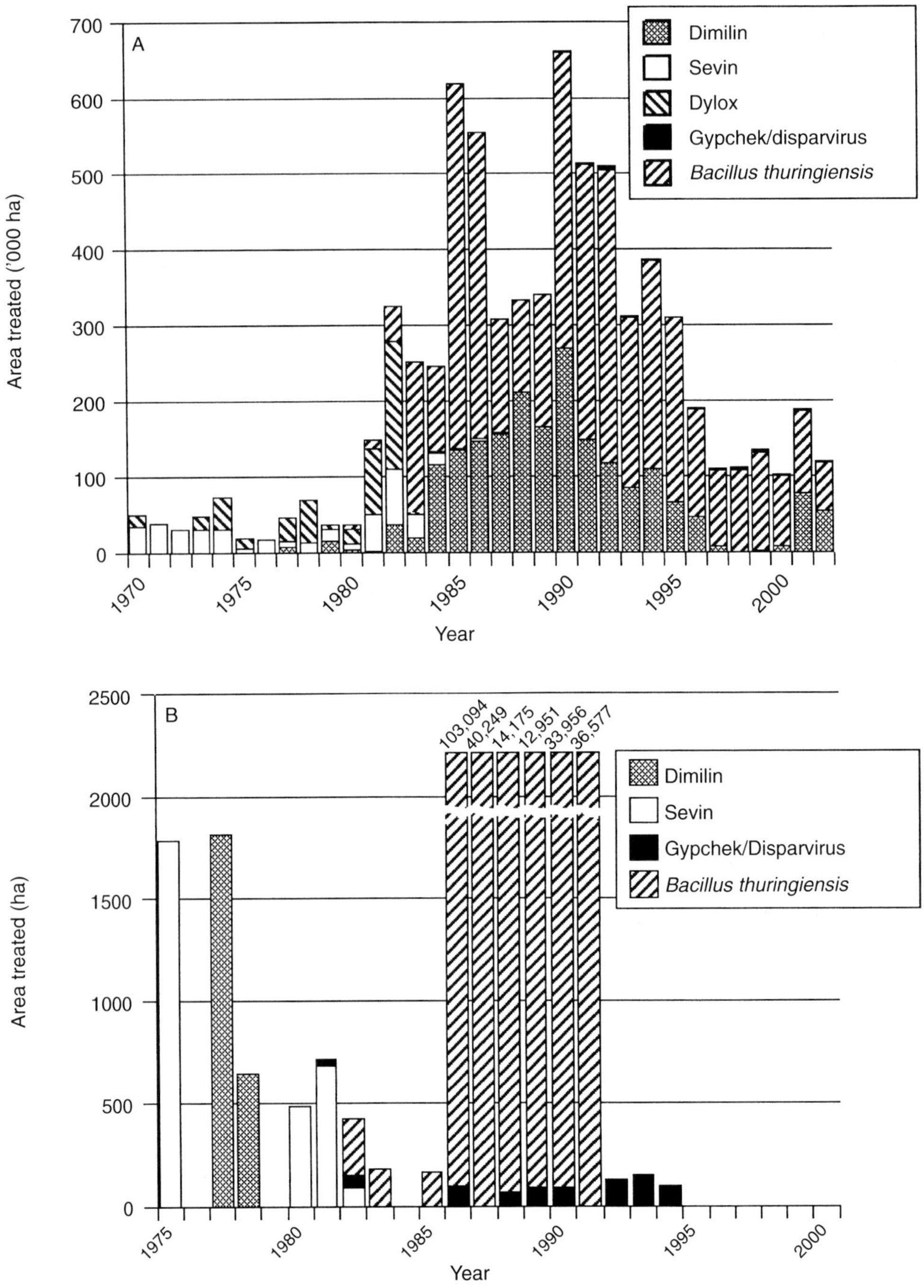

Fig. 10.5. Total area treated with insecticides during suppression and eradication programmes to control the gypsy moth (A) in the USA, 1970–2002, (from http://fhpr8.srs.fs.fed.us/wv/gmdigest, 2003); (B) in Canada, 1975–2000 (from Armstrong and Cook, 1993; Mason and Huber, 2002).

Biological control

Natural enemies

Classical biological control (the introduction of natural enemies to control recently introduced pests), started in 1905, was the first line of defence. Importation of exotic parasitoids and predators for gypsy-moth control was one of the largest programmes in the history of biological control (Brown, 1961). About 150 species of parasitoids have been reared from the gypsy moth throughout its distribution in Eurasia (Doane and McManus, 1981). Over the years, about 60 of these natural enemies, mainly from Europe, Asia and Japan (obtained from high and moderately high gypsy-moth populations), were introduced into the USA.

Early attempts at parasitoid introductions (1905–1914 and 1922–1933) were successful, but later introductions (1960–1977) failed (Reardon, 1981). By 1933, 44 species of parasitoids and nine species of predators had been introduced and, of these, 13 parasitoids and one predator became established (Dowden, 1962). Most researchers agree that classical biological control did not prevent the spread of the gypsy moth, but it very probably contributed to the collapse of some outbreaks and possibly slowed the spread. The introduced parasitoids may have also kept the populations at endemic levels after the collapse of an outbreak (Ticehurst *et al.*, 1978). Augmentative releases of some of the established parasitoids resulted in increased parasitism in the release areas (treatment) compared with the controls in three different studies (Weseloh and Anderson, 1975; Hoy, 1976; Blumenthal *et al.*, 1978). However, the augmentative parasitoid releases did not appear to significantly reduce gypsy moth egg-mass density (a measure of success) after the releases.

Pathogens

During the late 1950s and early 1960s, in response to public concern over the use of chemical insecticides, efforts were intensified to develop environmentally acceptable control alternatives. Entomopathogenic microorganisms offered such alternatives and considerable work has been done on the use of pathogens for gypsy-moth control. The pathogens investigated include *Bacillus thuringiensis* (Dubois, 1981), an NPV (Lewis, 1981) and a fungus (*Entomophaga maimaiga*) (Hajek *et al.*, 1990a; Elkinton *et al.*, 1991; Hajek, 1997).

Research into the use of *Btk* for the suppression of gypsy moth began in earnest in the early 1960s (Cantwell *et al.*, 1961) and, by the 1980s, it was accepted as an effective alternative to chemical insecticide for gypsy-moth control (Dubois, 1981). *Btk* use began in 1980 and increased gradually, reaching a peak in 1985 when about 481,900 ha were treated with *Btk* (Fig. 10.5A).

According to Campbell and Podgwaite (1971), the gypsy-moth NPV (*Ld*NPV) was probably introduced inadvertently with the gypsy moth or with one of its parasitoids (Glaser, 1915). It was considered to be the most important natural control agent of the gypsy moth until 1980 (Campbell, 1963; Doane, 1970, 1976; Podgwaite and Campbell, 1972). The development of *Ld*NPV as a biopesticide began shortly after its effectiveness in reducing larval populations was realized. Considerable research into strain selection, production, and laboratory and field testing (Lewis, 1981) and other factors (see section 6.3 of the gypsy-moth compendium, edited by Doane and McManus (1981)) preceded registration of the gypsy-moth NPV in 1978 as Gypchek for gypsy-moth control (Lewis, 1981).

Unfortunately, commercial production of Gypchek did not follow registration. Its greatest attribute, high host specificity, is also its greatest shortfall, as it makes it economically unattractive for private companies to produce and sell it. Large companies are only interested in developing products that generate US$40–50 million in sales per year (Podgwaite, 1999), while the market for Gypchek is only about US$5 million per year. To date, every company that tried to produce Gypchek commercially has failed economically (Podgwaite, 1999). Consequently, the US Forest Service (USFS) continues to produce Gypchek and make it available on a limited basis through cooperative suppression programmes (Podgwaite,

1999). Since the registration of Gypchek in 1978, only about 20,200 ha have been treated with it in the USA. This represents less than 1% of the total area treated for gypsy-moth control (Podgwaite, 1999; Fig. 10.5A).

There is some confusion about the origin of the gypsy-moth fungus that causes epizootics in North America (Hajek *et al.*, 1995). A fungal pathogen of the gypsy moth was introduced from Japan (where it causes extensive epizootics) into the USA on two separate occasions. The early introductions and releases were in 1910 and 1911 (Speare and Colley, 1912) and then later in 1985 and 1986 via infected larvae (Hajek *et al.*, 1995). No evidence was found indicating establishment of the fungus from either of these releases within a few years following introduction (Hajek *et al.*, 1995). After its second release, the fungus introduced from Japan was described as a new species, *E. maimaiga* Humber, Shimazu & Soper (Soper *et al.*, 1988). An epizootic of this fungus was discovered in 1989 (Andreadis and Weseloh, 1990) and surveys in the same year showed the fungus occurring in seven states and in virtually all samples collected from areas that were infested by the gypsy moth prior to 1980 (Hajek *et al.*, 1990a). More intensive surveys in the following year showed that *E. maimaiga* occurs in ten states in the northeastern USA (Elkinton *et al.*, 1991) and in Ontario (Welton, 1991; Nealis *et al.*, 1999), where the fungus must have spread on its own as no intentional introduction of the fungus has occurred there. Large-scale epizootics caused by *E. maimaiga* were widespread in areas colonized by the gypsy moth prior to 1980 (Elkinton *et al.*, 1991), but not in areas invaded more recently (Hajek *et al.*, 1996). The wide geographical distribution of the fungal epizootic suggests that the fungus probably became established in the New England states from the first intentional releases of *E. maimaiga* near Boston in 1910 and 1911 (Andreadis and Weseloh, 1990; Hajek *et al.*, 1990a; Elkinton *et al.*, 1991). The first authors to report on the 1989 discovery of the *E. maimaiga* epizootic summarized it the best by saying, 'the current epizootic may have resulted from the survival and inapparent spread of an early introduction in 1910–1911' (Andreadis and Weseloh, 1990: 2461).

Before the discovery of *E. maimaiga* in North America in 1989, NPV was generally credited with causing the collapse of gypsy-moth outbreaks (Hajek *et al.*, 1996). *Ld*NPV epizootics generally occur at high host densities in dry years, when rainfall is low (Elkinton *et al.*, 1991), while *E. maimaiga* can cause high levels of infection across a wide range of host densities (Hajek *et al.*, 1996). Since 1989, *E. maimaiga* epizootics have been shown to decimate gypsy-moth populations at both low and high host densities in years with higher than normal precipitation (Elkinton *et al.*, 1991; Hajek and Roberts, 1991; Smitley *et al.*, 1995) and even when NPV is present (Smitley *et al.*, 1995). Several authors have noted a positive correlation between the intensity of the epizootic and rainfall (Andreadis and Weseloh, 1990; Elkinton *et al.*, 1991; Hajek *et al.*, 1993; D'Amico and Elkinton, 1995). This is not surprising, since *E. maimaiga*-killed larvae discharge conidia in the presence of dew or when relative humidity is greater than 90% (Hajek *et al.*, 1990b; Hajek and Soper, 1992). In dry weather, fungal-killed gypsy moth have the same appearance as NPV-killed larvae (Hajek and Roberts, 1992). This holds true for other insects and Entomophthorales fungi (I.S. Otvos, personal observation). For example, epizootics in the eastern hemlock looper, *Lambdina fiscellaria fiscellaria* (Guenee), in Newfoundland were attributed to an undetermined wilt (viral) disease for over 20 years. The true cause of the epizootics, infection by two fungal pathogens, was only determined in 1969 (Otvos, 1973; Otvos *et al.*, 1973). The absence of external fruiting bodies on the insect cadavers makes distinguishing between mortality caused by these pathogens in dry weather in the field difficult. Given this information, it is not hard to imagine that earlier identifications of viral infection in the field would have also included some fungal-infected insects.

The fungus has spread dramatically between 1989 and 1992 and is prevalent in areas not recently colonized by the gypsy moth (it does not occur in recently colonized infestations), causing up to 100% mortality of late-instar gypsy-moth larvae (Reardon and Hajek, 1993). The fungus is considered by some to be a more important pathogen than NPV (Hajek, 1997). Consequently, it has been intentionally 'spread' or introduced, especially to areas newly invaded by the gypsy moth (Hajek *et al.*, 1996; Smitley *et al.*, 1995).

E. maimaiga resting spores can be translocated or even introduced into new fungus-free areas and along the leading edge of the spreading gypsy-moth infestation by the purposeful introduction of resting spores (Smitley *et al.*, 1995; Hajek *et al.*, 1996) by moving soil with resting spores from areas where fungal epizootics have recently occurred. However, care must be taken to ensure that pathogens, such as *Armillaria mellea* rhizomorphs, are not spread unintentionally as well (Reardon and Hajek, 1993). '*E. maimaiga* is now a dominant natural enemy associated with gypsy moth in North America' (Hajek, 1997: 67) and might be considered more important than NPV in moister environments. The fungus might be an excellent candidate for introduction around lakes, on islands and on the west coast of North America should the gypsy moth become established in that region.

Combinations of biological control agents

In order to increase their control impact, some biological control agents have been used in combination, such as parasitoids and pathogens. The parasitoid *Apanteles melanoscelus* (Ratzeburg) was used in combination with *Btk*, and their combined use resulted in higher population reduction and foliage protection than when each was used alone (Reardon, 1981). Studies have also shown that parasitoids contaminated with NPV in the laboratory can transmit the virus to the gypsy moth, both in the laboratory and in the field (Reardon, 1981).

Dual or mixed infection by *Ld*NPV and *E. maimaiga* has been reported from the gypsy moth (Hajek and Roberts, 1991; Weseloh and Andreadis, 1992; Malakar *et al.*, 1999). Generally, *E. maimaiga* out-competes *Ld*NPV when both the fungus and the virus infect the host at about the same time, because the fungal infection develops faster.

Sex pheromones

Disparlure is used in pheromone traps to monitor (survey and detect) gypsy-moth populations (Cameron, 1974; Schwalbe, 1981) and to suppress populations through mating disruption (Beroza *et al.*, 1975; Reardon *et al.*, 1998). Decrease in mating success was found to be inversely dependent on the density of the moths and directly related to the amount of disparlure used. Greater mating disruption occurred in lower-density gypsy-moth populations than in higher density populations. Because of the cost of disparlure production, gypsy-moth population suppression by mating disruption is only practised when low doses are used against low-density insect populations (Webb *et al.*, 1990). A double application of 75 g disparlure/ha in 1 year delayed population increase by 1 to 4 years. In low-density populations, mating disruption with aerially applied disparlure (at 75 g/ha)[4] is an effective control of gypsy-moth populations (Leonhardt *et al.*, 1996), reducing mating and the number of fertile egg masses laid by 70–85% (Webb *et al.*, 1990). Work is continuing to evaluate the efficacy of lower dosages of disparlure to achieve mating disruption, making it more cost effective (Reardon *et al.*, 1998).

Use of the barrier concept in gypsy-moth management

Between 1920 and the 1950s, several large-scale 'barrier' programmes were implemented to prevent the westward expansion of the gypsy moth (McManus and McIntyre, 1981; McFadden and McManus, 1991; Sharov *et al.*, 1998). Until the early 1980s, the main goal of the barrier-zone management concept was eradication of infestations outside the barrier zone and suppression of the moderate- and high-density populations inside the barrier zone by various means (Sharov *et al.*, 1998; Liebhold and McManus, 1999). The largest outbreak on record occurred in 1981 when 5.2 million ha were defoliated by the gypsy moth, of which 150,000 ha were treated mostly with Dylox® (58.4%) and Sevin® (32.2%). The second largest outbreak was in 1990, when 2.9 million ha were defoliated and more than 660,000 ha of gypsy-moth-infested stands were treated with *Btk* (59.0%) and Dimilin® (40.8%) (Fig. 10.5A). The barrier zone management concept was abandoned when the USFS embraced the IPM concept and different alternative approaches were tried (Reardon, 1991) to prevent gypsy-moth populations from expanding. These alternative approaches were tested in two pilot projects.

The first alternative approach, tried in a pilot project (Maryland IPM 1983–1987) to manage low to moderate populations of gypsy moth using IPM, involved annual surveillance of insect densities using pheromone traps to determine gypsy-moth distribution and density to maximize natural control and the use of direct control measures when necessary in an environmentally acceptable way (Reardon *et al.*, 1987).

This approach was later modified in a second pilot project over a much larger area (Reardon, 1991). The Appalachian Integrated Pest Management Gypsy Moth Project (AIPM 1987–1992) was designed 'to demonstrate the effectiveness of new and existing management technology in an IPM approach to minimize the spread and adverse effect of the gypsy moth' (Reardon, 1991: 108). In the first phase of this second project, three products (*Btk*, Gypchek and disparlure) were tested on a small scale before applying them operationally (Reardon, 1991). During this project, areas with high gypsy-moth density were treated with *Btk*, Dimilin®[4] and virus. Areas with low-density gypsy moth were treated with synthetic flakes impregnated with sex pheromones (disparlure) to reduce the number of fertile egg masses through mating disruption.

Following these two projects, a third one, the Slow-the-Spread (STS) pilot project (1993–1998), was initiated by the USDA Forest Service with participating state agencies. This study determined the feasibility of using IPM strategies to slow the spread of the gypsy moth, with reduced pesticide use and management costs compared with AIPM, over large geographical areas (Leonard and Sharov, 1995; Sharov and Liebhold, 1998a). Extensive monitoring of low-level gypsy-moth populations was conducted with pheromone traps. Isolated populations, well in advance of the infestation front (100–150 km), were suppressed or preferably eradicated. *Btk*, Gypchek and Dimilin® were used to eradicate or suppress gypsy moth in high-density populations and Gyplure (mating disruption) to manage low-density populations (Leonard and Sharov, 1995; R.C. Reardon, March 2003, personal communication).

The effectiveness of the treatments applied during the STS gypsy-moth project conducted by the USDA Forest Service (on 188,064 ha) were analysed and compared (Sharov *et al.*, 2002b). Disparlure treatment (93 blocks) was significantly more effective against isolated low-density populations of gypsy moth than *Btk* treatments (173 blocks). A large-scale evaluation of operational disparlure treatment of gypsy-moth populations has shown that this method is effective in isolated, well-defined, low-density infestations, but does not appreciably disrupt mating in high-density populations (Sharov *et al.*, 2002b). The development of mating disruption has become the key element to the success of the STS programme (Sharov *et al.*, 2002a). Analysis of gypsy-moth spread data has also shown that the STS programme has reduced the spread of this insect by more than 50% (Sharov *et al.*, 2002a).

In the past, the tendency was to eradicate isolated infestations as soon as they were detected. However, analysis of the treatments of gypsy-moth populations during the STS programme, from 1993 to 2001, suggested that it is better to postpone treatment until the gypsy-moth population is well delineated with a dense grid of pheromone traps (Sharov *et al.*, 2002b).

Since 1999, the STS strategy has become a comprehensive long-term national programme to protect the trees and forests in the USA from gypsy moth along the entire length of the expanding gypsy-moth front. STS coordinates efforts by the USDA Forest Service, the Animal and Plant Health

Inspection Service (APHIS) and several states where the gypsy moth is not yet established (Sharov *et al.*, 2002a). It uses three strategies:

- Suppressing populations within generally infested areas.
- Slowing the spread of gypsy moth to delay the impacts and costs associated with managing gypsy-moth outbreaks.
- Eradicating isolated infestations outside generally infested areas.

In the USA, the approach currently used is to control the advancing front of increasing gypsy-moth populations to slow the expansion of this insect's range. Pheromone traps are used to detect new outbreaks on the leading edge of the infestation, and environmentally benign insecticides, such as *Btk*, *Ld*NPV or mating disruption, are used in the suppression or eradication programmes. Eradication is implemented in areas where the gypsy moth is detected well in front (100–150 km) of the spread. Suppression projects are conducted to reduce damage in areas where the gypsy moth is well established.

Analysis of the historical records of the gypsy-moth spread in the eastern USA, revealed three distinct time periods with different rates of spread or expansion (Liebhold *et al.*, 1992). From 1900 to 1915 the rate of spread was slow, about 9 km/year. The spread was reduced to *c.* 3 km/year from 1916 to 1965. This reduction was very probably due to the aggressively managed 'barrier zone' (e.g. detection, suppression and eradication) to reduce the gypsy moth's westward movement. From 1966 to 1989, the rate of spread was very high (*c.* 21 km/year) (Liebhold *et al.*, 1992, 1995).

The STS project demonstrated the feasibility of reducing the rate at which insect infestations spread. *Btk*, Dimilin®, Gypchek and disparlure were tested to determine their efficacy against different densities of gypsy-moth infestation. *Btk*, Dimilin® and Gypchek (*Ld*NPV) are used to eradicate outlier or to suppress moderate to high gypsy-moth populations, while mating disruption using Gyplure is used to manage low-density populations (Reardon *et al.*, 1994, 1998; Leonard and Sharov, 1995). Mating disruption with disparlure was significantly more effective than *Btk* in reducing isolated low-density gypsy-moth populations (Sharov *et al.*, 2002b).

The gypsy-moth virus (Gypchek), because of its long incubation period and inconsistent efficacy (Podgwaite, 1999), is not the best choice for defoliation prevention. Dimilin® would be a better choice for use in campsites and other areas where its non-target effects might be tolerated. Disparlure has been shown to substantially reduce gypsy-moth populations at low to medium densities (Reardon *et al.*, 1998), and its operational use is increasing (Sharov *et al.*, 2002a).

Success of the STS strategy

Sharov and Liebhold (1998b) developed a model predicting that the spread of gypsy moth could be slowed by as much as 50% utilizing barrier zones (Liebhold *et al.*, 1992; Sharov and Liebhold, 1998b). In practice, the actual rate of spread in the Appalachian Mountains was reduced by 59% (Liebhold *et al.*, 1992), from 21 km/year to 9 km/year (Sharov and Liebhold, 1998c). Sharov and Liebhold (1998b) developed a model that specifies optimal strategies for eradication and containment. Their analysis and containment model disagree with the statement by Dahlsten *et al.* (1989) that 'insects that have already colonized parts of the United States or any large land mass or continent, probably should not be the targets for eradication programs in other sections of the country because of their potential for recolonization'. Sharov and Liebhold (1998a) state that their 'analysis clearly demonstrates that this statement is wrong. Eradication of small, isolated colonies of the gypsy moth within barrier zones is not only feasible, but also economically justified because the model predicts positive net benefits under realistic assumptions.'

There is economic benefit to slowing the spread of gypsy-moth populations (Leuschner *et al.*, 1996; Sharov and Liebhold, 1998a). The costs of slowing the spread of gypsy moth have been estimated to be about

25% of the expected potential benefits (Sharov *et al.*, 1998). Since about two-thirds of the potential area in the USA containing highly susceptible host trees (mainly in the south-eastern USA) remains uninfested (Liebhold *et al.*, 1997), slowing the spread of gypsy moth into these areas is economically beneficial (Leuschner *et al.*, 1996; Sharov and Liebhold, 1998a). Estimated benefits associated with reducing the rate of spread outweighed the cost of implementing the STS programme (detecting and treating isolated infestations along and ahead of the expanding gypsy-moth population front) by at least 3 : 1 (Leuschner *et al.*, 1996). This perceived or real benefit is indicated by individual households[5] who, when surveyed, indicated a willingness to pay between US$13 and 57/ha for gypsy-moth control (Miller and Lindsay, 1993).

Gypsy moth in eastern Canada

The use of chemicals and biological insecticides for eradication, suppression and delaying the spread of gypsy-moth populations, as well as the introduction of natural enemies, followed the same general trend in Canada (Fig. 10.5B) as in the USA (Fig. 10.5A). Therefore, for the sake of brevity, only the highlights will be mentioned here.

In eastern Canada, the gypsy moth was first recorded in Quebec near the US border in 1924, and covered about 90 ha. Control measures using lead arsenate commenced the following year, and within 3 years the infestation was eradicated. A second invasion of gypsy moth, detected in New Brunswick in 1936, was successfully eradicated by 1940 (Jobin, 1995). No further introductions of gypsy moth were found in eastern Canada for the next 16 years.

A survey and detection programme using pheromone-baited traps was initiated in 1954 and resulted in the discovery of a third infestation in 1956, near the site of its first introduction in Quebec (Cardinal, 1967). Spray operations, initiated in 1960, to control this infestation failed. The gypsy moth was later detected in Ontario in 1969, the infestation originating from a separate invasion from the USA. After the failure to eradicate the gypsy moth in eastern Canada, only suppression programmes were conducted against the European strain of gypsy moth. The outbreak in Quebec peaked in 1977, with 518,000 ha of defoliated stands and in Ontario in 1985, with 246,000 ha defoliated. The area infested in Ontario decreased to almost 168,000 ha of moderate to severe defoliation in 1986 (Jobin, 1995), of which 103,094 ha were treated (Nealis *et al.*, 2002).

In the aerial-spray operations conducted between 1960 and 1974, chemical insecticides, mainly Sevin® and some DDT, were used (Brown, 1975). Until 1969, the main purposes of these control operations were twofold: eradication of small, incipient infestations and suppression of larger infestations. After 1969 the principal aims of these spray operations were, as in the USA, to suppress gypsy-moth larval populations and delay the spread of gypsy moth into uninfested areas (Jobin, 1995). Between 1975 and 1989 *Btk*, Sevin® and Dimilin® were mainly used operationally for suppression (Jobin, 1995). *Btk* was used almost exclusively operationally after 1981 (Nealis *et al.*, 2002; Fig. 10.5B).

Work on the use of pathogens (*Btk* and *Ld*NPV) for gypsy-moth suppression did not start in Canada until the 1970s (Griffiths and Quednau, 1984). *Btk* slowly gained acceptance as an effective suppression tool of gypsy-moth populations and, at its peak use in 1986, over 103,000 ha of gypsy-moth-infested stands were treated in Ontario (Jobin, 1995; Nealis *et al.*, 2002).

Although the virus (*Ld*NPV) is also registered and effective for suppression of gypsy-moth populations, it is not yet commercially available. Consequently, the virus was only used experimentally in Canada (Jobin, 1995; Nealis *et al.*, 2002), unlike in the USA where the USDA Forest Service produces and uses it in cooperative projects.

Since the establishment of the gypsy moth in North America, 26 species of parasitoids native to North America have been found to successfully attack and develop in this unwanted newcomer (Griffiths, 1976; Sabrosky and Reardon, 1976; Griffiths and Quednau, 1984). In addition, 17 native insect predators and four mammalian predators also prey on the gypsy moth (Griffiths, 1976). Of the 26 parasitoid species that attack gypsy moth in the USA, 14 species were also reared from other hosts in southern Ontario and Quebec near the US border (Griffiths and Quednau, 1984), indicating the generalist nature of most of these parasitoids.

A review of the world literature on the parasitoids and predators of the gypsy moth showed that there are close to 400 species of natural enemies associated with the gypsy moth (Griffiths, 1976). Some of these were originally introduced against the browntail moth, *Euproctis chrysorrhoea* (Linnaeus) (Hewitt, 1916). Of the over 50 species of exotic parasitoids and predators released in the USA, 13 species of exotic parasitoids and one predator became established, 29 parasitoids and eight predators did not (Griffiths, 1976). Of these 13 exotic species of parasitoids established in the USA, nine species spread into Canada on their own. In fact, four of these parasitoid species were recovered in southern Ontario and Quebec before the gypsy moth was recorded in the area and before any exotic parasitoids were introduced against the gypsy moth in Canada (Griffiths and Quednau, 1984). Based on the review of literature of gypsy-moth parasitoids and predators (Griffiths, 1976) and their own work (including egg parasitoids) in Canada, Griffiths and Quednau (1984) concluded, 'The establishment of exotic insect parasites [parasitoids] on the gypsy moth in Canada is proceeding well largely through natural dispersal' and 'there is little more to be done in the introduction of biological control agents because there are no more suitable candidates' (Griffiths and Quednau, 1984). However, most of the seven introduced parasitoids were collected in moderate to high gypsy-moth populations. Consequently, since 1980, work in Canada has focused on finding parasiotids in Europe that might be effective at low gypsy-moth densities. This resulted in the introduction and release of a little known tachinid fly in Canada (Mills and Nealis, 1992; Nealis and Quednau, 1996). However, it is too early to evaluate the impact of this latest introduction on the gypsy-moth population.

Gypsy moth on the west coast

In western North America, neither the European nor the Asian strains of the gypsy moth have become established to date. Effective detection grids using pheromone traps around ports and in suspect areas and immediate eradication are the policy on the Pacific coast of Canada and the USA. So far, only small spot infestations of these two strains (mainly the European strain) have been found in British Columbia, Washington and Oregon, and all have been successfully eliminated through aggressive aerial and ground application of *Btk*, as was the infestation of the Asian strain introduced into North Carolina in 1993 on US military equipment returning from Germany, at a cost of about US$9 million (Wallner, 1996).

The first interception of the gypsy moth on the west coast was in 1911 in British Columbia, when egg masses were found on ornamental *Thuja* trees from Japan (Brown, 1975). The egg masses were destroyed but not before a few larvae hatched (Humble and Stewart, 1994). The interceptions and repeated eradications of the gypsy moth in British Columbia (both European and Asian strains) have been summarized by Humble and Stewart (1994). In eradication programmes, application is generally from the air, although at times the less effective and more expensive ground treatments are used as a result of court challenges by environmental groups opposing the eradication programmes and/or aerial spraying of *Btk* or any other insecticidal product.

The Asian strain was introduced in 1991 by Russian ships coming from the Far East but was also eradicated successfully in 1992 (Humble and Stewart, 1994). During this eradication, 19,000 ha were treated, at an estimated cost of CAN$6.5 million (Nealis, 2002). Since this incident, federal inspectors have banned ships from inshore waters when gypsy-moth egg masses are discovered on the superstructure of freighters during larval hatch and development (Humble and Stewart, 1994).

Eradication gets complicated when non-infested countries or regions impose trade embargoes. Such was the case with the 1999 eradication programme in British Columbia, which resulted from an unusual situation arising among the various government agencies responsible for responding to the exotic-insect threat and trying to meet trade conditions imposed by regions that are threatened by *L. dispar* invasion. Until 1998, the federal government's Canadian Food Inspection Agency (CFIA) conducted eradication programmes against the gypsy moth (both European and Asian strains) in British Columbia. Following the eradication programme in 1998, CFIA announced that it would no longer consider eradication of the European gypsy moth from British Columbia and would only regulate infested areas. Consequently, the USA imposed trade restrictions on lumber and log exports from British Columbia, forcing the provincial government to pass an order-in-council to enable them to treat the 13,000 ha that had been delineated for treatment, based on pheromone-trap catches on southern Vancouver Island. This eradication programme, conducted in 1999, cost *c.* CAN$3.7 million (Nealis, 2002).

Management of Bark Beetles

Over 200 species of scolytid bark beetles occur in Canada and Alaska (Bright, 1976). Nine are economically important and seven of these attack conifers. The most destructive conifer-attacking bark beetles, in descending economic importance, are:

- mountain pine beetle, *Dendroctonus ponderosae* Hopkins;
- spruce beetle, *Dendroctonus rufipennis* (Kirby);
- Douglas fir beetle, *Dendroctonus pseudotsugae* Hopkins;
- western balsam bark beetle, *Dryocoetes confusus* Swaine.

These beetles breed in the inner bark and phloem of the main bole of their host trees. Needles of trees successfully attacked by bark beetles first fade and then turn to a reddish-brown colour (pines, firs, Douglas fir) or the faded needles fall off (spruce). The trees attacked by mountain pine beetle die within 3 or 4 weeks of attack, but the foliage may not change colour until the following year. For the host trees attacked by the other three destructive bark-beetle species, the process of tree death is similar to that caused by the mountain-pine beetle. Periodic bark-beetle outbreaks frequently cause catastrophic economic losses. During the last outbreak (1972 to 1985) the mountain pine beetle killed approximately 195.7 million pines in British Columbia, representing an estimated potential economic loss of $14.4 to 19.6 billion, of which about $4.0 to 5.4 billion worth of beetle-killed trees could not even be salvaged (Borden, 1990). During the peak of the outbreak in 1983, an estimated 43 million mature lodgepole pine (*Pinus contorta* Dougl. ex. Loud. var. *latifolia* Engelm.), representing enough lumber to build 270,000 three-bedroom homes (S.R. Whitney, 1985, personal communication), were killed in infestations that covered nearly half a million hectares of forests (Wood *et al.*, 1983). The current mountain pine beetle outbreak in British Columbia, which started around 1992, now (2003) covers an estimated 2.0 million ha. Infestations by *D. rufipennis and D. pseudotsugae* occurred over an additional 1.2 million ha (British Columbia Ministry of Forests, 2003). There is no indication of a decline in the mountain pine beetle outbreak. At present, bark-beetle outbreaks of this magnitude can only be terminated by unseasonably cold temperatures (–35°C or lower for several days) (Somme, 1964) with little or no snow cover around the infested bole.

The earliest control operations against bark beetles occurred in the late 1910s and were directed at *D. rufipennis* and *D. ponderosae* outbreaks in eastern and western Canada, respectively. These early control attempts involved harvesting the infested stands or performing individual tree treatments such as fell-and-burn, or peeling and burning the infested bark. By the late 1940s, the chemical insecticides ethylene dibromide (EDB), a fumigant, and benzene hexachloride (BHC (lindane)), were the most commonly used chemicals. BHC was formulated in fuel oil, usually as a 2% solution, and used as a bark-penetrating insecticide to kill broods under the bark or as a water emulsion for protecting trees from beetle attack. Trap trees, felled or standing, either treated or not treated with chemical insecticides, were commonly used after the 1940s, and aerial-detection surveys were used for delineating infestations. By the late 1970s, the use of BHC and EDB was phased out, owing to environmental concerns, and were replaced by a systemic pesticide, monosodium methane arsenate (MSMA).

Management methods

Management options are available for two of the most important beetles: the mountain pine beetle and the spruce beetle (McMullen *et al.*, 1986; Safranyik *et al.*, 1990; Maclauchlan and Brooks, 1994). Bark-beetle management is based on two approaches: direct control and prevention (McMullen *et al.*, 1986).

Direct control is currently used operationally, and is most effective when a beetle infestation is small or just beginning to develop. At this time the direct-control measures are implemented to lower beetle populations to endemic levels or until one of the other management options can be implemented. There are two main operational direct-control measures: sanitary logging of infested stands and treatment of individual trees, mainly using lethal and conventional trap trees or fell-and-burn, or using sevin (carbaryl) as a bark-penetrating chemical.

Silvicultural methods, such as sanitation–salvage logging or felling and burning of the infested trees (McMullen *et al.*, 1986), will reduce beetle populations if applied in a timely fashion, that is, before the new generation of adult beetles leaves the infested trees. Individual trees are 'treated' by the cut–pile–burn method or by injecting the infested trees with the herbicide MSMA to kill the new brood of larvae (McMullen *et al.*, 1986). MSMA has to be applied within a few weeks of the attack while the trees are still alive and can translocate the poison. Sanitation logging is generally used to treat larger, more diffuse infestations, and involves removing (logging) the attacked trees containing the beetle brood. In high-value or high-impact areas, such as parks and campsities, the lower boles of trees can be treated with 2% active ingredient (AI) carbaryl in water to protect trees from lethal attack (McMullen *et al.*, 1986).

Preventive management is the preferred option, and entails long-term planning and the use of forestry practices that reduce stand susceptibility. For *D. ponderosae*, preventive management is based on reducing the susceptibility of lodgepole pine stands (Shore and Safranyik, 1992; Shore *et al.*, 2000) and involves forestry practices such as spacing, age stratifying, stocking and stand conversion and species control, which may be combined with sanitary harvesting/logging. One or a combination of the following management options that prevent or delay outbreak development can reduce stand susceptibility, i.e. shorter rotation age, conversion of forest type to a less susceptible one, creation of mixed-age stands, creation of mixed-host-species stands and changing stand structure and stocking (Safranyik *et al.*, 1974). All of these options are used in British Columbia. The spacing and thinning of stands and partial cutting have been effective in reducing losses caused by mountain pine beetle in mature lodgepole pine (Mitchell *et al.*, 1983; Cole and McGregor, 1985) and second-growth ponderosa pine (Sartwell and Dolph, 1976) in the USA.

Semiochemicals – chemicals eliciting interactions between organisms – can be used to manipulate mountain pine beetle (Borden, 1989). During the past 25 years, the use of population-aggregating pheromones became standard practice to contain and concentrate infestations in order to increase the efficacy of the control operation. The aggregation pheromone can be used in surveys to monitor beetle populations or to concentrate flying beetles in trees where they can be treated with a systemic herbicide, such as MSMA. Baiting of individual trees and small isolated stands with bark-beetle aggregation pheromone concentrates beetle attacks, decreasing the spread of the infestation and increasing the cost-effectiveness of direct-control operations through sanitation logging (Borden *et al.*, 1983). Treating individual or small groups of trees with anti-aggrega-

Table 10.5. Major forestry weed species competing with the production of commercial forest crop trees in Canada (from Wall *et al.*, 1992).

Species	Common name	Region
Trees		
Acer macrophyllum Pursh	Big-leaf maple	Coastal British Columbia
Acer rubrum L.	Red maple	Ontario, Quebec, Maritimes
Acer spicatum Lambert	Mountain maple	Ontario, Quebec, Manitoba, Maritimes
Alnus incana (L.) Moench	Speckled alder	Transcontinental
Alnus rubra Bongard	Red alder	Coastal British Columbia
Corylus cornuta Marshall	Beaked hazelnut	Transcontinental
Populus tremuloides Michaux	Trembling aspen	Transcontinental
Salix spp.	Willows	Transcontinental
Shrubs		
Gaultheria shallon Pursh	Salal	Coastal British Columbia
Ribes spp.	Currants and gooseberries	British Columbia, Ontario, Quebec, Maritimes
Rubus idaeus L.	Red raspberry	Ontario, Quebec, Maritimes
Rubus parviflorus Nuttall	Thimbleberry	British Columbia
Rubus spectabilis Pursh	Salmonberry	Coastal British Columbia
Sambucus spp.	Elderberries	Transcontinental
Herbaceous plants		
Calamagrostis canadensis (Michx.) Palisot de Beauvois	Blue-joint grass	Alberta, interior British Columbia
Epilobium angustifolium (L.)	Fireweed	Transcontinental
Kalmia angustifolia L.	Sheep laurel	Newfoundland, Maritimes
Pteridium aquilinum (L.) Kuhn	Bracken fern	Transcontinental

tion pheromone is thought to repel beetles from the trees in the treated stand. Variation in the use of these aggregation and anti-aggregation pheromones, as proposed and tested by Borden (1995), has met with varying degrees of success.

Population-aggregating pheromones are used extensively to increase the efficacy of control programmes. Direct-control operations are guided by pretreatment population and damage assessments and an appropriate strategy and set of tactics that are based on the field assessment combined with consideration of environmental and socio-economic concerns. The management operations are guided by decision support systems that determine susceptibility and risk (Shore and Safranyik, 1992; Shore *et al.*, 2000) and feasible strategies and tactics to be used in particular situations (Maclauchlan and Brooks, 1994).

Vegetation Management in Forestry

Unwanted (weedy) vegetation is a major source of competition for water, nutrients, light and space in nurseries, plantations and areas of natural conifer regeneration (Watson and Wall, 1995). Some weeds, such as *Ribes* spp. and fireweed, *Epilobium angustifolium* (Linnaeus), are alternative hosts of important forest-tree rust diseases (Ziller, 1974; Hansen and Lewis, 1997). Competition from weedy vegetation can delay seedling establishment, reduce growth, decrease timber yield and delay harvest (increase rotation age) of conifer crop trees. The economic losses caused by weeds in forestry were not fully appreciated until recently (Walstad and Kuch, 1987).

Most forest weeds are native species. Among the major competitors of crop-tree species, nine are fast-growing hardwood

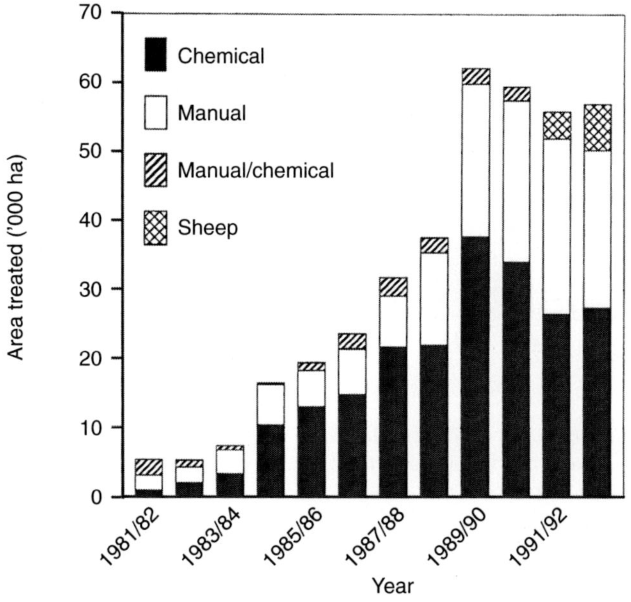

Fig. 10.6. Methods and trends of vegetation management in forest areas on crown-land in British Columbia, 1981–1993 (after Boateng, 1996).

trees, six are shrubs and four are herbaceous plants (Wall *et al.*, 1992; Table 10.5). In most cases, the unwanted vegetation can only be managed, not eliminated. In forestry, as in agriculture, competing vegetation has to be managed to favour the growth of the crop species. A number of methods have been, and are still, used for vegetation management, as illustrated in Fig. 10.6. In British Columbia, of the vegetation management treatments conducted during a 10-year period (1981/82–1992/93), 57% were done by using herbicides alone, 36% by using manual cutting, 4% by using a combination of herbicides and cutting and 3% by sheep grazing (Fig. 10.6). Manual removal of the competing vegetation is expensive and inefficient (Pendl and D'Anjou, 1990). Mechanical methods, using various kinds of machinery for site preparation after logging, are also expensive and, in addition, favour resprouting of some weedy species and tend to compact the soil (Prasad, 1996).

Prescribed fire is sometimes used in an attempt to eliminate weed seeds, unwanted seedlings and stumps. It also retards the regrowth of vegetation that would compete with the establishment and growth of crop-tree species (Feller, 1996). Fire does not eliminate the roots of perennial weeds and in fact stimulates sprouting of some perennial weeds and tree species (Prasad, 1996).

Grazing by animals, mainly sheep, has been tried experimentally on a relatively small scale in Australia, Canada, Ireland, New Zealand, Sweden and the USA (Sharrow *et al.*, 1989; Cayford, 1993). Sheep grazing is now used operationally in forestry to control weeds, but only in small plantations. Mulching with allelopathic plant material (i.e. chemicals that leach out of the mulch and suppress weed growth) has been tried experimentally (McDonald and Fiddler, 1996). Mulching, as well as brush and plastic blankets, works well in nurseries and in small intensively managed plantations, but operational use of these methods is not yet practical in forestry (Jobidon *et al.*, 1989; Jobidon, 1991a).

Some microorganisms and plants produce natural herbicides that are toxic to some weed species (Duke, 1986; Duke and Lydon, 1987). For example, bialophos (aminohydroxy-phospho-vinyl-butyryl-alanine), originally

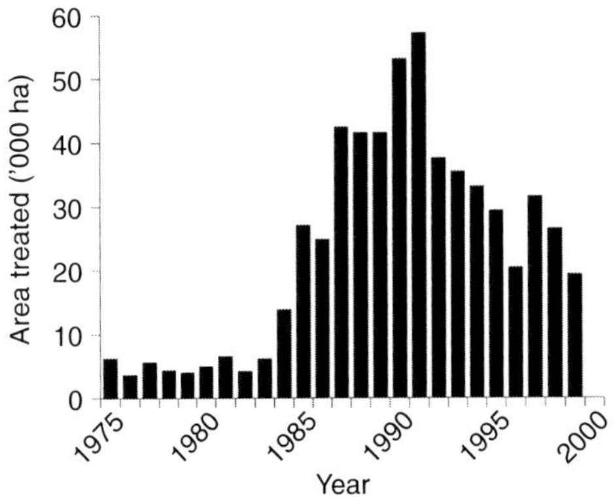

Fig. 10.7. Forest area treated with herbicides[a] for weed management in British Columbia by the Ministry of Forests and Forest Companies between 1975 and 2000 (from Humphreys, 1986, 1990; Canadian Council of Forest Ministers, 2003).
[a] Herbicides applied in large-scale operations include 2,4-D, 2,4-E, fosamine, glyphosate, hexazinone, MSMA, picloram and triclopyr. Glyphosate has been used preferentially since 1985.

isolated from the soil microbe *Streptomyces viridochromogens* (Jobidon, 1991b), controls red raspberry (*Rubus idaeus* L.) in forest plantations. However, it is expensive, is toxic to some conifers and has only limited use in forestry (Jobidon, 1991b; Prasad, 1992).

Competition by herbaceous and woody vegetation frequently requires treatment with commercial herbicides. Although the use of herbicides has been part of forest management since the mid-1940s, in British Columbia, Canada, the area treated annually was small (about 5000 ha/year) until the early 1980s (DeBoo and Prasad, 1997). Chemical herbicides were the choice of treatment (mainly 2,4-dichlorophenoxyacetic acid (2, 4-D) and 2, 4, 5-trichlorophenoxyethanoic acid (2, 4, 5-T)) and have been used for vegetation management in British Columbia since the mid-1970s (Fig. 10.7) and are the most cost-effective, especially if they can be applied from the air (Thorpe, 1996). After 1983, increased funding was available for forest management, resulting in generally increased pesticide use until 1992 (Fig. 10.7). After 1985, 2,4-D and 2,4,5-T were replaced by glyphosate because of health and environmental concerns (DeBoo and Prasad, 1997) and for political reasons (Halleran, 1990).

Restrictions on pesticide use for weed control in forestry will probably increase. Therefore, it is necessary to develop one or more cost-effective and environmentally acceptable methods of weed control (i.e. alternatives to chemical herbicides) in forestry. Biological control (especially in the case of introduced organisms) and the so-called bioherbicides may offer such an option. Bioherbicides are 'living entities' used to deliberately suppress the growth of weeds or reduce weed populations to economically acceptable levels in areas where they compete with the desired crop grown (Watson, 1989; Auld and Morin, 1995). These controls may include insects and pathogens (bacteria, fungi and viruses). The terms biological herbicides, bioherbicides and mycoherbicides are used interchangeably in the scientific literature to describe all pathogens (Winder and Shamoun, 1991). The term mycoherbicides, however, is more appropriate when referring to herbicides based on fungal pathogens. For the purposes of this chapter, the term originally used in the paper is retained when reference is made to a particular paper.

Biological control strategies in weed management in forestry include both classical (inoculative) and inundative (augmentative) inoculation with plant pathogens (Watson and Wall, 1995; Wall and Hasan, 1996). The inoculative (classical) approach is generally employed when the noxious weed was accidentally or intentionally introduced, usually from one continent to another. The natural enemy (usually a pathogen) or most promising enemies are introduced in the hope that they will spread throughout the 'pest' host population, resulting in control. The inundative approach is mainly employed against endemic weedy populations. The weeds are treated with a large amount of indigenous natural enemies, usually fungal pathogens of the weeds, which augment the local, naturally occurring pathogens. The remainder of this section will review the major accomplishments using these approaches for vegetation management in forestry.

The biological control of weeds in agriculture has a long history, with many notable successes (McLeod, 1962; Harris, 1984; Kelleher and Hulme, 1984), but it is relatively new in forestry (Fig. 10.1). Similar successes are also reported from the USA. The study of biological controls for vegetation and plant-pathogen management started only recently, in the last 20 years. In Canada between 1981 and 2000, there were 20 studies on the biocontrol of weeds of agricultural crops and only five on forest weeds and two on forest-tree diseases (Mason and Huber, 2002).

Inoculative strategy (classical biological control)

The classical biological control strategy (introducing insect enemies of exotic weeds) has been used successfully in agriculture (Harris, 1984; Julien and Griffiths, 1998; Harris and Shamoun, 2002).

The European blackberry (*Rubus fruticosus* L. agg.) has been introduced to many parts of the world and is regarded as an important weed pest in several countries, including Australia, where it invades pastures, forest and national parks (Amor and Richardson, 1980; Wall and Hasan, 1996). The rust fungus *Phragmidium violaceum* (Shultz) was introduced into Australia, where it controlled the European blackberry under the shady canopy of pine trees, thereby freeing pine seedlings from blackberry competition and allowing them to grow normally after the release. However, the rust was less effective on blackberries growing under the more open eucalyptus stands. This was not unexpected because the microclimate is probably drier in the more open stand of eucalyptus.

A gall-forming rust fungus, *Uromycladium tepperianum* (Sacc.) McAlp., was introduced into South Africa against the weed tree, *Acacia saligna* (Labill) Wendl. Both the fungus and the host tree originated in Australia. The fungus became established in South Africa and infected and formed galls on the flowers of *A. saligna*, killing some of the trees (Morris, 1991, 1997). In South Africa, the white-rot fungus *Cylindrobasidium laeve* (Pers.: Fr.) Chamois has been registered as Stumpout™ for the control of introduced wattle trees, *Acacia* spp. (Morris *et al.*, 1998).

However, some introductions can lead to controversy. When the gorse mite, *Tetranychus lintearius* Dufour, was imported to control the introduced gorse, *Ulex europaeus* L., in New Zealand, it became controversial because the introduction gave only variable control and the exotic mites interbred with one of the native mite species (Syrett *et al.*, 1985).

Bioherbicides

Despite three decades of research, the development of effective bioherbicides for weed control in forestry is still in its infancy. Many preliminary results indicate the great potential value derived from the use of biological control in vegetation management in forestry (Shamoun, 2000; Shamoun and DeWald, 2002; Shamoun *et al.*, 2002, and references therein). In fact, to date, only two mycoherbicides are commercially available for controlling weeds along hydroelectric power lines and for forestry use (Morris *et al.*, 1998; Shamoun and Hintz, 1998a,b). Of the bioherbicides tested, the manipulation of indigenous fungi probably offers the best chance for the development of mycoherbicides because they are native and generally persist at endemic levels in the environment. Research on plant pathogens as potential biocontrol agents for weed control started in the 1970s to suppress ericaceous shrubs on cut-over sites in eastern Canada (Wall, 1977) and later to control *Rubus* spp. (Wall, 1983). Of the biocontrol agents examined, *Chondrostereum purpureum* Fr./Pouzar was the most promising bioherbicide (Wall, 1990).

In western Canada, research has focused on weedy trees (*Acer macrophyllum* Pursh, *Alnus rubra* Bong., *Populus tremuloides* Michx.), and shrubs (*Rubus parviflorus* Nutt., *Rubus spectabilis* Pursh., *Gaultheria shallon* Pursh.). Pathogens from all of these target species have been identified and many of them have been tested as potential biocontrol agents (Dorworth, 1990, 1992; Sieber *et al.*, 1990a,b; Wall and Shamoun, 1990b).

When applied to cut stumps, *C. purpureum* provided spectacular results and gave 90% control of black cherry (*Prunus serotina* Ehrhert) regrowth in conifer plantations in Holland. This level of control matched the effectiveness of the chemical herbicide glyphosate (Scheepens and Hoogerbrugge, 1989). However, in Canada, the same treatment gave somewhat less spectacular results against *Populus* spp., *Acer* spp., *Betula* spp., *Prunus pensylvania* and *Alnus rubra* (Wall and Shamoun, 1990a; Wall, 1994). Considerable research has also been done on *C. purpureum* in Canada. Myco-Forestis Corporation (Quebec) has registered one strain (HQ-1) under the name Myco-Tech™ Paste for use east of the Rocky Mountains in Canada. Another strain (PFC 2139) is currently being registered for use in western Canada and the entire USA (Shamoun and Hintz, 1998a,b). After considerable testing, *C. purpureum* has been adopted, with a refined formulation, and patented for control of hardwoods in North American forests (Wall *et al.*, 1996). However, *C. purpureum* has not been registered as a mycoherbicide in Holland but is used only as a wood-decay promoter (Ravensberg, 1998). Presumably, it was not registered there as a herbicide because of the cost of registration and the expected return on this investment. The same economic considerations may hinder the commercial production and sale of mycoherbicide in forestry worldwide.

Some bioherbicides have been tested against several hardwood trees (Shamoun *et al.*, 2002), marsh reed grass (*Calamagrostis canadensis* (Michx.) Beauv) (Mallet *et al.*, 2002), *Rubus* spp. (Oleskevich *et al.*, 1998), fireweed (*Epilobium* spp.) (Winder, 2002), Scotch broom, *Cytisus scoparius* (Linnaeus) Link (Prasad, 2002a) and gorse (*Ulex europaeus* L.) (Prasad, 2002b). However, there are several obstacles to the development of commercial products, such as application technology, formulation to increase efficacy of the product so that it can be stored and readily applied in the field, optimum volume and droplet size to give the optimum efficiency, which are still unknown, and the scarcity of examples proving the commercial viability of this approach. The main reason for this is due to the fact that *C. purpureum* can only enter through an open wound of the host tree and at present this can only be achieved through manual inoculation of the wound (cut surface) of the target plant. In forestry, on a large scale, this is probably not feasible, but it may be practical to use under hydroelectric powerlines.

Silvicultural Management of Root Diseases

Root diseases caused by *Armillaria ostoyae* (Romagn.) Herink, *Phellinus weirii* (Murr.) Gilb. and *Inonotus tomentosus* (Fr.: Fr.) Teng. are among the most destructive pathogens of trees, mainly conifers grown for timber, in western North America. They cause significant growth loss and tree mortality in managed forests (Bloomberg and Reynolds, 1985; McDonald *et al.*, 1987; Morrison *et al.*, 1988, 1992; Bloomberg and Morrison, 1989; Ives and Rentz, 1993; Woods, 1994). Of these, *Armillaria* root disease caused by *A. ostoyae*, is the most important. It occurs in all forest regions of Canada. In the interior of British Columbia it is estimated to cause losses of 2–3 million m^3 annually (A. Van Sickle, 1999, personal communication). Even in undisturbed stands (Morrison and Mallet, 1996), up to 80% of the trees in a stand can be infected. All three root diseases reduce merchantable volume at the end of the planned rotation, change species composition over the rotation by killing the most susceptible hosts and may lengthen the rotation age (Morrison and Mallett, 1996; Sturrock, 2000).

The same management strategies can be applied to reduce the impact for all three root diseases. It is best to apply treatments at the time of harvest and at stand regeneration. The two most effective ways of reducing the root diseases and their impacts are the following:

1. Machine-assisted removal of inoculum source (infected stumps and large roots) at the time of harvesting or by push–pull logging of trees in root-diseased stands

(Morrison *et al.*, 1992; Sturrock *et al.*, 1994; Morrison and Mallett, 1996; Sturrock, 2000).
2. Planting resistant trees. Birch, aspen and poplar are tolerant of *A. ostoyae* (Morrison and Mallett, 1996). All hardwoods are immune to *P. weirii* (Morrison *et al.*, 1988, 1992) and birch is immune to *I. tomentosus* (Sturrock, 2000).

Both mechanical methods to remove sources of inoculum are called stumping (Morrison and Mallett, 1996). Stumping exposes infected roots to the sun, thus drying the roots and killing the fungi. Removal of the source of the inoculum from the soil reduces the possibility of infection in the next crop of trees. Economic analysis conducted for *P. weirii* in western Washington (Russell *et al.*, 1986) and for *A. ostoyae* in New Zealand (Shaw and Calderon, 1977) indicates that the high disease incidence and losses that occur on non-treated sites can be reduced by removing the inoculum. The associated costs are justified by increased rates of return in the next rotation: that is, the volume grown on the treated sites will more than pay for the inoculum removal.

Notes

[1] The remainder was treated with the chemical insecticides fenitrothion and aminocarb (Cunningham and van Frankenhuyzen, 1991).
[2] The virus has to be produced in living insects, therefore it is expensive to make. The cost of producing enough virus to treat 1 ha at the recommended label dose of 2.5×10^{11} PIB/ha is about CAN$40–50/ha.
[3] The cost of disparlure treatment at this dose was estimated at US$64/ha, while double application of *Btk* cost *c.* US$64–69/ha (Sharov *et al.*, 2002b).
[4] Dimilin® is a chitin synthesis inhibitor, and only affects arthropods. However, it can kill or negatively affect many non-target arthropods (Butler *et al.*, 1994).
[5] The survey was conducted in New Hampshire and included both households that had and had not experienced gypsy-moth damage to their properties.

References

Alfaro, R.I. and Maclauchlan, L.E. (1992) A method to calculate the losses caused by western spruce budworm in uneven-aged Douglas-fir forests of British Columbia. *Forest Ecology and Management* 55, 295–313.

Amor, R.L. and Richardson, R.G. (1980) The biology of Australian weeds 2. *Rubus fruticosus* L. agg. Includes distribution and ecological aspects. *Journal of the Australian Institute of Agricultural Science* 46, 87–97.

Andreadis, T.G. and Weseloh, R.M. (1990) Discovery of *Entomophaga maimaiga* in North American gypsy moth, *Lymantria dispar. Proceedings of the National Academy of Sciences of the USA* 87, 2461–2465.

Armstrong, J.A. and Cook, C.A. (1993) *Aerial Spray Applications on Canadian Forests: 1945–1990.* Information Report ST-X-2, Forestry Canada, Ottawa, Ontario, 266 pp.

Auld, B. and Morin, L. (1995) Constraints in the development of bioherbicides. *Weed Technology* 9, 638–652.

Balch, R.E. and Bird, F.T. (1944) A disease of the European spruce sawfly, *Gilpinia hercyniae* (Htg.), and its place in natural control. *Scientific Agriculture* 25, 65–80.

Baranchikov, Y.N. (1989) Ecological basis of the evolution of host relationships in Eurasian gypsy moth populations. In: Wallner, W.E. and McManus, K.A. (technical coordinators) *Proceedings, Lymantriidae: a Comparison of New and Old World Tussock Moths, June 26-July 1, 1988, New Haven, Connecticut.* General Technical Report NE-123, USDA Forest Service, Northeastern Forest Experiment Station, Broomall, Pennsylvania, pp. 319–338.

Beroza, M., Hood, C.S., Trefey, D., Leonard, D.E., Knipling, E.F. and Klassen, W. (1975) Field trials with disparlure in Massachusetts to suppress mating of the gypsy moth. *Environmental Entomology* 4, 705–711.

Bird, F.T. (1959) Polyhedrosis and granulosis virus causing single and double infections in the spruce budworm. *Journal of Insect Pathology* 1, 406–430.

Bird, F.T. (1961) Transmission of some insect viruses with particular reference to ovarial transmission and its importance in the development of epizootics. *Journal of Insect Pathology* 3, 352–380.

Blais, J.R. (1985) The ecology of the eastern spruce budworm: a review and discussion. In: Sanders, C.J., Stark, R.W., Mullins, E.J. and Murphy, J. (eds) *Recent Advances in Spruce Budworms Research: Proceedings of CANUSA Spruce Budworms Research Symposium, Bangor, Maine, September 16–20, 1984.* Canadian Forestry Service, Ottawa, Ontario, pp. 49–59.

Bloomberg, W.J. and Morrison, D.J. (1989) Relationship of growth reduction in Douglas-fir to infection by *Armillaria* root disease in southeastern British Columbia. *Phytopathology* 79, 482–487.

Bloomberg, W.J. and Reynolds, G. (1985) Growth loss and mortality in laminated root rot infection centres in second-growth Douglas-fir on Vancouver Island. *Forest Science* 31, 497–508.

Blumenthal, E.M., Fusco, R.A. and Reardon, R.S. (1978) Augmentative release of two established parasite species to suppress populations of the gypsy moth. *Journal of Economic Entomology* 72, 281–288.

Boateng, J. (1996) Past and future trends in forest vegetation management in BC. In: Coroneau, P.G., Harper, G.J., Blache, M.E., Boateng, J.O. and Gilkeson, L.A. (eds) *Integrated Forest Vegetation Management: Options and Applications. Proceedings of the Fifth British Columbia Forest Vegetation Management Workshop, November 29–30, 1993 Richmond, British Columbia.* Forest Resource Development Agreement Report No. 251, Canadian Forest Service and British Columbia Ministry of Forests, Victoria, British Columbia, pp. 1–4.

Borden, J.H. (1989) Semiochemicals and bark beetle populations: exploitation of natural phenomena by pest management strategists. *Holarctic Ecology* 12, 501–510.

Borden, J.H. (1990) Semiochemicals to manage coniferous tree pests in western Canada. In: Ridgway, R.L., Silverstein, R.M. and Inscoe, M.N. (eds) *Behavior-modifying Chemicals for Insect Management: Applications of Pheromones and Other Attractants.* Marcel Dekker, New York, pp. 281–315.

Borden, J.H. (1995) Development and use of semiochemicals against bark and timber beetles. In: Armstrong, J.A. and Ives, W.G.H. (eds) *Forest Insect Pests in Canada.* Natural Resources Canada, Canadian Forest Service, Ottawa, Ontario, pp. 431–449.

Borden, J.H., Chong, L.J., Pratt, K.E.G. and Gray, D.R. (1983) The application of behaviour-modifying chemicals to contain infestations of the mountain pine beetle, *Dendroctonus ponderosae* Hopk. *The Forestry Chronicle* 59, 235–239.

Boulton, T.J. (1999) Impacts of *Bacillus thuringiensis* subsp. *kurstaki* (*Btk*) on the population dynamics of non-target Lepidoptera on *Ribes cereum.* MSc thesis, University of Victoria, Victoria, British Columbia, 287 pp.

Boulton, T.J., Otvos, I.S. and Ring, R.A. (2002) Monitoring nontarget Lepidoptera on *Ribes cereum* to investigate side effects of an operational application of *Bacillus thuringiensis* subsp. *kurstaki. Environmental Entomology* 31, 903–913.

Bright, D.E. Jr (1976) *The Insects and Arachnids of Canada.* Part 2: *The Bark Beetles of Canada and Alaska. Coleoptera: Scolytidae.* Publication 1576, Biosystematics Research Institute, Research Branch, Canadian Department of Agriculture, Ottawa, Ontario, 241 pp.

British Columbia Ministry of Forests (2003) Forest Health Conditions 2002, Aerial Overview Survey Summary. Available at: www.for.gov.bc.ca/hfp/FORSITE/overview/2002table.htm

Brown, G.S. (1975) Gypsy moth *Porthetria dispar* (L.). In: Prebble, M.L. (ed.) *Aerial Control of Forest Insects in Canada.* Environment Canada, Ottawa, Ontario, pp. 208–212.

Brown, W.L. Jr (1961) Mass insect control programs: four case histories. *Psyche* 68, 75–111.

Buckner, C.H., Kingsbury, P.D., McLeod, B.B., Mortensen, K.L. and Ray, D.G.H. (1974) Impact of aerial treatment on non-target organisms, Algonquin Park, Ontario, and Spruce Woods, Manitoba. In: Chemical Control Research Institute (ed.) Bacillus thuringiensis – *Evaluation of Commercial Preparation of* Bacillus thuringiensis *With and Without Chitinase Against Spruce Budworm.* Information Report CC-X-59, Chemical Control Research Institute, Ottawa, Ontario, pp. F1-F72.

Butler, L., Chrislip, G.A., Kondo, V.A. and Townsend, E.C. (1974) Effect of diflubenzuron on nontarget canopy arthropods in closed, deciduous watersheds in Central Appalachian forest *Journal of Economic Entomology* 90, 784-779.

Cadogan, B.L. (1995) Jack pine budworm, *Choristoneura pinus.* In: Armstrong, J.A. and Ives, W.G.H. (eds) *Forest Insect Pests in Canada.* Natural Resources Canada, Canadian Forest Service, Ottawa, Ontario, pp. 123–126.

Cameron, E.A. (1974) Programs utilizing pheromones in survey or control: the gypsy moth. In: Birch, M.C. (ed.) *Pheromones.* American Elsevier Publication Company, New York, pp. 431–435.

Cameron, J.M. (1975a) Biological insecticides. In: Prebble, M.L. (ed.) *Aerial Control of Forest Insects in Canada.* Department of Environment, Ottawa, Canada, pp. 25–33.

Cameron, J.M. (1975b) Field trials of spruce budworm viruses, 1971–1973. In: Prebble, M.L. (ed.) *Aerial Control of Forest Insects in Canada*. Department of Environment, Ottawa, Canada, pp. 134–137.

Campbell, R.W. (1963) The role of disease and desiccation in the population dynamics of the gypsy moth *Porthetria dispar* (L.) (Lepidoptera: Lymantriidae). *The Canadian Entomologist* 95, 426–434.

Campbell, R.W. and Podgwaite, J.D. (1971) The disease complex of the gypsy moth. I. Major components. *Journal of Invertebrate Pathology* 18, 101–107.

Canadian Council of Forest Ministers (2003) National Forestry Database Programme. Available at: http://nfdp.ccfm.org

Canadian Department of Agriculture and Canadian Department of Fisheries and Forestry (1971) *Biological Control Programmes Against Insects and Weeds in Canada 1959–1968*. Technical Communication No. 4, Commonwealth Institute of Biological Control, Commonwealth Agricultural Bureaux, Slough, UK, 266 pp.

Canadian Forest Service (1975) *Report of the Annual Forest Pest Control Forum*. Environment Canada, Canadian Forest Service, Ottawa, Ontario, 225 pp.

Canadian Forest Service (1976) *Report of the Annual Forest Pest Control Forum*. Environment Canada, Canadian Forest Service, Ottawa, Ontario, 341 pp.

Canadian Forest Service (1977) *Report of the Annual Forest Pest Control Forum*. Environment Canada, Canadian Forest Service, Ottawa, Ontario, 337 pp.

Canadian Forest Service (1978) *Report of the Annual Forest Pest Control Forum*. Environment Canada, Canadian Forest Service, Ottawa, Ontario, 443 pp.

Cantwell, G.E., Dutky, S.R., Keller, J.C. and Thompson, C.G. (1961) Results of tests with *Bacillus thuringiensis* Berliner against gypsy moth larvae. *Journal of Invertebrate Pathology* 8, 228–233.

Cardinal, J.A. (1967) Lutte contre la spongieuse *Porthetria dispar* L. (Lepidoptera: Lymantriidae) au Québec. *Phytoprotection* 48, 92–100.

Carson, R.L. (1962) *Silent Spring*. Houghton Mifflin, Boston, Massachusetts, 368 pp.

Casida, S.E. and Quistad, G.B. (1998) Golden age of insecticide research: past, present or future? *Annual Review of Entomology* 43, 1–16.

Cayford, J.H. (1993) Sheep for vegetation management. *The Forestry Chronicle* 69, 27.

Cole, W.E. and McGregor, M.D. (1985) Reducing and preventing mountain pine beetle outbreaks in lodgepole pine stands by selective cutting. In: Safranyik, L. (ed.) *Role of the Host in the Population Dynamics of Forest Insects: Proceedings of IUFRO Conference, Banff, Alberta, September 4–7, 1983*. Canadian Forestry Service, Pacific Forestry Centre, Victoria, British Columbia, 240 pp.

Council of Forest Industries (2000) British Columbia Forest Industry Fact Book, Vancouver, British Columbia, 82 pp. Available at: www.cofi.org/reports/factbooks.htm

Cunningham, J.C. (1985a) Status of viruses as biocontrol agents for spruce budworms. In: Grimble, D.G. and Lewis, F.B. (eds) *Microbial Control of Spruce Budworms and Gypsy Moth*. General Technical Report, GTR-NE-100, Northeastern Forestry Experimental Station, US Department of Agriculture Forest Service, Broomall, Pennsylvania, pp. 61–67.

Cunningham, J.C. (1985b) Biorationals for control of spruce budworms. In: Sanders, C.J., Stark, R.W., Mullins, E.J. and Murphy, J. (eds) *Recent Advances in Spruce Budworm Research: Proceedings of CANUSA Spruce Budworm Research Symposium, September 16–20, 1984, Bangor, Maine*. Canadian Forestry Service, Ottawa, Ontario, pp. 320–349.

Cunningham, J.C. (1998) North America. In: Hunter-Fujita, F.R., Entwistle, P.F., Evans, H.F. and Cook, N.E. (eds) *Insect Viruses and Pest Management*. John Wiley and Sons, Chichester, UK, pp. 313–331.

Cunningham, J.C. and de Groot, P. (1984) *Neodiprion lecontei* (Fitch), redheaded pine sawfly (Hymenoptera: Diprionidae). In: Kelleher, J.S. and Hulme, M.A. (eds) *Biological Control Programmes Against Weeds and Insects in Canada 1969–1980*. Commonwealth Agricultural Bureaux, Slough, UK, pp. 323–329.

Cunningham, J.C. and Howse, G.M. (1984) Viruses: application and assessment. In: Kelleher, J.S. and Hulme, M.A. (eds) *Biological Control Programmes Against Weeds and Insects in Canada 1969–1980*. Commonwealth Agricultural Bureaux, Slough, UK, pp. 248–259.

Cunningham, J.C. and Kaupp, W.J. (1995) Insect viruses. In: Armstrong, J.A. and Ives, W.G.H. (eds) *Forest Insect Pests in Canada*. Natural Resources Canada, Canadian Forest Service, Ottawa, Ontario, pp. 327–340.

Cunningham, J.C. and McPhee, J.R. (1986) *Production of Sawfly Viruses in Plantations*. Technical Note No. 4, Canadian Forestry Service, Forest Pest Management Institute, Sault Ste Marie, Ontario, 4 pp.

Cunningham, J.C. and Shepherd, R.F. (1984) *Orgyia pseudotsugata* (McDunnough), Douglas-fir tussock moth (Lepidoptera: Lymantriidae). In: Kelleher, J.S. and Hulme, M.A. (eds) *Biological Control Programmes against Insects and Weeds in Canada, 1969–1980*. Commonwealth Agricultural Bureaux, Slough, UK, pp. 363–367.

Cunningham, J.C. and van Frankenhuyzen, K. (1991) Microbial insecticides in forestry. *The Forestry Chronicle* 67, 473–480.

Cunningham, J.C., de Groot, P. and Kaupp, W.J. (1986) A review of aerial spray trials with Lecontvirus for control of red-headed pine sawfly, *Neodiprion lecontei* (Hymenoptera: Diprionidae), in Ontario. *Proceedings of the Entomological Society of Ontario* 117, 65–72.

Dahlsten, D.L., Garcia, R. and Lorraine, H. (1989) Eradication as a pest management tool: concepts and contexts. In: Dahlsten, D.L. and Garcia, R. (eds) *Eradication of Exotic Pests: Analysis with Case Histories*. Yale University Press, New Haven, Connecticut, 296 pp.

D'Amico, V. and Elkinton, J.S. (1995) Rainfall effects on transmission of gypsy moth (Lepidoptera: Lymantriidae) nuclear polyhedrodis virus. *Environmental Entomology* 24, 1144–1149.

DeBoo, R.F. and Prasad, R. (1997) Pesticide use in forest management: the British Columbia experience since 1946. *Pesticide Outlook* 8, 9–14.

Doane, C.C. (1970) Primary pathogens and their role in the development of an epizootic in the gypsy moth. *Journal of Invertebrate Pathology* 15, 21–33.

Doane, C.C. (1976) Ecology of pathogens of the gypsy moth. In: Anderson, J.F. and Kaya, H.K. (eds) *Perspectives in Forest Entomology*. Academic Press, New York, pp. 285–293.

Doane, C.C. and McManus, M.L. (eds) (1981) *The Gypsy Moth: Research Toward Integrated Pest Management*. Technical Bulletin 1584, Forest Service, US Department of Agriculture, Washington, DC, 757 pp.

Dorworth, C. (1990) Use of indigenous microorganisms for forest weed biocontrol – the PFC enhancement process. In: Basset, C., Whitehouse, L.J. and Zabkiewicz, J.A. (eds) *Alternatives to the Chemical Control of Weeds: Proceedings of an International Conference held at the Forest Research Institute, Rotorua, New Zealand, July 25–27, 1989*. Forest Research Institute Bulletin 155, Forest Research Institute, Ministry of Forestry, Rotorua, New Zealand, pp. 116–119.

Dorworth, C.E. (1992) Biological control of weeds using indigenous pathogens. In: Shrimpton, G. (ed.) *Canadian Forest Nursery Weed Management Association: Proceedings of Annual Meeting. Victoria, British Columbia, July 6–8, 1992*. BC Ministry of Forests, Victoria, British Columbia, pp. 17–24.

Dowden, P.B. (1962) *Parasites and Predators of Forest Insects Liberated in the United States through 1960*. Agricultural Handbook 226, Forest Service, US Department of Agriculture, Washington, DC, 70 pp.

Duan, L. and Otvos, I.S. (2001) Influence of larval age and virus concentration on mortality and sublethal effects of a nucleopolyhedrovirus on the western spruce budworm (Lepidoptera: Tortricidae). *Environmental Entomology* 30, 136–146.

Dubois, N.R. (1981) *Bacillus thuringiensis*. In: Doane, C.C. and McManus, M.L. (eds) *The Gypsy Moth: Research Toward Integrated Pest Management*. Technical Bulletin 1584, Forest Service, US Department of Agriculture, Washington, DC, pp. 445–453.

Duke, S.O. (1986) Naturally occurring chemical compounds as herbicides. *Reviews of Weed Science* 2, 15–44.

Duke, S.O. and Lydon, J. (1987) Herbicides from natural compounds. *Weed Technology* 1, 122–128.

Dulmage, H.T. (1982) Distribution of *Bacillus thuringiensis* in nature. In: Kurstak, E. (ed.) *Microbial and Viral Pesticides*. Marcel Dekker, New York, pp. 209–237.

Dunlap, T.R. (1980) The gypsy moth: a study in science and public policy. *Journal of Forest History* 24, 116–126.

Elkinton, J.S. and Liebhold, A.M. (1990) Population dynamics of gypsy moth in North America. *Annual Review of Entomology* 35, 571–596.

Elkinton, J.S., Hajek, A.E., Boettner, G.H. and Simons, E.E. (1991) Distribution and apparent spread of *Entomophaga maimaiga* (Zygomycetes: Entomophthorales) in gypsy moth (Lepidoptera: Lymantriidae) populations in North America. *Environmental Entomolgy* 20, 1601–1605.

Embree, D.G. (1971) The biological control of the winter moth in eastern Canada by introduced parasites. In: Huffaker, C.B. (ed.) *Biological Control*. Plenum, New York, pp. 217–226.

FAO (Food and Agriculture Organization of the United Nations) (2002) FAO Statistics. Available at: www.fao.org/forestry/index.jsp

Feller, M. (1996) Use of prescribed fire for vegetation management. In: Comeau, P.G., Harper, G.J., Blache, M.E., Boateng, J.O. and Gilkeson, L.A. (eds) *Integrated Forest Vegetation Management: Options and Applications. Proceedings of the Fifth British Columbia Forest Vegetation Management Workshop, November*

29–30, 1993, Richmond, British Columbia. Forest Resource Development Agreement Report No. 251, Canadian Forest Service and British Columbia Ministry of Forests, Victoria, British Columbia, pp. 17–34.

Fleming, R. and Retnakaran, A. (1985) Evaluating single treatment data using Abbott's formula with reference to insecticides. *Journal of Economic Entomology* 78, 1179–1181.

van Frankenhuyzen, K. (1990) Development and current status of *Bacillus thuringiensis* for control of defoliating forest insects. *The Forestry Chronicle* 66, 498–507.

van Frankenhuyzen, K. (1993) The challenge of *Bacillus thuringiensis*. In: Entwistle, P.F., Cory, J.S., Bailey, M.J. and Higgs, S. (eds) Bacillus thuringiensis, *an Environmental Pesticide: Theory and Practice.* J. Wiley & Sons, New York, pp. 1–35.

van Frankenhuyzen, K. (1995) Development and current status of *Bacillus thuringiensis* for control of defoliating forest insects. In: Armstrong, J.A. and Ives, W.G.H. (eds) *Forest Insect Pests in Canada.* Natural Resources Canada, Canadian Forest Service, Ottawa, Ontario, pp. 315–325.

van Frankenhuyzen, K. (2000) Application of *Bacillus thuringiensis* in forestry. In: Charles, J.F., Delecluse, A. and Nielsen-LeRoux, C. (eds) *Entomopathogenic Bacteria: From Laboratory to Field Application.* Kluwer Academic Publishers, Dordrecht, The Netherlands, pp. 355–370.

Glaser, R.W. (1915) Wilt of gypsy-moth caterpillars. *Journal of Agricultural Research* 4, 101–128.

Griffiths, K.J. (1976) *The Parasites and Predators of the Gypsy Moth: a Review of the World Literature with Special Application to Canada.* Information Report O-X-243, Canadian Forestry Service, Great Lakes Forest Research Centre, Sault Ste Marie, Ontario, 92 pp.

Griffiths, K.J. and Quednau, F.W. (1984) *Lymantria dispar* (L.), gypsy moth (Lepidoptera: Lymantriidae). In: Kelleher, J.S. and Hulme, M.A. (eds) *Biological Control Programmes Against Weeds and Insects in Canada 1969–1980.* Commonwealth Agricultural Bureaux, Slough, UK, pp. 303–310.

Hajek, A.E. (1997) Fungal and viral epizootics in gypsy moth (Lepidoptera: Lymantriidae) populations in central New York. *Biological Control* 10, 58–68.

Hajek, A.E. and Roberts, E.W. (1991) Pathogen reservoirs as a biological control resource: introduction of *Entomophaga maimaiga* to North American gypsy moth, *Lymantria dispar*, populations. *Biological Control* 1, 29–34.

Hajek, A.E. and Roberts, E.W. (1992) Field diagnosis of gypsy moth (Lepidoptera: Lymantriidae) larval mortality caused by *Entomophaga maimaiga* and the gypsy moth nuclear polyhedrosis virus. *Environmental Entomology* 21, 706–713.

Hajek, A.E. and Soper, R.S. (1992) Temporal dynamics of *Entomophaga maimaiga* after death of gypsy moth (Lepidoptera: Lymantriidae) larval hosts. *Environmental Entomology* 21, 129–135.

Hajek, A.E., Humber, R.A., Elkinton, J.S., May, B., Walsh, S.R.A. and Silver, J.C. (1990a) Allozyme and RFLP analyses confirm *Entomophaga maimaiga* responsible for 1989 epizootics in North American gypsy moth populations. *Proceedings of the National Academy of Sciences of the USA* 87, 6979–6982.

Hajek, A.E., Carruthers, R.I. and Soper, R.S. (1990b) Temperature and moisture relations of sporulation and germination by *Entomophaga maimaiga* (Zygomycetes: Entomophthoraceae), a fungal pathogen of *Lymantria dispar* (Lepidoptera: Lymantriidae). *Environmental Entomology* 19, 85–90.

Hajek, A.E., Larkin, T.S., Carruthers, R.I. and Soper, R.S. (1993) Modeling the dynamics of *Entomophaga maimaiga* (Zygomycetes: Entomophthorales) epizootics in gypsy moth (Lepidoptera: Lymantriidae) populations. *Environmental Entomology* 22, 1172–1187.

Hajek, A.E., Humber, R.A. and Elkinton, J.S. (1995) Mysterious origin of *Entomophaga maimaiga* in North America. *American Entomologist* 41, 31–42.

Hajek, A.E., Elkinton, J.S. and Witcosky, J.J. (1996) Introduction and spread of the fungal pathogen *Entomophaga maimaiga* (Zygomycetes: Entomophthorales) along the leading edge of gypsy moth (Lepidoptera: Lymantriidae) spread. *Environmental Entomology* 25, 1235–1247.

Hall, P.J. (comp.) (1994) *Forest Insect and Disease Conditions in Canada 1992.* Forest Insect and Disease Survey, Forestry Canada, Ottawa, Ontario, 120 pp.

Hall, P.J. (comp.) (1995) *Forest Insect and Disease Conditions in Canada 1993.* Forest Insect and Disease Survey, Forestry Canada, Ottawa, Ontario, 133 pp.

Hall, P.J. (comp.) (1996) *Forest Insect and Disease Conditions in Canada 1994.* Forest Insect and Disease Survey, Forestry Canada, Ottawa, Ontario, 112 pp.

Halleran, M. (1990) Forest vegetation management and the politics of environment. *The Forestry Chronicle* 66, 369–371.

Hamel, D.R. (1977) The effects of *Bacillus thuringiensis* on parasitoids of the western spruce budworm, *Choristoneura occidentalis* (Lepidoptera: Tortricidae), and the spruce coneworm, *Dioryctria reniculelloides* (Lepidoptera: Pyralidae), in Montana. *The Canadian Entomologist* 109, 1409–1415.

Hansen, E.M. and Lewis, K.J. (eds) (1997) *Compendium of Conifer Diseases*. American Phytopathological Society Press, St Paul, Minnesota, 101 pp.

Harris, J.W.E., Alfaro, R.I., Dawson, A.F. and Brown, R.G. (1985a) *The Western Spruce Budworm in British Columbia 1909–1983*. Information Report BC-X-257, Canadian Forestry Service, Pacific Forestry Centre, Victoria, British Columbia, 32 pp.

Harris, P. (1984) Current approaches to biological control of weeds. In: Kelleher, J.S. and Hulme, M.A. (eds) *Biological Control Programmes Against Weeds and Insects in Canada 1969–1980*. Commonwealth Agricultural Bureaux, Slough, UK, pp. 95–103.

Harris, P. and Shamoun, S.F. (2002) Biological control of weeds in Canada: results, opportunities, and constraints. In: Claudia, R., Nantel, P. and Muckle-Jeffs, E. (eds) *Alien Invaders in Canada's Waters, Wetlands, and Forests*. Canadian Forest Service, Natural Resources Canada, Ottawa, Ontario, pp. 291–302.

Hewitt, G.C. (1916) The introduction and establishment in Canada of the natural enemies of the browntail and gypsy moth. *The Agricultural Gazette of Canada* 3, 20–21.

Hoy, M.A. (1976) Establishment of gypsy moth parasitoids in North America: an evaluation of possible reasons for establishment or non-establishment. In: Anderson, J.F. and Kaya, H.K. (eds) *Perspectives in Forest Entomology*. Academic Press, New York, pp. 215–232.

Hughes, K.M. and Addison, R.B. (1970) Two nuclear polyhedrosis viruses of the Douglas-fir tussock moth. *Journal of Invertebrate Pathology* 16, 196–204.

Hulme, M.A. (1988) The recent Canadian record in applied biological control of forest insect pests. *The Forestry Chronicle* 64, 27–31.

Humble, L. and Stewart, A.J. (1994) *Gypsy Moth*. Forest Pest Leaflet 75, Canadian Forest Service and British Columbia Ministry of Forests, Victoria, British Columbia, 8pp.

Humphreys, P. (1986) *Pesticide Use in British Columbia Forestry 1985*. Forest Service Internal Report PM–PB–26, British Columbia Ministry of Forests and Lands, Victoria, British Columbia, 31 pp.

Humphreys, P. (1990) *Pesticide Use in British Columbia Forestry 1989*. Forest Service Internal Report PM–PB–46, British Columbia Ministry of Forests, Victoria, British Columbia, 73 pp.

Ilnytzky, S., McPhee, J.R. and Cunningham, J.C. (1977) Comparison of field-propagated nuclear polyhedrosis virus from Douglas-fir tussock moth with laboratory virus. *Fisheries and Environment Canada, Forestry Service, Bi-monthly Research Notes* 33, 5–6.

Ives, W.G.H. and Rentz, C.L. (1993) *Factors Affecting the Survival of Immature Lodgepole Pine in the Foothills of West-Central Alberta*. Information Report NOR-X-330, Canadian Forest Service, Northwest Region, Northern Forestry Centre, Edmonton, Alberta, 49 pp.

Jobidon, R. (1991a) Some future directions for biologically based vegetation control in forestry. *The Forestry Chronicle* 67, 514–519.

Jobidon, R. (1991b) Potential use of bialaphos, a microbially produced phytotoxin, to control red raspberry in forest plantations and its effect on black spruce. *Canadian Journal of Forest Research* 21, 489–497.

Jobidon, R., Thibault, J.R. and Fortin, J.A. (1989) Phytotoxic effect of barley oat and wheat-straw mulches in eastern Quebec forest plantations 1. Effects on red raspberry (*Rubus idaeus*). *Forest Ecology and Management* 29, 277–294.

Jobin, L. (1995) Gypsy moth, *Lymantria dispar*. In: Armstrong, J.A. and Ives, W.G.H. (eds) *Forest Insect Pests in Canada*. Natural Resources Canada, Canadian Forest Service, Ottawa, Ontario, pp. 133–139.

Julien, M.H. and Griffiths, M.W. (1998) *Biological Control of Weeds: a World Catalogue of Agents and their Target Weeds*. CAB International, Wallingford, UK, 223 pp.

Keena, M.A., Grinberg, P.S. and Wallner, W.E. (1995) Asian gypsy moth genetics: biological consequences of hybridization [abstract]. In: Fosbroke, S.L.C. and Gottschalk, K.W. (eds) *Proceedings: U.S. Department of Agriculture Interagency Gypsy Moth Research Forum 1995, January 17–20, 1995, Loews Annapolis Hotel, Annapolis, Maryland*. General Technical Report NE-213, USDA Forest Service, Northeastern Forest Experiment Station, Radnor, Pennsylvania, p. 80.

Kelleher, J.S. and Hulme, M.A. (eds) (1984) *Biological Control Programmes Against Weeds and Insects in Canada 1969–1980*. Commonwealth Agricultural Bureaux, Slough, UK, 410 pp.

Kettela, E.G. (1983) *A Cartographic History of Spruce Budworm Defoliation from 1967 to 1981 in Eastern North America*. Information Report DPC-X-14, Canadian Forestry Service, Maritimes Research Centre, Fredericton, New Brunswick, 8 pp.

Kettela, E.G. (1995) Attempts to develop strategies to control spruce budworm, *Choristoneura fumiferana*, moth populations by spraying. In: Armstrong, J.A. and Ives, W.G.H. (eds) *Forest Insect Pests in Canada*. Natural Resources Canada, Canadian Forest Service, Ottawa, Ontario, pp. 113–117.

Kinghorn, J.M., Fisher, R.A., Angus, T.A. and Heimpel, A.M. (1961) Aerial spray trials against the black-headed budworm in British Columbia. *Department of Forestry, Forest Entomology and Pathology Branch, Bi-monthly Progress Report* 17, 3–4.

Leonard, D.E. (1981) Bioecology of the gypsy moth. In: Doane, C.C. and McManus, M.L. (eds) *The Gypsy Moth: Research Toward Integrated Pest Management*. Technical Bulletin 1584, Forest Service, US Department of Agriculture, Washington, DC, pp. 9–29.

Leonard, D.S. and Sharov, A.A. (1995) Slow the spread project update: developing a process for evaluation. In: Fosbroke, S.L.C. and Gottschalk, K.W. (eds) *Proceedings: US Department of Agriculture Interagency Gypsy Moth Research Forum 1995, January 17–20, 1995, Annapolis, Maryland*. General Technical Report NE-213, Northeastern Forest Experiment Station, USDA Forest Service, Radnor, Pennsylvania, pp. 82–85.

Leonhardt, B.A., Mastro, V.C., Leonard, D.S., McLane, W., Reardon, R.C. and Thorpe, K.W. (1996) Control of low-density gypsy moth (Lepidoptera: Lymantriidae) populations by mating disruption with pheromone. *Journal of Chemical Ecology* 22, 1255–1272.

Leuschner, W.A., Young, J.A., Waldon, S.A. and Ravlin, F.W. (1996) Potential benefits of slowing the gypsy moth's spread. *Southern Journal of Applied Forestry* 20, 65–73.

Lewis, F.B. (1981) Registration and cost effectiveness. In: Doane, C.C. and McManus, M.L. (eds) *The Gypsy Moth: Research Toward Integrated Pest Management*. Technical Bulletin 1584, Forest Service, US Department of Agriculture, Washington, DC, pp. 514–515.

Liebhold, A. and McManus, M. (1999) The evolving use of insecticides in gypsy moth management. *Journal of Forestry* 97, 20–23.

Liebhold, A., Mastro, V. and Schaefer, P.W. (1989) Learning from the legacy of Léopold Trouvelot. *Bulletin of the Entomological Society of America* 35, 20–22.

Liebhold, A.M., Halverson, J.A. and Elmes, G.A. (1992) Gypsy moth invasion in North America: a quantitative analysis. *Journal of Biogeography* 19, 513–520.

Liebhold, A.M., MacDonald, W.L., Bergdahl, D. and Mastro, V.C. (1995) *Invasion by Exotic Forest Pests: a Threat to Forest Ecosystems*. Forest Science Monograph 30, Society of American Foresters, Bethesda, Maryland, 49 pp.

Liebhold, A.M., Gottschalk, K.W., Mason, D.A. and Bush, R.R. (1997) Forest susceptibility to the gypsy moth. *Journal of Forestry* 95, 20–24.

Logie, R.R. (1975) Effects of aerial spraying of DDT on salmon populations of the Miramichi River. In: Prebble, M.L. (ed.) *Aerial Control of Forest Insects in Canada*. Department of the Environment, Ottawa, Canada, pp. 293–300.

Lowe, J.J., Power, K. and Gray, S.L. (1996) *Canada's Forest Inventory 1991: the 1994 Version. An Addendum to Canada's Forest Inventory 1991*. Information Report BC-X-362E, Natural Resources Canada, Canadian Forest Service, Pacific Forestry Centre, Victoria, British Columbia, 23 pp.

Maclauchlan, L.E. and Brooks, J.E. (eds) (1994) *Strategies and Tactics for Managing the Mountain Pine Beetle,* Dendroctonus ponderosae. British Columbia Forest Service, Kamloops Region, Forest Health, Kamloops, British Columbia, 60 pp.

MacLean, D.A. (1980) Vulnerability of fir-spruce stands during uncontrolled spruce budworm outbreaks: a review and discussion. *The Forestry Chronicle* 56, 213–222.

Malakar, R., Elkinton, J.S., Hajek, A.E. and Burand J.P. (1999) Within-host interactions of *Lymantria dispar* (Lepidoptera: Lymantriidae) nucleopolyhedrosis virus and *Entomophaga maimaiga* (Zygomycetes: Entomophthorales). *Journal of Invertebrate Pathology* 73, 91–100.

Mallet, K.I., Macey, D.E. and Winder, R.S. (2002) *Calamagrostis canadensis* (Michaux) Palisot de Beauvois, Marsh Reed Grass (*Poaceae*). In: Mason, P.G. and Huber, J.T. (eds) *Biological Control Programmes in Canada, 1981–2000*. CAB International, Wallingford, UK, pp. 298–301.

Marsden, J.S., Martin, G.E., Parham, D.J., Ridsill-Smith, T.J. and Johnston, B.G. (1980) *Returns on Australian Agricultural Research: Joint IAC-CSIRO Benefit Cost Study of the CSIRO Division of Entomology*. Commonwealth Scientific and Industrial Research Organization, Canberra, Australia, 107 pp.

Martin, P.A.W. and Travers, R.S. (1989) Worldwide abundance and distribution of *Bacillus thuringiensis* isolates. *Applied and Environmental Microbiology* 55, 2437–2442.

Mason, C.J. and McManus, M.L. (1981) Larval dispersal of the gypsy moth. In: Doane, C.C. and McManus, M.L. (eds) *The Gypsy Moth: Research Toward Integrated Pest Management*. Technical Bulletin 1584, Forest Service, US Department of Agriculture, Washington, DC, pp. 161–202.

Mason, P.G. and Huber, J.T. (eds) (2002) *Biological Control Programmes in Canada, 1981–2000*. CAB International, Wallingford, UK, 583 pp.

Mason, P.G., Huber, J.T. and Boyetchko, S.M. (2002) Introduction. In: Mason, P.G. and Huber, J.T. (eds) *Biological Control Programmes in Canada, 1981–2000*. CAB International, Wallingford, UK, pp. xi–xiv.

Mattson, W.J., Niemela, P., Millers, I. and Inguanzo, Y. (1994) *Immigrant Phytophagous Insects on Woody Plants in the United States and Canada: An Annotated List*. General Technical Report NC-169, Forest Service, US Department of Agriculture, North Central Forest Experiment Station, St Paul, Minnesota, 27 pp.

McDonald, G.I., Martin, N.E. and Harvey, A.E. (1987) Armillaria *in the Northern Rockies: Pathogenicity and Host Susceptibility on Pristine and Disturbed Sites*. Research Note INT-371, Forest Service, US Department of Agriculture, Intermountain Research Station, Ogden, Utah, 3 pp.

McDonald, P.M. and Fiddler, G.O. (1996) Mulching: a persistent technique for weed suppression. In: Comeau, P.G., Harper, G.J., Blache, M.E., Boateng, J.O. and Gilkeson, L.A. (eds) *Integrated Forest Vegetation Management: Options and Applications: Proceedings of the Fifth British Columbia Forest Vegetation Management Workshop, November 29–30, 1993, Richmond, British Columbia*. Forest Resource Development Agreement Report No. 251, Canadian Forest Service and British Columbia Ministry of Forests, Victoria, British Columbia, pp. 51–58.

McFadden, M.W. and McManus, M.E. (1991) An insect out of control? The potential for spread and establishment of the gypsy moth in new forest areas in the United States. In: Baranchikov, Y.N., Mattson, W.J., Hain, F.P. and Payne, T.L. (eds) *Forest Insect Guilds: Patterns of Interaction with Host Trees: Proceedings of a Joint IUFRO Working Party Symposium, Abakan, Siberia, USSR, August 13–17, 1989*. General Technical Report NE-153, Forest Service, US Department of Agriculture, Northeastern Forest Experiment Station, Radnor, Pennsylvania, pp. 172–186.

McGugan, B.M. and Coppel, H.C. (1962) Part II: Biological control of forest insects, 1910–1958. In: McLeod, J.H., McGugan, B.M. and Coppel, H.C. (compilers) *A Review of the Biological Control Attempts Against Insects and Weeds in Canada*. Commonwealth Agricultural Bureaux, Slough, UK, pp. 35–127.

McLeod, J.H. (1962) Part I: Biological control of pests of crops, fruit trees, ornamentals, and weeds in Canada up to 1959. In: McLeod, J.H., McGugan, B.M. and Coppel, H.C. (compilers) *A Review of the Biological Control Attempts Against Insects and Weeds in Canada*. Commonwealth Agricultural Bureaux, Slough, UK, pp. 1–33.

McManus, M.L. and McIntyre, T. (1981) Introduction: historical chronology. In: Doane, C.C. and McManus, M.L. (eds) *The Gypsy Moth: Research Toward Integrated Pest Management*. Technical Bulletin 1584, Forest Service, US Department of Agriculture, Washington, DC, pp. 1–7.

McMullen, L.H., Safranyik, L. and Linton, D.A. (1986) *Suppression of Mountain Pine Beetle Infestations in Lodgepole Pine Forests*. Information Report BC-X-276, Canadian Forestry Service, Pacific Forestry Centre, Victoria, British Columbia, 20 pp.

Miller, C.A. (1957) A technique for estimating the fecundity of natural populations of the spruce budworm. *Canadian Journal of Zoology* 35, 1–13.

Miller, C.A., Kettela, E.G. and McDougall, G.A. (1971) *A Sampling Technique for Overwintering Spruce Budworm and its Applicability to Population Surveys*. Information Report M-X-25, Department of Fisheries and Forestry, Canadian Forest Service, Maritimes Forest Research Centre, Fredericton, New Brunswick, 11 pp.

Miller, J.C. (1990) Field assessment of the effects of a microbial pest control agent on nontarget Lepidoptera. *American Entomologist* 36, 135–139.

Miller, J.D. and Lindsay, B.E. (1993) Willingness to pay for a state gypsy moth control program in New Hampshire: a contingent valuation case study. *Journal of Economic Entomology* 86, 828–837.

Mills, N.J. and Nealis, V.G. (1992) European field collections and Canadian releases of *Ceranthia samarensis* (Dipt.: Tachinidae), a parasitoid of the gypsy moth. *Entomophaga* 37, 181–191.

Mitchell, R.G., Waring, R.H. and Pitman, G.B. (1983) Thinning lodgepole pine increases tree vigor and resistance to mountain pine beetle. *Forest Science* 29, 204–211.

Montgomery, M.E. and Wallner, W.E. (1988) The gypsy moth: a westward migrant. In: Berryman, A.A. (ed.) *Dynamics of Forest Insect Populations: Patterns, Causes, Implications*. Plenum Press, New York, pp. 253–275.

Moody, B.H. (comp.) (1993a) *Forest Insect and Disease Conditions in Canada 1991*. Forest Insect and Disease Survey, Forestry Canada, Ottawa, Ontario, 99 pp.

Moody, B.H. (comp.) (1993b) *Forest Insect and Disease Conditions in Canada 1990*. Forest Insect and Disease Survey, Forestry Canada, Ottawa, Ontario, 115 pp.

Morris, M.J. (1991) The use of plant pathogens for biological weed control in South Africa. *Agriculture, Ecosystems and Environment* 37, 239–255.

Morris, M.J. (1997) Impact of the gall-forming rust fungus, *Uromycladium tepperianum* on the invasive tree *Acacia saligna* in South Africa. *Biological Control* 10, 75–82.

Morris, M.J., Wood, A.R. and den Breeyen, A. (1998) Development and registration of a fungal inoculant to prevent re-growth of cut wattle tree stumps in South Africa. In: Burge, M.N. (ed.) *IV International Bioherbicides Workshops – Programmes and Abstracts, 6–7 August, 1998*. University of Strathclyde, Glasgow, p. 15.

Morris, O.N. (1963) The natural and artificial control of the Douglas-fir tussock moth, *Orgyia pseudotsugata* McDunnough, by a nuclear-polyhedrosis virus. *Journal of Insect Pathology* 5, 401–414.

Morris, O.N. (1983) Comparative efficacy of Thuricide 16B and Dipel 88 against the spruce budworm, *Choristoneura fumiferana* (Lepidoptera: Tortricidae) in balsam fir stands. *The Canadian Entomologist* 115, 1001–1006.

Morris, R.F. (1954) A sequential sampling technique for spruce budworm egg surveys. *Canadian Journal of Zoology* 32, 302–313.

Morris, R.F. (ed.) (1963) *The Dynamics of Epidemic Spruce Budworm Populations*. Memoirs of the Entomological Society of Canada 31, Ottawa, Ontario, Canada, 332pp.

Morrison, D. and Mallett, K. (1996) Silvicultural management of armillaria root disease in western Canadian forests. *Canadian Journal of Plant Pathology* 18, 194–199.

Morrison, D.J., Merler, H. and Norris, D. (1992) *Detection, Recognition and Management of* Armillaria *and* Phellinus *Root Diseases in the Southern Interior of British Columbia*. Forest Resource Development Agreement Report 179, Canadian Forest Service and British Columbia Ministry of Forests, Victoria, British Columbia, 25 pp.

Morrison, D.J., Wallis, G.W. and Weir, L.C. (1988) *Control of* Armillaria *and* Phellinus *Root Diseases: 20-year Results from the Skimikin Stump Removal Experiment*. Information Report BC-X-302, Canadian Forestry Service, Pacific Forestry Centre, Victoria, British Columbia, 16 pp.

Mott, D.G., Angus, T.A., Heimpel, A.M. and Fisher, R.A. (1961) Aerial spraying of Thuricide against the spruce budworm in New Brunswick. *Department of Forestry, Forest Entomology and Pathology Branch, Bi-monthly Progress Report* 17, 2.

Nealis, V.G. (2002) Gypsy moth in Canada: case study of an invasive insect. In: Claudia, R., Nantel, P. and Muckle-Jeffs, E. (eds) *Alien Invaders in Canada's Waters, Wetlands and Forests*. Natural Resources Canada, Science Branch, Canadian Forest Service, Ottawa, Ontario, pp. 151–159.

Nealis, V.G. and Quednau, F.W. (1996) Canadian field releases and overwinter survival of *Ceranthia samarensis* (Villeneuve) (Diptera: Tachinidae) for biological control of the gypsy moth, *Lymantria dispar* (L.) (Lepidoptera: Lymantriidae). *Proceedings of the Entomological Society of Ontario* 127, 11–20.

Nealis, V.G. and van Frankenhuyzen, K. (1990) Interactions between *Bacillus thuringiensis* Berliner and *Apanteles fumiferanae* Vier. (Hymenoptera: Braconidae), a parasitoid of the spruce budworm, *Choristoneura fumiferana* (Clem.) (Lepidoptera: Tortricidae). *The Canadian Entomologist* 122, 585–594.

Nealis, V.G., Roden, P.M. and Ortiz, D.A. (1999) Natural mortality of the gypsy moth along a gradient of infestation. *The Canadian Entomologist* 131, 507–519.

Nealis, V.G., Carter, N., Kenis, M., Quednau, F.W. and van Frankenhuyzen, K. (2002) *Lymantria dispar* (L.), gypsy moth (Lepidoptera: Lymantriidae). In: Mason, P.G. and Huber, J.T. (eds) *Biological Control Programmes in Canada, 1981–2000*. CAB International, Wallingford, UK, pp. 159–168.

Niemela, P. and Mattson, W.J. (1996) Invasion of North American forests by European phytophagous insects – legacy of the European crucible? *BioScience* 46, 741–753.

Nigam, P.C. (1975) Chemical insecticides. In: Prebble, M.L. (ed.) *Aerial Control of Forest Insects in Canada*. Department of the Environment, Ottawa, Canada, pp. 8–24.

Nigam, P.C. (1980) Use of chemical insecticides against the spruce budworm in eastern Canada. *CANUSA Newsletter* 11, 1–3.

Niwa, C.G., Stelzer, M.J. and Beckwith, R.C. (1987) Effects of *Bacillus thuringiensis* on parasites of western spruce budworm (Lepidoptera: Tortricidae). *Journal of Economic Entomology* 80, 750–753.

Oleskevich, C., Shamoun, S.F., Vesonder, R.F. and Punja, Z.K. (1998) Evaluation of *Fusarium avenaceum* and other fungi for potential as biological control agents of invasive *Rubus* species in British Columbia. *Canadian Journal of Plant Pathology* 20, 12–18.

Otvos, I.S. (1973) *Biological Control Agents and Their Role in the Population Fluctuation of the Eastern Hemlock Looper in Newfoundland.* Information Report N-X-102, Environment Canada, Newfoundland Forest Research Centre, St John's, Newfoundland, 34 pp.

Otvos, I.S. and Raske, A.G. (1980a) *The Effects of Fenitrothion, Matacil, and* Bacillus thuringiensis *plus Orthene on Larval Parasites of the Spruce Budworm.* Information Report N-X-184, Canadian Forestry Service, Newfoundland Forest Research Centre, St John's, Newfoundland, 24 pp.

Otvos, I.S. and Raske, A.G. (1980b) *Effects of Aerial Application of Matacil on Larval and Pupal Parasites of the Eastern Spruce Budworm,* Choristoneura fumiferana *(Lepidoptera: Tortricidae).* Information Report N-X-189, Canadian Forest Service, Newfoundland Forest Research Centre, St John's, Newfoundland, 12 pp.

Otvos, I.S. and Vanderveen, S. (1993) *Environmental Report and Current Status of* Bacillus thuringiensis *var.* kurstaki *Use for Control of Forest and Agricultural Insect Pests.* Joint report by Forestry Canada, Pacific Forestry Centre and British Columbia Ministry of Forests, Victoria, British Columbia, 81 pp.

Otvos, I.S., MacLeod, D.M. and Tyrrell, D. (1973) Two species of *Entomophthora* pathogenic to the eastern hemlock looper (Lepidoptera: Geometridae) in Newfoundland. *Canadian Entomologist* 105, 1435–1441.

Otvos, I.S., Cunningham, J.C. and Friskie, L.M. (1987) Aerial application of nuclear polyhedrosis virus against Douglas-fir tussock moth, *Orgyia pseudotsugata* (McDunnough) (Lepidoptera: Lymantriidae): I. Impact in the year of application. *The Canadian Entomologist* 119, 697–706.

Otvos, I.S., Cunningham, J.C. and Kaupp, W.J. (1989) Aerial application of two baculoviruses against the western spruce budworm (Lepidoptera: Tortricidae) in British Columbia. *The Canadian Entomologist* 121, 209–217.

Otvos, I.S., Cunningham, J.C. and Shepherd, R.F. (1995) Douglas-fir tussock moth, *Orgyia pseudotsugata.* In: Armstrong, J.A. and Ives, W.G.H. (eds) *Forest Insect Pests in Canada.* Natural Resources Canada, Canadian Forest Service, Ottawa, Ontario, pp. 127–132.

Pearce, P.A. (1975) Effects on birds. In: Prebble, M.L. (ed.) *Aerial Control of Forest Insects in Canada.* Department of the Environment, Ottawa, Canada, pp. 306–313.

Pendl, F. and D'Anjou, B. (1990) *Effect of Manual Treatment Timing on Red Alder Regrowth and Conifer Response.* Forest Resource Development Agreement Report 112, Canadian Forestry Service and British Columbia Ministry of Environment, Victoria, British Columbia, 21 pp.

Perlman, F.D., Press, E., Googins, J.A., Malley, A. and Poarea, H. (1976) Tussockosis: reactions to Douglas-fir tussock moth. *Annals of Allergy* 36, 302–307.

Podgwaite, J.D. (1999) Gypchek: biological insecticide for the gypsy moth. *Journal of Forestry* 97, 16–19.

Podgwaite, J.D. and Campbell, R.W. (1972) The disease complex of the gypsy moth. II. Aerobic bacterial pathogens. *Journal of Invertebrate Pathology* 20, 303–308.

Prasad, R. (1992) Some aspects of biological control of weeds in forestry. In: Richardson, R.G. (compiler) *Proceedings of the First International Weed Control Congress, February 17–21, 1992, Monash University, Melbourne, Australia,* Vol. 2. Weed Science Society of Victoria Inc., Melbourne, Australia, pp. 398–402.

Prasad, R. (1996) Development of bioherbicides for integrated weed management in forestry. In: Brown, H. (ed.) *Proceedings, Second International Weed Control Congress, Copenhagen, Denmark, June 25–28 1996.* Department of Weed Control and Pesticide Ecology, Slagelse, Denmark, pp. 1197–1203.

Prasad, R. (2002a) *Cytisus scoparius* (L.) Link, Scotch Broom (Fabaceae). In: Mason, P.G. and Huber, J.T. (eds) *Biological Control Programmes in Canada, 1981–2000.* CAB International, Wallingford, UK, pp. 343–345.

Prasad, R. (2002b) *Ulex europaeus* L., Gorse (Fabaceae). In: Mason, P.G. and Huber, J.T. (eds) *Biological Control Programmes in Canada, 1981–2000.* CAB International, Wallingford, UK, pp. 431–433.

Prasher, D. and Mastro, V.C. (1995) Genotype analyses of 1994 port specimens. In: Hilburn, D., Johnson, K.J.R. and Mudge, A.D. (eds) *Proceedings of the 1994 Annual Gypsy Moth Review: Held at the Portland Hilton, Portland, Oregon, October 30-November 2, 1994.* US National Gypsy Moth Management Board, Salem, Massachusetts, pp. 61–63.

Prebble, M.L. (ed.) (1975) *Aerial Control of Forest Insects in Canada.* Environment Canada, Ottawa, Ontario, 330 pp.

Ravensberg, W.J. (1998) *BioChon: Effective Biological and Environmentally Friendly Product.* Pest Leaflet, Koppert Biological Systems, The Netherlands, 2 pp.

Reardon, R.C. (1981) Summary. In: Doane, C.C. and McManus, M.L. (eds) *The Gypsy Moth: Research Toward Integrated Pest Management.* Technical Bulletin 1584, Forest Service, US Department of Agriculture, Washington, DC, pp. 412–413.

Reardon, R.C. (1991) Appalachian gypsy-moth integrated pest-management project. *Forest Ecology and Management* 39, 107–112.

Reardon, R. and Hajek, A. (1993) Entomophaga maimaiga *in North America: a Review.* AIPM Technology Transfer, US Department of Agriculture Forest Service, Northeastern Area, Morgantown, West Virginia, 22 pp.

Reardon, R.C., Johnson, D.R., Narog, M.G., Banash, S.E. and Hubbard, H.B. Jr (1982) Efficacy of two formulations of *Bacillus thuringiensis* on populations of spruce budworm (Lepidoptera: Tortricidae) on balsam fir in Wisconsin. *Journal of Economic Entomology* 75, 509–514.

Reardon, R., McManus, M., Kolodny-Hirsch, D., Tichenor, R., Raupp, M., Schwalbe, C., Webb, R. and Meckley, P. (1987) Development and implementation of a gypsy moth integrated pest management program. *Journal of Arboriculture* 13, 209–216.

Reardon, R.C., Dubois, N.R. and McLane, W. (1994) Bacillus thuringiensis *for Managing Gypsy Moth: a Review.* FHM-NC-01–94, National Centre of Forest Health Management, US Department of Agriculture Forest Service, Morgantown, West Virginia, 32 pp.

Reardon, R.C., Leonard, D.S., Mastro, V.C., Leonhardt, B.A., McLane, W., Talley, S., Thorpe, K.W. and Webb, R.E. (1998) *Using Mating Disruption to Manage Gypsy Moth: A Review.* USDA Forest Service, Morgantown, West Virginia, 85 pp.

Reeks, W.A. and Cameron, J.M. (1971) Current approach to biological control of forest insects. In: Canadian Department of Agriculture and Canadian Department of Fisheries and Forestry (ed.) *Biological Control Programmes Against Insects and Weeds in Canada 1959–1968.* Technical Communication No. 4, Commonwealth Institute of Biological Control, Commonwealth Agricultural Bureaux, Slough, UK, pp. 105–113.

Regniere, S. and Lysyk, T.J. (1995) Population dynamics of the spruce budworm, *Choristoneura fumiferana.* In: Armstrong, J.A. and Ives, W.G.H. (eds) *Forest Insect Pests in Canada.* Natural Resources Canada, Canadian Forest Service, Ottawa, Ontario, pp. 95–105.

Russell, K., Johnsey, R. and Edmonds, R. (1986) Disease and insect management for Douglas-fir. In: Oliver, C.D., Hanley, D.P. and Johnson, J.A. (eds) *Douglas-fir: Stand Management for the Future.* Contribution No. 55, College of Forest Resources, Institute of Forest Resources, University of Washington, Seattle, Washington, pp. 189–207.

Sabrosky, C.W. and Reardon, R.C. (1976) *Tachinid Parasites of the Gypsy Moth,* Lymantria dispar, *with Keys to Adults and Puparia.* Miscellaneous Publications of the Entomological Society of America 10, Lanham, Maryland, 126 pp.

Safranyik, L., Shrimpton, D.M. and Whitney, H.S. (1974) *Management of Lodgepole Pine to Reduce Losses from the Mountain Pine Beetle.* Forestry Technical Report 1, Canadian Forestry Service, Pacific Forestry Centre, Victoria, British Columbia, 24 pp.

Safranyik, L., Simmons, C. and Barclay, H.J. (1990) *A Conceptual Model of Spruce Beetle Population Dynamics.* Information Report BC-X-316, Forestry Canada, Pacific Forestry Centre, Victoria, British Columbia, 13 pp.

Sanders, C.J. (1988) Monitoring spruce budworm population density with sex pheromone traps. *Canadian Entomologist* 120, 175–183.

Sanders, C.J., Stark, R.W., Mullins, E.J. and Murphy, J. (eds) (1985) *Recent Advances in Spruce Budworms Research: Proceedings of CANUSA Spruce Budworms Research Symposium, Bangor, Maine, September 16–20, 1984.* Canadian Forestry Service, Ottawa, Ontario, 527 pp.

Sartwell, C. and Dolph, R.E. Jr (1976) *Silvicultural and Direct Control of Mountain Pine Beetle in Second-Growth Ponderosa Pine.* Research Note PNW-268, US Department of Agriculture Forest Service, Pacific Northwest Forest and Range Experiment Station, Portland, Oregon, 8 pp.

Scheepens, P.C. and Hoogerbrugge, A. (1989) Control of *Prunus serotina* in forests with the endemic fungus *Chondrostereum purpureum.* In: Delfosse, E.S. (ed.) *Proceedings, VII International Symposium on Biological Control of Weeds, March 6–11, 1988, Rome, Italy.* Commonwealth Scientific and Industrial Research Organization Publications, East Melbourne, Australia, pp. 545–551.

Schwalbe, C.P. (1981) Disparlure-baited traps for survey and detection. In: Doane, C.C. and McManus, M.L. (eds) *The Gypsy Moth: Research Toward Integrated Pest Management.* Technical Bulletin 1584, Forest Service, US Department of Agriculture, Washington, DC, pp. 542–548.

Shamoun, S.F. (2000) Application of biological control to vegetation management in forestry. In: Spence, N.R. (ed.) *Proceedings of the X International Symposium on Biological Control of Weeds, 4–14 July 1999.* Montana State University, Bozeman, Montana, pp. 87–96.

Shamoun, S.F. and DeWald, L.E. (2002) Management strategies for dwarf mistletoes: biological, chemical, and genetic approaches. In: Geils, B.W., Tovar, J.C. and Moody, B. (technical coordinators) *Mistletoes of North American Conifers*. US Department of Agriculture Forest Service General Technical Report RMRS-GTR-98, US Department of Agriculture, Rocky Mountain Research Station, Fort Collins, Colorado, pp. 75–82.

Shamoun, S.F. and Hintz, W.E. (1998a) Development and registration of *Chondrostereum purpureum* as a mycoherbicide for hardwood weeds in conifer reforestation sites and utility rights-of-way. In: Burge, M.N. (ed.) *IV International Bioherbicide Workshop – Programme and Abstracts, 6–7 August, 1998*. University of Strathclyde, Glasgow, p. 14.

Shamoun, S.F. and Hintz, W.E. (1998b) Development of *Chondrostereum purpureum* as a biological control agent for red alder in utility rights-of-way. In: Wagner, R.G. and Thompson D.G. (compilers) *Third International Conference on Forest Vegetation Management, Popular Summaries*. Forest Research Information Paper No. 141, Ontario Ministry of Natural Resources, Ontario Forest Research Institute, pp. 308–310.

Shamoun, S.R., Macey, D.E., Prasad, R. and Winder, R.S. (2002) *Acer, Alnus, Betula, Populus* and *Prunus* spp., weedy hardwood trees (Aceraceae, Betulaceae, Salicaceae, Rosaceae). In: Mason, P.G. and Huber, J.T. (eds) *Biological Control Programmes in Canada, 1981–2000*. CAB International, Wallingford, UK, pp. 283–289.

Sharov, A.A. and Liebhold, A.M. (1998a) Bioeconomics of managing the spread of exotic pest species with barrier zones. *Ecological Applications* 8, 833–845.

Sharov, A.A. and Liebhold, A.M. (1998b) Model of slowing the spread of gypsy moth (Lepidoptera: Lymantriidae) with a barrier zone. *Ecological Applications* 8, 1170–1179.

Sharov, A.A. and Liebhold, A.M. (1998c) Quantitative analysis of gypsy moth spread in the Central Appalachians. In: Baumgartner, J., Brandmayr, P. and Manly, B.F.J. (eds) *Population and Community Ecology for Insect Management and Conservation: Proceedings of the Ecology and Population Dynamics Section of the 20th International Congress of Entomology, Florence, Italy, 25–31 August, 1996*. A.A. Balkema, Rotterdam, pp. 99–110.

Sharov, A.A., Liebhold, A.M. and Roberts, E.A. (1998) Optimizing the use of barrier zones to slow the spread of gypsy moth (Lepidoptera: Lymantriidae) in North America. *Journal of Economic Entomology* 91, 165–174.

Sharov, A.A., Leonard, D., Liebhold, A.M., Roberts, E.A. and Dickerson, W. (2002a) 'Slow the Spread': a national program to contain the gypsy moth. *Journal of Forestry* 100, 30–36.

Sharov, A.A., Leonard, D., Liebhold, A.M. and Clemens, N.S. (2002b) Evaluation of preventative treatments in low-density gypsy moth populations using pheromone traps. *Journal of Economic Entomology* 95, 1205–1215.

Sharrow, S.H., Leininger, W.C. and Rhodes, B. (1989) Sheep grazing as silvicultural tool to suppress brush. *Journal of Range Management* 42, 2–4.

Shaw, C.G. III and Calderon, S. (1977) Impact of *Armillaria* root rot in plantations of *Pinus radiata* established on sites converted from indigenous forest. *New Zealand Journal of Forest Science* 7, 359–373.

Shepherd, R.F. and Otvos, I.S. (1986) *Pest Management of Douglas-fir Tussock Moth: Procedures for Insect Monitoring, Problem Evaluation and Control Actions*. Information Report BC-X-270, Canadian Forestry Service, Pacific Forestry Centre, Victoria, British Columbia, 14 pp.

Shepherd, R.F., Otvos, I.S. and Chorney, R.J. (1984a) Pest management of the Douglas-fir tussock moth (Lepidoptera: Lymantriidae): a sequential sampling method to determine egg mass density. *The Canadian Entomologist* 116, 1041–1049.

Shepherd, R.F., Otvos, I.S., Chorney, R.J. and Cunningham, J.C. (1984b) Pest management of the Douglas-fir tussock moth (Lepidoptera: Lymantriidae): prevention of a Douglas-fir tussock moth outbreak through early treatment with a nuclear polyhedrosis virus by ground and aerial applications. *The Canadian Entomologist* 116, 1533–1542.

Shepherd, R.F., Gray, T.G., Chorney, R.J. and Daterman, G.E. (1985) Pest management of the Douglas-fir tussock moth: monitoring endemic populations with pheromone traps to detect incipient outbreaks. *The Canadian Entomologist* 117, 839–848.

Shepherd, R.F., Cunningham, J.C. and Otvos, I.S. (1995) Western spruce budworm, *Choristoneura occidentalis*. In: Armstrong, J.A. and Ives, W.G.H. (eds) *Forest Insect Pests in Canada*. Natural Resources Canada, Canadian Forest Service, Ottawa, Ontario, pp. 119–121.

Shore, T.L. and Safranyik, L. (1992) *Susceptibility and Risk Rating Systems for the Mountain Pine Beetle in Lodgepole Pine Stands*. Information Report BC-X-336, Forestry Canada, Pacific Forestry Centre, Victoria, British Columbia, 12 pp.

Shore, T.L., Safranyik, L. and Lemieux, J.P. (2000) Susceptibility of lodgepole pine stands to the mountain pine beetle: testing of a rating system. *Canadian Journal of Forest Research* 30, 44–49.

Sieber, T.N., Sieber-Canavesi, F. and Dorworth, C.E. (1990a) *Identification of Key Pathogens of Major Coastal Forest Weeds.* Forest Resource Development Agreement Report No. 113, Canadian Forest Service and British Columbia Ministry of Forests, Victoria, British Columbia, 54 pp.

Sieber, T.N., Sieber-Canavesi, F. and Dorworth, C.E. (1990b) Simultaneous stimulation of endophytic *Cryptodiaporthe hystrix* and inhibition of *Acer macrophyllum* callus in dual culture. *Mycologia* 82, 569–575.

Smirnoff, W.A. and Morris, O.N. (1984) Field development of *Bacillus thuringiensis* Berliner in eastern Canada, 1970–80. In: Kelleher, J.S. and Hulme, M.A. (eds) *Biological Control Programmes Against Weeds and Insects in Canada 1969–1980.* Commonwealth Agricultural Bureaux, Slough, UK, pp. 238–247.

Smith, S.M., van Frankenhuyzen, K., Nealis, V.G. and Bourchier, R.S. (2002) *Choristoneura fumiferana* (Clemens), eastern spruce budworm (Tortricidae). In: Mason, P.G. and Huber, J.T. (eds) *Biological Control Programmes in Canada, 1981–2000.* CAB International, Wallingford, UK, pp. 58–68.

Smitley, D.R., Bauer, L.S., Hajek, A.E., Sapio, F.J. and Humber, R.A. (1995) Introduction and establishment of *Entomophaga maimaiga*, a fungal pathogen of gypsy moth (Lepidoptera: Lymantriidae) in Michigan. *Environmental Entomology* 24, 1685–1695.

Somme, L. (1964) Effects of glycerol on cold hardiness in insects. *Canadian Journal of Zoology* 42, 87–101.

Soper, R.S., Shimazu, M., Humber, R.A., Ramos, M.E. and Hajek, A.E. (1988) Isolation and characterization of *Entomophaga maimaiga* sp. nov., a fungal pathogen of gypsy moth, *Lymantria dispar*, from Japan. *Journal of Invertebrate Pathology* 51, 229–241.

Speare, A.T. and Colley, R.H. (1912) *The Artificial Use of the Brown-tail Fungus in Massachusetts, with Practical Suggestions for Private Experiment, and a Brief Note on a Fungus Disease of the Gypsy Caterpillar.* Wright and Potter, Boston, Massachusetts, 31 pp.

Statistics Canada (2002) International Trade. Available at: http://www.statcan.ca/ english/Pgdb/intern.htm

Stelzer, M.J., Neisess, J., Cunningham, J.C. and McPhee, J.R. (1977) Field evaluation of baculovirus stocks against Douglas-fir tussock moth in British Columbia. *Journal of Economic Entomology* 70, 243–246.

Sturrock, R.N. (2000) *Management of Root Diseases by Stumping and Push-falling.* Technology Transfer Note, Canadian Forest Service, Pacific Forestry Centre, Victoria, British Columbia, 8 pp.

Sturrock, R.N., Phillips, E.J. and Fraser, R.G. (1994) *A Trial of Push-Falling to Reduce* Phellinus weirii *Infection of Coastal Douglas-fir.* Forest Resource Development Agreement Report No. 217, Canadian Forest Service and British Columbia Ministry of Forests, Victoria, British Columbia, 22 pp.

Syrett, P., Hill, R.L. and Jessep, C.T. (1985) Conflict of interest in biological control of weeds in New Zealand. In: Delfosse, E.S. (ed.) *Proceedings of the VI International Symposium on the Biological Control of Weeds, August 19–25, 1984, Vancouver, British Columbia.* Agriculture Canada, Ottawa, Ontario, pp. 391–397.

Talerico, R.L. (1984) General biology of the spruce budworm and its hosts. In: Schmitt, D.M., Grimble, D.G. and Searcy, J.L. (eds) *Managing the Spruce Budworm in Eastern North America.* Agriculture Handbook No. 620, US Department of Agriculture Forest Service, Washington, DC, pp. 1–10.

Thorpe, S. (1996) Chemical and manual treatment in the northern interior. In: Comeau, P.G., Harper, G.J, Blache, M.E., Boateng, J.O. and Gilkeson, L.A. (eds) *Integrated Forest Vegetation Management: Options and Applications. Proceedings of the Fifth British Columbia Forest Vegetation Management Workshop, November 29–30, 1993, Richmond, British Columbia.* Forest Resource Development Agreement Report No. 251, Canadian Forest Service and British Columbia Ministry of Forests, Victoria, British Columbia, pp. 61–66.

Ticehurst, M., Fusco, R.A., Kling, R.P. and Unger, J. (1978) Observations on parasites of gypsy moth in first cycle infestations in Pennsylvania from 1974–1977. *Environmental Entomology* 7, 355–358.

Varty, I.W. (1978) *1977 Environmental Surveillance of Insecticide Spray Operations in New Brunswick's Budworm-infested Forests.* Information Report M-X-87, Canadian Forestry Service, Maritimes Forest Research Centre, Fredericton, New Brunswick, 24 pp.

Wall, R.E. (1977) *Ericaceous Ground Cover on Cutover Sites in Southwestern Nova Scotia.* Information Report M-X-71, Canadian Forestry Service, Maritimes Forest Research Centre, Fredericton, New Brunswick, 55 pp.

Wall, R.E. (1983) Fireblight of wild raspberry on clear-cut forest areas. *Environment Canada, Canadian Forestry Service, Research Notes* 3, 2–3.

Eco-labels

An eco-label differentiates environmentally preferable products based on an environmental impact assessment of the product compared with other products in the same category (Loureiro *et al.*, 2001). The eco-label may emphasize any of a number of environmentally related themes, including biodiversity, social justice, energy conservation or agricultural sustainability.

While the purpose of an eco-label is ultimately to market one product over another, labels carry significant educational content as well as implicit and sometimes explicit, guarantees that the labelled product has improved quality or benefits compared with those that are not labelled. For example, 'Dolphin Safe' tuna, one of the first eco-labels to achieve widespread distribution in the USA, became the industry standard because of consumer awareness and concern that dolphins were being killed as a 'by-catch' by commercial tuna fishers.

One of the best-known food-related eco-labels is the organic food designation. This label has achieved success in the market-place by combining explicit environmental guarantees (e.g. that practitioners strive to maintain biological diversity and to maintain and improve soil health on their farms) and implicit quality guarantees (e.g. nutritional superiority, absence of toxic chemical residues). Organic labels cite the promise of sustainable agriculture and health of the soil, in part, by avoiding the use of synthetic fertilizers and pesticides, though naturally occurring pesticides such as rotenone, ryania and copper are generally allowed. The 'Certified Organic' label and its many relatives are governed by various organizations throughout the world. In the USA, organic standards have been codified into federal law (Anon., 2000).

IPM and IP programmes, with their focus on understanding ecological aspects of crop production and on reducing or eliminating the use of pesticides, particularly those with negative non-target effects, are natural candidates for eco-labels. Compliance with IPM/IP systems is similar to those in organic agriculture, though there is generally less emphasis placed on proscribed chemicals and more emphasis placed on monitoring pest populations. When compliance with IPM standards or guidelines is documented, presence of 'Certified IPM' crops in the market-place can assure concerned consumers that the producer is taking measures to protect the environment while growing crops. This presents an opportunity to educate customers about the realities of farming and pest damage and the rationale behind judicious pesticide use.

Challenge: assessing IPM practices

For consumers to give credibility to an eco-label, the labelling programme must have a basis in ecological principles, criteria to judge the progress towards those principles, standards for practical measurement of those criteria and certification documenting that the standards are followed (Kurki and Matheson, 2001).

If IPM is to be used as a marketing label, IPM practices must be documented and certified. Self-certification is not adequate. McDonald and Glynn (1994) found that, while 85% of apple-growers surveyed described themselves as practising IPM, many of them did not follow general IPM practices. For example, only 17% of self-described IPM growers waited for pest action thresholds to be reached before applying pesticides. Hamilton *et al.* (1997) found similar contradictions when growers were asked to classify their own farming systems.

A wide array of IPM evaluation systems exists, including output-oriented systems and input-oriented systems (Swinton and Williams, 1998). Output-oriented systems measure such variables as pesticide use, residues or hazard indices. Input-oriented systems relate to completion of IPM practices and are commonly used to assess farm compliance, as in programmes in Massachusetts (Hollingsworth and Coli, 1999), New York (Petzoldt and Kovach, 1996) and those under the umbrella of the International Organization of Biological Control (IOBC) (Cross and Dickler, 1994). Wisconsin's *Protected Harvest* guidelines (Benbrook *et al.*, 2002) combine outputs (pesticide-hazard indices) and inputs (specific farmer practices).

Consumers interested in purchasing goods with environmental labels indicate a strong preference for an independent or government-associated organization to verify completion of practices. New England consumers, farmers, food processors and retailers supported the role of state departments of agriculture as a certifying agency over private or farmer organizations (Paschall *et al.*, 1992; Pan Atlantic, 1997).

Challenge: marketing

Several challenges to the marketing of IPM-grown produce have been identified by Shelton *et al.* (1990): the term integrated pest management or IPM is perceived negatively by consumers; the terms and concepts associated with IPM (e.g. cultural techniques, biological controls, genetic resistance, pest-population levels, etc.) are technical and complex; there are negative attitudes associated with pests and chemical applications; and there is a lack of personal experience with pesticides among urban consumers. However, as shown above (Bruhn *et al.*, 1992), consumers are receptive to the ideas behind IPM and, by educating consumers through marketing, potential obstacles can be overcome. One approach to educating the public is through eco-labels, discussed below. Another approach, through schools, can provide both passive and active education of IPM. In the USA, many states have passed legislation that requires schools to practise and document their practice of IPM. Thus, individuals associated with schools are learning its terms and principles. Schools are also teaching IPM: the Pennsylvania Department of Education (2002) mandates that children in grades 4, 7, 10 and 12 demonstrate knowledge of increasingly complex concepts of IPM. Various extension programmes and independent agencies have developed numerous IPM curricula (http://www.ipminstitute.org).

Reluctance to promote IPM labels has also formed institutional barriers within grower groups, universities and government organizations. Reasons for opposition to IPM labels include: reluctance to differentiate produce from that of a large clientele of conventional growers; a sense of increasing government regulation; reluctance to discuss the use of pesticides with consumers; and potential costs associated with a programme (e.g. Acuff, 1997). While farmers have expressed reluctance to promote IPM produce as 'good', because it might taint their conventional produce as 'bad', in our experience this is not an issue of concern to consumers (Hollingsworth, 1994).

IPM-grown goods also face many of the same challenges to market penetration that all goods encounter. Beamer and Preston (1991), exploring the process of allocation of shelf space in stores, confirmed that stores employ different strategies in selecting brands and labels. Factors that could affect the acceptance of IPM-grown produce include: variety, in that some stores provide more consumer options than others; uniformity, the need for a store chain to provide very similar products among its stores; seasonality, for stores to provide produce throughout the year; and firm image. A firm's image may emphasize low prices, wide variety, high quality, or, perhaps, awareness of social and environmental causes. Another barrier in the penetration of some markets is the 'loyalty factor' between established growers and store managers or buyers.

Examples of IPM labels

A number of IPM certification programme labels have been promoted in recent years. The following are examples of programmes that promote IPM farms and products through the use of farm certification. These examples do not include labels for organic products, discussed above, but do include those labels that include IPM as a significant promotional trait. The field of IPM certification is dynamic: new IPM labels are developed and others are phased out. Hence, the following examples are not comprehensive, but are provided for the purpose of illustration. Current information on these and other eco-labels can be found in Barstow (2002) and through websites of the IPM Institute of North America (http://www.ipminstitute.org) and the Consumers Union (http://www.ecolabels.org).

Integrated production (IOBC)

Since the late 1980s, the IOBC has provided standards for commercial labelling of pome fruit grown under 'integrated fruit production' (IFP) practices (Malavolta *et al.*, 1998). Now under the label of IP, IOBC standards for stone fruit, soft fruit, viticulture and subtropical fruit have been promulgated. These generalized standards are used to develop regional IP guidelines by local or national organizations or agencies.

Currently, over 50 regional or national agencies in more than 20 countries, in Europe, North America, South America and the South Pacific have IP certification programmes, with the greatest adoption in Western Europe (Table 11.1). Switzerland has the most extensive adoption of IP standards, including 85% of Swiss pome fruit, 81% of stone fruit, 73% of soft fruit and 81% of wine grapes (Dickler, 1999).

Responsible Choice

Administered through Stemilt Growers Inc., a fruit-grower's cooperative in Washington State, Responsible Choice was the first IPM certification programme in the USA (Reed, 1995). Certification requirements were based on those of the IFP programme of the IOBC, though it also incorporated an environmental impact scoring system for pesticide use. Two hundred and fifty growers of apple, cherries and pears participated in the programme.

Wegmans supermarkets

Under the motto, 'Food You Can Feel Good About', Wegmans Food Markets of Rochester, New York has promoted fruit and vegetable products grown using IPM. Farmers who grow for the Wegmans programme adhere to a point-system evaluation developed in cooperation with Cornell University. Although the programme started initially with fresh produce, processed foods (e.g. sweet corn, beans) bearing the Wegmans brand also bear a New York IPM logo. IPM information, including video loops, is prominently displayed in Wegmans stores. IPM-labelled foods do not command a higher price at Wegmans (W. Poole, 2001, personal communication), but support the company's theme, 'Food You Can Feel Good About,' by supporting local

Table 11.1. Regional integrated fruit production (IFP) certification of pome fruit (1000 ha) (from Dickler, 1999).

Country	IP	Conventional	Per cent
Argentina	396	35,104	1.1
Australia	12,000	3,000	80.0
Austria	6,030	1,061	85.0
Croatia	820	540	60.3
Denmark	670	852	44.0
Germany	26,042	5,859	81.6
Great Britain	10,184	3,289	75.6
Italy	32,191	22,699	58.6
Norway	66	1,874	3.4
Poland	5,100	136,900	3.6
Portugal	350	7,650	4.4
Slovenia	1,200	1,868	39.1
Spain	600	16,643	3.5
Switzerland	4,316	778	84.7
Uruguay	183	4,010	4.4
USA (Hood River, Oregon)	105	745	12.4
Total	100,253	242,904	29.2

farmers, providing healthy food and caring about the environment

Partners with Nature

A collaboration of the Massachusetts Department of Food and Agriculture, the University of Massachusetts Extension IPM Programme and the US Department of Agriculture (USDA) Farm Service Agency, Partners with Nature (PWN) certified Massachusetts-grown IPM vegetables and small fruits from 1993 to 1999. Certification was based on best management practices, which were weighted and assigned points (Hollingsworth and Coli, 1999). The programme focused on consumer and grower education, certifying 109 different growers over 6 years. In 1999, the programme certified 53 growers of 13 different crops. When polled, 92% of growers enrolled in the programme agreed that educating the public was the most important role of the programme. Nearly 90% agreed that the programme provided them with a greater understanding of IPM and encouraged greater adoption of IPM practices. A third of the growers noted an increase in profits through the programme. An independent survey (Bonanno, 1997) demonstrated that certified PWN growers used less than half the amount of pesticides than self-described IPM growers (http://www.umass.edu/umext/ipm/ipm_projects/education/partners_with_nature.html).

Eco-OK

Originally founded by the Rain Forest Alliance in 1991, Eco-OK certifies coffee, banana, cocoa and orange farms according to social and environmental standards, including IPM. As of June 2001, 218 farms/cooperatives were certified in Costa Rica, Colombia, Ecuador, Guatemala, Honduras, Panama, Mexico, El Salvador and Hawaii. The programme's guidelines regulate agrochemicals: pesticides approved in the USA are specified, though any pesticides designated by the Pesticide Action Network as in the 'dirty dozen' are not permitted. Other regulations address land use, water treatment, worker conditions, wildlife conservation and community relations.

The Rain Forest Alliance claims that the certification process benefits farmers by increasing production efficiency, reducing costly inputs and improving farm management and that certified farmers have better access to speciality buyers, contract stability, favourable credit options, publicity, technical assistance and niche markets. Most farmers receive a price premium on average from 5 to 20% (http://www.rainforest-alliance.org/programs/cap/index.html).

Food Alliance-approved

The Food Alliance was founded in the north-western USA in 1994 and expanded to the Midwest in 2002. It recognizes farmers who seek alternatives to conventional pesticides, protect soil and water and promote the well-being of farm workers (http://www.thefoodalliance.org). Approved commodities include fruits, vegetables, wheat, livestock, dairy products and wine. In 2002, nearly 200 farms and ranches in eight states participated in the programme and approved products are sold in retail outlets in 20 states. Marketing and promotional campaigns resulted in a peak consumer awareness of the programme of 24% in 2001. The programme plans to expand to the north-eastern USA (Kane and Ennis, 2002).

Core Values Northeast (CVN)

Founded by the consumer education organization, Mothers and Others for a Livable Planet, in 1995 to promote IPM-grown and locally grown apples, CVN certified tree fruit, small fruits and some vegetables. In 2002, they listed 18 participating farms from six states (http://www. corevalues.org).

Linking Environment and Farming (LEAF)

Promoting awareness and adoption of integrated farm management (IFM) in Britain,

LEAF relies on farmers' self-assessment of the economic and environmental consequences of current and potential farm practices. Public awareness is promoted through press releases and tours at participating farms. Thirty-six cooperating farms throughout Britain are currently listed and a label for fresh produce is under development (http://www.leafuk.org).

Protected Harvest

A collaboration of the World Wildlife Fund, the Wisconsin Potato and Vegetable Growers Association and the University of Wisconsin, Protected Harvest certifies potatoes grown using IPM. Guidelines have been under development since 1996 and the certification programme began in 2001. Certification is based on 'preventive practice points', which measure the position of the grower along the 'IPM continuum' (Benbrook, 1996) and an additive index of the toxicity of pesticides used in crop production (Benbrook *et al.*, 2002). Certain pesticides and genetically engineered cultivars are prohibited from use (Sexson and Dlott, 2001). In 2001, 20 growers were enrolled in the programme. Currently, Protected Harvest certifies only 'Healthy Grown' brand potatoes.

Conclusions

While a significant number of consumers are interested in supporting environmentally friendly products, the concepts of IPM are complex and difficult to communicate to the harried shopper whose goal is to purchase the requisite amount of groceries in the minimum amount of time. The success of an IPM label is dependent on the success of its supporting marketing programme and ecolabels with strong marketing support (e.g. Food Alliance) have shown significant growth. Labels are only part of the educational process. Other media, such as brochures, posters, news releases and media events, help support the educational process. Clearly, IPM is much more than a marketing incentive, but marketing efforts not only promote the product, but also educate the consumer about agriculture, affecting attitudes and perceptions about farming in general (Bruhn *et al.*, 1992; Hollingsworth, 1994).

References

Acuff, G. (1997) Labels send wrong message. *Fruit Grower* September, 29.

Anderson, M.D., Hollingsworth, C.S., Van Zee, V., Coli, W.M. and Rhodes, M. (1996) Consumer response to integrated pest management and certification. *Agriculture, Ecosystems and Environment* 60, 97–106.

Anon. (2000) National organic program; final rule. *US Federal Register* 65(245); 80547–80596.

Barstow, C. (2002) *The Eco-Foods Guide,* 1st edn. New Society Publishers, Gabriola Island, British Columbia, Canada, 271 pp.

Beamer, B.G. and Preston, W.P. (1991) Shelf space allocation in the produce section: implications for marketing specialty produce. *Journal of Food Distribution Research* September, 23–35.

Benbrook, C.M. (1996) *Pest Management at the Crossroads.* Consumers Union, Yonkers, New York, 272 pp.

Benbrook, C.M., Sexon, D.L., Wyman, J.A., Stevenson, W.R., Lynch, S., Wallendal, J., Diercks, S., Van Haren, R. and Granadino, C.A. (2002) Developing a pesticide risk assessment tool to monitor progress in reducing reliance on high-risk pesticides. *American Journal of Potato Research* 79, 183–199.

Blend, J. and van Ravenswaay, E.O. (1998) *Consumer Demand for Ecolabelled Apples: Survey Methods and Descriptive Results.* Staff Paper No. 98–20, Department of Agricultural Economics, Michigan State University, East Lansing, Michigan.

Bonanno, R. (1997) *New England Vegetable Growers Association Pesticide Use.* Report to US Environmental Protection Agency Region I.

Bruhn, C., Peterson, S., Phillips, P. and Sakovidh, K. (1992) Consumer response to information on integrated pest management. *Journal of Food Safety* 12, 315–326.

Bunn, D., Feenstra, G.W., Lynch, L. and Sommer, R. (1990) Consumer acceptance of cosmetically imperfect produce. *Journal of Consumer Affairs* 24, 268–279.

Burgess, R., Kovach, J., Petzoldt, C., Shelton, A. and Tette, J. (1989) Results of IPM marketing survey. In: *Proceedings of the Fifty-First Annual New York State Pest Management Conference.* Cornell University, Ithaca, New York.

Cartwright, B., Collins, J.K. and Cuperus, G.W. (1993) Consumer influences on pest control strategies for fruits and vegetables. In: Leslie, A.R. and Cuperus, G.W. (eds) *Successful Implementation of Integrated Pest Management for Agricultural Crops.* Lewis Publishers, Boca Raton, Florida, pp. 151–171.

Collins, J.K., Cuperus, G.W., Cartwright, B., Stark, J.A. and Ebro, L.L. (1992) Consumer attitudes on production systems for fresh produce. *Journal of Sustainable Agriculture* 3, 67–77.

Cross, J.V. and Dickler, E. (1994) Guidelines for integrated production of pome fruits in Europe. Technical Guideline III. *International Organization of Biological Control WPRS Bulletin* 17, 9.

Cuperus, G.W., Kendall, P., Rehe, S., Sachs, S., Frisbie, R., Hall, K., Bruhn, C., Deer, H., Woods, F., Branthaver, B., Weber, G., Poli, P., Buege, D., Linker, M., Andress, E., Wintersteen, W., Dost, F., Damicone, J., Herzfeld, D., Collins, J., Cartwright, B. and McNeal, C.D. (1991) *Integration of Food Safety and Water Quality Concepts Throughout the Food Production, Processing and Distribution Educational Programmes: Using Hazard Analysis Critical Control Point (HACCP) Philosophies.* Oklahoma Cooperative Extension Service Circular E-903, Stillwater, Oklahoma.

Dickler, E. (1999) Encuesta sobre producciòn integrada de frutas en el Periodo 1995–1997. Producciòn integrada de frutas en Europa y en el mundo. In: *Curso international de Producciòn integrada y organica de frutas mayo, General Roca, Rio Negro, Argentina,* Vol. 2, pp. 1–7.

Govindasamy, R. and Italia, J. (1997) *Consumer Response to Integrated Pest Management and Organic Agriculture: an Econometric Analysis.* New Jersey Agricultural Experiment Station Publication P-02137–2-97, New Brunswick, New Jersey.

Govindasamy, R. and Italia, J. (1998) *Consumer Concerns about Pesticide Residues.* New Jersey Agricultural Experiment Station Publication FS896, New Brunswick, New Jersey.

Govindasamy, R., Italia, J. and Rabin, J. (1998) *Consumer Response and Perceptions to Integrated Pest Management Produce.* New Jersey Agricultural Experiment Station Publication P-02137–5-98, New Brunswick, New Jersey.

Grant, J., Tette, J., Petzoldt, C. and Kovach, J. (1990) *Feasibility of an IPM-Grower Recognition Program in New York State.* IPM No. 3, New York Agricultural Experiment Station, Geneva, New York.

Hamilton, G.C., Robson, M.G., Ghidiu, G.M., Samulis, R. and Prostko, E. (1997) 75% adoption of integrated pest management by 2000? A case study from New Jersey. *American Entomologist* 43, 74–78.

Hartman, H. (1996) *The Hartman Report on Food and the Environment: a Consumer's Perspective, Phase I.* The Hartman Group, Bellevue, Washington.

Hartman, H. (1997) *The Hartman Report on Food and the Environment: a Consumer's Perspective, Phase II.* The Hartman Group, Bellevue, Washington.

Hollingsworth, C.S. (1994) Integrated pest management certification: a sign by the road. *American Entomologist* 40, 74–75.

Hollingsworth, C.S. and Coli, W.M. (eds) (1999) *Massachusetts Integrated Pest Management Guidelines: Crop Specific Definitions.* University of Massachusetts Extension Publication IP-IPMA, Amherst, Massachusetts, 66 pp.

Hollingsworth, C.S., Paschall, M.J., Cohen, N.L. and Coli, W.M. (1993) Support in New England for certification and labeling of produce grown using integrated pest management. *American Journal of Alternative Agriculture* 8, 78–84.

Kane, D.J. and Ennis, J.F. (2002) The Food Alliance: transforming a regional success story into a national network. In: *Proceedings of the Conference on Ecolabels and the Greening of the Food Market. Boston, Massachusetts, 7–9 November 2002.* Tufts University, Medford, Massachusetts.

Kurki, A. and Matheson, N. (2001) Eco-labelled foods: profit or problems? *Appropriate Technology Transfer for Rural Areas News* 8(3).

Loureiro, M.L., McCluskey, J.L. and Mittelhammer, R.C. (2001) *Willingness to Pay for Sustainable Agricultural Products. Report to U.S. Federal State Marketing Improvement Programme.* Department of Agricultural Economics, Washington State University, Pullman, Washington, 55 pp.

McDonald, D.G. and Glynn, C.J. (1994) Difficulties in measuring adoption of apple IPM: a case study. *Agriculture, Ecosystems and Environment* 48, 219–230.

Malavolta, C., Avilla, J., Boller, E.F., Gendrier, J.P. and Jorg, E. (1998) Integrated production, recent environmental policy and market trends: the role of IOBC? *International Organization of Biological Control WPRS Bulletin* 21, 23–27.

Morris, P.M., Rosenfield, A. and Bellinger, M. (1993) *What Americans Think about Agrichemicals. A Nationwide Survey on Health, the Environment and Public Policy*. Public Voice for Food and Health Policy, Washington, DC.

Ott, S.L., Huang, C.L. and Misra, S.K. (1991) Consumers' perceptions of risk from pesticide residue and demand for certification of residue-free produce. In Caswell, J.A. (ed.) *Economics of Food Safety*. Elsevier, New York, pp. 176–188.

Pan Atlantic Consultants (1997) *Report to the Massachusetts Department of Food and Agriculture on Consumer Preferences and Attitudes Regarding Massachusetts Grown Agricultural Products*. Massachusetts Department of Food and Agriculture, Boston, Massachusetts.

Paschall, M.J., Hollingsworth, C.S., Coli, W.M. and Cohen, N.L. (1992) Attitudes and perceptions of New England consumers and the food industry toward a certification program for integrated pest management. *Fruit Notes (University of Massachusetts Extension)* 57, 3–11.

Pennsylvania Department of Education (2002) *Academic Standards for Environment and Ecology. 22 Pennsylvania Code*. Chapter 4, Appendix B.

Petzoldt, C. and Kovach, J. (1996) *New York IPM Elements*. New York IPM Program, New York Agricultural Experiment Station, Geneva, New York.

Pool, W.M. (1996) The influence of consumer attitudes and perceptions about pesticides and produce quality on technology transfer. MS thesis, Rochester Institute of Technology, Rochester, New York, 160 pp.

van Ravenswaay, E.O. and Blend, J.R. (1999) Measuring consumer demand for eco-labelled apples. *American Journal of Agricultural Economics* 81, 1078–1083.

Reed, A.N. (1995) Responsible choice: a systems approach to growing, packing and marketing fruit. In: Hull, J. Jr and Perry, R. (eds) *The 125th Annual Report of the Secretary of the State Horticultural Society of Michigan for the Year 1995*. Hartford, Michigan, pp. 68–78.

Sexson, D.L. and Dlott, J. (2001) Companion documentation for the eco-potato standards. Available at: http://ipcm.wisc.edu/bioIPM

Shelton, A.M., Burgess, R., Lanier, J., Petzoldt, C.H., Kovach, J., Grant, J. and Tette, J. (1990) Market research in consumer attitudes to IPM. In: *Proceedings of the Fifty-Second Annual New York State Pest Management Conference*. Cornell University, Ithaca, New York.

Swinton, S.M. and Williams, M.B. (1998) *Assessing the Economic Impacts of Integrated Pest Management: Lessons from the Past, Directions for the Future*. Staff Paper No. 98–12, Department of Agricultural Economics, Michigan State University, East Lansing, Michigan.

Underhill, S.E. and Figueroa, E.E. (1993) *Consumer Preferences for Non-conventionally Grown Produce*. Agricultural Economics Staff Paper 93–07. Department of Applied Economics and Management, Cornell University, Ithaca, New York.

12 The Essential Role of IPM in Promoting Sustainability of Agricultural Production Systems for Future Generations

G.W. Cuperus,[1] R.C. Berberet[1] and R.T. Noyes[2]
[1]*Department of Entomology and Plant Pathology,*
[2]*Department of Biosystems and Agricultural Engineering,*
Oklahoma State University, Stillwater, Oklahoma, USA
E-mail: gcuperus@sbcglobal.net

Introduction

Two of the most common phrases used since the early 1980s in relation to systems designed for production of food and fibre are 'sustainable agriculture' and ' integrated pest management' (IPM). These phrases almost invariably appear in publications that stress the efficiency and profitability of production systems and, more emphatically, the necessity of protecting soil, water and the human food supply from contamination by agrochemicals. Our goal in this chapter is to support the concept that improving the sustainability of production systems and implementation of IPM must be linked. Our basic premise is that employment of principles of IPM is essential to optimizing sustainability of agricultural systems. We believe that the future development and success of IPM are quite important to the sustainability of agriculture for the coming centuries.

Many definitions have been proposed to describe sustainable agriculture and IPM, and we realize the necessity of presenting those that we intend to use in assessing the merits of our basic premise. We consider sustainable agriculture to be 'an agriculture that can evolve indefinitely toward greater human utility, greater efficiency of resource use, and a balance with the environment that is favourable both to humans and to most other species' (Harwood, 1990). The following conditions (modified from Benbrook, 1990) must be satisfied for agricultural systems to be sustainable:

- Soil resources must not be degraded through erosion, salination or contamination with toxic compounds (e.g. pesticides).
- Water resources must be managed to meet needs for irrigation and to prevent degradation with silt and toxic compounds.

The biological and ecological integrity of systems must be preserved through careful management of genetic resources (for both crops and livestock), nutrient cycles and pest species.

- Production systems must be economically viable, returning an acceptable profit to farmers.

- Social expectations must be satisfied, and food and fibre needs must be met in terms of quality and quantity of commodities available at reasonable prices to consumers.

These attributes give a clear emphasis for two primary goals of sustainable systems. These systems must be economically viable, and they must contribute to desirable environmental qualities over the long term.

The basic approach for pest regulation that we envision is consistent with these primary goals for sustainability, as is clearly evident in a recent definition proposed for IPM:

> The judicious use and integration of various pest control tactics in the context of the associated environment of the pests in ways that complement and facilitate the biological and other natural controls of pests to match economic, public health, and environmental goals.
>
> (Anon., 2000)

This definition implies that IPM employs ecologically based management processes developed with an understanding of natural cycles and natural regulators of those species that compete with humans for resources in agricultural production systems (Cate and Hinkle, 1994). Successful IPM programmes, by this definition, are those that will enhance the profitability of the agricultural enterprise and protect the environment for the indefinite future.

Our basic premise that IPM is essential to sustainability stems from our contention that insect pests, pathogenic microorganisms and weeds pose substantial threats to yields and quality of agricultural commodities. It is essential to the productivity and profitability of agriculture that effective means for regulating these species be employed. Production systems that do not include effective pest regulation cannot remain profitable over the long term; stated another way, they cannot be sustainable. Also of great concern have been the increasing difficulties experienced over the last 50 years resulting from reliance on single control agents, particularly chemical pesticides. One of the first to voice these concerns was Rachel Carson in her classic commentary on pesticides entitled *Silent Spring* (Carson, 1962). In the years since 1962, it has become increasingly apparent that employing chemical controls unilaterally will not provide safe and effective regulation of pests over the long term (Cuperus *et al.*, 1990). Problems ranging from pesticide resistance in target species (resulting in control failures) to environmental degradation and contamination of food products by pesticide residues have proved that reliance on unilateral controls seriously detracts from sustainability. The IPM philosophy has evolved through approaches designed to solve these types of problems. Clearly, an attribute that must be added to those previously stated for sustainable systems is that these systems 'employ integrated management programs for safe and effective regulation of pest species'.

Mandates for Sustainability

In our view, there are three mandates to be addressed in assuring sustainability of agricultural production systems. Contributions of IPM are critical to meeting these economic, environmental and social mandates.

Economic mandate

Economic considerations basic to IPM are consistent with the requirement for profitability in sustainable agricultural systems. Rather than insisting upon eradication of pests, an understanding has evolved with the development of IPM that low population densities of pests usually do not threaten the profitable production of agricultural commodities. While this concept is accepted most readily in the discipline of entomology, it has support in plant pathology and weed science as well. This aspect of IPM philosophy has resulted from the development of 'economic injury level' and 'economic threshold' concepts (Stern *et al.*, 1959). The economic injury level concept is based on the assumption that a pest species must be present at some minimum population level before losses resulting from that species will exceed the cost of available controls, typically the cost of applying chemical pesticides. Derived from the economic injury level is the economic threshold, the operative concept for decision making regarding pesticide applications, in particular. The economic threshold is defined as 'the pest population level at which the potential loss exceeds the cost of control'. When control applications are made at the time that pest populations have reached this defined population density, as determined by systematic sampling in commodity-production areas, maximum profits from the pest-control enterprise accrue to farmers.

The economic injury level/economic threshold concept is employed in regulation of most key insect pests and is being adopted with some modifications in control programmes for plant-pathogenic microorganisms and weeds, for which sampling methods and timing of controls may differ from what are usual for insect pests. This sequence of sampling to assess the prevalence of pests followed by decision making using criteria of economic thresholds supports the basic tenet of both IPM and sustainable agriculture, i.e. production systems must be economically viable over the long term. In this regard, the economic injury level and economic threshold are parameters that not only serve as criteria for decision making, but are also important for defining the contribution of effective pest management to the sustainability of production systems. As the frequency of pest occurrence at population densities exceeding the cost of available controls and the profitability of pest-control measures are summarized over time, the value of pest management in production systems can be estimated.

Environmental mandate

The idea that profitability in agricultural production and protection of the environment are jointly attainable goals is central to the philosophy of IPM and sustainable agriculture. It is critical to sustainability that the use of off-farm inputs such as agrochemicals, which may be potential pollutants, and farming practices such as tillage operations, which may contribute to soil erosion, be employed in a manner that does not result in the degradation of soil and water. In addition, there is increasing concern about farming inputs and operations that threaten wildlife species, either by direct mortality or by disruption of habitats. The contention has been made that there is an overemphasis on removing pests from agricultural systems, even when concepts of IPM are employed. Perhaps more attention should be paid to integrated 'habitat' management, addressing in a broad sense the influence of habitat modifications in agriculture on pests and other species that may be present (Zorner, 2000).

The appropriate use of chemical pesticides in agricultural production systems has been central to the development of IPM since its inception (Newsom, 1967; Smith, 1970). Through IPM, the judicious use of artificial pest controls, particularly chemical pesticides, is emphasized to: (i) preserve natural control agents, such as entomophagous insects, and beneficial microorganisms; (ii) decrease the potential for mortality of an array of non-target organisms, such as wildlife species; and (iii) limit the accumulation of toxic residues in the environment. Much more work is needed to develop the same level of comprehensive understanding of effects of habitat modifications typical of agricultural ecosystems on the whole array of plant and animal species in residence. Greater efforts to address concerns about habitat preservation for natural enemies of pests and for wildlife species are being initiated.

Social mandate

Health and well-being are highly valued in societies around the world, resulting in demands for a safe, wholesome food supply that is produced without harm to the environment or hazards to those who work in agriculture. Meeting this mandate for the indefinite future may be quite challenging because of the added expectation in many countries that food must remain relatively inexpensive and be free of any traces of damage by pests. To the present time, pesticides have provided the primary means for limiting pest populations so that abundant supplies of unblemished produce are made available. However, it seems apparent that consumers in the USA may regard the use of pesticides as the most critical hazard to a safe food supply (Pomerantz, 1995). Clearly, the demands of people around the world for a safe, abundant food supply can be met only by the development of sustainable agricultural systems that can effectively combine pest controls with profitability and the maintenance of a safe environment and human food supply. While pesticides are most often cited as the most serious threat to food safety, there is a great need for improved understanding of the hazards posed by bacterial and fungal contaminants in food commodities. In addition to providing a safe and effective means for reducing damage to commodities by all types of pests, IPM is also the primary means by which hazards of both chemical and microbial contamination of food commodities may be greatly decreased.

Public understanding must be extended beyond somewhat vague misgivings about the safety of the environment and the human food supply. Strong negative reactions by people in several countries to the production of food commodities using genetically modified plants or animals have left little doubt of the need to educate the public regarding risks and benefits associated with the products of biotechnology. It is critical, as well, that the public be informed about the potential for pests of all types to limit the availability of foods and certainly to cause food prices to increase. The perception of an unlimited supply of cheap food that exists in many countries could rapidly be proved an illusion if effective means of pest management are not maintained through judicious use of existing control technologies and consistent investments in research to develop new avenues for pest control for the future. Through deliberate and persistent educational efforts, the public must come to appreciate that the development and implementation of IPM, even when programmes involve the use of the products of biotechnology such as genetically modified crops, are a sound investment of resources in support of sustainable agricultural production for the future.

Developing Resources for IPM in the Future

Our intent in this section is not to develop an exhaustive review of all resources that may possibly contribute to more effective pest management for the future, but to select several topic areas that will make essential contributions to sustainable agricultural systems. Among the disciplines of entomology, plant pathology and weed science, there is a wide array of ecological considerations and pest controls that may have utility in the design and implementation of IPM programmes. We have selected several types of resources for our discussions, realizing that there are other resources with which those we describe will be integrated in developing IPM programmes. Also important is the concept that specific resources may have limited value for pest control until they are combined in comprehensive programmes, and we have included a commentary relating to the integration of resources in the sections to follow.

Diagnostic tools

Effective management decisions depend greatly on accurate diagnosis of pest species or biotypes/races in association with symptoms observed in crops or livestock. The need for accurate diagnosis is steadily becoming more acute as particular biotypes/races within pest species adapt to pesticides, crop cultivars or genetically modified strains. While identification of pest species by anatomical or structural features was once sufficient for most instances of pest diagnosis, the more typical case now and in the future will involve more complex procedures involving immunological or nucleic-acid analyses. The most common immunological approaches involve one of three methods for labelling antibodies: (i) with enzymes, such as in the enzyme-linked immunosorbent assay (ELISA); (ii) with coloured particles; and (iii) with materials such as radioisotopes or fluorophores. These approaches have great utility for detecting the presence of specific pathogens in plants or for testing insect-species populations that may serve as vectors for plant pathogens for the presence of specific microorganisms. Their use is increasing greatly for detection of specific microbial products, such as toxins of the bacterium, *Bacillus thuringiensis* (*Bt*), that are present in genetically modified crop cultivars. Diagnostic kits are available for identifying several viruses that infect plants, such as cucumber mosaic virus and tomato spotted-wilt virus, in the field. Also available are field kits for differentiating between eggs of the cotton bollworm, *Helicoverpa zea* (Boddie), and those of the tobacco budworm, *Heliothis virescens* (Fabricius) (Agdia Inc., Elkhart, Indiana).

The same concept can be used in the preparation and labelling of nucleic-acid probes developed for polymerase chain reaction (PCR) processes that are being used to detect viruses and other types of organisms that may be targeted in IPM programmes. It is remarkable that many of these diagnostic processes have been adapted for field use or for field collection of samples for rapid laboratory processing when response time is critical, as is often the case for decision making regarding the application of pest-control agents. It is clear that the availability of these techniques will continue to increase and greatly enhance more traditional diagnostics that employ morphological examinations and bioassays.

Other important technologies that will enhance the value and application of information made available through improved diagnostic capabilities are geographical information systems (GIS) and global positioning systems (GPS). These systems have a great capacity for organizing, mapping and applying information from a variety of diagnostic processes. The capability exists for farmers to map soil types, topography, soil fertility and spatial patterns for pest infestations within individual fields for use in planning appropriate agronomic and pest-control operations. These presentations, when compared over time can provide important insights on patterns of weed interference and areas most likely to be infested by insects or plant pathogens. For soil insects such as the lesser cornstalk borer, *Elasmopalpus lignosellus* (Zeller), that exhibit spatial patterns in fields that are greatly influenced by soil texture and drainage (Berberet *et al.*, 1986), the application of GIS technology can provide valuable maps to assist in scouting and decision making for controls. Mapping over large areas can provide important information on broad migrational patterns of insects and movement from field to field. For both targeted sampling and area-wide management, GIS/GPS technologies will be critical for the incorporation of IPM into precision agriculture for the future (Ellsbury *et al.*, 2000).

Weather forecasting

Accurate weather forecasting and record keeping are essential for IPM programmes. Weather data will continue to be important in two primary ways. The first of these is to provide basic information that, when coupled with species population data, is used in preparing models of seasonal life histories or life systems for crops or pests. For example, insects and plants are poikilothermic, and temperature conditions are usually the primary determinant for population growth in these species. In comparison, for fungal pathogens of crops or of insects, rainfall and relative humidity conditions are major determinants for the prevalence of infections. The models generated from weather data and results of sampling are based on events that have occurred and are often used in simulations to develop predictions about future events, such as population increases in insect species or increased prevalence of infection by a fungal pathogen, given certain weather conditions.

The second major application of weather data is in decision making for current or future pest-control activities. By coupling current weather data with the predictions based on models, forecasts for crop development or pest activity are prepared. These forecasts often have great value in allowing farmers to conduct pest-control activities in a more timely manner, which is often critical – with applications of fungistatic compounds for limiting infections of pathogens, for example. Weather parameters such as degree-day accumulations for insect development or relative humidity conditions will be increasingly important in regard to improved decision making for pest-control activities.

Major improvements in data collection for IPM programmes have occurred with the establishment of weather networks, such as the Oklahoma Mesonet system. This system collects weather data at 117 sites in Oklahoma, which are used to develop comprehensive summaries of current and past conditions. Integrated with this system are a number of programmes that make important contributions to IPM in the state, such as calculations of degree-day accumulations for development of the lucerne weevil, *Hypera postica* (Gyllenhall). These calculations are used in conjunction with current field-sampling data for decision making relative to the need for insecticide applications (Berberet and Mulder, 1993). Increasingly, site-specific weather-data systems are being developed that will enhance decision making on farm-by-farm or field-by-field bases in the future. One such service currently operating is SkyBit, Inc. (Boalsburg, Pennsylvania, USA), a company that has pioneered the implementation of automated weather services at the farm scale (Russo, 2000). Improved means for obtaining and applying weather data are essential for both agronomic and pest-management decisions in crop production.

Transgenic plants

Traditional plant-breeding approaches have provided high-yielding, pest-resistant crops that have made major contributions to modern agricultural systems, in terms of both overall productivity and limiting losses due to insect pests and plant pathogens. In a very real sense, these cultivars have been basic to IPM programmes because they not only directly reduce losses due to pests but have also performed well in combination with pesticides, biocontrols or other approaches in comprehensive management programmes. An excellent example of the contributions of traditional breeding exists in the lucerne crop, for which there are currently over 200 cultivars registered in the USA, covering all ranges of winter dormancy, from those well adapted for production in Canada to those suited for southern California. These cultivars have varying degrees of host resistance to over ten insect pests and plant pathogens that occur in different regions of the world (Anon., 1999).

While products of traditional plant breeding will continue to be important for improved productivity of crops and applications in IPM, the advent of biotechnology has over the last 10 years resulted in the development and release of crop germplasms that are already making remarkable contributions to agriculture. It is not possible at this time to even estimate realistically the great potential of transgenic plants, not only for pest management but also for the production of foods with enhanced nutritional qualities, pharmaceuticals for human health and a variety of industrial products, such as plastics. Two major contributions for IPM to date have involved insect-protected cultivars and those with tolerance to herbicides. Insect-protected cultivars, such as BOLLGARD® (INGARD®) cotton, into which genes controlling production of endotoxins in the bacterium *Bt* have been placed are already showing excellent results in limiting damage by the cotton bollworm, *H. zea* (*armigera*), and tobacco budworm, *H. virescens*, in the USA and several other countries, such as Australia, China and Argentina (Fitt and Wilson, 2000). The efficacy of BOLLGARD® cotton has permitted reductions of >50% in insecticide applications and has greatly enhanced IPM programmes through increased populations of beneficial insects and ready integration with selective insecticides and cultural controls. The technology used to transfer genes from *B. thuringiensis* into crop plants can also be used to incorporate genes controlling pest-resistance factors from a variety of plant sources into cultivated crops.

To date, the greatest number of US Department of Agriculture (USDA) permits issued for regulated technologies relating to transgenic plants have been for herbicide-tolerant crops. Of over 30 million ha of these crops planted worldwide in 1998, about 80% were located in the USA. The most widely grown have been maize, soybeans and cotton having tolerance to glyphosate (Hess and Duke, 2000). However, there are many cultivars available each with tolerance for one of several herbicides in addition to glyphosate. These cultivars hold great promise for IPM because they allow farmers to replace pre-emergence herbicide applications made without assessing weed interference with post-emergence applications that are need-based. Further, the herbicide-tolerance trait can be combined with insect or disease resistance in cultivars to form the foundation for pest-management programmes (Hess and Duke, 2000). As with insect-protected cultivars, the contributions of herbicide-tolerant cultivars can be great, especially when coupled with considerations such as reduced soil erosion in crops, made possible because reduced tillage systems are enhanced with these cultivars.

The potential is great for contributions by transgenic plants to enhance productivity and pest management in all types of crops. Currently, it appears that this potential may be limited by concerns of the general public about the acceptability of transferring genes among plant and animal species. This concern has been fuelled by relatively small groups of people having great fears about application of gene-transfer technology in agriculture. Clearly, the future of transgenic crops and the future availability of this tremendous resource for sustainable agricultural production are greatly dependent upon public education and acceptance around the world.

Biological control

Biological control was regarded as an important tool for managing pests before the inception of the IPM concept. Initially, the approach that generated the greatest interest was classical biological control, i.e. introducing beneficial species to control introduced pests. There have been many instances of successful control of pest species by the classical approach. This approach will clearly remain quite important for the future, especially with the problem of introduced pests becoming more serious around the world, and the critical need to identify and introduce natural enemies to regulate the populations of these pests. For both insect (and other arthropod) pests and weeds, there is a great need worldwide for increased exploration, identification of species (and biotypes/races) of potentially effective natural enemies and importation/establishment of these natural enemies in regions where introduced pests are causing damage. Along with emphases on classical approaches, the potential for effective biological control by both augmentation and conservation of natural enemies for a large number of important pest species, including insects, plant-pathogenic microorganisms and weeds, must be realized.

Although there have been extensive programmes for the mass rearing and release of beneficial insects, such as *Trichogramma* spp., to control a variety of insect pests, perhaps the greatest potential for successful biological control through augmentation of natural enemies exists with pathogenic microorganisms, those infecting insects or weeds and those that compete with plant-pathogenic agents. Currently, there are about 20 microorganisms used in formulating registered microbial insecticides in the USA. While most of these are bacteria, there are also preparations of several viruses and fungi. These products are applied on about 1 million ha annually (Federici, 2000). The utility of these agents has been limited by difficulties in the production and formulation of products, particularly with viruses and fungi, and by limited spectra of activity against pest complexes in comparison with chemical insecticides. However, recent research has indicated that the efficacy of bacteria and viruses may be improved through recombinant DNA technology

(Baum *et al.*, 1998; Treacy, 1998). The advantages of microbial insecticides over chemical insecticides in terms of reduced threats to non-target species and limited potential for environmental degradation are compelling reasons for increasing efforts to improve these agents and increase their utility for IPM programmes.

Although the use of such products has not covered large areas, the integration of biological control agents for plant pathogens into IPM programmes has become more common since the early 1990s. Currently, there are more than 35 commercial products available for the control of plant-pathogenic organisms (Loper and Stockwell, 2000). Notable success attained with two agents in the USA – *Agrobacterium radiobactor* K84, for control of crown gall in nursery tree crops caused by *Agrobacterium tumefaciens*, and *Pseudomonas fluorescens* A506, for control of fire blight in pome fruits caused by *Erwinia amylovora* – has clearly demonstrated that there is great potential for the biological control of plant pathogens. The value of these and other biological controls for plant pathogens is enhanced by their utility in IPM programmes where they have been used in combination with cultural controls and even some types of chemical pesticides (Loper and Stockwell, 2000).

Seven products that are formulations containing microorganisms used as bioherbicides have been registered since 1980. These products have emerged from a total of about 250 agents having proved efficacy for weed control. There has been relatively little interest from the agrochemical industry in these agents because of their host (target) specificity and limited sales potential in comparison with chemical herbicides (Charudattan, 2000). The first of these products to be registered in the USA, a preparation of *Phytophthora palmivora* called DeVine®, has been in use since 1980 for the control of strangler-vine, *Morrenia odorata*, in Florida citrus-production areas (Kenney, 1986). Another product named Collego®, a formulation of *Colletotrichum gloeosporioides* f. sp. *aeschynomene*, has been registered for the control of northern joint-vetch, *Aeschynomene virginica*. At least 12 other potential bioherbicidal agents are currently being investigated for possible registration. Clearly these products could make valuable contributions in the future to IPM programmes where it is desirable and profitable to target particular weed species with agents that do not have adverse effects on non-target species. In reviewing the range of products and applications that have been identified using microbial formulations for the control of insects, plant pathogens and weeds, it seems apparent that these agents will make important contributions to IPM programmes in the future.

Habitat management to enhance the survival and increase the effectiveness of natural enemies is the aim of conservation biological control (Barbosa, 1998). While it is not possible to avoid all disruptions to beneficial species in farming systems, the goal of habitat management in IPM is to develop systems that favour natural enemies over pest species. The relatively newly defined discipline of landscape ecology examines structure, function and change in ecosystems. To enhance the conservation of natural enemies, disturbance regimes in agricultural production systems must be understood (Landis *et al.*, 2000). For successful conservation programmes, resources must be made available in cropland areas and in adjacent riparian areas to sustain beneficial species throughout the year. An excellent example of habitat management for the enhancement of natural enemies are the so-called 'beetle banks' employed in Europe, where raised strips have been planted to grasses within fields to provide overwintering sites for beneficial insects (Thomas *et al.*, 1992). Studies in the USA have also demonstrated the value of providing 'refuge strips' for beneficial insects within fields or at field edges (Landis *et al.*, 2000). In many instances, habitats for natural enemies can be provided within areas that are established in fields or adjacent to fields for purposes other than the conservation of beneficial species, such as grass waterways designed to reduce soil erosion and conservation headlands established for soil, water and wildlife conservation. In the coming years, studies in landscape ecology will make great contributions to IPM, aiding in the design of habitat management systems that enhance the activity of the natural enemies of pests while conserving soil and water for the increased productivity of crops. Simply put, these studies will be essential for the sustainability of crop production systems.

Also essential to conservation and the enhancement of beneficial species in crop production is an improved understanding of the potential effects of pest-resistant cultivars developed by traditional breeding approaches and transgenic, insect-protected cultivars on successful biological control with parasitoids and predators. Knowledge of the multitrophic effects of these cultivars may offer opportunities to reduce potential deleterious effects on beneficial species and/or allow opportunities for the incorporation of traits that will enhance the effectiveness of these species.

Chemical pesticides

Since the early 1950s, chemical pesticides have been the predominant input for pest control in agricultural production systems. The role of chemical pesticides as the most commonly used tool for the regulation of pests has not changed greatly since the inception of IPM. Although the use of pesticides is expected to decline with the adoption of IPM, it is important to state that elimination of chemical pesticides is not regarded as a goal of IPM. Still, there are several objectives for IPM programmes that relate to these compounds, including: (i) reduced reliance on chemical pesticides as unilateral controls for pests; (ii) decreasing the potential for the disruption of non-target species, such as natural enemies of pests; and (iii) limiting contamination of the environment by residues of pesticides. Progress has been made in addressing these objectives through improved pest monitoring and decision-making processes relating to pesticide applications. Use of more efficient and accurate sampling procedures and employing economic thresholds in treatment decisions has resulted in reductions in pesticide use, particularly in insecticide applications in some crops, such as cotton, lucerne and apples. It is essential that research continues in the future to develop even more accurate and efficient sampling/decision-making processes.

Additionally, improving the effectiveness of alternative controls has resulted in reduced reliance on chemical pesticides, as has occurred in lucerne production in the USA. The release of a broad array of multiple-pest resistant cultivars has greatly reduced the need for insecticide applications against species such as the spotted lucerne aphid, *Therioaphis trifolii* f. *maculata* (Buckton), and the pea aphid, *Acyrthosiphon pisum* (Harris). Also, successful biological control of the lucerne weevil with several species of hymenopteran parasites has greatly reduced the need for insecticide applications in the eastern and central areas of the USA (Kingsley *et al.*, 1993).

Much greater efficiency and sustainability in the use of chemical pesticides for the future will be related to the development and registration of new types of products. Since 1980, several new classes of insecticides, fungicides and herbicides have been developed and registered that require the application of but a small fraction of the active ingredient that has been typical of pesticides registered previously to give highly effective pest control. Among these compounds are the strobilurin fungicides (e.g. azoxystrobin), sulphonylurea herbicides (e.g. prosulphuron, triasulphuron) and insecticides such as neonicotinoids (e.g. imidacloprid), spinosyns (e.g. spinosad) and avermectins (e.g. emamectin benzoate). These compounds and other new chemistries yet to be discovered will greatly enhance the contributions of chemical pesticides to IPM programmes by providing a high degree of efficacy while posing limited threats of non-target toxicity and environmental contamination.

New types of compounds and improved application equipment must be combined with improved decision-making processes to promote the safest and most efficient use of chemical pesticides for the future. The use of pesticides in IPM programmes in a manner that contributes to the sustainability of agricultural systems for the indefinite future will require the adoption of the fundamental principles proposed below (adapted from Dennehy, 2000):

- Limiting the use of pesticides to conditions where the potential for losses due to pests exceeds the cost of control (application of economic thresholds) and/or there is no effective alternative to these compounds.
- Selecting pesticides that are most compatible with other components of the IPM programme. The development and registration of new types of compounds holds great promise for more effective integration of chemical and non-chemical controls.
- Applying appropriate rates of active ingredients, taking into account the required efficacy, the potential non-target effects of compounds and the safety of food commodities for consumers.
- Using application technology that provides the greatest safety for those who are applying pesticides, the lowest probability of drift into non-target areas and the best protection against unwarranted environmental contamination.

Meeting the Challenges Posed by Pest Adaptation

One of the greatest challenges to the long-term contributions of IPM programmes to the sustainability of agricultural systems is adaptation by pests, a phenomenon with demonstrated potential to render ineffectual several important types of control agents, including chemical pesticides, crop cultivars developed by traditional plant breeding and transgenic, insect-protected cultivars. It is essential to continued progress for the development and implementation of IPM programmes that the efficacy of these agents be preserved through the adoption of strategies that allow them to be used indefinitely without selecting for resistant populations of pests. If adaptation by key pests to these essential resources for IPM continues at the rate, that has occurred since the early 1970s, the sustainability of agricultural production systems could be placed at risk. Resistance to chemical pesticides is described by the World Health Organization as 'the development of an ability in a strain of some organism to tolerate doses of a toxicant which would prove lethal to the majority of individuals in a normal population of the same species' (World Health Organization, 1957). We define resistance to transgenic, insect-protected cultivars in a similar manner as 'the ability of an arthropod species or plant pathogenic organism to grow and reproduce on plants that would not normally serve as suitable hosts for the species'. Since the report of resistance in the San José scale, *Quadraspidiotus perniciosus* (Comstock), to lime sulphur (Melander, 1914), the problem of pest adaptation to chemical insecticides and/or acaricides has become increasingly more serious. Currently, a total of over 500 species of Arthropoda have developed resistance to chemical pesticides (Pedigo, 1999). The efficacy of many chemical insecticides and acaricides has been compromised and many have been rendered completely ineffectual against certain pests. For some species, such as the Colorado potato beetle, *Leptinotarsa decemlineata* (Say), effective control with chemical insecticides has become nearly impossible because of resistance to chemical toxicants, including even the most recently registered materials, such as imidacloprid (Ferro, 2000). Laboratory studies have shown the potential of pest adaptation to other new forms of insecticides, as in the case of the tobacco budworm, *H. virescens*, exhibiting a resistance ratio greater than 355-fold after 11 generations of selection with spinosad (Bailey *et al.*, 1999).

Before the registration of fungicides having systemic properties and fairly specific modes of action began in the late 1960s, there had been relatively little evidence of resistance in plant pathogens to the conventional dithiocarbamate (e.g. maneb) and phthalimide (e.g. captan). These fungicides exhibit a non-specific mode of action, to which the pathogens apparently cannot adapt (Koller, 1991). However, the resistance of pathogens to newer classes of compounds, such as the benzimidazoles (e.g. benomyl) and phenylamides (e.g. metalaxyl), has resulted in many control failures and the need to apply mixtures of fungicides and to introduce alternative compounds in some crops to achieve consistent disease control (Koller, 1991).

Before 1970, there were no proven cases of resistance in weeds to chemical herbicides (Georghiou, 1986). In part, the absence of resistance could have been due to traditional emphases on non-chemical controls, such as tillage and crop rotation. Also, many weeds complete only one reproductive cycle (generation) per year and have long seed-dormancy periods, which may reduce the rate of adaptation in comparison with insects and plant pathogens. However, proven instances of herbicide resistance in weeds are increasing rapidly with reduced-tillage systems and greater usage of these compounds. Over 50 species have been proved to be resistant to triazine herbicides and, more recently, some species with resistance have been reported to have evolved resistance to the sulphonylurea compounds (e.g. chlorsulphuron).

Adaptation to resistant crop cultivars developed through traditional breeding approaches has been proved for many insect pests (Nielson *et al.*, 1970; Zarrabi *et al.*, 1995; Porter *et al.*, 2000; Ratcliffe *et al.*, 2000) and plant pathogens (Masterbroek, 1984; Kolmer and Dyck, 1994; Young *et al.*, 1994). The potential for pest adaptation to transgenic, insect-protected cultivars such as those that produce the toxins of *B. thuringiensis*, cannot be denied. Resistance to endotoxins produced by *B. thuringiensis* when the bacterium is applied as a microbial insecticide has already been proved (Tabashnik *et al.*, 1990). It is critical to IPM programmes for the future that protocols be developed to guide the appropriate integration of innovative types of chemical pesticides and transgenic plants into IPM programmes. The efficacy of these valuable resources must be protected to the greatest extent possible for the long term. This will necessitate the wide-scale adoption of aggressive methods for the detection of resistance in pests and the implementation of resistance management plans to reduce the possibility of pest adaptation and limit the spread of resistant strains of pests should they evolve in the field.

Resistance detection should include an array of biochemical, immunological, molecular and bioassay approaches. The objective of these methods is to detect/identify the presence of members of pest populations having specific enzymes or other proteins that may be important in detoxifying active ingredients in pesticides or bacterial endotoxins (Roe *et al.*, 2000). Several of these approaches are similar to those described in the section on diagnostics for purposes of pest identification and the delineation of races or biotypes. Bioassays could be used at regular intervals to test samples of pests for increasing levels of tolerance/resistance to toxicants. The 'feeding-disruption assay' is a specific type of bioassay that has been developed to detect changes in susceptibility to the endotoxins of *B. thuringiensis* in populations of the cotton bollworm and tobacco budworm (Roe *et al.*, 2000).

The management of resistance in pest species to active ingredients in synthetic chemical pesticides or microbial toxins is a major challenge to the sustainability of agriculture. Basically, resistance-management programmes require first that an effective plan be formulated with input from researchers, extension specialists, consultants, representatives of agribusiness and clientele groups that are to be the end-users. A basic premise for these plans should be the limitation of pesticide use in accordance with appropriate decision-making criteria for applications. The types of toxicants used should be diversified, with excessive reliance on any particular mode of action being eliminated. There should be serious attempts to utilize alternatives to pesticides and to harmonize pesticide usage with all other types of controls that are included in the IPM programme for the particular crop(s) involved. Secondly, the management plan must be implemented through extensive education programmes with all potential end-users. Crop consultants and representatives of the agrochemical and seed industries must be closely involved in implementation. This step relies heavily on aspects of communication and adoption, which are discussed in the next section of this chapter. Thirdly, as the plan is implemented, monitoring procedures must be used to allow continual evaluation of the efficacy of the management strategies. Regular sampling of pest populations, with testing to detect the presence of resistant individuals in pest populations, must be conducted. Finally, the plan must be flexible enough to allow modifications when they are required based on results of monitoring or of ongoing research.

An ongoing programme following this model for resistance management has operated successfully in Arizona dealing with the silver-leaf whitefly, *Bemisia argentifolii* Bellows & Perring, in winter vegetables, melons, lucerne and cotton (Dennehy, 2000). The primary purpose of this plan has been to maintain the effectiveness of several types of insecticides, including insect growth regulators (buprofezin and pyriproxyfen), synthetic pyrethrins and imidacloprid, in attaining effective, long-term control of the whitefly and other insect pests, such as *Lygus* bugs.

A similar design has been used in formulating a resistance-management plan designed to sustain the effectiveness of the transgenic cotton, INGARD®, against the cotton bollworm/tobacco budworm in Australia under the direction of the Transgenic and Insecticide Management Strategy (TIMS) Committee (Fitt and Wilson, 2000). The primary elements of this plan include:

- Placement of refuges (susceptible cotton) on each farm where INGARD® cotton is grown.
- Enforcement of a defined planting schedule for INGARD® cotton.
- Mandatory cultivation of crop residues to destroy overwintering insect pupae.
- Use of defined economic thresholds to control surviving cotton bollworms.
- Regular monitoring of *Bt* resistance levels in field populations.

Resistance-management plans signal a radical departure from what has been the usual approach for pesticide applications since the early 1950s. We have abundant information and experience regarding the limited potential for long-term efficacy of these agents when there are no management plans designed to reduce pest adaptation. The record shows literally hundreds of cases of pest resistance to these compounds. There is no clear reason to suggest that experience with transgenic plants would be different from what has already occurred with products of traditional plant breeding and with chemical pesticides. In fact, instances of pest adaptation to cultivars developed by traditional plant breeding are abundant. It is in formulating plans to promote the lasting effectiveness of these essential resources for agriculture that the principles of IPM will have great value. These plans will rely on the effective integration of control measures and realistic analyses of benefits, risks and costs relative to all aspects of pest-control programmes, considerations that are viewed as strengths of IPM.

Communication is Critical to the Adoption of IPM by End-users

To provide the greatest contributions towards the sustainability of agricultural systems, the improvement in resources for IPM resulting from innovative research efforts must be matched by more effective efforts to promote the adoption of pest-management programmes. Our discussion to this point has addressed several types of resources that are essential to the enhancement of IPM programmes. For these resources to be utilized effectively there is a critical need for improved information transfer among all groups who work in the development, implementation and application of IPM. Our comments will pertain primarily to means for improving communications among researchers, extension specialists, consultants, representatives of agribusiness concerns and farmers, all of whom must work cooperatively to achieve the greatest success in the adoption of IPM.

A critical aspect that must be addressed is more effective communication and cooperation among scientific disciplines. Our intention is not to suggest that such cooperation does not exist or that there are no examples of excellent interdisciplinary programmes. However, with the passage of over 30 years since cooperation among scientists was first promoted under the banner of IPM, disciplines such as entomology, plant pathology, weed science, and economics continue to use different vocabularies when addressing concepts such as the definition of IPM, formulation and application of economic thresholds and problems/solutions relating to races/strains/biotypes within pest species that have adapted to pesticides or host-plant resistance. In the process of planning all aspects of IPM programmes, scientists from these disciplines must concentrate not only on the critical resources and components needed but also on more effective means of communicating among the disciplines and with clientele. Often, potential end-users of information are required to interpret 'mixed messages' coming from specialists within disciplines that may actually pertain more to different vocabularies than to disagreement on basic pest-management approaches. While progress has certainly been made, much work remains for the delivery of more fully integrated programmes communicated to farmers with a more integrated vocabulary for pest management.

A second issue to be addressed regarding information transfer relates to more effective interaction of researchers and extension specialists with crop consultants and representatives of agribusiness in delivering programmes to farmers. The consultants and business representatives have a vested interest in the success of farmers and must be fully engaged in communicating the most cost-effective and safest means of management to end-users. Pest management is becoming an increasingly complex aspect of agricultural production, and farmers often require assistance in developing the technical skills necessary to adopt the current methods and technologies employed in IPM. It is in the best interest of farmers and consumers that cooperative relationships exist among all parties involved in the development and marketing of products and services for agriculture. And such cooperation can assist greatly in further reducing the volume of 'mixed messages' with which farmers must contend.

For IPM to make the greatest possible contribution to the sustainability of agriculture, a critical aspect of information transfer must be an effective dialogue between farmers and those who develop the programmes and market the technology and services. The utility of resources developed for pest management depends greatly on how well these resources can be adapted to the needs and priorities of individual farmers. While concepts of IPM are often expressed as general principles, applications in IPM must be site-specific. Those who develop IPM programmes must constantly monitor success in the application of these programmes and be prepared to make adjustments as these are needed. One of the best ways to monitor the effectiveness and to improve the adoption of IPM is to seek feedback from the end-users. Information transfer is not intended to be solely from those in academia, consultancies, and industry to the farmers.

Although fact sheets, farmer meetings and field days will continue to provide important avenues for communication for the adoption of IPM, the Internet is a new resource that offers great potential for improving information transfer. The Center for Integrated Pest Management located at North Carolina State University (http://cipm.ncsu.edu) was founded in 1991 to enhance programme development and communication for IPM. This centre was among the first to promote IPM information sharing on the Worldwide Web and now supports an Internet system with cooperators in over 35 states within the USA. Another effort, involving federal–state cooperation in the USA, to make the latest and most accurate pest-management information available to users is the National IPM Network (http://www.ipm.ncsu.edu). Many additional websites have been established to address IPM in particular cropping systems, such as cotton (http://www.gaipm.org/cotton/) and lucerne (http://www.agr.okstate.edu/alfalfa). And, the establishment of websites is by no means restricted to the USA, as sites are appearing around the world using the Internet to promote information transfer for IPM, including those in Australia (http://www.cse.csiro.au/

research), China (http://www.ipmchina.net/) and Europe (http://www.ipmeurope.org) (Stinner, 2000).

It is difficult to predict the overall contributions of the Internet to enhanced information transfer among those who develop, market and employ resources for IPM, even within just the next 10 years. The potential of Internet-driven information sharing worldwide is virtually limitless. As technology improves to make Internet sites more interactive and hardware resources are developed to make Internet access available to greater numbers of end-users worldwide, the adoption of IPM programmes will increase dramatically.

In a world that is rapidly becoming more and more urbanized, it is critical that consensus be developed in the general population to support needs attendant to production and processing of food and fibre commodities. Recent experiences involving negative reactions of the public to research into and the registration of transgenic crops have served to emphasize the necessity for public education and acceptance of methods used in crop production. Effective communication with the public regarding the goals and approaches for IPM is critical to the sustainability of agriculture. Consumers must be educated to appreciate the fact that safer, more effective means of pest management are constantly being developed. In our opinion, the current turmoil regarding transgenic crops has resulted, in large measure, from a failure to communicate the benefits and risks associated with the technology used in the development of these crops. A major challenge for the effective adoption of IPM is communication with the general public. The Internet must be employed to meet this challenge just as it is employed to inform and educate farmers who make the decisions to use IPM.

References

Anon. (1999) *Fall Dormancy and Pest Resistance Ratings for Alfalfa Varieties.* Certified Alfalfa Seed Council, Davis, California

Anon. (2000) Definition for integrated pest management. USDA CSREES, Washington, DC. Available at: http://www.reeusda.gov

Bailey, W.D., Young, H.P. and Roe, R.M. (1999) Laboratory selection of a Tracer-resistant strain of the tobacco budworm and comparisons with field strains from the southeastern US. In: *Proceedings of the Beltwide Cotton Conference.* National Cotton Council, Memphis, Tennessee, pp. 1221–1224.

Barbosa, P. (1998) *Conservation Biological Control.* Academic Press, San Diego, California.

Baum, J.A., Johnson, T.B. and Carlton, B.C. (1998) *Bacillus thuringiensis*, natural and recombinant bioinsecticide products. In: Hall, F.R. and Menn, J.J. (eds) *Methods in Biotechnology,* Vol. 5, *Biopesticides: Use and Delivery.* Humana Press, Totowa, New Jersey, pp. 189–209.

Benbrook, C.M. (1990) Society's stake in sustainable agriculture. In: Edwards, C.A., Lal, R., Madden, P., Miller, R.H. and House, G. (eds) *Sustainable Agricultural Systems.* St Lucie Press, Delray Beach, Florida, pp. 68–76.

Berberet, R. and Mulder, P. (1993) *Scouting for the Alfalfa Weevil in Oklahoma.* OSU Current Report CR-7177, Oklahoma Cooperative Extension Service, Stillwater, Oklahoma.

Berberet, R.C., Wall, R.G. and Peters, D.C. (1986) *The Lesser Cornstalk Borer: a Key Pest of Peanuts in Oklahoma.* Bulletin B-778, Oklahoma Agricultural Experiment Station.

Carson, R. (1962) *Silent Spring.* Fawcett Crest, New York.

Cate, J.R. and Hinkle, M.K. (1994) *Integrated Pest Management: the Path of a Paradigm.* National Audubon Society, Washington, DC.

Charudattan, R. (2000) Current status of biological control of weeds. In: Kennedy, G.G. and Sutton, T.B. (eds) *Emerging Technologies for Integrated Pest Management.* American Phytopathological Society Press, St Paul, Minnesota, pp. 269–288.

Cuperus, G.W., Noyes, R.T., Fargo, W.S., Clary, B.L., Arnold, D.C. and Anderson, K. (1990) Successful management of a high risk stored wheat system in Oklahoma. *American Entomologist* 36, 129–134.

Dennehy, T.J. (2000) Fulfilling the role of resistance management to preserve effectiveness of new insecticide technologies. In: Kennedy, G.G. and Sutton, T.B. (eds) *Emerging Technologies for Integrated Pest Management*. American Phytopathological Society Press, St Paul, Minnesota, pp. 400–417.

Ellsbury, M.M., Clay, S.A., Fleischer, S.J., Chandler, L.D. and Schneider, S.M. (2000) Use of GIS/GPS systems in IPM: progress and reality. In: Kennedy, G.G. and Sutton, T.B. (eds) *Emerging Technologies for Integrated Pest Management*. American Phytopathological Society Press, St Paul, Minnesota, pp. 419–438.

Federici, B.A. (2000) Genetically engineered pathogens of insects for IPM: concepts and status. In: Kennedy, G.G. and Sutton, T.B. (eds) *Emerging Technologies for Integrated Pest Management*. American Phytopathological Society Press, St Paul, Minnesota, pp. 163–176.

Ferro, D.N. (2000) Success and failure of *Bt* products: Colorado potato beetle – a case study. In: Kennedy, G.G. and Sutton, T.B. (eds) *Emerging Technologies for Integrated Pest Management*. American Phytopathological Society Press, St Paul, Minnesota, pp. 177–189.

Fitt, G.P. and Wilson, L.J. (2000) Genetic engineering in IPM: Bt cotton. In: Kennedy, G.G. and Sutton, T.B. (eds) *Emerging Technologies for Integrated Pest Management*. American Phytopathological Society Press, St Paul, Minnesota, pp. 108–125.

Georghiou, G.P. (1986) *Pesticide Resistance: Strategies and Tactics for Management*. National Academy Press, Washington, DC.

Harwood, R.R. (1990) A history of sustainable agriculture. In: Edwards, C.A., Lal, R., Madden, P., Miller, R.H. and House, G. (eds) *Sustainable Agricultural Systems*. St Lucie Press, Delray Beach, Florida, pp. 3–19.

Hess, F.D. Sr and Duke, S.O. (2000) Genetic engineering in IPM: a case study: herbicide tolerance. In: Kennedy, G.G. and Sutton, T.B. (eds) *Emerging Technologies for Integrated Pest Management*. American Phytopathological Society Press, St Paul, Minnesota, pp. 126–140.

Kenney, D.S. (1986) De-Vine – the way it was developed – an industrialist's view. *Weed Science* 34 (suppl. 1), 15–16.

Kingsley, P.C., Bryan, M.D., Day, W.H., Burger, T.L., Dysart, R.J. and Schwalbe, C.P. (1993) Alfalfa weevil (Coleoptera: Curculionidae) biological control: spreading the benefits. *Environmental Entomology* 22, 1234–1250.

Koller, W. (1991) Fungicide resistance in plant pathogens. In: Pimentel, D. (ed.) *Handbook of Pest Management in Agriculture*, Vol. II. CRC Press, Boca Raton, Florida, pp. 679–720.

Kolmer, J.A. and Dyck, P.L. (1994) Gene expression in the *Triticum aestivum–Puccinia recondita* fsp. *tritici* gene-for-gene system. *Phytopathology* 84, 437–440.

Landis, D.A., Menalled, F.B., Lee, J.C., Carmona, D.M. and Perez-Valdez, A. (2000) Habitat management to enhance biological control in IPM. In: Kennedy, G.G. and Sutton, T.B. (eds) *Emerging Technologies for Integrated Pest Management*. American Phytopathological Society Press, St Paul, Minnesota, pp. 226–239.

Loper, J.E. and Stockwell, V.O. (2000) Current status of biological control of plant diseases. In: Kennedy, G.G. and Sutton, T.B. (eds) *Emerging Technologies for Integrated Pest Management*. American Phytopathological Society Press, St Paul, Minnesota, pp. 240–256.

Masterbroek, H.D. (1984) Utility of defeated resistance genes to powdery mildew, *Erysiphe graminis* fsp. *hordei*, in spring barley variety mixtures. *Netherlands Journal of Plant Pathology* 90, 257–265.

Melander, A.L. (1914) Can insects become resistant to sprays? *Journal of Economic Entomology* 7, 167.

Newsom, L.D. (1967) Consequences of insecticide use on non-target organisms. *Annual Review of Entomology* 12, 257–286.

Nielson, M.W., Don, H., Schonhorst, M.H., Lehman, W.F. and Marble, V.L. (1970) Biotypes of the spotted alfalfa aphid in western United States. *Journal of Economic Entomology* 63, 1822–1825.

Pedigo, L.P. (1999) *Entomology and Pest Management*. Prentice Hall, Upper Saddle River, New Jersey.

Pomerantz, M.L. (1995) A profile of the fresh produce consumer. *The Packer* 54, 30–39.

Porter, D.R., Burd, J.D., Shufran, K.A. and Webster, J.A. (2000) Efficacy of pyramiding greenbug (Homoptera: Aphididae) resistance genes in wheat. *Journal of Economic Entomology* 93, 1315–1318.

Ratcliffe, R.H., Cambron, S.E., Flanders, K.L., Bosque-Perez, N.A., Clement, S.L. and Ohm, H.W. (2000) Biotype composition of hessian fly (Diptera: Cecidomyiidae) populations from the southeastern, midwestern, and northwestern United States and virulence to resistance genes in wheat. *Journal of Economic Entomology* 93, 1319–1328.

Roe, R.M., Bailey, W.D., Gould, F., Sorenson, C.E., Kennedy, G.G., Bacheler, J.S., Rose, R.L., Hodgson, E. and Sutula, C.L. (2000) Detection of resistant insects and IPM. In: Kennedy, G.G. and Sutton, T.B.

(eds) *Emerging Technologies for Integrated Pest Management*. American Phytopathological Society Press, St Paul, Minnesota, pp. 67–84.

Russo, J.M. (2000) Weather forecasting for IPM. In: Kennedy, G.G. and Sutton, T.B. (eds) *Emerging Technologies for Integrated Pest Management*. American Phytopathological Society Press, St Paul, Minnesota, pp. 453–473.

Smith, R.F. (1970) Pesticides: their use and limitations in pest management. In: Rabb, R.L. and Guthrie, F.E. (eds) *Concepts of Pest Management*. North Carolina State University, Raleigh, North Carolina, pp. 103–118.

Stern, V.M., Smith, R.F., van den Bosch, R. and Hagen, K.S. (1959) The integrated control concept. *Hilgardia* 29, 81–101.

Stinner, R.E. (2000) Information management: past, present, and future. In: Kennedy, G.G. and Sutton, T.B. (eds) *Emerging Technologies for Integrated Pest Management*. American Phytopathological Society Press, St Paul, Minnesota, pp. 474–481.

Tabashnik, B.E., Cushing, N.L., Finson, N. and Johnson, M.W. (1990) Field development of resistance to *Bacillus thuringiensis* in diamondback moth (Lepidoptera: Plutellidae). *Journal of Economic Entomology* 83, 1671–1676.

Thomas, M.B., Wratten, S.D. and Sotherton, N.W. (1992) Creation of 'island' habitats in farmland to manipulate populations of beneficial arthropods: predator densities and species composition. *Journal of Applied Ecology* 29, 524–531.

Treacy, M.F. (1998) Recombinant baculoviruses. In: Hall, F.R. and Menn, J.J. (eds) *Methods in Biotechnology*, Vol. 5, *Biopesticides: Use and Delivery*. Humana Press, Totowa, New Jersey, pp. 321–367.

World Health Organization (1957) *Seventh Report of the Expert Committee on Insecticides*. Technical Report Series No, 125, 31 pp.

Young, B.A., St Martin, S.K., Schmitthenner, A.F., Buzzell, R.I. and McBlain, B.A. (1994) Absence of residual effects of defeated resistance genes on *Phytophthora* rot of soybean. *Crop Science* 34, 409–414.

Zarrabi, A.A., Berberet, R.C. and Caddel, J.L. (1995) New biotype of *Acyrthosiphon kondoi* (Homoptera: Aphididae) on alfalfa in Oklahoma. *Journal of Economic Entomology* 88, 1461–1465.

Zorner, P.S. (2000) Shifting agricultural and ecological context for IPM. In: Kennedy, G.G. and Sutton, T.B. (eds) *Emerging Technologies for Integrated Pest Management*. American Phytopathological Society Press, St Paul, Minnesota, pp. 32–41.

13 Opportunities and Challenges for IPM in Developing Countries

David Bergvinson
International Maize and Wheat Improvement Center (CIMMYT), El Batán, Mexico, CP 56130, Mexico
E-mail: d.bergvinson@cgiar.org

Introduction

Dramatic increases in rice and wheat production during the green revolution were achieved through the introduction of short-stature, high-yielding wheat and rice varieties into Asia as part of a production package that included fertilizers and pesticides to meet the urgent food demands during the 1960s and 1970s. During the next two decades, the world will have to feed approximately 2.5 billion more people, with less arable land, fewer non-renewable and renewable resources and a smaller percentage of people working in the agricultural sector. This challenge will be particularly acute in the less developed countries (LDCs), where food production will largely come from production increases rather than expansion into marginal lands. This will take place against the backdrop of climate change and its poorly understood impact on agroecosystems.

Today, the rate of production gains from conventional breeding for staple food crops has slowed from 2.9% (green revolution era) to 1.9% in LDCs, with gains frequently being offset by environmental stresses and declining soil fertility. In order to address the looming production shortfall, crop production must increase as much as it did during the first green revolution (Rosegrant *et al.*, 2001). To meet this challenge, integrated crop management (ICM) technologies must be applied to overcome the production constraints faced by resource-poor farmers in LDCs. A central component of ICM is integrated pest management (IPM). To realize the full potential of IPM, sufficient government support for IPM research and a favourable policy environment must be maintained (Pinstrup-Andersen, 2002). Researchers, extension providers and farmers will need to collaborate to develop, adapt and adopt IPM technologies and also to influence the formulation of policies that favour the adoption of sustainable cropping technologies generally.

Using a strengths, weaknesses, opportunities and threats (SWOT) analysis, this chapter highlights factors that will bear directly on the acceptance and use of IPM within the context of sustainable cropping systems in LDCs. Because each region of the world has unique constraints and advantages related to the adoption and deployment of IPM, sections of the chapter briefly look at such particulars for Africa, Latin America and Asia. A short exploration of what the future might hold for IPM in LCDs is explored through the use of four scenarios, which were developed by International

 Integrated Pest Management: Potential, Constraints and Challenges (eds O. Koul, G.S. Dhaliwal and G.W. Cuperus)

Maize and Wheat Improvement Center (Spanish acronym CIMMYT) scientists to assist them in mapping out directions for crop research in coming years. The chapter finishes with concluding observations and thoughts about how to ensure that best-practice pest-management technologies and dissemination strategies meet the needs of LDC farmers in the 21st century.

Strengths: Factors Encouraging IPM Use in Developing Countries

IPM enjoys the support of the international development community because it is recognized as a key component to sustainable rural development in LDCs. It is uniquely positioned for deployment in LDCs, thanks to the increasing trend towards participatory research methods that enable farmers to fully engage in the development and deployment of IPM. Through socio-economic research in LDCs, our knowledge of indigenous IPM practices has increased, as has our understanding of cropping constraints posed by diverse agroecological and socio-economic conditions. LDCs host a wealth of diversity in the form of cropping systems and species diversity in tropical and subtropical ecologies, which will serve as a tremendous asset to the development and adoption of IPM technologies. Centres of origin for the major crops found in LDCs already serve the world by providing genetic diversity for crop-improvement programmes, in both industrial countries and LDCs, through their own breeding efforts and those of international agricultural research centres (IARCs). This section discusses in greater detail how all these factors serve as a positive foundation for the expanded use of IPM in the farm fields of developing countries.

The role of IPM in the context of integrated crop management

Although IPM has a long history in both industrial countries and LDCs (Morse and Buhler, 1997), it is now being viewed as a key component within the wider context of sustainable cropping systems. Due to the complexity of constraints that farmers face in commercial and small-scale agriculture – including marketing, economic use of inputs and complex biological interactions – IPM programmes have evolved to become part of ICM (Meerman *et al.*, 1996). The objective of ICM is to manage a production system in a way that optimizes the use of natural resources, protects the environment and maximizes output in a sustainable manner, taking into consideration the socio-economic and technological framework and the interaction of management components. The management components focus on different production constraints, with control of pests assuming a prominent position, especially in developing countries, where insect and disease pressures tend to be severe due to the subtropical and tropical climates found there. With ICM gaining widespread acceptance, the importance of and preference for IPM have also grown.

Geier described the spectrum of IPM philosophies. At the tactical end of the spectrum is pest control, which 'amounts to hardly more than bulldozing nature without thought to consequences and frequently creates more problems than it solves' (Geier, 1966: 471). At the strategic end of the spectrum is integrated 'pest management', which 'is intended to convey the idea of intelligent manipulation of nature for man's lasting benefit, as in wildlife management' (Geier, 1966: 474). Morse and Buhler (1997) summarized the differences between control and management as a control beating the pest into submission using direct interventions, with little knowledge of the pest population and community interactions and usually operating at a local level (i.e. field). Management is considered a knowledge-intensive control strategy that takes into account the pest complex and social issues to achieve a balanced ecosystem, using a diverse range of interventions to bring the pest population below an economic threshold over a large area. The management philosophy has been supported in developing countries, in part, through loans by the World Bank and their Operational Directive 4.03, which states: 'The Bank's policy is to promote effective and environmentally sound pest management practices in Bank-supported agricultural development' (World Bank, 1992). This policy was reinforced by funding in the USA for IPM, totalling more than US$180 million annually, to promote IPM research and education as part of a campaign to implement IPM on 75% of the nation's crop area by the year 2000 (Jacobsen, 1996). Clearly there is political will and support for IPM in both the developing and developed world, but how this will play out in the future and what types of technologies are likely to be promoted under different policy environments remain open questions.

Participatory learning and technology development

Historically, technology development and transfer has followed the so-called 'transfer of technology' (TOT) approach, in which extension played a pivotal role in communicating information between the farmer and researcher (Fig. 13.1A). This model is a 'top-down' approach to technology development in which information flows predominantly from the research to the farmer, as new crop-management interventions are developed. Farmer input has been largely confined to fine-tuning the intervention. Under the best of circumstances, information is fed back to research scientists through extension, although, in actual practice, little, if any, feedback occurs. This model has worked for many agricultural interventions in industrial countries and has evolved to include the private sector as a major player in technology delivery with improved feedback from both farmers and consumers on the types of technologies that should be developed. Consumer feedback has largely been channelled through the popular press to influence government policy, which sets research mandates. Although this model has worked well for simple interventions, it has achieved less success with knowledge-intensive technologies, such as IPM. Morse and Buhler (1997) point out that, under the TOT model, blame for the lack of IPM adoption falls on the shoulders of extension providers, with their shortcoming attributed to being understaffed or inadequately trained. This is often the case for LDCs, where extension services are often poorly staffed and equipped and

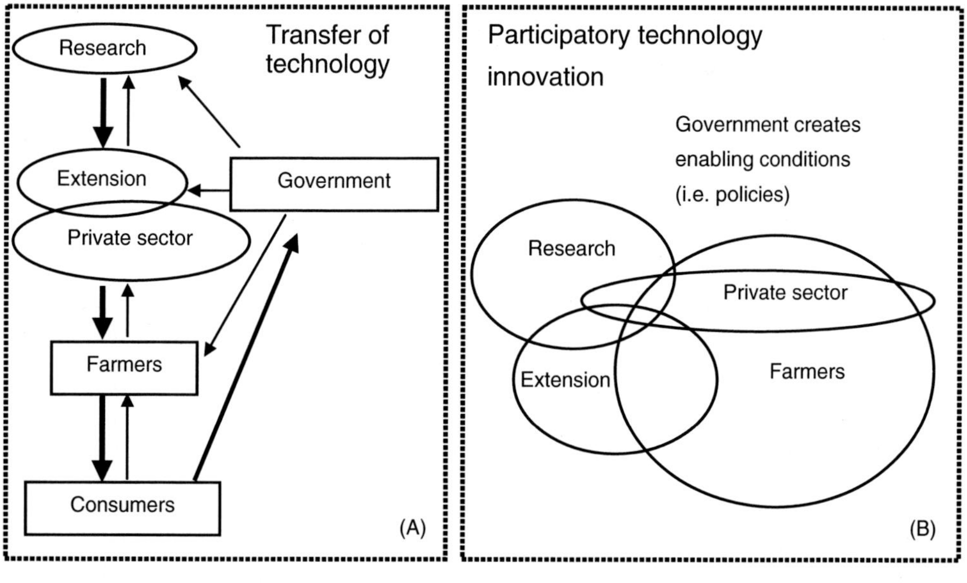

Fig. 13.1. Technology dissemination models for IPM. (A) Transfer of technology model and (B) participatory technology innovation model. (Adapted from Bruin and Meerman, 2001:22.)

operating funds for transferring technologies to farmers are limited or non-existent.

During the last two decades, an increasing body of evidence has shown that participatory methods for technology development and transfer result in high adoption rates, due to a sense of ownership of the process by those for whom the technology is targeted. Morse and Buhler (1997) have called for a paradigm shift from the 'technology first' approach, in which farmers play a subordinate role, to a 'farmer first' approach. This view places IPM within the broad context of the needs of the local people, with the farmers' constraints being clearly identified and indigenous technical knowledge (ITK) about crop management being recognized as a starting point for further technology development. This was promoted earlier by Matteson *et al.* (1984), who stated: 'Useful pest control characteristics of traditional systems must be preserved or augmented whenever possible and suggested changes should be tested to make sure they are consonant with the farmers' environment and socio-economic circumstances.'

Under this participatory model (Fig. 13.1B), farmers are seen as equal partners in the development of IPM technologies through their contribution of information about ITK, socio-economic factors and ecological conditions. Scientists and extension providers contribute basic IPM principles, assist in a better understanding of the agroecosystem, offer experimental know-how to assess new IPM options and facilitate interaction and effective feedback. The private sector is an important partner in delivering input technologies (i.e. improved seed, appropriate pesticides) and marketing. Practical examples of the effective application of the participatory model may be found in various IPM and ICM development projects in Zanzibar on rice, cassava, banana and vegetables (Bruin and Meerman, 2001). Another example is found in Zambia where participatory methods were used to assess ITK and production constraints for maize farmers (Nkunika, 2002). Using a farmer field school (FFS) approach, demonstration plots integrated improved fertilizer and cultural practices (early weeding) with ITK for pest control. Researchers found that farmers employed effective ITK to manage maize pests – mainly storage pests and termites – and when these were integrated with ICM techniques it led to a 37% increase in productivity.

Tools for participatory research and technology development have evolved to address interrelated steps that cover the following areas: defining the problem, understanding the problem, developing and testing solutions making the interventions available to farmers to choose from and monitoring adoption to define impact and refine the intervention. The key first participatory step – defining the problem – is achieved using rapid rural appraisal (RRA) or participatory rural appraisal (PRA) in the targeted communities to characterize the social organization, agricultural system and pest constraints. An RRA is less participatory than a PRA, in which the survey results are discussed and a joint plan of action is formulated within the community (Bruin and Meerman, 2001). Both these techniques, however, enable researchers to document ITK used by farmers to address constraints in crop production. Participatory technology development (PTD) looks for practical solutions to the problems, which include the use of ITK. Farmers and researchers work together to characterize the agroecosystem followed by farmer-led testing, monitoring and evaluation of different interventions. During the process, farmers develop the analytical capacity to sustain technology development to address future constraints.

FFSs are another form of PTD, whereby farmers learn by doing and eventually become 'IPM experts in their own field' (Gallagher, 1992). Training of trainers is the first step in which IPM training, if required, is provided for extension providers, along with facilitation skills and the principles of adult education. Trainers then engage farmers in an intensive 3-month course during the cropping season to learn basic IPM principles and develop a better understanding of the agroecosystem, including pests, biological controls, environmental hazards and good cropping practices. The four goals of FFSs for IPM are to grow a healthy crop, to conserve natural enemies, to conduct regular field observations and to make farmers IPM experts. Farmers are encouraged to share their knowledge with their neighbours to disseminate lessons learned from the FFSs. This participatory approach has been adopted by various development organizations to develop IPM programmes for different crops around the world.

Another participatory technique, recently employed in southern Africa, is the 'mother–baby' testing system, used specifically to evaluate experimental and commercial maize varieties (Bänziger and de Meyer, 2002). In this testing system, a complete trial, called the 'mother', is grown under recommended conditions at a school or agricultural training centre to serve as the focal point for community discussions about the best varieties. Individual farmers in the community are given four of these varieties, each called a 'baby', to grow under their own management practices. At harvest, farmers evaluate the crops for different traits and the relative importance of these traits is determined through farmer interviews. This approach is very relevant to what Brundtland (1987) calls 'complex, diverse and risk-prone' (CDR) agriculture, which is practised by the majority of resource-poor farmers in developing countries. Varieties undergoing testing have been improved for resistance to important biotic stresses and tolerance to drought and low soil fertility – production constraints often faced by resource-poor farmers. This scheme may well lend itself to future testing of localized IPM interventions.

Communication methods

Understanding how farmers perceive and classify pests, as well as their terminology for ITK is a basic starting-point for effective communication between scientists and farmers (Bentley and Rodríguez, 2001). The importance of this communication linkage has received renewed attention with the emergence of farmer participatory learning and technology development. Communication is a two-way street, in which scientists need to communicate concepts that may not be apparent to farmers based on observation (e.g. metamorphosis of insects or spores of diseases), while farmers need to convey ITK to scientists, based on years of farmer experimentation. Looking strictly at IPM, comprehension of the 'folk' language and classification of the farmers' biotic world based on cultural experiences is essential for effective communication about simple concepts (tactical IPM and pesticide handling) and knowledge-intensive technologies (strategic IPM and managing the agroecosystem).

Innovative thinking about how best to communicate messages related to reducing pesticide use and exposure in developing countries has advanced considerably since the early 1990s. Calendar spraying serves as a useful illustration. In some parts of the developing world, those who do not spray their crops at the first sign of insect damage are perceived as lazy or poor agriculturalists. This practice, however, often conflicts with those advocated by IPM tactics. In this situation, the question arises, how does one change the perceptions and consequently the practices of whole farming communities?

One approach, called 'conflict information', is to shock the farmers and create a state of dissonance that challenges them to evaluate and change their perceptions in respect of a simple rule of thumb, or heuristic, that is scientifically supported but in conflict with their prevailing views (Heong and Escalada, 1997). This approach was tested on rice farmers in the Philippines, with the heuristic being that leaf-folder control is not necessary during the first 30 days after transplanting. Research had shown, contrary to farmers' beliefs, that rice plants tolerate early defoliation by the rice leaf-folder, *Cnaphalocrocis medinalis* (Guenée) (Heong *et al.*, 1994). Farmers were asked to measure out a 500 m^2 plot that would not be sprayed with insecticide for the first 30 days after transplanting to later compare with their plots that were sprayed. It was quickly observed that yields from the unsprayed plot were equal to those of the sprayed plots. Farmer opinion on the need to spray for leaf-feeding insects early in the season changed from 62% supporting it down to 10%.

Furthermore, related to benefits of adopting the new practice, 94% recognized that there would be financial savings, 35% cited reduced labour requirements, 26% perceived reduced exposure to health hazards, 11% noted reduced insecticide residues in rice and 2% said it would help conserve biological control agents. This approach holds considerable promise in reaching a larger audience of farmers where the overuse of insecticides is prevalent.

Another innovative approach enjoying wide use is 'social marketing', which employs marketing principles and techniques to advance a social cause, idea or behaviour (Kotler and Zaltman, 1971). Marketing a commercial product is achieved by setting a measurable objective, marketing research, developing a product that meets a genuine need, creating awareness and demand through advertising, and establishing prices that generate a return and meet the original marketing objective. Atkin and Leisinger (2000) report on the impact of social marketing to promote the adoption of safe and effective practices for insecticide use for maize in Mexico and for cotton in India and Zimbabwe. Surveys were conducted to assess knowledge, attitudes and practices (KAP) as a baseline for identifying pesticide-related illnesses and safety practices. Following the KAP survey, the prominent handling and application health hazards were addressed through communication campaigns using different media. The cost–benefit analysis of this study identified, measured and validated the benefits (improved health, reduced use of sprays based on economic injury level (EIL)) and costs (protective gear, equipment maintenance, lost opportunity due to time spent scouting and washing after pesticide applications). Critical to this approach are choosing the most effective media to promote the message and couching the message in socially and culturally acceptable terms. In Mexico, comic books and posters proved effective, while local radio and theatrical plays for parents and schoolchildren worked well in India and Zimbabwe, respectively. The lessons learnt from these case studies were: (i) the message must be simple; (ii) costly interventions, if not subsidized, will not be maintained; and (iii) the media used should be an accepted form of communication for the whole farm family, because social obligations (family) often enforced the pesticide safety message.

Biodiversity

Developing countries predominantly have subtropical or tropical agroecosystems, which generally have high levels of biodiversity. Altieri (1995) described biodiversity as the key element in sustainable agriculture, not only in the maintenance of diversified cropping systems, but also in the maintenance of diverse varieties of a given crop. Within the spectrum of LDC farming systems, small-scale farming systems generally employ mixed cropping systems and subsequently promote greater diversity than large-scale, commercial agriculture. Several examples of the beneficial effects of increased biodiversity are reported in this book (see Verkerk, Chapter 4, this volume) and in recent reviews (Landis *et al.*, 2000).

Since the early 1990s, the concept of agroecology has been promoted by many groups working in LDCs as a means of achieving sustainable crop production through the promotion of environmentally and socially sensitive technologies, especially for small-scale or subsistence farming systems (Altieri, 1993). Altieri (1995) depicts an agroecological approach as the convergence of three circles: social, economic and environmental goals. Under the environmental goal, biodiversity and the stability that results are important objectives in low-input cropping systems. Altieri's approach emphasizes the social and environmental components in cropping systems to promote equity and the use of local resources while sustaining yields and economic viability.

In relation to IPM, management of the environment overlaps with habitat management, which plays an important role in the conservation of biological control agents, namely, parasitoids and predators. Maintenance of suitable conditions within the agricultural landscape is necessary to ensure food (pollen and nectar sources), shelter, hosts and alternative prey for the biological control agents, in order to increase their survival, fecundity, longevity and ultimate effectiveness (Landis *et al.*, 2000). Knowledgeable use of habitat management holds considerable promise both for buffering the environment for biological control agents and, as we shall see later, to serve as a refugia in managing insect resistance to genetically engineered (GE) plants.

Several good examples of biological control through habitat management have been documented in LDCs. One particularly innovative strategy, the 'push–pull strategy', is currently being exploited in eastern Africa to control stem borers in small-scale maize farming systems (Khan *et al.*, 1997). The two predominant stem borers, *Chilo partellus* (Swinhoe) and *Busseola fusca* (Fuller), are 'pulled' from the maize plots by planting highly susceptible trap crops around the field perimeter – in this case, napier grass (*Pennisetum purpureum* Schumach) and Sudan grass (*Sorghum vulgare sudanense* Stapf.). The 'push' component is provided by intercropping maize with grasses that repel the stem borers, such as molasses grass (*Melinis minutiflora* Beauv.) or species of *Desmodium* (*D. uncinatum* Jacq. or *D. intortum* Urb). Although the repellent crop does not provide complete control, *M. minutiflora* produces a volatile, nonatriene, which attracts the parasitoid *Cotesia sesamiae* (Cameron), thereby increasing the rate of parasitism fourfold (Khan *et al.*, 1997). Recent on-farm studies confirmed the efficacy of the push–pull strategy in reducing stem borer populations to 25% of the levels found in control plots (Khan *et al.*, 2001). The strategy yields other benefits as well. In the predominantly mixed crop–livestock farming systems of eastern Africa, the napier grass, a popular livestock forage, contributes directly to farm income. Meanwhile, the desmodium, a leguminous plant, when sown in alternating rows with maize, increased soil fertility, which resulted in the suppression of *Striga* spp., a parasitic weed that is prevalent in low-fertility soils in Africa. An economic analysis of the push–pull strategy generated from this study found the cost : benefit ratio to be 1 : 1.4 for improved maize varieties grown with no intervention and 1 : 2.3 for the push–pull strategy (using napier grass and desmodium).

Secondary plant compounds are also an important part of the diversity equation in LDC ecologies. There has been a growing interest since the inception of IPM to exploit natural plant products for pest control. The most popular examples are pyrethrum and neem oil. Pyrethrum, derived from chrysanthemum plants, originated in Africa and has served as the model molecule for the synthesis of pyrethroid insecticides which have proved to be effective and safe. The neem tree (*Azadirachta indica*) from India produces a mixture of insecticidal tetranortriterpenoids, with the most active being azadirachtin (Schmutterer, 1995). Over the centuries, Indians have processed neem seeds, bark and leaves for their insecticidal and medicinal properties (Koul, 1996). Several other plants belonging to *Meliaceae*, apart from *A. indica*, have been characterized for their biological activity. Given the wealth of knowledge derived from a single genus, one wonders how much phytochemical diversity remains to be characterized and utilized as control treatments in IPM. Clearly, LDCs are home to most of the world's phytochemical diversity, which holds great promise for the future development of botanical pesticides. It is likely that botanical pesticides will become increasingly important in LDCs for a number of reasons: (i) they provide effective control of insects that have become resistant to synthetic insecticides; (ii) they generally pose a low risk to non-target organisms; (iii) they are naturally occurring and so are sometimes accepted by organic certification programmes; and (iv) they are often more accessible than synthetic insecticides in the LDCs where they are grown (Weinzierl, 1999).

Genetic diversity within crop species is another important contribution developing countries have made to IPM. As is current practice, plant breeding in the future will utilize genetic diversity to develop crops with unique and desirable traits, including conventional host-plant resistance (HPR) (see Smith, Chapter 7, this volume). When one considers the centres of origin, and hence diversity, for the major crops of the world, developing countries have already made a tremendous contribution to plant improvement, both *in situ* and *ex situ* (in the form of germplasm bank collections held in trust around the world). Centres of diversity for selected crops include Mexico and Central America for maize (*Zea mays*), beans (*Phaseolus* spp.), cotton (*Gossypium hirsutum*) and tomato (*Lycopersicon esculentum*); South America for sweet potato (*Ipomoea batatas*), rubber (*Hevea brasiliensis*), cotton (*Gossypium barbadense*), potato (*Solanum tuberosum*) and cassava (*Manihot esculenta*); East Africa for finger millet (*Eleusine coracan*), sorghum (*Sorghum bicolor*), cowpea (*Vigna unguiculata*) and coffee (*Coffea* spp.); West Africa for bulrush millet (*Pennisetum americanum*) and rice (*Oryza glaberrima*); North Africa for rye (*Secale cereale*); south-west Asia for barley (*Hordeum vulgare*), South-east Asia for rice (*Oryza sativa*) and citrus (*Citrus* spp.); New Guinea for sugarcane (*Saccharum* spp.); the Fertile Crescent in the Iran/Iraq/Syria/Turkey region for wheat (*Triticum* spp.); and China for soybeans (*Glycine max*) (Simmonds, 1976). This long list of basic food crops highlights the important role developing countries have played and will continue to play in the future stewardship of genetic resources, which could contribute to IPM strategies both locally and around the world.

Arthropod diversity is also a tremendous resource to be utilized more fully in the future. Historically, some of the limitations for using introduced or exotic biological control agents in classical biological control have been misidentification of species, collection of biotypes not adapted to the target agroecology and quarantine capacity to facilitate the introduction (Neuenschwander, 1993). With the establishment of research organizations such as EcoPort, BioNET and the International Organization for Biological Control, information management and exchange will be enhanced to enable the identification and localization of potential control agents and to ensure the highest probabilities of success by using site-similarity tools to target collection activities for classical biological control programmes.

Research networks: Consultative Group on International Agricultural Research Centres

The green revolution achieved its success in large part through the networks consisting of IARCs and national agricultural research and extension systems (NARES). The first IARC, the International Rice Research Institute (IRRI), was established in 1960 and its sister centre, CIMMYT, was formally established in 1966. These two centres took the lead in developing and releasing high-yielding varieties for developing countries, especially in south Asia where real and projected food shortages placed a sense of urgency on increasing rice and wheat production. In 1971, the Consultative Group on International Agricultural Research (CGIAR) was formed to serve as an umbrella organization to coordinate funding to the then five-member group of IARCs. Today, the CGIAR includes 16 IARCs, each having a mandate region, crop or cropping system (CGIAR, 2003). Many of these IARCs hold genetic resources of the main cereal and pulse crops in germplasm banks. These collections of global accessions are held 'in trust' under a Food and Agriculture Organization (FAO) agreement signed in 1994. Under this agreement, the IARCs are responsible for the long-term storage and regeneration of holdings for the benefit of the international community. Most sources of conventional HPR originated from materials now held in germplasm banks, which will continue to serve as a valuable resource for HPR development.

Morse and Buhler (1997) assert that the green revolution laid the foundation for the later adoption of IPM. It enabled higher production and lower grain prices, which benefited both farmers and consumers. Productivity gains also benefited the environment, as they largely arrested the further expansion of agriculture into marginal lands in many parts of Asia. However, the green revolution has also been criticized for harming the environment by reducing genetic crop diversity and beneficial insect populations in agroecosystems and adversely affecting farmers' health through the overuse and mishandling of pesticides. Across Asia, insecticide consumption increased from US$347 million in 1980 to a peak of US$1.08 billion in 1990 (Pingali *et al.*, 1997). The resurgence of secondary pests, such as the brown planthopper, *Nilaparvata lugens* (Stål) in rice (Heinrichs *et al.*, 1982), was a driving force for the development and promotion of IPM technology. The goal was simple: maintain established yield gains with reduced pesticide use.

This unique network of IARCs has served as a hub for coordinating research within and between developing countries around the world. Although in its early days the CGIAR focused on pressing food demands, in more recent years the CGIAR centres have aimed to respond more actively to environmental, social and small-farmer concerns as part of the quest to further sustainable rural development. In that spirit, the CGIAR established a system-wide programme on IPM, which brought together ten IARCs to address issues related to the development, promotion and adoption of IPM technologies in different cropping systems. During the late 1990s, membership was expanded and current stakeholders now include various national agricultural research systems (NARS), diverse promoters of IPM, such as FAO and the World Bank, advocacy groups, such as the Pesticide Action Network and other non-governmental organizations (NGOs) and the private sector (CropLife International) (James *et al.*, 2003). Together, these parties have developed and promoted IPM technologies for a range of global pests including the whitefly, *Trialeurodes vaporariorum* (Westwood); stem borers, *C. partellus* and *B. fusca*; *Striga* species, *S. hermonthica* and *S. asiatica*; leaf-miners, *Liriomyza trifolii* (Burgess), *Liriomyza sativae* Blanchard and *Liriomyza huidobrensis* (Blanchard); and white grubs (e.g. *Phyllophagus* spp.). The system-wide programme partners are also addressing broader themes that bear on IPM, including farmer participatory research, invasive species, impact assessment and the use of GE crops in LDCs.

Weaknesses: Factors Working against IPM Use in LDCS

In contrast to industrial countries, LDCs have fewer financial resources to address a broad range of development objectives. As a result, funding for ICM development and deployment is generally low, which results in low to moderate levels of financial support for agricultural research and extension. Government funding for improved infrastructure, telecommunications in particular, needs to continue, as it will enable IPM practitioners to keep abreast of IPM advances within their region and around the world. IPM is also challenged by the economic trend of farmers pursuing off-farm employment when possible, which often results in labour shortages and less active engagement in agricultural production. These factors are reviewed in more detail below.

LDC regulatory agencies

LDCs have established regulatory bodies similar in function to the Environmental Protection Agency, the US Department of Agriculture (USDA) Animal and Plant Health Inspection Service (APHIS) and the Food and Drug Administration in the USA. The mandate for these agencies is to ensure that agricultural products, including agrochemicals and GE crops, are safe for consumers and the environment and improve the quality and quantity of food produced. The high cost associated with the commercialization of agrochemical products has encouraged agrochemical companies to place a high priority on product stewardship and the use of modern science to promote sustainable agriculture (CropLife, 2003). However, such vigilance is often lacking in LDCs because the small agrochemical companies, which are often major suppliers, often have little interest in product stewardship. Commonly they sell products based on older chemistry that have a low profit margin and are generally more damaging to human health and the environment. The ability of the regulatory agencies to enforce their mandates and ensure safe and sustainable crop production range from that demonstrated in recently industrialized countries, as seen by the exemplary Taiwan Agricultural Chemicals and Toxic Substances Research Institute (TACTRI) in Taiwan

(TACTRI, 2002), to countries that have yet to establish regulatory bodies for various reasons. As the profitability of agriculture increases and export markets develop, it is anticipated that these maturing agencies will better regulate pesticide use and enforce policies that favour IPM technologies. Where profitability is expected to decline, one fears the opposite will happen.

Government investment in agricultural research

In most LDCs, government funding for agricultural research and extension is low, which is not surprising given the demands placed on their limited treasuries by health, education and other competing development fields. Public-sector agricultural research is not keeping pace with the private sector in terms of facilities, operational funds and compensation. This results in many of the most capable scientists being lured away to the private sector or to positions abroad and, for those who remain, the prospect of not being given the resources to perform their jobs properly. As LDCs build up their agricultural and industrial sectors and, with them, their export markets, it is hoped that more funding will be directed towards agricultural research, but, unfortunately, this cannot be assumed at this time.

International development organizations, which recognize the importance of agriculture in development, complement national funding through projects, but these funding sources carry a finite time link, typically 3–5 years. For IPM, a project may not have sufficient time to characterize, identify, test, scale up and monitor management interventions. The problem is exacerbated by a high turnover rate of research and technical staff in some NARES, due to the reasons cited above. Another important problem receiving more attention of late is gender imbalance as it relates to extension and researchers working in communities where women are engaged in crop management (Bruin and Meerman, 2001). Organizations such as the International Service for National Agricultural Science (ISNAR) are looking at these and other issues related to capacity building in agriculture within LDCs (ISNAR, 2003).

Telecommunications

Although the Internet and other forms of telecommunication are taken for granted in the industrial countries, their limited availability (including affordability) and poor reliability in some LDCs are a constraint to their extensive use. Poor telecommunication infrastructure impedes the use of Internet-oriented systems that link interdisciplinary research and extension teams and users and providers of new technology within a country and between countries. Frequently we find adequate telecommunication capabilities in a LDC's capital city, but poor or no communication system serving the countryside, where the researchers and extensionists conduct their work and farmers reside. Between the LDCs and the industrial countries, many refer to a widening of the 'digital divide' (Kates *et al.*, 2001), which leaves developing countries in the technological wake of their more developed colleagues; this would also apply to IPM. The Internet will become increasingly important for IPM networks in LDCs. It will enable scientists and policy makers to access IPM expertise through various databases, including one recently offered through the Global IPM Facility (http://www.wisard.org/wisard/clients/ippm), which lists more than 400 IPM experts around the developing world.

Off-farm employment

In industrial countries, agriculture has increasingly become a capital-intensive enterprise which achieves high production but often requires off-farm employment to supplement farm revenue. In many industrialized countries less than 25% of the population works in agriculture, while in many developing countries this figure is generally above 60% (Bongaarts, 2002). Off-farm employment is also a reality for many farm families in LDCs, but the trend is for young adults, frequently male, to leave the farm to seek urban employment, leaving behind an agricultural workforce that is either older or dominated by young women (Pingali, 2001). Income forecasting by farmers in LDCs is becoming increasingly difficult as imported grain can upset local grain prices due to the absence of effective marketing boards that enable farmers to plan farm activities based on expected future income. Ethiopia and Malawi are two examples of price fluctuations cited by Pinstrup-Andersen (2002): 2000–2001 maize prices in Ethiopia fell to 20% of the 1999–2000 price and, in the course of the following year, maize prices in Malawi increased 400%. The lack of infrastructure that results in such price fluctuations leave resource-poor farmers in a dilemma in terms of both investing in inputs to improve crop performance and securing income from outside sources to cover basic expenses, such as education fees.

Opportunities

Scientific advances in IPM and supporting disciplines have been tremendous in the past two decades in industrial countries and, to a lesser extent, in LDCs. This section highlights recent successes and opportunities for further development of IPM component research in LDCs.

Past successes and the lessons learned

Since its inception in the early 1960s, many examples of successful IPM programmes have been reported, mainly in industrial countries, where the overuse of pesticides had reached a critical level. Several relatively recent reviews have pointed to similar successes in developing countries (Mengech *et al.*, 1995; Guan Soon, 1996; Bruin and Meerman, 2001). Crop-specific studies – for instance, on rice (Matteson, 2000) and cassava (Bellotti *et al.*, 1999) – lend further weight to the efficacy of IPM in LDCs. Guan Soon (1996) examined ten IPM programmes in developing countries to identify common features that they shared. The cropping systems included banana (rust thrips, *Chaetanaphothrips orchidii* (Moulten) and banana weevil, *Cosmopolites sordidus* (Germar), Costa Rica); cotton (secondary pests such as *Argyrotaenia sphaleropa* (Fernald) and *Platynota* sp., Peru); mangoes (mealy-bug, *Drosicha stebbingi* (Green), Pakistan); cassava (mealy-bug, *Phenacoccus manihoti* Matile-Ferrero, Congo); coconut (rhinoceros beetle, *Oryctes rhinoceros* (Linnaeus), Malaysia); cocoa (bark-boring caterpillars, Malaysia); crucifers (diamondback moth, *Plutella xylostella* (Linnaeus), Asia); and rice (brown planthopper, *N. lugens*, Asia), among others. The key features shared by some of the programmes included recognition that chemical control eliminates natural enemies, which results in pest outbreaks; failure of chemical control leads to IPM adoption; IPM strategies emphasized the reduction and/or use of selective pesticides, which in turn, restored natural enemies; pest problems declined after reducing pesticide use; research focused on field problems; there was close collaboration between researchers, extension providers and farmers; IPM had a strong biological control component; and the presence of strong government and management support.

Biological control

Biological control programmes in LDCs offer some of the most dramatic examples of impact and high cost:benefit ratios realized by IPM. One of the most celebrated cases of classical biological control involved the cassava mealy bug *P. manihoti* and the parasitoid *Epidinocarsis lopezi* (De Santis). The mealy-bug was introduced from South America to Central Africa in the early 1970s and, in the absence of its natural enemies, it rapidly established itself as a devastating pest of cassava, a staple food in sub-Saharan Africa (Herren *et al.*, 1987). The use of insecticides was judged to be an impractical control strategy for the mealy bug given the low crop value, economic constraints and farmers' lack of experience in handling insecticides. Biological control was seen as a promising alternative and options were sought through the formation, in 1979, of a network headed by the International Institute for Tropical Agriculture (IITA), which included the Center for Tropical Agriculture (CIAT), the Brazilian Agricultural Research Corporation (EMBRAPA), the International Institute of Biological Control (IIBC) in London, the Inter-African Phytosanitary Council (IAPSC) and the Nigerian quarantine service (Neuenschwander, 1993). Identification of suitable foreign biological controls required that national biological control programmes be established, with national scientists trained in the handling and testing of candidate control agents within their country. Releases were made in collaboration with national institutions and often accompanied by media coverage, which popularized the concept of biological control with both farmers and consumers.

In 1981, the parasitoid *Apoanagyrus* (*Epidinocarsis*) *lopezi* De Santis (Hymenoptera: Encyrtidae) was imported from Paraguay into Nigeria for use in the biological release programme for cassava mealy-bug control. Three years after the parasitoid's initial series of releases, it had spread over 200,000 km^2 in south-western Nigeria. By 1985, more than 50 releases in 12 African countries had been made, which by 1990 resulted in *E. lopezi* being established in 24 countries covering more than 12 million km^2 (Neuenschwander *et al.*, 1990).

With ever-increasing transborder movement of people and international trade, the incidence and hazards posed by invading alien species will also increase – and, with them, the demand for solutions using classical biological control and other IPM components. The Convention on Biological Diversity recognized the 'taxonomic impediment' for the management of biodiversity, in large part due to knowledge gaps and shortage of trained taxonomists. Networks such as BioNET-International focus on increasing taxonomic infrastructure in LDCs by providing training and technologies that enable a country to be self-reliant (BioNET-International, 2003). A global taxonomic platform to standardize the inventory of living organisms would enable the development of searchable databases to accelerate the implementation of classical biological control programmes in the future, especially using south-to-south exchanges.

Diversified cropping systems and reduced rates of targeted pesticides will encourage the conservation of biological control agents. Cropping practices can be developed and implemented that allow natural enemies to realize their full potential to suppress pests. This can be done through identification and remediation of negative influences that suppress natural enemies or by enhancing habitats for resident natural enemies (i.e. providing pollen and nectar sources) (Landis *et al.*, 2000). This approach has been encouraged in many IPM programmes, especially the FFSs, in which farmers are introduced to indigenous biological control agents and are encouraged to adopt practices that do not have an adverse impact on non-target insects.

Host-plant resistance

Farmers, especially small-scale farmers, have selected directly or indirectly for traits of interest, including resistance to pests, over the millennia. One reason farmers cite for not adopting improved, high-yielding varieties is because of their perceived susceptibility to local pests and diseases. HPR is an economically viable and environmentally appealing technology for LDCs. Resistant varieties provide a number of advantages: they lessen the need for costly imported pesticides, reduce hazards posed by exposure to pesticides and misuse of recycled pesticide containers and minimize pesticide residues on produce, with subsequent benefits to consumer health and safety. The economic impact of HPR on farmers can be significant, not just in reduced purchases of pesticides but also in increased insect biodiversity through reduced applications of synthetic insecticides. The estimated value of insect resistance in wheat at a global level is over US$250 million/year (Smith, 1999). The rice variety IR36, with resistance to brown planthopper, green leafhopper, striped stem-borer and gall midges, saves rice farmers an estimated US$1 billion (Khush and Brar, 1991). It should be noted, however, that impact assessment for conventional host-plant resistance is scarce and should be promoted in the future.

Huge strides have been made in the development of methodologies to select and incorporate HPR (Panda and Khush, 1995). Inheritance of HPR falls into two broad categories: major gene resistance and polygenic (quantitative) resistance. Major gene resistance involves a single gene, usually dominant, that confers resistance to a pest or disease that is usually specialized on that crop. Polygenic resistance involves several genes, each contributing a small level of resistance, which is often influenced by the environment. Given these characteristics, more plant varieties with major gene resistance have been deployed; however, the single-gene approach has also resulted in the development of resistant pest biotypes, such as brown planthopper resistance to *Bph* genes in rice and Hessian fly biotypes in wheat. One can observe from the summary by Khush and Brar (1991) that major gene resistance is more prevalent against Homoptera, Hemiptera and Diptera, while polygenic resistance is more prevalent for Lepidoptera, Coleoptera and mites. Through the use of modern molecular tools (see Smith, Chapter 7, this volume) and an understanding of the biochemical or biophysical basis of resistance, the process of incorporating resistance into accepted, high-yielding varieties has accelerated and been refined as only the genomic regions of interest are incorporated into the improved varieties without the extra genetic 'baggage' from the resistant source. Molecular markers have been used in conventional insect resistance in several crops (Yencho *et al.*, 2000; Willcox *et al.*, 2002) and are being adopted by NARS in LDCs. Future HPR efforts will continue to play a major role in developing resistance to pests, with gene sequencing of dominant genes potentially serving as constructs for future transformation efforts.

As was noted earlier, most crops have their centres of diversity in developing countries. Much remains to be done to characterize and utilize this diversity in order to incorporate conventional resistance into crop-breeding programmes. HPR has tremendous potential in LDCs, as the technology is packaged in the seed and does not require additional knowledge to utilize it, but it may require education about its management to maintain a high level of resistance in recycled or 'saved' seed, a practice that is commonly used by subsistence farmers.

Genetically engineered crops

Much debate has taken place regarding the use, benefits and risks of GE crops in industrial countries, in particular Europe. In fact, GE crops have been the predominant topic in integrated crop protection literature in the past 5 years. In the USA, the Council for Agricultural Science and Technology (CAST) undertook a review of the impacts of agricultural biotechnology to assess its impact in cotton, maize and soybean on environmental and sustainability issues, such as pesticide use, soil management and conservation tillage, crop weediness, gene flow, pest resistance, pest population shifts (secondary pests) and non-target organisms (Carpenter *et al.*, 2002). The report recommended the use of GE technologies as a means of sustainable intensification of crop production as the results showed an increase in insect biodiversity (*Bacillus thuringiensis* (*Bt*)-based technology) and enhanced adoption of soil-conservation technologies (herbicide-resistance genes to facilitate no-till in soybean) associated with GE crops. The best expression of utility of GE crops is by farmer adoption, which shows that there are economically tangible benefits with GE crops that are reflected in the dramatic increase in area planted to genetically modified (GM) crops in both industrial and developing countries (Fig. 13.2).

Although investment in agricultural biotechnology is concentrated in industrial countries, three developing countries have committed significant levels of public funding to biotechnology, namely China (US$112 million), India (US$15 million) and Brazil (US$2 million) (Huang *et al.*, 2002). China has

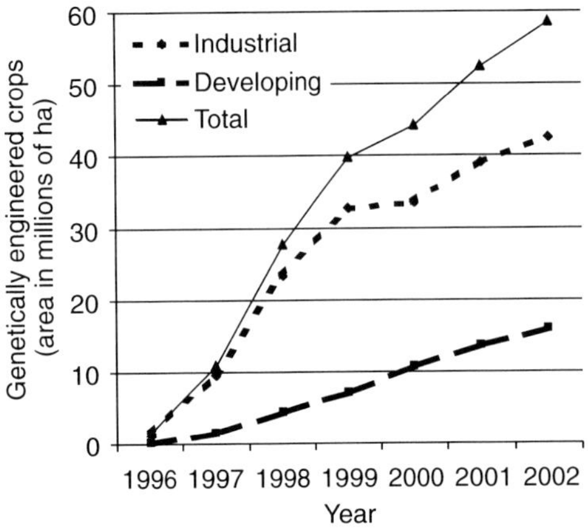

Fig. 13.2. Area planted to genetically engineered crops in developing and industrial countries between 1996 and 2002 (from James, C., 2003b).

announced plans to raise plant-biotechnology budgets by 400% before 2005, which would then account for approximately one-third of the world's public investment in plant biotechnology. This level of investment by LDCs is driven by expected returns, which for IPM include the reduction in insecticide use by both large-scale and small-scale farmers.

Recent studies show that small-scale farmers in developing countries stand to benefit more from GE technology than commercial farmers in industrial countries, due to the limited access to agrochemicals and greater losses associated with insect pests in tropical and subtropical agroecologies (Qaim and Zilberman, 2003). The best example of GE crop-protection technology in LDCs is *Bt* cotton containing the *cry*1Ac gene from *Bacillus thuringiensis,* which is being grown in on-farm trials in India (Qaim and Zilberman, 2003) and commercially in China (Huang *et al.*, 2002). In China, *Bt* cotton was approved for commercial release in 1997, due to a sharp reduction in cotton production caused by losses and control costs associated with the cotton bollworm, *Helicoverpa armigera* (Hübner). The response by China's poor farmers to *Bt* cotton has been astonishing, increasing from 2000 ha in 1997 to 2 million ha in 2002, accounting for approximately half the cotton area in China (Huang *et al.*, 2002; James, C., 2003a). Between 1999 and 2001, the reduction in insecticide applied to cotton was 123,000 t of formulated product, which resulted in a 75% reduction in insecticide poisonings of cotton farmers using knapsack sprayers (James, 2002). Economic benefit has been the main driving force for adoption, with an average reduction of 13 sprays (49 kg)/ha, amounting to a cost savings of US$762/ha, excluding labour costs (Huang *et al.*, 2002). The net effect has been a 28% reduction in production costs and reduced exposure and poisonings by farmers to produce a product of the same value.

Scientists like Florence Wambugu (1999) have highlighted the economic and health benefits derived from *Bt* crops to promote their use in Africa. Wambugu states, 'Africa missed the green revolution, which helped Asia and Latin America achieve self-sufficiency in food production. Africa cannot afford to be excluded or to miss another major global "technology revolution". It must join the biotechnology endeavour.' A Kenyan project entitled Insect Resistant Maize for Africa (IRMA) is looking at the issues surrounding *Bt* maize to control stem borers in both commercial and subsistence maize cropping systems (Mugo *et al.*, 2001). Yield-loss studies from the project established that stem borer losses in Kenya amount to US$72 million per annum (De Groote, 2002). PRAs were used to establish farmer constraints, which, apart from stem borers, include other biotic (storage pests, maize streak virus) and abiotic (drought and low soil fertility) stresses. Expected benefits from *Bt* maize in Kenya will include increased profitability to farmers and increased insect biodiversity within the mixed maize cropping system of small-scale farmers.

Although *Bt* cotton and maize have reduced insecticide use, a central tenet in IPM, the use of *Bt* crops begs the question of its impact on other IPM components, such as biological control, and the agroecosystem in general. The large-scale deployment of *Bt* crops has generated public concern regarding its ecological, food-safety and social consequences. These concerns were heightened by a laboratory study showing mortality of the monarch butterfly, *Danaus plexippus*

(Linnaeus) after consuming pollen from *Bt* maize expressing *Cry*IAb toxin (Losey *et al.*, 1999). Concern was also raised regarding impacts on biological controls when Hilbeck *et al.* (1998) demonstrated adverse effects of *Bt* maize, mediated through the herbivore prey, on lacewing larvae, *Chrysoperla carnea* (Stephens). While laboratory studies are useful in pointing to potential risks, it is difficult to translate these results to a field setting. In the case of the monarch butterfly, *Bt* maize had a negligible impact due to low exposure rates under field conditions (Sears *et al.*, 2001). Field trials in maize comparing different control interventions on insect diversity found that *Bt* (Thuricide®)-sprayed plots hosted greater diversity and abundance of non-target insects than insecticide (Dimethoate and Bulldock)-sprayed plots, with both treatments controlling stem borer attack (Songa, 2002). Baseline surveys for insect abundance and diversity in different maize cropping systems in Kenya have also been established to monitor the future impact of *Bt* maize on non-target insects (Songa *et al.*, 2002). Such an approach should be considered for future GE projects to enable monitoring and more conclusive statements to be made about the impact of *Bt* crops on non-target organisms.

Recent reviews compiled on the impact of *Bt* crops on non-target organisms (including soil biota), outcrossing to weedy relatives, horizontal gene transfer (gene movement between plants and bacteria) and resistance by targeted pests (Carpenter *et al.*, 2002; Shelton *et al.*, 2002) have concluded that the risks are low. However, the issue of resistance management remains a serious concern for *Bt* crops (Shelton *et al.*, 2002). Currently, the only insect-resistance management (IRM) strategy available for commercial GE crops is the high-dose/refugia strategy. The central principle of this strategy involves the use of a refugia to maintain susceptible alleles in insect populations by not imposing a selection pressure on a portion of the pest population (Gould, 1996). In the USA, farmers are expected to plant a structured refugia. In developing countries, enforcement of a structured refugia will be difficult as this often runs counter to the growers' objective of controlling insect pests and would be virtually impossible to enforce given the large number of small-scale farmers relative to extension agents available for monitoring compliance. China is one example where resistance management is not mandatory for commercial *Bt* cotton (Zhao *et al.*, 2000). Developing viable IRM strategies will require participatory research involving socio-economists, entomologists and policy makers to develop strategies that are socially, culturally and economically viable for small-scale farmers in LDCs.

Establishing IRM strategies and conveying IRM's importance to small-scale farmers are a challenge currently being undertaken by the IRMA project in Kenya. Farm surveys in different agroecological zones quantified the area planted to maize and other crops that are known hosts of stem borers. Field trials were conducted by planting plots of these alternative hosts bordered by commercial maize varieties to establish moth oviposition preference and productivity by counting exit holes from the different crops. Using the surveys and moth production data, the area of effective refugia can be estimated for each agroecology. Using geographical information systems (GIS), regions were identified that contained a 'natural' refugia equivalent to 20% maize. Lowland tropical ecologies contain a diverse range of alternative hosts for stem borers (mostly forage grasses) and have an adequate natural refugia; however, commercial maize areas will require a structured refugia (Mulaa *et al.*, 2001). This approach would complement the 'push–pull' strategy outlined earlier for stem borer control in maize, as napier grass would serve as a refugia while *Bt* maize would provide more complete protection than that currently being offered by desmodium. In the end, IRM strategies for use on small-scale farms in LDCs must be economically attractive and must fit into the farmers' cropping system.

In the future, monitoring for insect resistance may be achieved through molecular markers for resistance genes, such as the recently identified cadherin alleles associated with resistance in the pink bollworm, *Pectinophora gossypiella* (Saunders) (Morin *et* al., 2003). Clearly, IRM in LCDs will be challenging but not impossible if: (i) robust varieties containing two or more complementary resistance genes are developed; (ii) the abundance of alternative hosts to serve as a refuge are well characterized; (iii) resistance-management programmes can be integrated into established cropping systems; and (iv) GIS is used to target monitoring efforts.

As with synthetic pesticides, which were perceived to be a panacea for pest control over half a century ago, transgenic technology should be considered as an important component (HPR) in an IPM programme. *Bt* crops should not displace but rather should complement indigenous technologies for pest control in LDCs, to avoid overreliance on one technology, which could break down if alternative control strategies are not utilized.

Resource conservation technology

Resource conservation technologies (RCTs), such as no-till practices, have experienced rapid adoption in recent years. No-till agriculture started to be adopted by large-scale farmers in Brazil and Argentina in the early 1990s, associated with price reductions in glyphosate (broad-spectrum herbicide) and the development of more suitable planters and sprayers by machine manufactures (Ekboir, 2002). Monsanto supported a project in Brazil to refine the technology package for small-scale farmers and, within 2 years, 90% of the farmers (820,000 ha) were using no-till.

The popularity of no-till with farmers has increased dramatically with the advent of herbicide-resistant soybean, which accounted for the largest area planted to GE crops (62% of the global total) in 2002 (James, C., 2003a). It also accounted for the largest area of GE crops in LDCs, with Argentina growing 13.5 million ha of Round-up Ready® soybean. The high adoption rate of this technology has been driven by economic returns; it has been estimated that farmers using no-till management obtain savings, due to reduced labour, fuel and pesticide use, of about US$56/ha. This resulted in a nationwide savings of US$356 million in 2000 (James, 2002). Environmental benefits include reduced soil erosion and runoff of pesticide residues, improved air and water quality and reduced exposure to pesticides (Carpenter *et al.*, 2002), which make this technology attractive for other crops in LDCs. Herbicide consumption will continue to increase in regions where no-till is being adopted, such as Latin America and the USA (Fig. 13.3), with herbicide consumption accounting for 45% of the global pesticide market (FAOSTAT, 2002). The role of

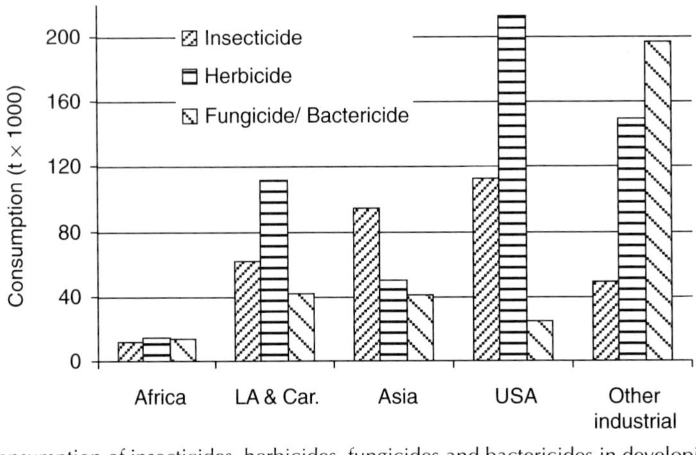

Fig. 13.3. Consumption of insecticides, herbicides, fungicides and bactericides in developing countries within Africa, Latin America–Caribbean, Asia and industrial countries. Based on 89 countries, not including China. (Based on FAOSTAT, 2002, for 1997.)

herbicide-resistant products will increase as the rural workforce continues its urban migration in the coming years (Pingali, 2001). However, IPM has not kept pace with the practice of zero-till in Argentina. Monitoring of pests, diseases and biological control agents should be undertaken, with proper baseline studies being conducted to help in the development of IPM programmes tailored for no-till cropping systems.

The rice–wheat cropping systems (RWCS) of the Indo-Gangetic Plain occupy 13.5 million ha and account for one-third of the total rice and wheat production in this heavily populated region. A falling water-table, excessive soil erosion and low organic-matter content in the soil are major production constraints in the region, with farmers using elevated nitrogen rates to compensate for reduced soil fertility, resulting in increased losses from disease and insect attack (Srivastava, 2000). Given these problems, the region is an excellent candidate for RCTs and associated IPM strategies.

In 1994, the Rice–Wheat Consortium was established to tackle these constraints by introducing RCTs into farmers' fields. After harvesting rice, wheat is direct-drilled into rice residue to avoid ploughing, utilize the residual moisture in the soil and conserve the straw residue. Labour, fuel and water savings have driven the rapid adoption of no-till, but concern about the impact of residues on insect pests and diseases has been expressed. Insect control was traditionally achieved using cultural-control practices, such as removing rice stubble to reduce stem borer populations (Srivastava, 2000). Conversely, under the no-till system, reduced disturbance of the agroecosystem increases the abundance and diversity of biological control agents that regulate stem borer populations (P. Hobbs, Mexico, 2003, personal communication). Understanding the interactions of conservation-tillage practices with agroecosytems will become increasingly important as more farmers in LDCs adopt RCTs, such as zero-till, in order to increase the profitability and sustainability of farm enterprises. Since RCT and IPM are both key components of future ICM programmes, it will be important to characterize the interactions (positive and negative) to develop intensified, sustainable cropping systems in LDCs.

Geographical information systems to target technologies and monitor insect pest and disease movement

Geographical Information Systems (GIS) will probably play an ever-increasing role in IPM, in both developed and developing countries. In developing countries, GIS can be used to assess and predict the distribution of weed and insect pests, by enabling scientists to visualize geographically large data sets, provided the collection sites are geographically referenced (latitude and longitude). Databases for GIS include FAO's classification of soils, temperature, rainfall, potential evapotranspiration, population, rivers, roads and specialized databases such as insect population density and species distribution. A good example of specialized databases is that developed at the International Centre for Insect Physiology and Ecology (ICIPE) to characterize the distribution of maize stem-borers in East Africa (Irungu *et al.*, 2003). Several native species of stem borer are found in Africa, but the spotted stem-borer, *C. partellus*, has expanded into new agroecologies since its introduction into the region about 50 years ago. GIS are being used to predict the distribution and potential impact of classical biological control efforts following the introduction of the parasitic wasp, *Apanteles flavipes* (Cameron) (Emana *et al.*, 2002). They have also been used to define which climatic factors favour the development of *C. partellus* and the location of those agroecologies to better target the release of biological control agents in the future. GIS may also help researchers target locations for the collection of biological control agents to better reflect the agroecology in which the biological control agent is to be released, thus ensuring that the collected species/biotype is already conditioned to the target ecology.

The application of appropriate GIS tools and databases can also help scientists in LDCs to make better decisions regarding where to target their pest-control interventions and what control interventions used in other countries may be most suited to their targeted agroecology. However, for most LDC scientists, this technology has remained largely in the domain of GIS experts, due to impediments such as cost, complexity and the availability of standardized data sets that have sufficient resolution to be useful in IPM. Some of these constraints are now being addressed by the development of new software platforms that are easy to learn and are able to integrate GIS tools with extensive databases relevant to agriculture (Irungu *et al.*, 2003).

Another promising area is the marriage of GIS and global positioning systems (GPS) along with satellite images and detailed soil mapping, prerequisites for precision agriculture. This cutting-edge package of technological tools allows farmers to apply agrochemicals, such as fertilizers and pesticides, only as needed. From the perspective of tactical pest control, precision IPM would mean the application of selective pesticides to control weeds, insects and diseases when and where it is necessary to maximize returns in commercial crop production. The prospect of using precision agriculture in LDCs may seem a distant dream, but this may come sooner than we think, particularly in the large-scale commercial operations found in some of these countries. Expected technology cost decreases, particularly for the imaging and GPS components, should spur greater adoption. Browsing the Internet, one can find several on-line degree-day models for several insects and diseases in industrial countries. The availability of comparable databases and models for LDCs will depend on demonstrating the utility to commercial farmers and on collaborative field research efforts between NARS, advanced research institutes and IARCs.

Internet access to IPM technology and networks

As illustrated in the previous section, the Internet has opened up information and technology exchange between the North and the South. Internet cafés are now commonly found in larger towns in LDCs and increasingly in smaller villages. This means that a researcher, extension provider or farmer can now access information regarding new technologies being used on the other side of the world. For the time being, most of these 'exchanges' are in one direction, North to South. However, as Internet-based networks for IPM become established, exchanges between IPM practitioners in LDCs will increase as scientists and farmers look to each other in similar agroecologies to find common solutions for crop-production constraints.

All the IARCs have websites that provide useful introductions to the research centre, but, as the Internet has matured, so has the power of those websites in enabling NARS scientists to access technologies recently developed by the IARCs, including germplasm, ICM and IPM for different staple crops. These sites are accessed through the CGIAR (http://www.cgiar.org) or directly by substituting the centre's acronym (e.g. IRRI for the International Rice Research Institute) for 'cgiar'. The IRRI website is a good example of how the Internet can serve LDC farmers by helping them identify crop-production constraints and address them with solutions, including IPM technologies. The website is called the Rice Knowledge Bank (http://www.knowledgebank.irri.org/) and it provides the latest information on ICM for rice cropping systems in LDCs. The website contains decision tools for crop management and pest identification, which enables identification of pests versus non-target insects during field surveys, resulting in fewer insecticide applications. NARS can access research materials such as manuals on hybrid rice seed production, IPM, rice taxonomy, technology-transfer methods and support skills such as consensus building, e-learning and presentation and training skills. Similar websites are being developed at other IARCs to accelerate the devolution of technology to NARS and other research partners. The IARCs are also in the process of developing websites that can serve as a forum for ICM in developing countries. The most advanced example of IPM networking in developing countries is the System-wide Programme on IPM (SP–IPM) (James *et al.*, 2003). As discussed earlier, this network represents a diverse set of IPM stakeholders in both the public and private sectors and focuses on addressing the most pressing biotic constraints of the resource-poor through IPM. Noteworthy websites that link to additional resources include FAO, the Technical Centre for Agricultural and Rural Cooperation (CTA) and PEST CABWeb (http://pest.cabweb.org/), among others. The Internet, through such sites as the SP-IPM, has been a valuable tool for disseminating information, thereby increasing public awareness of the benefits of IPM and raising the profile of IPM within communities. We can expect these trends to grow exponentially in the years to come.

Threats

A threat is a perceived imminent danger and, for IPM, these imminent dangers can be averted by advocacy on the part of scientists, farmers and the general public in shaping the political and social environment to favour the adoption of sustainable technologies. Many of these threats are well recognized and are now being addressed by some, but not all, LDC governments. The purpose of highlighting some selected threats is to encourage strategic thinking about advocacy measures and new directions in public awareness.

Policy

Public policy, perhaps more than any other factor, holds sway over the course of IPM. During the early stages of the green revolution, intensified rice production was visibly and sometimes severely damaged by insects, which created apprehension among some scientists and policy makers about a major outbreak. To address this threat, policies were implemented that made pesticides readily available and affordable to farmers (Pingali *et al.*, 1997). Unfortunately, farmers received little training at that time about the safe and judicious use of pesticides. Today, the aggregated decline in pesticide use in Asia during the past decade (Fig. 13.4) has largely come from IPM efforts on several fronts: host-plant resistance (insects and diseases), promotion (by IPM programmes) of the judicious use of pesticides and withdrawal of government-subsidized pesticide programmes in most Asian countries (Pingali *et al.*, 1997). Indonesia, which subsidized up to 85% of the cost of pesticides during the early 1980s, banned by presidential decree the use of 57 of the 66 broad-spectrum pesticides used on rice. By 1989, pesticide subsidies were no longer offered (Gallagher, 1992). The impact these policies, among others, had on Indonesian rice production from 1984 to 1990 was dramatic; production rose from 38 to 45 million t while Indonesia went from being a net importer of pesticides (US$47 million) to a net exporter (US$15 million) (FAOSTAT, 2002). By 2002, rice production reached 51 million t, providing a good example of what progressive policies combined with massive farmer training can accomplish.

Among international agencies, the World Bank is one of the most influential in promoting IPM. For loans to be granted, project proposals must conform to the World Bank Operational Policy 4.09, which defines IPM as:

> a mix of farmer-driven, ecologically based pest control practices that seeks to reduce reliance on synthetic chemical pesticides. It involves a) managing pests (keeping them below economically damaging levels) rather than seeking to eradicate them; b) relying, to the extent possible, on nonchemical measures to keep pest populations low; and c) selecting and applying pesticides, when they have to be used, in a way that minimizes adverse effects on beneficial organisms, humans and the environment.
>
> (World Bank, 1998)

The policy does not allow the use of class Ia and Ib pesticides and promotes the reduction of synthetic pesticide use in developing countries. However, organizations such as the Pesticide Action Network North America have called for tighter monitoring of the policy during project implementation to ensure that the policy is adhered to (Ishii-Eiteman *et al.*, 2002).

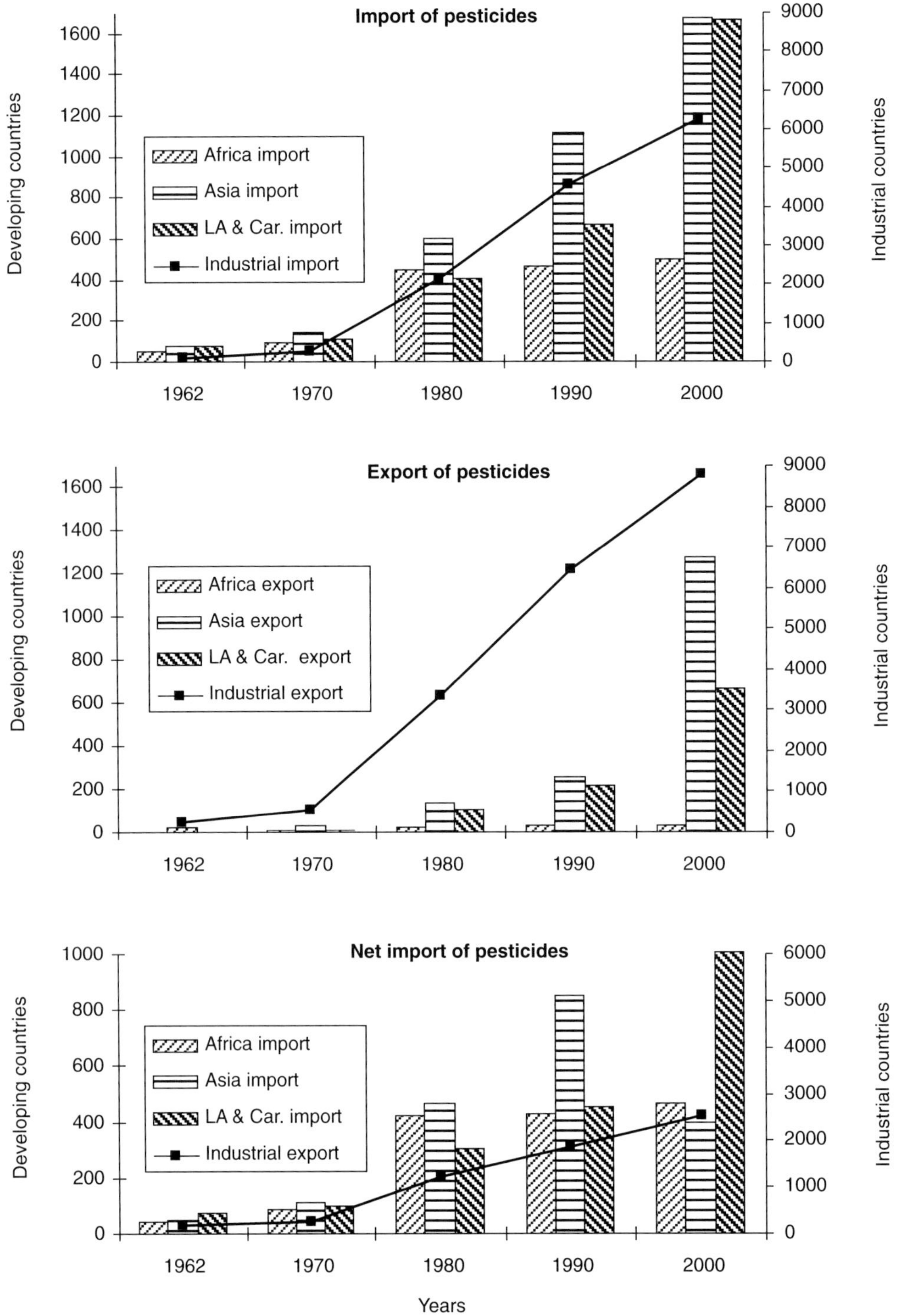

Fig. 13.4. Import, export and net import of pesticides (millions of US$) over the past 40 years in developing countries within Africa, Latin America–Caribbean, Asia and industrial countries (based on FAOSTAT, 2002, for years 1962, 1970, 1980, 1990 and 2000).

Globalization

Globalization is a trend that will probably continue due to the benefits accrued from trade liberalization, increased exchange of knowledge, technology and culture and increased mobility of capital and labour. If correct policies and institutions are in place, globalization could enhance economic growth and poverty alleviation in developing countries (Pinstrup-Andersen, 2002). One potential outcome would be increased competition in agricultural markets, which would promote the cultivation of crops that are best adapted to a given agroecology and are thus most profitable.

However, globalization also presents challenges to IPM, perhaps the most notable being the introduction of exotic pests, associated with increased trade liberalization and movement of labour. Making the problem worse, many developing countries do not have the human and financial resources needed to operate effective quarantine facilities. Moreover, the number of taxonomists available worldwide to conduct the necessary monitoring for exotic pests is inadequate. Through networks such as BioNET-International, taxonomists can pool their expertise through regional networks and common databases to address the challenges in monitoring pest introductions and prescribing biological control agents that can be used to regulate introduced pest species.

Agricultural subsidies in industrial countries further complicate world food security by increasing production in developed countries which depresses international grain markets, often to the detriment of the rural poor in developing countries, whose sole source of income is agriculture (Pinstrup-Andersen, 2002). Depressed prices are exacerbated by the inability of small-scale farmers to store their production, forcing them to sell their harvest quickly into saturated local markets, which further reduces their farm revenue. This points to the need for increased research on reducing postharvest losses for resource-poor farmers through improved storage systems at both local and national levels to buffer the effects of the variable production often experienced in semi-arid regions of the developing world.

Intensification of agriculture

Crop production in developed and developing countries should use all crop-management tools that have met biosafety, environmental and health standards to be applied to the challenge of meeting the food demands of the next 50 years. Pesticides have been and will continue to be an important component of modern agriculture in both developed and developing countries. Crop-protection technology, along with fertilizers and improved crop varieties, has resulted in unprecedented increases in productivity over the past half-century. However, these gains have been built on the intensification of cropping systems, such as rice production systems in which two or three crops are grown annually, resulting in pest outbreaks attributed to the absence of dry-fallow to break the pest cycle (Litsinger, 1989). Most intensified cropping systems were not designed with sustainability or the maintenance of biological diversity in mind. The upshot of this is narrow-based agroecologies that are susceptible to pest invasions.

Sustainable crop management systems must be maintained in the forefront of technology development to ensure the future viability of production systems. Agricultural technologies that have a high potential for sustainability include intercropping (increases soil fertility when legumes are used and reduces pest attack, as in the push–pull strategy); rotations; agroforestry (trees can exploit water and nutrients and provide mulch, while herbaceous crops prevent erosion); sylvo-pasture (trees and pasture for mixed livestock systems); green manuring (legumes to fix nitrogen for incorporation to improve soil fertility); conservation tillage (reduced soil erosion); biological control; and IPM (Conway, 1998).

Reduced international funding for agricultural development in LDCs

A decade-long decline in funding for agricultural development in LDCs by countries of the North has seriously slowed progress in this vital area, with some projects being discontinued before their benefits can reach farmers' fields. IPM projects have not gone unaffected and, should support continue in this downward direction, the consequences are very worrying. To counter this trend, the potential and actual impact of IPM must be documented in a scientifically sound manner and presented in simple terms to government officials and policy makers in developing and industrial countries. An impact assessment of IPM programmes should be implemented at the beginning of projects to document the impact and enable mid-course adjustments to better meet the socio-economic needs of farmers and to validate the investments made in the technology. Emphasis should also be given to IPM's contribution to sustainable cropping systems and improved quality of life (better human health and reduced pesticide loading in the environment).

Climate change

Greenhouse gases (CO_2, CH_4 and N_2O) have risen dramatically during the past century due to combustion of fossil fuels, changes in land use, etc., with CO_2 levels increasing 31% (Gitay *et al.*, 2002). In the same period, global mean surface temperature has increased by 0.6°C, with the largest increases occurring in the mid- and high latitudes of northern continents. The Intergovernmental Panel on Climate Change (IPCC) estimates that the mean temperature increase for the band encompassing most developing countries (30°S to 30°N) will increase by 3°C and changes in precipitation patterns will vary among continents (IPCC, 2001). The impacts of climate change on people's livelihoods will be greatest in the tropics and subtropics, in particular Africa, as many poor small-scale farmers will have few alternatives to agriculture (Gitay *et al.*, 2002). Using scenarios, the IPCC has described 'ecosystem movements', in which entire ecosystems migrate to new locations, and 'ecosystem modification', in which *in situ* changes in species composition and dominance occur. Overall, biodiversity is expected to decrease because of increased land-use intensity and destruction of natural habitats. Insect pest and disease pressures are projected to increase, potentially resulting in larger agricultural and forestry losses in Africa (Gitay *et al.*, 2002). A contributing factor would be the possible decoupling of biological control agents from their prey and new pests being controlled with increased use of pesticides, which would negatively affect non-target species. Reduced tillage, continual ground cover, agroforestry and IPM promote biodiversity and offer a buffer against these effects of climate change (Gitay *et al.*, 2002).

In order to position IPM to meet these environmental changes, scientists need to think ahead to the types of abiotic–biotic interactions that will have an impact on pest management, especially in LDCs, where these interactions are not well defined. Without any biotic stresses, maize production in Africa and Latin America is expected to decline by 10% and by as much as 30% in some parts of Brazil, Venezuela and South Africa (Jones and Thorton, 2003). Plants grown under environmental stress conditions associated with climate change are generally more susceptible to insects (Heinrichs, 1988) and diseases (Coakley *et al.*, 1999). However, the change in pest/disease status would depend on the specific insect– or disease–host complex as well as technological and socio-economic changes that occur within a given agroecological region. Modelling and GIS will be essential complementary tools for simulating and mapping expected changes in biotic stresses and for providing LDC policy makers and stakeholders with comprehensive advice on impacts and mitigation strategies, including IPM programmes.

Regional IPM Issues

Although LDCs worldwide have much in common when it comes to IPM, they also have considerable differences, based on socio-economic, cultural, political, economic and climatic conditions, which will affect IPM development and adoption. The following sections briefly highlight some of the unique issues that are likely to influence IPM practices in each region and areas of research emphasis to meet the changing political and physical climate in which IPM operates.

Africa

African agriculture is predominantly small-scale, mixed crop/livestock farming systems, consisting of holdings of less than 1 ha to a few hectares. Often, farmers do not own the land but have the right to farm the land their ancestors cultivated, under the regulation of village headmen or local government officials. Indigenous crops include: finger millet (*E. coracan*), bulrush millet (*P. americanum*), rice (*O. glaberrima*), sorghum (*S. bicolor*), cowpea (*V. unguiculata*), coffee (*Coffea* spp.) and rye (*S. cereale*), with sorghum and millet being well adapted to arid ecologies (Simmonds, 1976). The most important introduced crops include maize, cassava, sweet potato and Irish potato. Successful IPM interventions in Africa were recently reviewed (Zethner, 1995; Dabrowski, 1997; Abate *et al.*, 2000). Pesticide usage in Africa has been slowly increasing (Fig. 13.4), but remains much lower than in other regions. However, sustainable crop protection that improves farm income is a main driving force for African development projects; some of these have focused on research and development of biological control (cassava, sweet potato, maize, sorghum, stored products, fruit production, coffee, cotton and cowpea) and, to a lesser extent, host-plant resistance (maize, sweet potato, cassava); and chemical control, mainly in rice and cotton (Zethner, 1995). Although pests are a problem, farmers face other, often more pressing, constraints, such as poor soil fertility, erosion, drought and severe fluctuations in market prices. An ICM approach is required to address these wide-ranging issues. The 'push–pull' strategy for stem borer control in maize provides a good example of ICM used to address several constraints (stem borers, *Striga*, soil fertility) using economically viable crops (napier grass for livestock).

As in the past, successful IPM will need to show a clear economic benefit to the farmer for adoption to occur. This means that socio-economic studies should be incorporated into all IPM projects to ensure that the right technology is being targeted to the farmers' needs and is economically attractive. Biological control programmes have generally been extremely successful and well accepted in Africa, given their sustainability (no cost to farmers) once established and they have not suffered from the overuse of pesticides, given the low application rates commonly used by subsistence farmers. Development of host-plant resistance for diseases and insects will become more important as the technologies to efficiently incorporate resistance into local varieties become established in Africa. Already, conventional sources of resistance are being promoted in several crops.

GE crops with genes for insect, disease and herbicide resistance will play an important role in sustainable cropping systems for Africa. Herbicide-resistant maize with a seed treatment containing imazapyr and pyrithiobac has been identified as an effective control for the parasitic weeds *S. hermonthica* and *S. asiatica* (Kanampiu *et al.*, 2002) in on-farm trials in Kenya. This technology would be particularly suited to low-fertility soils, where the yield losses to *Striga* are most severe. The use of *Bt* maize and virus-resistant sweet potato is also expected to play an important part in boosting crop production to meet future demands in Africa (Wambugu, 1999). An important constraint facing the large-scale adoption of GE crops will be the harmonization of biosafety protocols within Africa, which are now under discussion at a regional level in East Africa (C. Ngichabe, Nairobi, 2003, personal communication).

Latin America

Latin America and the Caribbean are home to numerous agroecologies and cropping systems, with farms ranging from subsistence plots to large-scale commercial operations growing several crops. Indigenous crops in this region include maize (*Z. mays*), beans (*Phaseolus* spp.), cotton (*Gossypium hirsutum, G. barbadense*), tomato (*L. esculentum*), sweet potato (*I. batatas*), rubber (*H. brasiliensis*), potato (*S. tuberosum*) and cassava (*M. esculenta*) (Simmonds, 1976). Given the large number of indigenous crops, this region represents an important source of biodiversity for providing new genes for host-plant resistance as well as biological control agents for classical biological control or augmented release within the region. The overuse and misuse of pesticides have been a motivating force for IPM in this region, with Brazil accounting for approximately half of the pesticide consumption in Latin America (FAOSTAT, 2002). The impact of IPM in South American countries has recently been compiled by Campanhola *et al.* (1995), with most IPM projects focusing on the judicious use of selective pesticides (including biopesticides), crop rotations and cultural methods. Augmented release of parasitoids – *A. flavipes*, *Trichogramma pretiosum* Riley, *Paratheresia claripalpis* (Wulp) and *Metagonistylum minense* Townsend – has also been used successfully to control the sugarcane borer, *Diatraea saccharalis* Fabricius, in Brazil and Colombia.

The adoption of IPM has been largely driven by economics, but also by a growing public awareness of the hazards pesticides pose to human health and the environment. As with the other regions, IPM fits well within the context of ICM strategies. In Argentina and Brazil, the adoption of minimum and no-till has increased herbicide consumption (Fig. 13.3), but it has also reduced soil erosion and made farming more profitable and sustainable, once the appropriate planting equipment was made available (Ekboir, 2002). The development of herbicide-resistant soybean, which is now grown on approximately 13 million ha in Argentina (James, C., 2003a), has greatly accelerated adoption of RCTs. Given the expansive area planted to herbicide-resistant soybean, IRM strategies should be enforced for the long-term efficacy of this soil-conservation technology. As with Africa, regional biosafety protocols should be established to enable the trade of GE crops in the future.

Latin America has several strong NARS capable of taking the lead in the development of alternative control strategies for the principal food and fibre crops. EMBRAPA and the Argentinean National Institute for Agricultural Technology (INTA) have provided host-plant resistance, commercial capacity for rearing biological control agents and sustainable crop management methods (i.e. no-till). Exploration for new biological controls will be an important contribution from this region to the international community and it should be coordinated through a central platform such as BioNET-International, to ensure that passport data on putative control agents is available to IPM practitioners and government quarantine facilities. Using a central platform will also accelerate the use of these assets in classical biological control programmes.

Asia

Asia was the home of the green revolution and will probably be the major player in the 'doubly green revolution'. Prior to the green revolution, Filipino farmers grew local rice varieties with only 7.5% using synthetic pesticides in 1954; this grew to 90% by 1994, with a concomitant doubling of yield to 2.7 t/ha (Teng, 1994). However, by the 1970s, this was recognized as an unsustainable trend and, since then, rice scientists have refocused their research to IPM for rice pests, including the development of insect- and disease-resistant rice varieties.

In Asia, however, pesticides have become part of the rice production culture, a tendency that is slowly being changed through regional IPM programmes using an FFS approach. However, only a relatively small portion of the large Asian farming community has been exposed to IPM courses, but new communication techniques, as outlined in the Strengths/Communication section, show signs of accelerating the dissemination process.

Market price also plays a key role in pesticide use in parts of Asia. As cropping systems increasingly diversify into high-value commodities, in which aesthetics determine market price, farmers will probably return to prophylactic sprays as a way of ensuring high market value for their crops (Pingali,

2001). Although FAO is already working at addressing this issue through FFS programmes, such issues might best be handled though policy and pesticide registration procedures that are tightly enforced through monitoring of pesticide residues and subsequent penalties for violators. This system has been used successfully in Taiwan to reduce pesticide residues in vegetables, through a programme known as Good Agricultural Practice (GAP). Farmers obtain a higher market price when their products are inspected and labelled with the GAP logo, enabling consumers to recognize products that conform to safety regulations, which often results in a higher price for the farmer (TACTRI, 2002).

China and India have been dubbed 'super NARS' by some involved in agricultural development, based on their immense capacities in agricultural research, which is now expanding into biotechnology. Biotechnology will probably assume an increasingly important role in delivering crop varieties with greater tolerance to abiotic and biotic stresses, which will serve as the cornerstone of ICM programmes of the future. Already, the use of *Bt* cotton in China has demonstrated the economic, health and environmental benefits to be realized by commercial and small-scale farmers (Huang *et al.*, 2002). Future biotechnology applications will probably include the development of drought-tolerant crops and the use of herbicide resistance to address weed control in intensified cropping systems, as labour costs increase (Pingali, 2001). As with pesticide policies, the use of natural resources such as water, will probably require legislation to deter abuse of water resources and to make water access more equitable within and between countries. These policies may affect IPM as reduced irrigation may cause a shift in the pest complex and require new cultural management strategies to regulate new pest complexes.

A Look to the Future: Impact of Four Socio-economic Scenarios on IPM

IPCC has used scenarios as a tool to paint the backdrop of the future world under different socio-economic systems to generate 'alternative futures' (IPCC, 2001). The four divergent socio-economic circumstances they developed fall into four quadrants based on a regional/global axis and an economic/environmental axis (Fig. 13.5). For the purpose of a strategic planning exercise in CIMMYT, these different socio-economic environments were labelled by cropping system to capture the 'flavour' of the agricultural impact of each scenario (D. Watson and M. van Ginkel, Mexico, 2003, personal communication). These scenarios are driven by diverse forces, including globalization, rate of technology change, degradation of natural resources, water scarcity, climate change, changing farming structure and agricultural marketing, consumer behaviour/preferences, health issues and nutrition, urbanization, national and international instability and conflict, changing roles of key actors, demographics, environment and food trade policies, governance and intellectual property rights (IPCC, 2001; Pinstrup-Andersen, 2002).

The 'Monoculture' (global/economic) scenario predicts rapid economic growth and rapid development of technologies due to global markets and large financial investment of the private sector in technology development (IPCC, 2001). One example of this would be pesticide research to target major crops in both industrialized and developing countries. Increased polarization of wealth, population growth and climate change will have an adverse effect on natural resources, including reduced biodiversity. Production systems would scale up to compete in global markets, which would probably result in the increased use of pesticides as part of a tactical control strategy in order to meet the increased production demands.

The 'Intercropping' (global/environmental) scenario would see the interaction and interdependence of countries to attain more social equity and environmental sustainability, at some cost to economic development. Internationally agreed-upon policies would be enforced in a more uniform manner to reduce environmental degradation and regulate the management of renewable resources (IPCC, 2001). Crop diversity would increase, as crops would be targeted to agroecologies

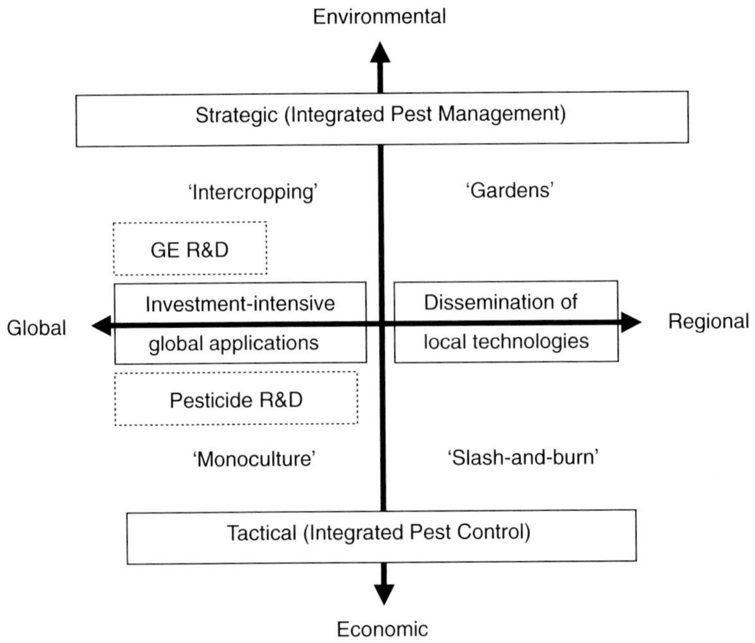

Fig. 13.5. Socio-economic scenarios based on global versus regional markets and economic (tactical IPM) versus environmental (strategic IPM) emphasis for technology development in crop management. R&D, research and development. (Based on scenarios developed by the Intergovernmental Panel on Climate Change (IPCC, 2001).)

in which they are most suited for cost-effective production, with production subsidies being reduced. Under this scenario, ICM would be promoted to deliver sustainable, high-yielding cropping systems. Pesticide use would be reduced in favour of alternative control strategies. GE crops would be the focal point for private sector investment to deliver sustainable cropping technologies. IPM technologies would be developed to operate at a global level for a common good, with such examples being the global monitoring of pests and diseases using GIS and enhanced cooperation and exchange of biological control agents for 'classical' biological control programmes.

The 'Gardens' (regional/environmental) scenario would place less emphasis on globalization and more on local/regional markets. In some ways this resembles an agricultural version of E.F. Schumacher's vision put forward in *Small is Beautiful* (1975). Agricultural policies would be developed for sustainable agriculture, at a possible cost to production efficiency and international competitiveness (IPCC, 2001). Agroecology would be the predominant agricultural philosophy to promote biodiversity and focus on indigenous IPM technologies, which would be exchanged between communities or within a region as described by Altieri (1995). ICM would be promoted using low-cost technologies, with capital-intensive technologies, such as genomics or GE crops, not having a large enough market to stimulate commercial interest in developing countries.

The final scenario, 'Slash-and-Burn' (regional/economic), would lead to a heterogeneous world in which the main theme would be self-reliance and the preservation of local identity. This scenario would see the degradation of the environment and misuse of natural resources, as socio-economic systems are focused on local exploitation and not sustainability (IPCC, 2001). Resource-poor nations would not be able to support agricultural research and development, resulting in low adoption of knowledge-intensive technologies, such as IPM, and the use of old and possibly obsolete pesticides (when economically viable), or they may resort to using no external inputs, resulting in low and variable yields, especially in marginal environments.

Examples of all four scenarios exist in the world today and this mosaic will probably continue for the foreseeable future given the divergent development and trade policies supported by different governments. However, the 'Monoculture' scenario is the most prevalent socio-economic scenario today, one that largely focuses on crop-production efficiency. Under this scenario, IPM would not be a high priority unless it came at little additional cost to the producer or was mandated through national/international policies. Increased net import of pesticides by developing countries since the early 1960s (Fig. 13.4) is evidence that this scenario will probably predominate at least in the short term. The political will to move towards the 'Intercropping' scenario appears to exist, as seen in US policy on IPM (Jacobsen, 1996), the European Union (EU) promoting ICM in agricultural policy (Wattiez *et al.*, 2003) and the World Bank Operational Policy 4.09 (World Bank, 1998). ICM is strongly promoted in the 'Intercropping' scenario, which would see the use of all control forms, including GE crops, to develop sustainable cropping systems. When it comes to IPM, there is an apparent dichotomy between the policies supporting IPM and the economic realities, as reflected by the global increase in pesticide use.

Conclusions

During the next two decades, it is estimated that the world will have to feed approximately 2.5 billion more people, with most of this growth occurring in LDCs, resulting in a projected shortfall of cereal production in LDCs of 197 million t (Rosegrant *et al.*, 2001). Most of this production will need to come from LDCs, which are showing stagnation in per cent yield gains, dropping from 2.9% during the green revolution to a predicted 1.2% during the next 20 years (Rosegrant *et al.*, 2001). Increasing the rate of yield gain will rely heavily on ICM and IPM in particular. In order to address this looming production demand, a 'doubly green revolution' will be required that exploits two areas of modern science: biotechnology to understand and manipulate living organisms and modern ecology to understand the dynamics of agricultural and natural-resource ecosystems (Conway, 1998). Such technologies can only be developed in partnership between farmers, extension, researchers and policy makers.

Intensification of cropping systems must be approached from a sustainable-agriculture perspective. If care is not used in increasing production, biodiversity, which has buffered many mixed cropping systems in LDCs, may be compromised and, with it, the diversity of organisms associated with tropical agroecologies. Management strategies and technologies that minimize disturbances to agroecosystems should be promoted, such as conservation-tillage technology and the strategic use of habitat management to enhance the efficacy of biological controls and reduce the use of broad-spectrum pesticides.

Pesticide use in LDCs has increased over the last 30 years, with Asia becoming a major exporter of pesticides and Latin America becoming the largest consumer of pesticides, in large part due to conservation-tillage practices. Pesticides will continue to be an important component in ICM, but the discontinuance of class 1a and 1b pesticides should be strongly promoted through the World Bank Operational Policy 4.09 and pesticide subsidies should be eliminated to allow market forces to exert economic pressure to minimize their use. Chemical manufacturers have become active in product stewardship, either directly or through a consortium like CropLife International. The private sector will probably expand its role in technology dissemination and training, by providing educational material to farmers to promote product safety and stewardship to enhance the efficacy, safety and long-term effectiveness of pesticide products in the developing world.

The use of biotechnology to produce insect-, disease- and herbicide-resistant plants has already become established in North America, Australia, Argentina and China, along with a steady rate of increased adoption in several developing countries (Fig. 13.2). The use of genetic engineering holds considerable promise in reducing the environmental and health costs associated with pesticide use, as well as increasing farm revenue. Since GE crops contain the pest control trait in the seed, distribution of the control technology to farmers should be accomplished in an easy and equitable manner. In order to realize the full potential of GE crops, biosafety and intellectual property legislation, at both national and international levels, must be established in LDCs. Traits to be incorporated should be decided in consultation with farmers, consumers and scientists to ensure that GE crops meet the needs of these groups and pose no significant risk to the environment.

Digital technology and high-speed telecommunications have enabled information to be accessed from around the world via the Internet. Although there was a great discrepancy in digital technology between industrial and LDCs at the end of the 20th century, this divide is rapidly narrowing as LDCs develop the telecommunication capacity to fully participate in the Internet. Most IPM websites are currently hosted in industrial countries; however, this trend is changing, as more researchers in developing countries are able to establish websites that promote local IPM technologies. The use of GPS will complement web networks by providing researchers and extension providers with tools that will enable them to define regions where production constraints are most acute, develop targeted technologies for those regions and monitor their use. The use of GIS will also become more important as increased effort is made to define the impact of IPM on crop production in LDCs; this is particularly true for the spread of classical biological control agents and resistant varieties. It will also provide a visual aid for researchers and policy makers to define priority areas for ICM and to better define production constraints and assess the impacts of policies *ex ante*. Using these tools to communicate the positive impact IPM has made will be critical in generating public and political support to increase the adoption of proven IPM technologies.

Given the environmental and social complexity of farming systems in developing countries, participatory methods have proved to be the most successful approach for disseminating IPM practices to farmers in LDCs. Viewing farmers as an equal partner in technology development and testing will foster ownership of IPM technologies and increase adoption. Farmers should be offered a 'basket of options' to test, given the socio-economic complexity of farming systems in LDCs. Dissemination of proven techniques that have clear economic and health benefits to farmers must be promoted, using various media and containing a simple message that is couched in a socially sensitive way to inform and persuade the farm family.

The green revolution, which addressed the food-security crisis of the 1960s and 1970s, involved scientific innovation and active promotion of high-yielding varieties that could transform fertilizer inputs into grain production – but it did not involve IPM. For the 'doubly green revolution' to succeed in feeding the world through sustainable cropping systems, a concerted and integrated effort on the part of scientists, farmers and policy makers will be required to create awareness and a political and regulatory environment in which sustainable technologies like IPM are embraced to meet the food and fibre demands of the 21st century.

Acknowledgements

Support by the Syngenta Foundation for Sustainable Agriculture is gratefully acknowledged. The author wishes to thank D. Poland for editing the manuscript and M. van Ginkel and D. Beck for reviewing the manuscript.

References

Abate, T., van Huis, A. and Ampofo, J.K.O. (2000) Pest management strategies in traditional agriculture: an African perspective. *Annual Review of Entomology* 45, 631–659.

Altieri, M.A. (ed.) (1993) *Crop Protection Strategies for Subsistence Farmers.* Westview Press, Boulder, Colorado, 197 pp.

Altieri, M.A. (1995) *Agroecology: the Science of Sustainable Agriculture.* Westview Press, Boulder, Colorado, 433 pp.

Atkin, J. and Leisinger, K.M. (2000) *Safe and Effective Use of Crop Protection Products in Developing Countries.* CAB International, Wallingford, UK, 163 pp.

Bänziger, M. and de Meyer, J. (2002) Collaborative maize variety development for stress-prone environments in southern Africa. In: Cleveland, D.A. and Soleri, D. (eds) *Farmers, Scientists and Plant Breeding.* CAB International, Wallingford, UK, pp. 269–296.

Bellotti, A.C., Smith, L. and Lapointe, S.L. (1999) Recent advances in cassava pest management. *Annual Review of Entomology* 44, 343–370.

Bentley, J.W. and Rodríguez, G. (2001) Honduran folk entomology. *Anthropology Research* 42, 285–301.

BioNET-International (2003) The Global Network for Taxonomy. Available at: http://www.bionet-intl.org

Bongaarts, J. (2002) Demography. In: *Sustainable Food Security for All by 2020. Proceedings of an International Conference, 4–6 September, Bonn, Germany.* IFPRI, Washington, DC, pp. 53–55.

Bruin, C.A. and Meerman, F. (2001) *New Ways of Developing Agricultural Technologies: the Zanzibar Experience with Participatory Integrated Pest Management.* Wageningen University and Research Centre/CTA, Wageningen, The Netherlands, 167 pp.

Brundtland, G.H. (1987) *Our Common Future: World Commission on Environment and Development.* Oxford University Press, Oxford, UK, 383 pp.

Campanhola, C., de Moraes, G.J. and deSá, L.A.N. (1995) Review of IPM in South America. In: Mengech, A.N., Saxena, K.N. and Gopalan, H.N.B. (eds) *Integrated Pest Management in the Tropics: Current Status and Future Prospects.* John Wiley & Sons, New York, pp. 117–152.

Carpenter, J., Felsot, A., Goode, T., Hammig, M., Onstad, D. and Sankula, S. (2002) *Comparative Environmental Impacts of Biotechnology-derived and Traditional Soybean, Corn and Cotton Crops.* Council for Agricultural Science and Technology, Ames, Iowa, 189 pp.

CGIAR (2003) Consultative Group on International Agricultural Research: Who we are. Available at: http://www.cgiar.org/who/index.html

Coakley, S.M., Scherm, H. and Chakraborty, S. (1999) Climate change and plant disease management. *Annual Review of Phytopathology* 37, 399–426.

Conway, G. (1998) *The Doubly Green Revolution: Food for All in the 21st Century.* Cornell University Press, Ithaca, New York , 335 pp.

CropLife (2003) CropLife International. Available at: http://www.croplife.org

Dabrowski, Z.T. (1997) *Integrated Pest Management in Vegetables, Wheat and Cotton in the Sudan: a Participatory Approach.* ICIPE Science Press, Nairobi, 245 pp.

De Groote, H. (2002) Maize yield losses from stem borers in Kenya. *Insect Science and its Application* 22, 89–96.

Ekboir, J. (2002) *CIMMYT 2000–2001 World Wheat Overview and Outlook: Developing No-till Packages for Small-scale Farmers.* CIMMYT, Mexico, DF, 65 pp.

Emana, G., Overholt, W.A., Kairu, E., MacOpiyo, L. and Zhou, G. (2002) Predicting the distribution of *Chilo partellus* (Swinhoe) and its parasitoid *Cotesia flavipes* Cameron in Ethiopia using correlation, step-wise regression and geographic information systems. *Insect Science and its Application* 22, 131–137.

FAOSTAT (2002) FAOSTAT Agricultural Data. Rome. Available at: http://apps.fao.org/page/collections?subset=agriculture

Gallagher, K.D. (1992) IPM development in the Indonesian National IPM Programme. In: Aziz, A., Kadir, S.A. and Barlow, H.S. (eds) *Pest Management and the Environment in 2000.* CAB International, Wallingford, UK, pp. 82–89.

Geier, P.W. (1966) Management of insect pests. *Annual Review of Entomology* 11, 471–490.

Gitay, H., Suàrez, A., Watson, R.T. and Dokken, D.J. (2002) *Climate Change and Biodiversity.* IPPCC Technical Paper V, Intergovernmental Panel on Climate Change, WMO, UNEP, Geneva, Switzerland, 77 pp. Available at: http://www.ipcc.ch/pub/tpbiodiv.pdf

Gould, F. (1996) Deploying pesticidal engineered crops in developing countries. In: Persley, G.J. (ed.) *Biotechnology and Integrated Pest Management*. Biotechnology in Agriculture Series No. 15, CAB International, Wallingford, UK, pp. 264–293.

Guan Soon, L. (1996) Integrated pest management in developing countries. In: Persley, G.J. (ed.) *Biotechnology and Integrated Pest Management*. Biotechnology in Agriculture Series No. 15, CAB International, Wallingford, UK, pp. 61–75.

Heinrichs, E.A. (ed.) (1988) *Plant Stress–Insect Interactions*. John Wiley & Sons, New York, 492 pp.

Heinrichs, E.A., Aquino, G.B., Chelliah, S., Valencia, S.L. and Reissig, W.H. (1982) Resurgence of *Nilaparvata lugens* (Stål) populations as influenced by method and timing of insecticide applications in lowland rice. *Environmental Entomology* 11, 78–84.

Heong, K.L. and Escalada, M.M. (1997) Perception changes in rice pest management: a case study of farmers' evaluation of conflict information. *Journal of Applied Communications* 81, 3–17.

Heong, K.L., Escalada, M.M. and Mai, V. (1994) An analysis of insecticide use in rice: case studies in the Philippines and Vietnam. *International Journal of Pest Management* 40, 173–178.

Herren, H.R., Neuenschwander, P., Hennessey, R.D. and Hammond, W.N.O. (1987) Introduction and dispersal of *Epidinocarsis lopezi* (Hom., Pseudococcidae), in Africa. *Agriculture, Ecosystems and Environment* 19, 131–144.

Hilbeck, A., Baumgartner, M., Freid, P.M. and Bigler, F. (1998) Effects of *Bacillus thuringiensis* corn-fed prey on mortality and development time of immature *Chrysoperla carnae*. *Environmental Entomology* 27, 480–487.

Huang, J., Rozelle, S., Pray, C. and Wang, Q. (2002) Plant biotechnology in China. *Science* 295, 674–677.

IPCC (2001) *Climate Change 2001: Mitigation – Contribution of Working Group III to the Third Assessment Report*. Intergovernmental Panel on Climate Change, World Meteorological Organization, Geneva. 700 pp. Available at http://www.grida.no/climate/ipcc_tar/wg3/pdf/TAR-total.pdf

Irungu, R.W., Hodson, D. and Muchugu, E.I. (2003) Awhere-ACT: predicting pest outbreaks in Africa. ICT Update, CTA, Issue 11. Available at: http://ictupdate.cta.int/index.php/article/articleview/193/1/34/

Ishii-Eiteman, M., Hamburger, J. and Lee, C. (2002) *Monitoring the World Bank's Pest Management Policy, a Guide for Communities*. Pesticide Action Network North America, San Franscisco, California, 20 pp. Available at: http://www.panna.org/ resources/documents/monitoringWB.pdf

ISNAR (2003) International Service for National Agricultural Science. Available at: http://www.isnar.cgiar.org/topics.htm

Jacobsen, B. (1996) USDA integrated pest management initiative. National IPM Network. Available at: http://ipmworld.umn.edu/ chapters/jacobsen.htm

James, B., Neuenschwander, P., Markham, R.H., Anderson, P., Braun, A., Overholt, W., Khan, K., Makkouk, K. and Emechebe, A. (2003) Bridging the gap with the CGIAR Systemwide Program on Integrated Management. In: Maredia, K., Dakouo, D. and Mota-Sanchez, D. (eds) *Integrated Pest Management in the Global Arena*. CAB International, Wallingford, UK, pp. 419–434.

James, C. (2002) Global review of commercialized transgenic crops: 2001. Feature: Bt cotton. ISAAA Briefs No. 26, Ithaca, New York. Available at: http://www.isaaa.org/publications/briefs/Brief_26.htm

James, C. (2003a) Global status of commercialized transgenic crops: 2002. ISAAA Briefs No. 27, International Service for the Acquisition of Agri-biotech Applications. Available at: http://www.isaaa.org

James, C. (2003b) Global status of GM crops. *ISAAA Global Status*. International Service for the Acquisition of Agri-biotech Applications. Available at: http://www.isaaa.org/kc/Bin/gstats/index.htm

Jones, P.G. and Thornton, P.K. (2003) The potential impacts of climate change on maize production in Africa and Latin America in 2055. *Global Environmental Change* 13, 51–59.

Kanampiu, F., Ransom, J., Gressel, J., Jewell, D., Friesen, D., Grimanelli, D. and Hoistington, D. (2002) Appropriateness of biotechnology to African agriculture: *Striga* and maize as paradigms. *Plant Cell, Tissue and Organ Culture* 69, 105–110.

Kates, R.W., Clark, W.C., Corell, R., Hall, J.M., Jaeger, C.C., Lowe, I., McCarthy, J.J., Schellnhuber, H.J., Bolin, B., Dickson, N.M., Faucheux, S., Gallopin, G.C., Grübler, A., Huntley, B., Jäger, J., Jodha, N.S., Kasperson, R.E., Mabogunje, A., Matson, P., Mooney, H., Moore, B. III, O'Riordan, T. and Svedlin, U. (2001) Environment and development: sustainability science. *Science* 292, 641–642.

Khan, Z.R., Ampong-Nyarko, K., Chiliswa, P., Hassanali, A., Kimani, S., Lwande, W., Overholt, W.A., Pickett, J.P., Smart, L.E., Wadhams, L.J. and Woodcock, C.M. (1997) Intercropping increases parasitism of pests. *Nature* 388, 631–632.

Khan, Z.R., Pickett, J.A., Wadhams, L. and Muyekho, F. (2001) Habitat management strategies for the control of cereal stemborers and striga in maize in Kenya. *Insect Science and its Application* 21, 375–380.

Khush, G.S. and Brar, D.S. (1991) Genetics of resistance to insects in crop plants. *Advances in Agronomy* 45, 223–274.

Kotler, P. and Zaltman, G. (1971) Social marketing: strategies for changing public behavior. *Journal of Marketing* 35, 3–12.

Koul, O. (1996) Neem research and development: present and future scenario. In: Handa, S.S. and Koul, M.K. (eds) *Supplement to Cultivation and Utilization of Medicinal Plants*. PID, CSIR, New Delhi, pp. 583–611.

Landis, D.A., Wratten, S.D. and Gurr, G.M. (2000) Habitat management to conserve natural enemies of arthropod pests in agriculture. *Annual Review of Entomology* 45, 175–201.

Litsinger, J.A. (1989) Second generation insect pest problems on high yielding rices. *Tropical Pest Management* 35, 235–242.

Losey, J., Raynor, L. and Carter, M.E. (1999) Transgenic pollen harms monarch larvae. *Nature* 399, 214.

Matteson, P.C. (2000) Insect pest management in tropical Asian irrigated rice. *Annual Review of Entomology* 45, 549–574.

Matteson, P.C., Altieri, A. and Gagne, W.C. (1984) Modification of small farmer practices for better pest management. *Annual Review of Entomology* 29, 383–402.

Meerman, F., van de Ven, G.W.J., van Keulen, H. and Breman, H. (1996) Integrated crop management: an approach to sustainable agricultural development. *International Journal of Pest Management* 42, 13–24.

Mengech, A.N., Saxena, K.N. and Gopalan, H.N.B. (1995) *Integrated Pest Management in the Tropics: Current Status and Future Prospects*. John Wiley & Sons, New York, 171 pp.

Morin, S., Biggs, R.W., Sisterson, M.S., Shriver, L., Ellers-Kirk, C., Higginson, D., Holley, D., Gahan, L.J., Heckel, D.G., Carriere, Y., Dennehy, T.J., Brown, J.K. and Tabashnik, B.E. (2003) Three cadherin alleles associated with resistance to *Bacillus thuringiensis* in pink bollworm. *Proceedings of the National Academy of Science* 100, 5004–5009.

Morse, S. and Buhler, W. (1997) *Integrated Pest Management: Ideals and Realities in Developing Countries*. Lynne Rienner, Boulder, Colorado, 171 pp.

Mugo, S., Bergvinson, D. and Hoisington, D. (2001) Options in developing stemborer-resistant maize: CIMMYT's approaches and experiences. *Insect Science and its Application* 21, 409–415.

Mulaa, M., Bergvinson, D. and Mugo, S. (2001) IRMA researchers work to develop insect resistance management strategies. *IRMA Updates* 2(1), 3. Available at: http://www.cimmyt.org/ABC/investin-insectresist/pdf/IRMA_2&1.pdf

Neuenschwander, P. (1993) Human interactions in classical biological control of cassava and mango mealybugs on subsistence farms in tropical Africa. In: Altieri, M.A. (ed.) *Crop Protection Strategies for Subsistence Farmers*. Westview Press, Boulder, Colorado, 197 pp.

Neuenschwander, P., Hammond, W.N.O., Ajuonu, O., Gado, A., Echendu, N., Bokonon-Ganta, A.H., Allomasso, R. and Okon, I. (1990) Biological control of the cassava mealybug, *Phenacoccus manihoti* (Hom., Pseudococcidae) by *Epidinocarsis lopezi* (Hym., Encyrtidae) in West Africa, as influenced by climate and soil. *Agriculture, Ecosystems and Environment* 32, 39–55.

Nkunika, P.O. (2002) Smallholder farmers' integration of indigenous technical knowledge (ITK) in maize IPM: a case study in Zambia. *Insect Science and its Application* 22, 235–240.

Panda, N. and Khush, G.S. (1995) *Host-plant Resistance to Insects*. CAB International, Wallingford, UK, 431 pp.

Pingali, P.L. (2001) Environmental consequences of agricultural commercialization in Asia. *Environmental and Development Economics* 6, 483–502.

Pingali, P.L., Hossain, M. and Gerpacio, R.V. (1997) *Asian Rice Bowls: the Returning Crisis?* CAB International, Wallingford, UK, 341 pp.

Pinstrup-Andersen, P. (2002) Food and agricultural policy for a globalizing world: preparing for the future. *American Journal of Agricultural Economics* 84, 1201–1214.

Qaim, M. and Zilberman, D. (2003) Yield effects of genetically modified crops in developing countries. *Science* 299, 900–902.

Rosegrant, M.W., Paisner, M.S., Meijer, S. and Witcover, J. (2001) *Global Food Projections to 2020: Emerging Trends and Alternative Futures*. IFPRI, Washington, DC, 206 pp. Available at: http://www.ifpri.org/pubs/ books/gfp/gfp.pdf

Schmutterer, H. (1995) *The Neem Tree: Source of Unique Natural Products for Integrated Pest Management, Medicine, Industry and Other Purposes*. VCH Publishers, New York, 696 pp.
Schumacher, E.F. (1975) *Small is Beautiful: Economics as if People Mattered*. Harper and Row, New York, 305 pp.
Sears, M.K., Hellmich, R.L., Stanley-Horn, D.E., Oberhauser, K.S., Pleasants, J.M., Mattila, H.R., Siegfried, B.D. and Dively, G.P. (2001) Impact of Bt corn pollen on monarch butterfly populations: a risk assessment. *Proceedings of the National Academy of Science* 98, 11937–11942.
Shelton, A.M., Zhao, J.-Z. and Roush, R.T. (2002) Economic, ecological, food safety and social consequences of the deployment of Bt transgenic plants. *Annual Review of Entomology* 47, 845–881.
Simmonds, N.W. (1976) *Evolution of Crop Plants*, 1st edn. Longman, New York, 339 pp.
Smith, C.M. (1999) Plant resistance to insects. In: Rechcigl, J. and Rechcigl, N. (eds) *Biological and Biotechnological Control of Insects*. Lewis Publishers, Boca Raton, Florida, pp. 171–205.
Songa, J. (2002) Study compares effect of Bt biopesticide and conventional insecticides on nontarget arthropods. *IRMA Updates* 3(3), 5–6. Available at: http://www.cimmyt.org/ABC/InvestIn-InsectResist/pdf/IRMA_ 3&3.pdf
Songa, J., Bergvinson, D.J. and Mugo, S (2002) A reference collection for non-target arthropods of Bt maize in Kenya. *IRMA Updates* 4(3), 8. Available at: http://www.cimmyt.org/ABC/ InvestIn-InsectResist/pdf/IRMA_3_I4.pdf
Srivastava, S.K. (2000) Effect of farm practices on natural resources and strategy of transfer of ecofriendly technologies. In: *International Conference on Managing Natural Resources for Sustainable Agricultural Production in the 21st Century, 14–18 February*. Indian Society of Soil Science, New Delhi, pp. 1134–1135.
TACTRI (2002) Taiwan Agricultural Chemicals and Toxic Substances Research Institute, Council of Agriculture. Available at: http://www.tactri.gov.tw
Teng, P.S. (1994) Integrated pest management in rice. *Experimental Agriculture* 30, 115–137.
Wambugu, F. (1999) Why Africa needs agricultural biotech. *Nature* 400, 15–16.
Wattiez, C., Scheuer, S. and Iwasaki-Riss, J. (2003) European Parliament passes a resolution calling for a reduction in pesticide use. Available at: http://www.icmfocus.com/inbrief.asp?Section=Display&NewsID=34
Weinzierl, R.A. (1999) Botanical insecticides, soaps and oils. In: Recheigl, J.E. and Recheigl, N.A. (eds) *Biological and Biotechnological Control of Insect Pests*. Agriculture and Environment Series, CRC Press, Boca Raton, Florida, pp 101–121.
Willcox, M.C., Khairallah, M.M., Bergvinson, D., Crossa, J., Deutsch, J.A., Edmeades, G.O., Gonzalez-de-Leon, D., Jiang, C., Jewell, D.C., Mihm, J.A., Williams, W.P. and Hoisington, D. (2002) Selection for resistance to southwestern corn borer using marker-assisted selection and conventional backcrossing. *Crop Science* 42, 1516–1528.
World Bank (1992) *Agricultural Pest Management*. Operational Directive 4.03, World Bank, Washington, DC, 5 pp.
World Bank (1998) Operational Policy 4.09. World Bank, Washington, DC, 2 pp. Available at: http://wbln0018.worldbank.org/Institutional/Manuals/OpManual.nsf/bytype/665DA6CA847982168525672C007D07A3?OpenDocument
Yencho, G.C., Cohen, M.B. and Byrne, P.F. (2000) Applications of tagging and mapping insect resistance loci in plants. *Annual Review of Entomology* 45, 393–422.
Zethner, O. (1995) Practice of integrated pest management in tropical and sub-tropical Africa: an overview of two decades. In: Mengech, A.N., Saxena, K.N. and Gopalan, H.N.B. (eds) *Integrated Pest Management in the Tropics: Current Status and Future Prospects*. John Wiley & Sons, New York, pp. 1–67.
Zhao, J.Z., Rui, C.H., Lu, M.G., Fan, X.L., Ru, L.J. and Meng, X.Q. (2000) Monitoring and managing of *Helicoverpa armigera* resistance to transgenic Bt cotton in Northern China. *Research on Pest Management Newsletter* 11, 28–31.

Index